Methods in Molecular Biology

Series Editor
John M. Walker
School of Life and Medical Sciences
University of Hertfordshire
Hatfield, Hertfordshire, UK

For further volumes:
http://www.springer.com/series/7651

For over 35 years, biological scientists have come to rely on the research protocols and methodologies in the critically acclaimed *Methods in Molecular Biology* series. The series was the first to introduce the step-by-step protocols approach that has become the standard in all biomedical protocol publishing. Each protocol is provided in readily-reproducible step-by step fashion, opening with an introductory overview, a list of the materials and reagents needed to complete the experiment, and followed by a detailed procedure that is supported with a helpful notes section offering tips and tricks of the trade as well as troubleshooting advice. These hallmark features were introduced by series editor Dr. John Walker and constitute the key ingredient in each and every volume of the *Methods in Molecular Biology* series. Tested and trusted, comprehensive and reliable, all protocols from the series are indexed in PubMed.

Golden Gate Cloning

Methods and Protocols

Edited by

Daniel Schindler

Max Planck Institute for Terrestrial Microbiology, Marburg, Germany; Center for Synthetic Microbiology (SYNMIKRO), Philipps-University Marburg, Marburg, Germany

Humana Press

Editor
Daniel Schindler
Max Planck Institute for Terrestrial
Microbiology
Marburg, Germany

Center for Synthetic Microbiology (SYNMIKRO)
Philipps-University Marburg
Marburg, Germany

ISSN 1064-3745 ISSN 1940-6029 (electronic)
Methods in Molecular Biology
ISBN 978-1-0716-4094-4 ISBN 978-1-0716-4220-7 (eBook)
https://doi.org/10.1007/978-1-0716-4220-7

Cover Caption: Artwork by Ehmad Chehrghani Bozhcheloe.

This Humana imprint is published by the registered company Springer Science+Business Media, LLC, part of Springer Nature.
The registered company address is: 1 New York Plaza, New York, NY 10004, U.S.A.

Preface

Golden Gate cloning is one of the most versatile DNA assembly strategies and has been widely used in all disciplines since its invention. It is based on the properties of type IIS restriction endonucleases, which recognize a defined, directed DNA motif but cut any DNA sequence at a fixed distance. This property led to the development of a one-pot, single-step molecular cloning technique by combining DNA restriction and DNA ligation in a single reaction, now referred to as Golden Gate cloning.

The principle is based on the loss of the type IIS recognition site in the cloning product, which accumulates over time in the reaction. Golden Gate cloning has proven to be highly efficient, modular, and capable of high throughput. It allows for the standardization of biological parts, which is why Golden Gate cloning is widely used in synthetic biology and enables laboratory automation approaches.

Synthetic biologists use Golden Gate cloning to create reusable libraries of parts for complex pathways, genetic circuits, or combinatorial assemblies. This book in the Methods in Molecular Biology series cannot fully cover the broad spectrum of Golden Gate cloning. However, this book provides an overview of well-established protocols that cover the full range of in silico, in vitro, and in vivo approaches to enable researchers to plan, design, and perform Golden Gate cloning experiments.

Marburg, Germany *Daniel Schindler*

Contents

Contributors

SARAH ABDUL-MAWLA • *Molecular Plant Sciences, Center for Synthetic Microbiology (SYNMIKRO), Philipps-University Marburg, Marburg, Germany*

FLORIAN ALTEGOER • *Institute of Microbiology, Heinrich-Heine University, Düsseldorf, Germany*

EBBE SLOTH ANDERSEN • *Department of Molecular Biology and Genetics, Aarhus University, Aarhus, Denmark; Interdisciplinary Nanoscience Center (iNANO), Aarhus University, Aarhus, Denmark*

ANKE BECKER • *Center for Synthetic Microbiology, Philipps-Universität Marburg, Marburg, Germany; Department of Biology, Philipps-Universität Marburg, Marburg, Germany*

BORK A. BERGHOFF • *Institute for Microbiology and Molecular Biology, Justus-Liebig University Giessen, Giessen, Germany*

SONJA BILLERBECK • *Molecular Microbiology, Groningen Biomolecular Sciences and Biotechnology Institute, University of Groningen, Groningen, The Netherlands*

CEDRIC K. BRINKMANN • *Max Planck Institute for Terrestrial Microbiology, Marburg, Germany*

JOHN A. BRYANT JR • *Department of Biological Systems Engineering, Virginia Polytechnic Institute and State University, Blacksburg, VA, USA*

ELLIOTT CHAPMAN • *Edinburgh Genome Foundry, Institute of Quantitative Biology, Biochemistry and Biotechnology, School of Biological Sciences, University of Edinburgh, Edinburgh, UK*

DAVID M. CHIASSON • *Department of Biology, Saint Mary's University, Halifax, Canada*

M. WAYNE DAVIS • *School of Biological Sciences, HHMI, University of Utah, Salt Lake City, UT, USA*

STIJN T. DE VRIES • *Max Planck Institute for Terrestrial Microbiology, Marburg, Germany*

THIJS DE VROET • *Molecular Microbiology, Groningen Biomolecular Sciences and Biotechnology Institute, University of Groningen, Groningen, The Netherlands*

SOPHIE DITTMAR • *Institute for Microbiology and Molecular Biology, Justus-Liebig University Giessen, Giessen, Germany*

PETER EMELIN • *Molecular Plant Sciences, Center for Synthetic Microbiology (SYNMIKRO), Philipps-University Marburg, Marburg, Germany*

ANNA FABER • *Center for Synthetic Microbiology, Philipps-Universität Marburg, Marburg, Germany; Department of Biology, Philipps-Universität Marburg, Marburg, Germany; School of Molecular Sciences, The University of Western Australia, Crawley, Australia*

RENNOS FRAGKOUDIS • *Edinburgh Genome Foundry, Institute of Quantitative Biology, Biochemistry and Biotechnology, School of Biological Sciences, University of Edinburgh, Edinburgh, UK; Department of Biochemistry and Biotechnology, University of Thessaly, Larissa, Greece*

XIAN FU • *BGI Research, Shenzhen, China; Guangdong Provincial Key Laboratory of Genome Read and Write, BGI Research, Shenzhen, China; BGI Research, Changzhou, China*

AUKSE GAIZAUSKAITE • *Max Planck Institute for Terrestrial Microbiology, Marburg, Germany; Center for Synthetic Microbiology (SYNMIKRO), Philipps-Universität Marburg, Marburg, Germany; Synthetic Biology, TUM Campus Straubing for Biotechnology and Sustainability, Technical University of Munich, Straubing, Germany*
EVA GARCÍA-RUIZ • *Departamento de Biocatálisis, Instituto de Catálisis y Petroleoquímica, ICP-CSIC, Madrid, Spain*
SARA GILLES • *Max Planck Institute for Terrestrial Microbiology, Marburg, Germany; Center for Synthetic Microbiology (SYNMIKRO), Philipps-Universität Marburg, Marburg, Germany*
CRAIG GILMOUR • *Edinburgh Genome Foundry, Institute of Quantitative Biology, Biochemistry and Biotechnology, School of Biological Sciences, University of Edinburgh, Edinburgh, UK*
RAMONA GRÜTZNER • *Leibniz Institut of Plant Biochemistry, Department of Cell and Metabolic Biology, Halle, Germany*
DAVID ANGELO V. GUANZON • *Microbial Biology, Ruhr University Bochum, Bochum, Germany*
JOHAN GUILLÉN • *Brightmont Academy, Seattle, WA, USA*
MELODY HAARMANN • *Microbial Biotechnology, Ruhr University Bochum, Bochum, Germany*
KATHARINA HÖFER • *Max Planck Institute for Terrestrial Microbiology, Marburg, Germany; Center for Synthetic Microbiology (SYNMIKRO), Marburg, Germany*
JOSEF HOFF • *Center for Synthetic Microbiology, Philipps-Universität Marburg, Marburg, Germany; Department of Biology, Philipps-Universität Marburg, Marburg, Germany; Microbial Biotechnology, Campus Straubing for Biotechnology and Sustainability, Technical University of Munich, Straubing, Germany*
STEFAN A. HOFFMANN • *Systems and Synthetic Biology, Wageningen University & Research, Wageningen, Netherlands; Manchester Institute of Biotechnology, University of Manchester, Manchester, UK*
AMANDA L. HUGHES • *European Molecular Biology Laboratory (EMBL), Genome Biology Unit, Heidelberg, Germany*
RENÉ INCKEMANN • *Max Planck Institute for Terrestrial Microbiology, Marburg, Germany*
ERIK M. JORGENSEN • *School of Biological Sciences, HHMI, University of Utah, Salt Lake City, UT, USA*
LAURA KLEY • *Max Planck Institute for Terrestrial Microbiology, Marburg, Germany*
TANIA S. KÖBEL • *Max Planck Institute for Terrestrial Microbiology, Marburg, Germany*
JESÚS LABORDA-MANSILLA • *Departamento de Biocatálisis, Instituto de Catálisis y Petroleoquímica, ICP-CSIC, Madrid, Spain*
FRANÇOIS-XAVIER LEHR • *Max Planck Institute for Terrestrial Microbiology, Marburg, Germany; Center for Synthetic Microbiology (SYNMIKRO), Philipps-Universität Marburg, Marburg, Germany*
ALEXANDER LEPAK • *Center for Synthetic Microbiology, Philipps-University Marburg, Marburg, Germany*
VIVIAN LINKE • *Microbial Biotechnology, Ruhr University Bochum, Bochum, Germany*
KATARZYNA ELŻBIETA LIPIŃSKA • *Max Planck Institute for Terrestrial Microbiology, Marburg, Germany; Center for Synthetic Microbiology (SYNMIKRO), Philipps-Universität Marburg, Marburg, Germany*
GREGORY J. S. LOHMAN • *Research Department, New England Biolabs, Ipswich, MA, USA*
ALICIA MACIÁ VALERO • *Molecular Microbiology, Groningen Biomolecular Sciences and Biotechnology Institute, University of Groningen, Groningen, The Netherlands*

SYLVESTRE MARILLONNET • *Leibniz Institut of Plant Biochemistry, Department of Cell and Metabolic Biology, Halle, Germany*
GABRIEL MENDOZA-ROJAS • *Institute of Microbiology, Heinrich-Heine University, Düsseldorf, Germany*
ANNA CHRISTINA R. NGO • *Microbial Biotechnology, Ruhr University Bochum, Bochum, Germany*
MICHAEL T. A. NGUYEN • *Department of Molecular, Cellular, and Developmental Biology, Yale University, New Haven, CT, USA; Systems Biology Institute, Yale University, West Haven, CT, USA*
HENRIKE NIEDERHOLTMEYER • *Max Planck Institute for Terrestrial Microbiology, Marburg, Germany; Center for Synthetic Microbiology (SYNMIKRO), Philipps-Universität Marburg, Marburg, Germany; Synthetic Biology, TUM Campus Straubing for Biotechnology and Sustainability, Technical University of Munich, Straubing, Germany*
NIELS N. OEHLMANN • *Max Planck Institute for Terrestrial Microbiology, Marburg, Germany*
GEORGIOS POTHOULAKIS • *Interdisciplinary Nanoscience Center (iNANO), Aarhus University, Aarhus, Denmark*
NADIIA POZHYDAIEVA • *Max Planck Institute for Terrestrial Microbiology, Marburg, Germany*
RIANNE C. PRINS • *Molecular Microbiology, Groningen Biomolecular Sciences and Biotechnology Institute, University of Groningen, Groningen, The Netherlands*
ADÁN A. RAMÍREZ ROJAS • *Max Planck Institute for Terrestrial Microbiology, Marburg, Germany*
LENNART RANDAU • *Department of Biology, Philipps Universität Marburg, Marburg, Germany; SYNMIKRO, Center for Synthetic Microbiology, Marburg, Germany*
JOHANNES G. REBELEIN • *Max Planck Institute for Terrestrial Microbiology, Marburg, Germany; Center for Synthetic Microbiology (SYNMIKRO), Philipps-University Marburg, Marburg, Germany*
SELINA RUST • *Department of Biology, Philipps Universität Marburg, Marburg, Germany*
ARPITA SAHOO • *Max Planck Institute for Terrestrial Microbiology, Marburg, Germany; Center for Synthetic Microbiology (SYNMIKRO), Philipps-Universität Marburg, Marburg, Germany*
DANIEL SCHINDLER • *Max Planck Institute for Terrestrial Microbiology, Marburg, Germany; Center for Synthetic Microbiology (SYNMIKRO), Philipps-University Marburg, Marburg, Germany*
SARINA SCOVILLE • *Department of Biology, Saint Mary's University, Halifax, Canada*
YUE SHEN • *BGI Research, Shenzhen, China; Guangdong Provincial Key Laboratory of Genome Read and Write, BGI Research, Shenzhen, China; BGI Research, Changzhou, China; BGI Research, China National GeneBank (CNGB), Shenzhen, Guangdong, China*
JOE SMITKA • *Microbial Biotechnology, Ruhr University Bochum, Bochum, Germany*
LARS M. STEINMETZ • *European Molecular Biology Laboratory (EMBL), Genome Biology Unit, Heidelberg, Germany; Stanford Genome Technology Center, Stanford University, Palo Alto, CA, USA; Department of Genetics, School of Medicine, Stanford University, Stanford, CA, USA*
DANIEL STUKENBERG • *Center for Synthetic Microbiology, Philipps-Universität Marburg, Marburg, Germany; Department of Biology, Philipps-Universität Marburg, Marburg, Germany*

NICHOLAS M. THOMSON • *Quadram Institute Bioscience, Norwich Research Park, Norwich, UK*
DIRK TISCHLER • *Microbial Biotechnology, Ruhr University Bochum, Bochum, Germany*
PETER VEGH • *Edinburgh Genome Foundry, Institute of Quantitative Biology, Biochemistry and Biotechnology, School of Biological Sciences, University of Edinburgh, Edinburgh, UK*
GONZALO VIDAL • *Interdisciplinary Computing and Complex Biosystems, School of Computing, Newcastle University, Newcastle upon Tyne, UK; Department of Electrical, Computer, and Energy Engineering, University of Colorado Boulder, Boulder, CO, USA*
CAROLUS VITALIS • *Department of Electrical, Computer, and Energy Engineering, University of Colorado Boulder, Boulder, CO, USA*
NILS WEINDORF • *Microbial Biotechnology, Ruhr University Bochum, Bochum, Germany*
STEFAN WERNER • *Icon Genetics GmbH, Halle, Germany*
FELIX WILLMUND • *Molecular Plant Sciences, Center for Synthetic Microbiology (SYNMIKRO), Philipps-University Marburg, Marburg, Germany*
R. CLAY WRIGHT • *Department of Biological Systems Engineering, Virginia Polytechnic Institute and State University, Blacksburg, VA, USA*
YU ZHANG • *BGI Research, Shenzhen, China; BGI Research, Hangzhou, China*
JU ZHENG • *BGI Research, Shenzhen, China; Guangdong Provincial Key Laboratory of Genome Read and Write, BGI Research, Shenzhen, China*
SONJA ZWENG • *Institute of Microbiology, Heinrich-Heine University, Düsseldorf, Germany*

Chapter 1

Golden Gate Cloning of MoClo Standard Parts

Ramona Grützner and Sylvestre Marillonnet

Abstract

Efficient DNA assembly methods are an essential prerequisite in the field of synthetic biology. Modular cloning systems, which rely on Golden Gate cloning for DNA assembly, are designed to facilitate assembly of multigene constructs from libraries of standard parts through a series of streamlined one-pot assembly reactions. Standard parts consist of the DNA sequence of a genetic element of interest such as a promoter, coding sequence, or terminator, cloned in a plasmid vector. Standard parts for the modular cloning system MoClo, also called level 0 modules, must be flanked by two BsaI restriction sites in opposite orientations and should not contain internal sequences for two type IIS restriction sites, BsaI and BpiI, and optionally for a third type IIS enzyme, BsmBI. We provide here a detailed protocol for cloning of level 0 modules. This protocol requires the following steps: (1) defining the type of part that needs to be cloned, (2) designing primers for amplification, (3) performing polymerase chain reaction (PCR) amplification, (4) cloning of the fragments using Golden Gate cloning, and finally (5) sequencing of the part. For large standard parts, it is preferable to first clone sub-parts as intermediate level −1 constructs. These sub-parts are sequenced individually and are then further assembled to make the final level 0 module.

Key words Modular cloning, Synthetic biology, Biological parts, DNA assembly, Multigene constructs

1 Introduction

Golden Gate cloning is a DNA assembly method designed for assembly of multiple DNA fragments (up to 52 fragments) in a recipient vector using a one-pot assembly reaction [1–4]. Golden Gate cloning is based on the use of type IIS enzymes such as BsaI, BpiI, or BsmBI (*see* **Note 1**). Modules flanked by restriction sites for a type IIS enzyme are pipetted together with a recipient vector containing compatible type IIS enzyme restriction sites. Assembly is performed with a restriction-ligation using the type IIS restriction enzyme and ligase. Golden Gate cloning takes advantage of the fact that restriction sites for the type IIS enzyme used are not present in the ligated product of interest, but are still present in re-ligated original vector or donor inserts plasmids. A

Daniel Schindler (ed.), *Golden Gate Cloning: Methods and Protocols*, Methods in Molecular Biology, vol. 2850,
https://doi.org/10.1007/978-1-0716-4220-7_1,

restriction-ligation allows repeated cycles of digestion and ligation to ensure that fragments that were ligated in their initial vector are made available for further rounds of ligation, until they are correctly assembled in the recipient vector, resulting in the desired construct. Golden Gate cloning is used in a multitude of DNA cloning systems developed for a variety of host organisms, including bacteria, yeasts, fungi, plants, microalgae, cyanobacteria, and animal cells [5–20]. One of these systems is the modular cloning system MoClo that was first designed to assemble multigene constructs for plants [5, 6, 21, 22], but that can also be adapted for other types of organisms. The MoClo system relies on libraries of standard parts. Each part consists of the DNA sequence of a genetic element of interest, such as a promoter, coding sequence, or terminator. Parts are flanked by two BsaI restriction sites in opposite orientations (Fig. 1). In addition, all parts should not contain restriction sites for BsaI and BpiI within internal sequences. Optionally, parts can be designed to lack restriction sites for a third type IIS enzyme, BsmBI (*see* **Note 2**). Generation of domesticated parts (lacking type IIS enzymes restriction sites in internal sequences) is the most labor-intensive part of cloning projects that use the MoClo system, as this process requires polymerase chain reaction (PCR) amplification of DNA, subcloning of the PCR products in a plasmid vector, and sequencing. The easiest solution for generating basic parts is to order the sequence from a gene synthesis company. This is useful when DNA material is not available for PCR amplification, as is the case, for example, for sequences from metagenomics projects. Even when a DNA template is available, gene synthesis allows sequences to be made directly lacking internal type IIS restriction sites, avoiding the domestication process. With a trend in decreasing prices for gene synthesis, ordering sequences from companies is likely to become the preferred solution. At present, however, gene synthesis still represents a substantial cost for most labs, and we still clone some basic parts ourselves. When ordering parts from a gene synthesis provider, it is important to be sure that the plasmid in which the part will be cloned has an antibiotic selection marker that is compatible with the cloning system with which this part needs to be used. It is also important to be sure that the sequence that one wants to order has been obtained from high-quality sequencing data, as what is ordered is what the user will get. In this chapter, we provide a detailed protocol for cloning of MoClo level 0 basic parts. This protocol can also be used to clone parts for other modular cloning systems, provided that cloning vectors with appropriate antibiotic resistance markers are used. This protocol requires the following steps: (1) defining the type of basic part that needs to be cloned, (2) designing primers for amplification, (3) performing PCR amplification, (4) cloning of the fragments using Golden Gate cloning, and finally (5) sequencing of the part. For large basic parts, it is

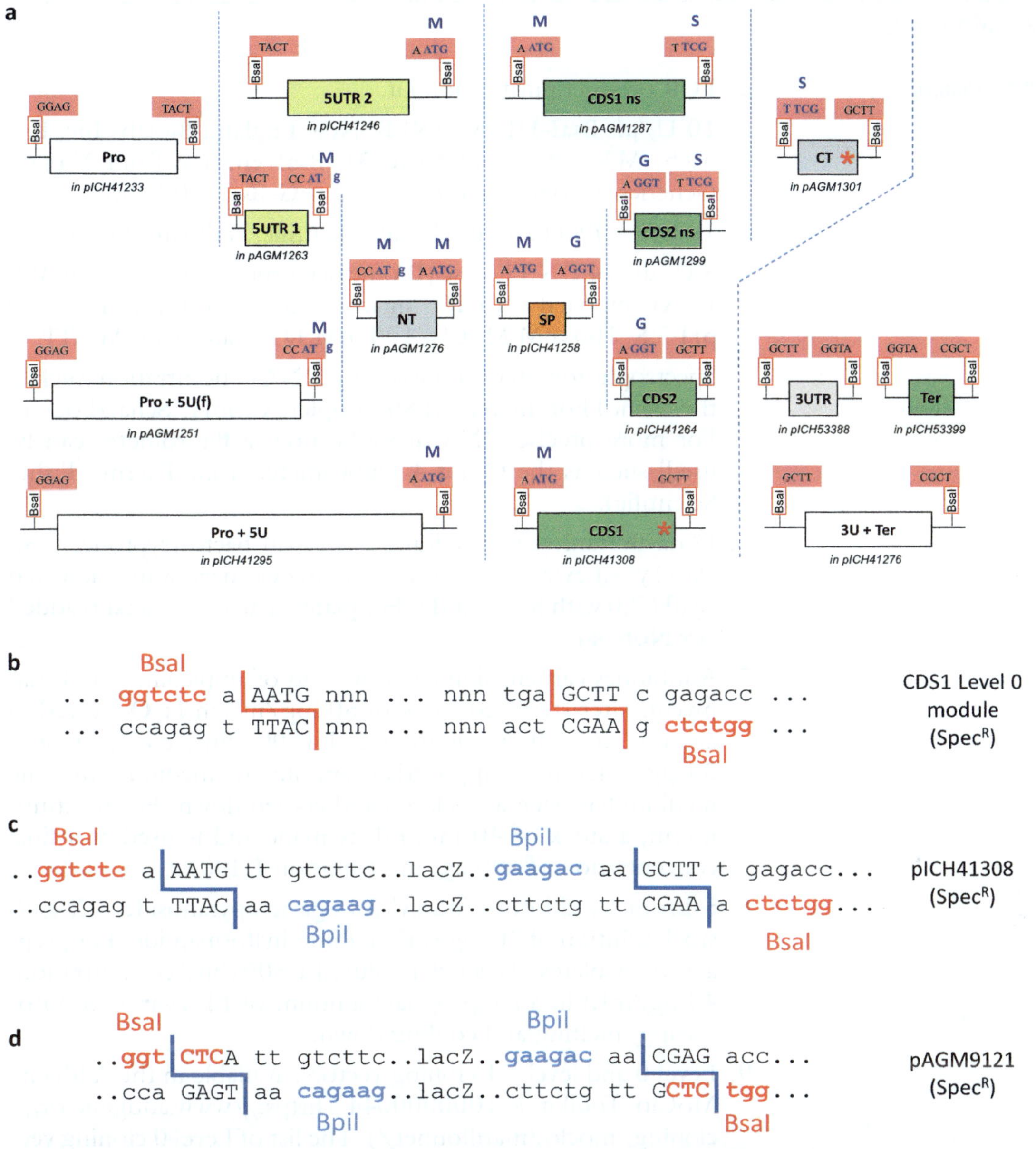

Fig. 1 Standard MoClo module types and cloning vectors. (**a**) Standard MoClo module types. Each level 0 module type has a special cloning vector, indicated below each module type. Pro promoter; 5 U/5UTR: 5′ untranslated leader; 3 U/3UTR 3′ untranslated sequence; NT N-terminal fusion tag or sequence; CT C-terminal fusion tag or sequence; SP signal peptide; CDS coding sequence; Ter terminator; M methionine; S serine; G glycine; * stop codon. (**b**) Sequence details of the CDS1 module (**b**) and its cloning vector (**c**) are given as examples. (**d**) Universal level 0 cloning vector pAGM9121

preferable to first clone sub-parts as intermediate level −1 (level minus one) constructs.

These sub-parts are sequenced individually and are then further assembled to make the final level 0 module.

2 Materials

2.1 Cloning

1. PCR product purification kit.
2. 10 U/μL BsaI-HF®v2 (NEB, New England Biolabs Inc, Ipswich, MA, USA) (*see* **Note 3**). BsaI enzyme from Nippon Genetics Europe can also be used (FG-BsaI, 10 U/μL).
3. 10 U/μL *Bpi*I (Thermo Fisher Scientific, Waltham, MA, USA).
4. 3 U/μL T4 DNA Ligase (Promega Corporation, Madison, WI, USA), supplied with 10× ligation buffer: 300 mM Tris–HCl pH 7.8, 100 mM $MgCl_2$, 100 mM DTT, and 10 mM ATP.
5. Spectrophotometer for measuring DNA concentration, such as the NanoPhotometer NP80 (Implen GmbH, Schatzbogen). For more precise DNA quantification, a fluorometer can be used, such as the Qubit 4 Fluorometer from Thermo Fisher Scientific).
6. Luria-Bertani (LB) medium: 1% (*w/v*) bacto-tryptone, 0.5% (*w/v*) yeast extract, 1% (*w/v*) NaCl in deionized water, adjusted to pH 7.0 with 5 N NaOH. For plates, 1.0% (*w/v*) agar is added (*see* **Note 4**).
7. Antibiotics carbenicillin (used instead of ampicillin) and kanamycin: filter-sterilized stock of 50 mg/mL in H_2O (stored in aliquots at −20 °C) is diluted 1:1000 (final concentration: 50 μg/mL) in an appropriate amount of medium after the medium has been autoclaved and cooled down. For spectinomycin, a stock of 40 mg/mL is made and is used at a final concentration of 100 μg/mL (dilution 1:400).
8. 5-Bromo-4-chloro-3-indolyl-β-D-galactopyranoside (X-gal): stock solution of 20 mg/mL in dimethylformamide. For preparation of plates, the stock is diluted 1:500 (final concentration: 40 μg/mL) in an appropriate amount of LB agar after autoclaving/melting and cooling down.
9. Level 0 and level −1 cloning vectors, available in the Addgene MoClo Toolkit # 1000000044 (https://www.addgene.org/cloning/moclo/marillonnet/). The list of Level 0 cloning vectors is shown in Fig. 1a. The universal level −1 cloning vector is pAGM1311.

2.2 Screening of Colonies

1. Plasmid extraction kit for preparation of miniprep DNA.
2. A DNA ladder, such as the GeneRuler™ 1 kb Plus DNA Ladder as an example, for gel electrophoresis.
3. 50× Tris Acetate-EDTA (TAE) buffer: 242.0 g Tris, 57.1 mL of acetic acid, and 100 mL of 0.5 M EDTA, pH 8.0, in 1 L of deionized water. Running buffer for agarose gels is 1× TAE.

4. For preparation of gels for electrophoresis, agarose (0.7–1.5%) in 1× TAE is melted in a microwave oven and one drop of a 0.025% (w/v) ethidium bromide solution is added per 100 mL of melted agarose.
5. Gels are checked visually using a transilluminator, such as the Fusion FX transilluminator (Vilber Lourmat, Eberhardzell, Germany), and pictures are taken by using the gel analysis software included.
6. DNA analysis software for generation of plasmid maps, such as Geneious Prime software or SnapGene software, as examples.

3 Methods

3.1 Defining the Part Type

Cloning of standard parts requires to first define the type of part that needs to be made. Parts can consist of promoters, coding sequences, or terminators (Fig. 1a). Each part type is flanked by two BsaI sites in opposite orientations (example for a CDS1 part, Fig. 1b). The 4-nucleotide sequence of the cleavage sites, called fusion sites, is defined for each part type. For example, coding sequences parts (CDS1) are flanked by fusion sites AATG at the start codon and GCTT (an arbitrary fixed sequence that is placed after the stop codon). Coding sequence parts that a user may want to fuse to other sequences at the C-terminal end need to be cloned without a stop codon (CDS1 no stop, or CDS1 ns). In this case, the fusion site at the end of the sequence is defined as TTCG, with TCG coding for a serine. For each part type, a specific cloning vector is available with fusion sites compatible with the part type, for example pICH41308 for cloning of CDS1 parts (Fig. 1c). A universal level 0 cloning vector, pAGM9121 (Fig. 1d), is also available to make parts with non-standard sites, which is sometimes needed for special applications. pAGM9121 can also be used for cloning any of the standard parts.

Summary of actions:

1. Determine the part type, and, as a consequence, determine the two fusion sites that should flank the part.

3.2 Domestication of the Part and Addition of Fusion Sites

Domestication consists of removing internal restriction sites for the type IIS enzymes that are used with a modular cloning system: BsaI, BpiI, and, optionally, BsmBI for the MoClo system. This step is first performed in silico. Removal of internal sites in coding sequences can always be done with a silent mutation. In contrast, removal of sites in non-coding sequences is more problematic since it is generally not known whether the site overlaps with an enhancer sequence important for promoter function. Therefore, after domesticating promoter sequences, it will be necessary to validate experimentally

a

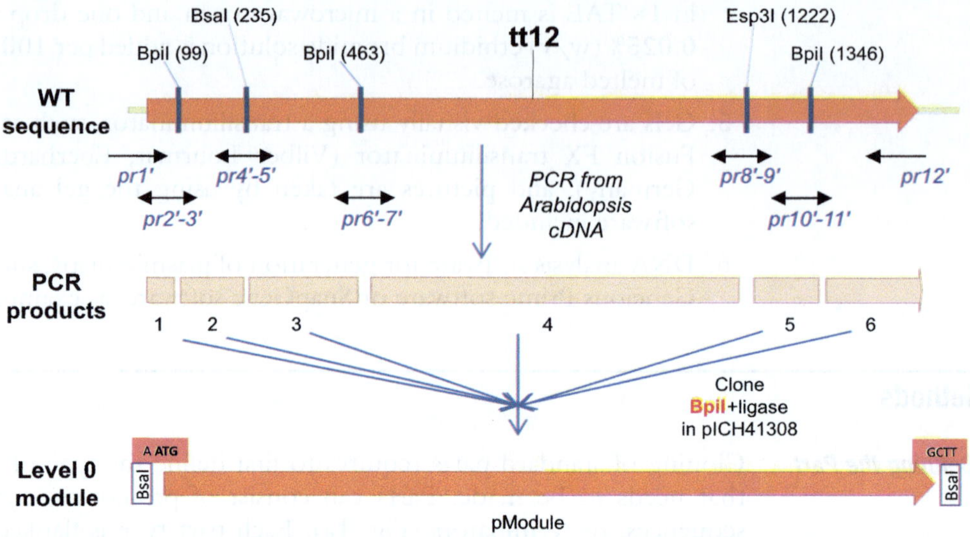

b

pr1'	tt gaagac aa aatg agctccacagagacatacgag
pr2'	tt gaagac aa cgat gaccgttcagttatctgagaatccg
pr3'	tt gaagac aa atcg ccagagatagaggagtttctc
pr4'	tt gaagac aa tacc actatagaggctccagagag
pr5'	tt gaagac aa ggta tctgttctgaattacatgctcagcttc
pr6'	tt gaagac aa gaac actgcagctgcaaggtgcaag
pr7'	tt gaagac aa gttc ctcacgttcctctactggtac
pr8'	tt gaagac aa cacg gctgctataacttctgcatcg
pr9'	tt gaagac aa cgtg tctgacctctttcctcttctcgc
pr10'	tt gaagac aa gtcc aatgacatagtacgtaacaagattcac
pr11'	tt gaagac aa ggac ttcctattggctgtgtccttggc
pr12'	tt gaagac aa aagc ttaaacacctgcgttagccatctcttg

Fig. 2 Cloning the *Arabidopsis* TT12 gene as an example of a CDS1 module. (**a**) The WT sequence contains five type IIS restriction sites that need to be removed. This requires the design of 12 primers (pr1′ to pr12′). Six PCR fragments are amplified and cloned using BpiI and ligase in level 0 cloning vector pICH41308. (**b**) Sequence of the primers

in the host organism that the promoter is working as intended. As an example for part domestication, we use here the *Arabidopsis* TT12 gene (GenBank accession NM_115765). This gene contains one BsaI site, three BpiI sites, and one BsmBI site (Fig. 2a). Therefore, five silent mutations are introduced to remove these sites (shown in green in Fig. 3). The next step consists of adding sequences for the standard fusion sites that will flank the part. Since this module will be a coding sequence with a stop codon (CDS1), we add one A before the start codon (resulting in sequence AATG) and the sequence GCTT after the stop codon (both fusion sites are highlighted in yellow in Fig. 3).

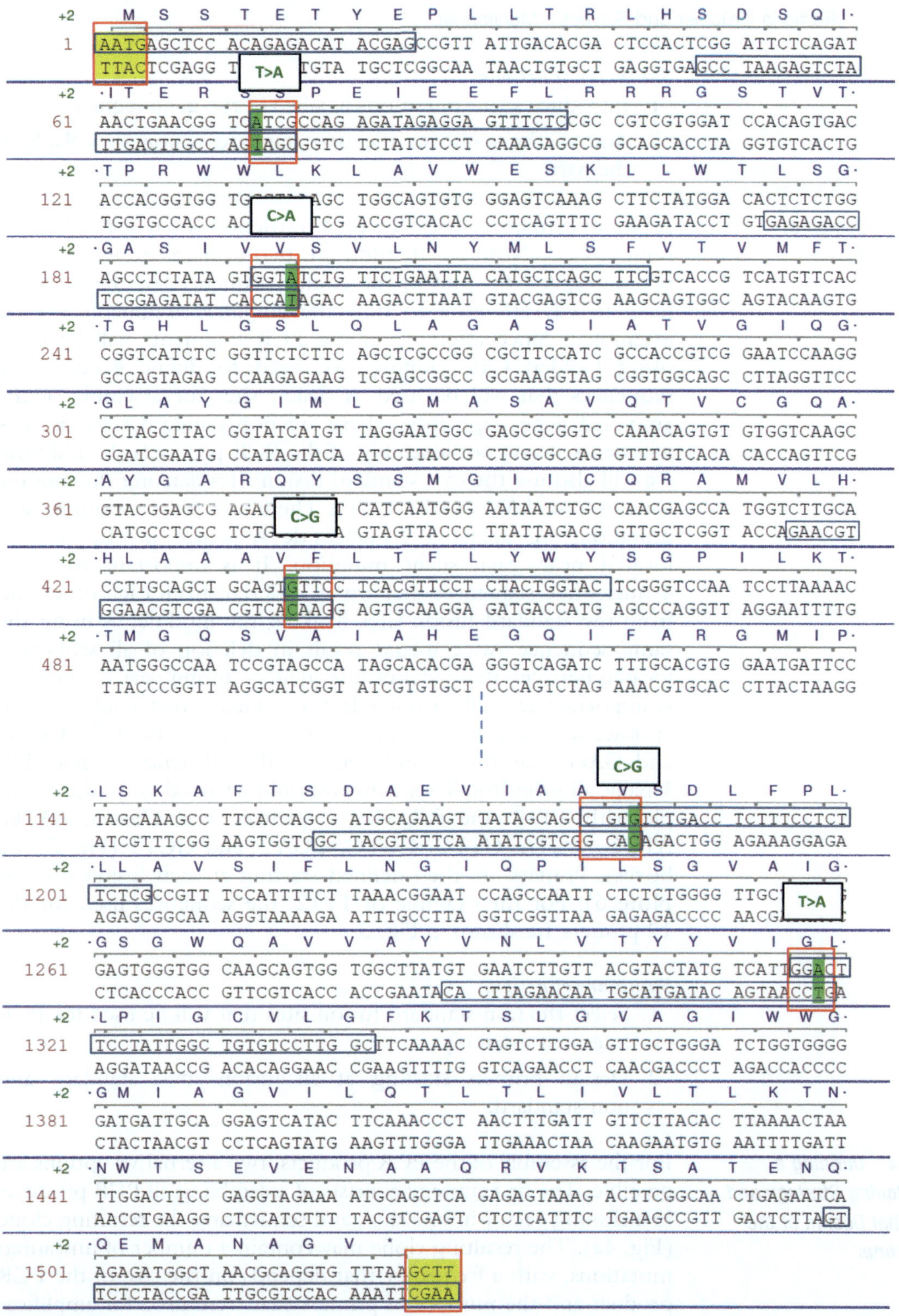

Fig. 3 Domestication of the TT12 gene. Nucleotide substitutions made to eliminate the Type IIS enzymes restriction sites are highlighted in green. The 4-bp sequences chosen as fusion sites are boxed in red. Standard fusion sites are highlighted in yellow. Primer sequences selected for amplification of the fragments are boxed in black

Summary of actions:

1. Introduce silent mutations in silico in the gene sequence.
2. Add to the sequence the two standard fusion sites flanking the part.

3.3 Design of Primers

A pair of primers is made for each restriction site that needs to be removed (*see* **Note 5**). In addition, two primers are made at the beginning and the end of the coding sequence. At this step, we need to pay attention to how the amplified fragments will be assembled. Since assembly of the PCR fragments will be done using Golden Gate cloning, we need to define 4-nucleotide sequences that will be used as fusion sites for assembly of the amplified gene fragments. These are non-standard sites, as they will be used only for assembly of the PCR products. In addition, we will also use the two standard fusion sites flanking the module (in this case AATG and GCTT). Therefore, for each introduced mutation, we define a 4-nucleotide sequence overlapping, or located near, each silent mutation. It is important to select 4-nucleotide sequences that are all different from each other and from the standard fusion sites flanking the module, as using the same sequence twice would result in deletion of all sequences located between these identical sequences during part assembly. It is important as well to not select a sequence that is identical to another sequence in the opposite orientation (for example, GCTC and GAGC) as this would decrease the efficiency of assembly. Finally, all palindromic sequences should be avoided. In the example with TT12, the following sequences were chosen: AATG, ATCG, GGTA, GTTC, CGTC, GGAC, and GCTT (Fig. 3, red boxes). Primers are then defined starting at each fusion site (*see* **Note 6**). For the example of TT12, the sequences used for the 12 primers are shown in Fig. 3.

Summary of actions:

1. Select the non-standard fusion sites that will be used for PCR fragment assembly.
2. Design primers starting at all fusion sites, standard and non-standard.

3.4 Defining a Cloning Strategy and Final Design of the Primer

For the assembly of the PCR products, two alternative options are possible. The first strategy consists of assembling all PCR products in a cloning vector in one step, and sequencing the resulting clone (Fig. 4a). The resulting clone may contain a number of unwanted mutations, with a frequency that depends on the size of the PCR product and the number of primers that were used for amplification. Unwanted mutations can come from the DNA polymerase during PCR amplification but may also come from random errors in the primers. Therefore, a part that has been assembled from a

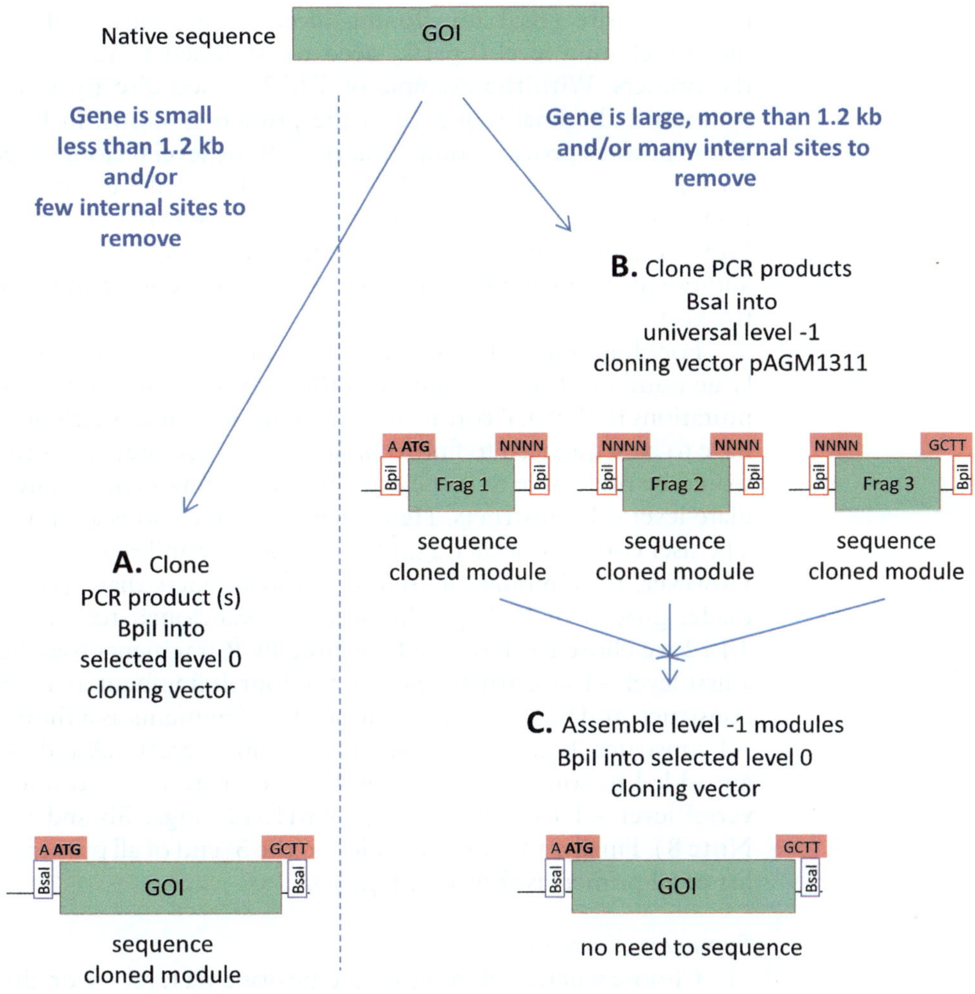

Fig. 4 Alternative strategies for cloning level 0 modules

larger number of PCR products will likely contain more unwanted mutations and may require sequencing more clones to get a correct one. A second problem that occurs when sequencing large modules (more than 1.2 kb) is that the module will be too large to get the entire sequence by sequencing from both sides. In this case, internal primers will be required to sequence the entire module. Such primers, located in part-specific sequences, will need to be designed and ordered specifically for each part. Therefore, cloning parts that are large (more than 1.2 kb) or that require assembly of a large number of PCR products is not a very efficient option. The alternative solution is to clone sub-fragments in an intermediate vector, sequence these intermediate constructs, and finally assemble the final part from these sub-parts (Fig. 4b, c). For these two alternatives, the sequence of the primers used for amplification will need to be different, as extensions containing a different type IIS enzyme restriction site (BsaI for cloning level −1 modules or BpiI for directly cloning level 0 parts) need to be added to the 5′ end of the primers. With the example of TT12 cloned directly as a level 0 module, the final sequence of the primers is shown in Fig. 2b. These primers provide compatibility with the level 0 cloning vector for CDS1 shown in Fig. 1c. Level 0 modules could also be cloned using the universal level 0 cloning vector pAGM9121 (Fig. 1d). However, in that case, the primers would need to contain an additional 4-nucleotide sequence to provide compatibility with the vector (*see* **Note** 7).

For cloning of TT12 CDS part, 12 primers are necessary. This large number of primers increases the chance of getting unwanted mutations in the final construct, which means that more clones may have to be sequenced to find a correct one. Therefore, in this case, it would be preferable to first clone several sub-fragments as intermediate level −1 constructs. Here again, several options are possible. The user could clone all fragments individually and sequence them. However, to minimize the number of constructs that need to be made, groups of PCR products can be subcloned together. For TT12 we chose to clone the first three PCR fragments together in a first level −1 construct, to clone the fourth fragment as a second construct, and finally to clone the last two fragments as a third level −1 construct (Fig. 5a). To do that, the sequences ACAA and ACAT are added to some primers to provide compatibility with the universal level −1 cloning vector pAGM1311 (Figs. 5b and 6) (*see* **Note 8**). Finally, a BsaI site is added to the 5′ end of all primers. The list of all primers is shown in Fig. 5b.

Summary of actions:

1. Choose whether cloning should be done via level −1 or directly as level 0.
2. If cloning is done as level −1 construct, add the sequences ACAT and ACAA to the 5′ end of selected primers.
3. Add the type IIS recognition sequence corresponding to the chosen cloning strategy (BsaI or BpiI) to the 5′ end of the primers. Now all primers are ready for PCR amplification.

3.5 PCR Amplification of the Fragments

The template for PCR depends on the module type that is made. For promoters, genomic DNA will be used. For coding sequences lacking introns, first-strand cDNA will usually be used as a template, but any plasmid containing this sequence could be used as well. For cloning the coding sequence of the TT12 gene, we use *Arabidopsis* cDNA as a template.

1. A PCR mix is set up following the manufacturer's instructions. For example, using KOD polymerase, the following conditions are used: 1 μL cDNA (5–20 ng/μL), 5 μL of 10× buffer, 3 μL of 25 mM MgSO4, 5 μL of 2 mM dNTPs, 1.5 μL each of

a

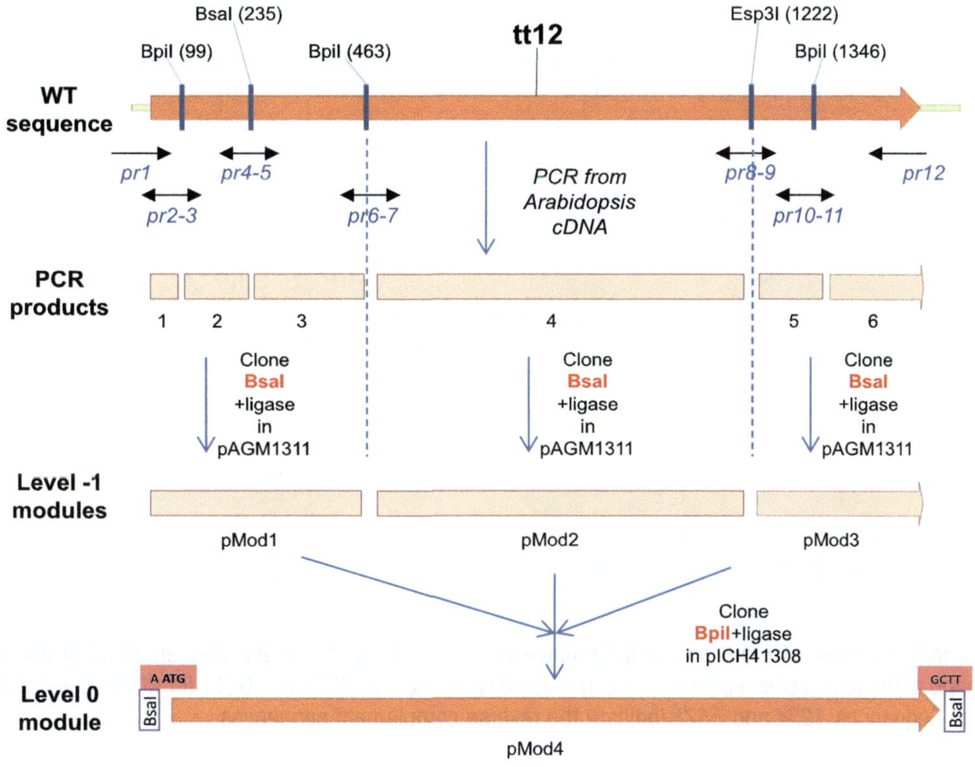

b

pr1	tt ggtctc a ACAT aatg agctccacagagacatacgag
pr2	tt ggtctc a cgat gaccgttcagttatctgagaatccg
pr3	tt ggtctc a atcg ccagagatagaggagtttctc
pr4	tt ggtctc a tacc actatagaggctccagagag
pr5	tt ggtctc a ggta tctgttctgaattacatgctcagcttc
pr6	tt ggtctc a ACAA gaac actgcagctgcaaggtgcaag
pr7	tt ggtctc a ACAT gttc ctcacgttcctctactggtac
pr8	tt ggtctc a ACAA cacg gctgctataacttctgcatcg
pr9	tt ggtctc a ACAT cgtg tctgacctctttcctcttctcgc
pr10	tt ggtctc a gtcc aatgacatagtacgtaacaagattcac
pr11	tt ggtctc a ggac ttcctattggctgtgtccttggc
pr12	tt ggtctc a ACAA aagc ttaaacacctgcgttagccatctcttg

Fig. 5 Cloning the *Arabidopsis* TT12 gene via level −1 cloning as an example for cloning of large modules. (**a**) Six PCR fragments amplified from WT cDNA are cloned by groups of 3, 1, and 2 PCR products in the universal level −1 cloning vector pAGM1311. These clones are sequenced and then assembled in the level 0 cloning vector pICH41308. (**b**) Sequence of the primers used in (**a**)

10 μM sense and antisense primers, and 1 μL of KOD Hot Start DNA polymerase (10 U/μL, final concentration 0.02 U/μL) in a total reaction volume of 50 μL.

2. PCR is performed using the following cycling conditions: (1) incubation at 95 °C for 2 min for polymerase activation, (2) denaturation at 95 °C for 20 s, (3) annealing at 60 °C for 10 s (the temperature for the annealing step can be adjusted for

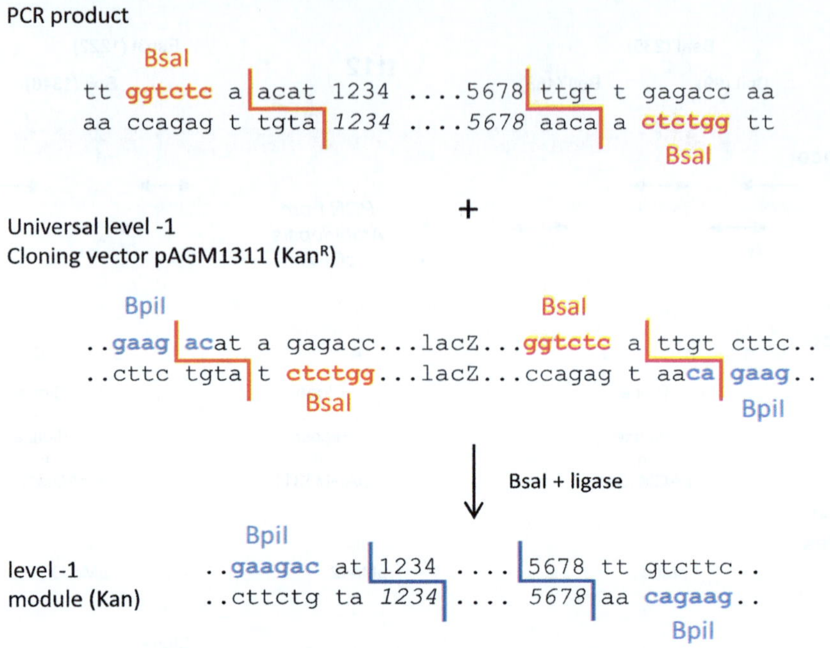

Fig. 6 Cloning of level −1 modules using the universal level −1 cloning vector. Sequences 1234 and 5678 correspond to the fusion sites specific for each module type, such as AATG and GCTT for CDS1 modules as an example. Sequences *1234* and *5678* indicate the reverse complement sequences

specific primers, but the temperature of 58 °C usually works well for most primers), and (4) extension at 70 °C, and the duration depends on the length of the expected fragment (from 10 s/kb for fragments smaller than 500 bp up to 25 s/kb for fragments larger than 3 kb; see manufacturer's instructions); **steps 2–4** are repeated 35–40 times and are followed by a final extension step at 70 °C for 20 s to 2 min (depending on fragment length). The reaction is then kept at 12 °C until taken out of the thermal cycler.

3. Of the PCR product obtained, 5 μL is analyzed by gel electrophoresis to make sure that a product of the correct size has been amplified.
4. The amplified fragment is purified from remaining primers, potential primer dimers, and remaining polymerase enzyme by using a PCR purification kit following the kit protocol. DNA is eluted from the column with 30–50 μL of elution buffer (5 mM Tris–HCl, pH 8.5). In case several bands were amplified rather than only the expected fragment, the same kit can also be used to cut and extract the appropriate DNA fragment from an agarose gel.

3.6 Cloning of Level −1 Constructs

1. Set up a restriction-ligation by pipetting into a tube 20 fmol (approximately 50 ng; *see* **Note 9**) of each purified PCR product and of the level −1 vector pAGM1311 (Addgene plasmid #47983), 1.5 μL 10× ligation buffer, 5 U (0.5 μL) of *Bsa*I, and 3 U (1 μL) of ligase (final volume of 15 μL).
2. Incubate the restriction-ligation mix in a thermal cycler for 1–4 h at 37 °C.
3. Restriction-ligation is followed by a digestion step (5 min at 50 °C) and then by heat inactivation for 5 min at 80 °C. The final incubation step at 80 °C is very important and is needed to inactivate the ligase at the end of the restriction-ligation. Omitting this step would lead to re-ligation of some of the insert and plasmid backbone fragments when the reaction vessel is taken out of the thermal cycler, and would lead to a higher proportion of colonies containing incorrect constructs.
4. Transform the entire ligation into chemically competent DH10B cells (*see* **Note 10**).
5. Thaw frozen chemically competent cells (50 μL per tube) on ice.
6. Add the entire Golden Gate reaction to the cells, and incubate on ice for 30 min.
7. Incubate 90 s at 42 °C in a water bath.
8. Let the cells recover on ice for 2 min.
9. Add 0.5 mL of LB medium to the cells and incubate the tube at 37 °C in a shaker-incubator (150 rpm) for 45 min.
10. After incubation, plate 10–20 μL of the transformation on LB agar plates containing Kanamycin and X-gal.
11. Incubate the plates overnight at 37 °C. Many white and very few blue colonies should be obtained.
12. Pick two white colonies from the plate and transfer to liquid TB medium containing Kanamycin.
13. Grow bacterial culture over night at 37 °C and use for isolation of plasmid DNA.
14. Check plasmids by restriction digest using *Bpi*I (*see* **Note 11**).
15. Sequence the clones with vector primers moclof (agcgaggaagcggaagagcg) and moclor (gccacctgacgtctaagaaacc) if cloning was done in the MoClo level −1 vector pAGM1311 (*see* **Notes 12** and **13**).

3.7 Assembly of Level 0 Modules

Level 0 modules are cloned from either PCR products (for parts that are not too large), or from sequenced level −1 modules. The Golden Gate reaction is virtually the same as described in Subheading 3.6 except that the enzyme *Bpi*I is used instead of *Bsa*I.

1. Set up a restriction-ligation by pipetting into a tube 20 fmol of each level −1 module (or PCR products) and of the selected level 0 cloning vector (in the present example, pICH41308), 1.5 μL 10× ligation buffer, 5 U (0.5 μL) of *Bpi*I, and 3 U (1 μL) of ligase (final volume of 15 μL).
2. Incubate the restriction-ligation mix in a thermal cycler. For assembly of two to four level −1 modules, incubation for 1–4 h at 37 °C is sufficient. If more modules are ligated together, ligation is done in a volume of 25 μL (2.5 μL 10× ligation buffer, 1 μL of 10 U/μL BpiI, and 1.5 μL of 3 U/μL ligase) and a cycling program is used as following: 2 min at 37 °C followed by 5 min at 16 °C, both repeated 50 times.
3. Restriction-ligation is followed by a digestion step (5 min at 37 °C) and then by heat inactivation for 10 min at 80 °C.
4. Transformation of the reaction mix into *Escherichia coli*. Cells are spread on LB plates containing Spectinomycin and X-gal. Colonies containing correct constructs should be white, while colonies containing uncut or re-ligated vector should be blue. Plasmids should be isolated from a few colonies. For clones assembled from level −1 modules, two colonies are generally more than enough. Plasmid miniprep DNA extracted from colonies should be analyzed by a restriction digest with BsaI. The final clones need to be sequenced if they were assembled from PCR products, but do not need to be sequenced if they were assembled from level −1 parts, as PCR was not involved.

Level 0 modules are then ready for assembly into level 1 constructs (usually containing complete transcription units) and then to level 2 [13], M or P [14] for assembly of multigene constructs.

4 Notes

1. The most commonly used type IIS enzymes used are BsaI, BpiI, and BsmBI. All of these have a 6 bp recognition site sequence and a 4 nt cleavage site located 1 or 2 bp away from the recognition site. SapI (isoschizomer, LguI) and AarI have a 7 bp recognition site sequence. This can be useful as these enzymes will cleave DNA on average every 8 kb (for DNA with an average GC content of 50%) versus 2 kb for 6 bp cutters.
2. Removing BsmBI restriction sites, in addition to BsaI and BpiI sites, requires more work but has some advantages. For example, one may want to make a construct containing several transcription units, but would like to test many variants of a given coding sequence in one of the transcription units. Rather than making a completely new construct from parts for each

new sequence variant, it is possible to make a construct that contains all transcription units but lacks the coding sequence (or promoter or terminator) in one of the transcription units. Instead of the coding sequence, a dummy sequence with two restriction sites is cloned. These sites should allow the multigene construct to be digested to insert the various coding sequences of interest in a final cloning step. These sites cannot be BsaI or BpiI, as these are needed for construction of the multigene construct. A third enzyme is necessary, and this can be BsmBI. All cloning vectors present in the MoClo kit have been domesticated for BsaI and BpiI and BsmBI, and are therefore suitable for this purpose.

3. BsaI is sold by NEB, but isoschizomers of BsaI (with the same recognition site sequence) are also available from other companies, for example Eco31I from Thermo Fisher Scientific. These enzymes do not necessarily have optimal cleavage efficiency in the same buffer. With the conditions that we are using, we have obtained better cloning efficiency with BsaI than with Eco31I, but it is possible that other restriction-ligation buffers may work well for Eco31I as well. NEB initially sold the standard BsaI enzyme, but has now replaced it by BsaI-HFv2 (version 2) that is claimed to have higher specificity than standard BsaI. This enzyme seems to work quite well, as it was reported that 24 fragments could be assembled with good efficiency using this enzyme [3], and more recently 52 fragments [4].
4. LB medium can be easily made using commercially available ready mixes. For example, we prepare liquid LB by adding 25 g of LB Broth High salt mix (from Duchefa) to 1 L of water. For LB solid medium, add 35 g of LB Agar High salt (Duchefa) to 1 L of water. Media is then autoclaved.
5. Two mutations that need to be introduced in close proximity can be eliminated with the same pair of primers. For example, mutations that need to be introduced 40 bp apart may be eliminated by selecting a fusion site in the middle of the sequence between these two sites. The two primers selected for amplification in different directions will each introduce a mutation. It is important that the primers are long enough so that enough sequence located after the mismatch is able to bind the template sequence. Note that long primers up to 100 nt can easily be synthesized by companies. We usually select for the lower price available (no other purification other than salt free). Such primers work just fine.
6. We usually design primers containing 12 G or C and with a length of at least 20 nt. For very AT-rich sequences, a longer primer might work with fewer G or C, but with a minimum of at least 9 G or C. Alternatively, the user can use one of many

online programs to select a primer with a desired melting temperature. It is important to calculate the Tm by considering only the sequence located downstream of the last introduced mutation, since what counts is annealing of the primer to its non-mutated template at the first amplification cycle.

7. Cloning a standard part in the universal level 0 cloning vector pAGM9121 (Addgene Plasmid #51833) requires to add sequences CTCA and CTCG to the primers, just upstream of the standard fusion sites, to provide compatibility to the vector. In the example with TT12 in Fig. 2b, pr1′ should be tt gaagac aa CTCA AATG agctccacagagacatacgag and pr12′ should be tt gaagac aa CTCG AAGC ttaaacacctgcgttagccatctcttg. The advantage of using the cloning vectors specific for each part rather than pAGM9121 is that the primers needed for amplification are 4 nucleotides shorter. But no matter which type of vector is used, the resulting part will be exactly the same. pAGM9121 can also be used to clone non-standard parts. This is because any non-standard fusion site of choice can be included in the primer. The forward primer for amplification will have the sequence tt gaagac aa CTCA, followed by 4 nucleotides of the upstream non-standard fusion site, followed by several nucleotides for annealing to the sequences interest. The reverse primer will have the sequence tt gaagac aa, CTCG, followed by 4 nucleotides complementary to the downstream non-standard fusion site, followed by several nucleotides for annealing to the sequences interest.
8. When cloning PCR fragments in the level −1 universal vector, the fusion sites ACAT and TTGT (complementary of ACAA) will be used for cloning. In cases where the level −1 modules are assembled from more than one PCR fragment, it is important to check that the non-standard fusion sites used for assembling the PCR fragments are all different from ACAT and TTGT.
9. In practice, if all module plasmids and the vector have approximately the same size (4–5 kb), simply adding 50 ng of DNA of each module and of the vector will work relatively well. However, when plasmids with widely different sizes are used, calculating an equimolar amount should provide a higher cloning efficiency. We have previously recommended using 40 fmol of each fragment and of the vector for cloning. It, however, appears that using less DNA leads to higher cloning efficiency. We therefore now recommend using 20 fmol for each insert and vector. The following formula (from the NEB catalog) can be used: 1 μg of a 1000 bp DNA fragment corresponds to 1.52 pmol. Therefore, the volume of DNA to pipet (in μL) to have 20 fmol is given by the equation: 20 (fmol) × size (bp) of the DNA fragment/(concentration (ng/μL) × 1520).

Frequently, this formula will lead to volumes <1 μL with standard miniprep DNA. In these cases, we recommend diluting the DNA in order to avoid pipetting errors.

10. Commercially produced cells can be used, but usually lab-made competent cells are sufficient.
11. White colonies may contain inserts smaller than the expected size. This is because PCR amplification often leads to at least some amount of primer dimer production. Primer dimers are normally removed by column purification of the products. However, it sometimes happens that some primer dimers are not completely removed by column purification. Some primer dimers are flanked by BsaI restriction sites and can therefore be cloned as efficiently as the full-length products. A digestion of the clones with BpiI allows the identification of the clones containing full-length inserts.
12. Usually, sequencing of one clone may be enough, especially when the sequence of interest was amplified from plasmid sequences. However, when cloned sequences were amplified from DNA or cDNA prepared from an organism of interest, it is quite common to find sequence polymorphisms that differ from the expected sequence. If the polymorphism leads to an amino-acid change, it is important to know whether we are dealing with a polymorphism or a PCR mutation. Therefore, a second clone needs to be sequenced. If the same polymorphism is seen twice, one can assume that we are dealing with a polymorphism and not a PCR mutation.
13. Another issue that may arise when sequencing level −1 clones (or alternatively, level 0 clones) assembled from multiple PCR fragments is the presence of chimeric sequences. This can occur when the sequence of interest is part of a gene family. In such instances, PCR amplification can yield sequences from multiple homologous genes, leading to the cloning of chimeric sequences that originate from several homologues. In such cases, it can be beneficial to consider revising the primer design and clone each amplified fragment individually, rather than in groups of several fragments. After sequencing, the correct fragments can be chosen for assembly of final level 0 module.

References

1. Engler C, Gruetzner R, Kandzia R, Marillonnet S (2009) Golden gate shuffling: a one-pot DNA shuffling method based on type IIs restriction enzymes. PLoS One 4(5):e5553. https://doi.org/10.1371/journal.pone.0005553
2. Engler C, Kandzia R, Marillonnet S (2008) A one pot, one step, precision cloning method with high throughput capability. PLoS One 3(11):e3647. https://doi.org/10.1371/journal.pone.0003647
3. Potapov V, Ong JL, Kucera RB, Langhorst BW, Bilotti K, Pryor JM, Cantor EJ, Canton B, Knight TF, Evans TC Jr, Lohman GJS (2018) Comprehensive profiling of four base overhang ligation Fidelity by T4 DNA

ligase and application to DNA assembly. ACS Synth Biol 7(11):2665–2674. https://doi.org/10.1021/acssynbio.8b00333

4. Pryor JM, Potapov V, Bilotti K, Pokhrel N, Lohman GJS (2022) Rapid 40 kb genome construction from 52 parts through data-optimized assembly design. ACS Synth Biol 11(6):2036–2042. https://doi.org/10.1021/acssynbio.1c00525
5. Weber E, Engler C, Gruetzner R, Werner S, Marillonnet S (2011) A modular cloning system for standardized assembly of multigene constructs. PLoS One 6(2):e16765. https://doi.org/10.1371/journal.pone.0016765
6. Werner S, Engler C, Weber E, Gruetzner R, Marillonnet S (2012) Fast track assembly of multigene constructs using Golden Gate cloning and the MoClo system. Bioeng Bugs 3(1): 38–43. https://doi.org/10.4161/bbug.3.1.18223
7. Binder A, Lambert J, Morbitzer R, Popp C, Ott T, Lahaye T, Parniske M (2014) A modular plasmid assembly kit for multigene expression, gene silencing and silencing rescue in plants. PLoS One 9(2):e88218. https://doi.org/10.1371/journal.pone.0088218
8. Sarrion-Perdigones A, Vazquez-Vilar M, Palaci J, Castelijns B, Forment J, Ziarsolo P, Blanca J, Granell A, Orzaez D (2013) GoldenBraid 2.0: a comprehensive DNA assembly framework for plant synthetic biology. Plant Physiol 162(3):1618–1631. https://doi.org/10.1104/pp.113.217661
9. Lampropoulos A, Sutikovic Z, Wenzl C, Maegele I, Lohmann JU, Forner J (2013) GreenGate---a novel, versatile, and efficient cloning system for plant transgenesis. PLoS One 8(12):e83043. https://doi.org/10.1371/journal.pone.0083043
10. Terfruchte M, Joehnk B, Fajardo-Somera R, Braus GH, Riquelme M, Schipper K, Feldbrugge M (2014) Establishing a versatile Golden Gate cloning system for genetic engineering in fungi. Fungal Genet Biol 62:1–10. https://doi.org/10.1016/j.fgb.2013.10.012
11. Lee ME, DeLoache WC, Cervantes B, Dueber JE (2015) A highly characterized yeast toolkit for modular, multipart assembly. ACS Synth Biol 4(9):975–986. https://doi.org/10.1021/sb500366v
12. Agmon N, Mitchell LA, Cai Y, Ikushima S, Chuang J, Zheng A, Choi WJ, Martin JA, Caravelli K, Stracquadanio G, Boeke JD (2015) Yeast Golden Gate (yGG) for the efficient assembly of S. Cerevisiae transcription units. ACS Synth Biol 4(7):853–859. https://doi.org/10.1021/sb500372z
13. Iverson SV, Haddock TL, Beal J, Densmore DM (2016) CIDAR MoClo: improved MoClo assembly standard and new E. coli part library enable rapid combinatorial design for synthetic and traditional biology. ACS Synth Biol 5(1):99–103. https://doi.org/10.1021/acssynbio.5b00124
14. Moore SJ, Lai HE, Kelwick RJ, Chee SM, Bell DJ, Polizzi KM, Freemont PS (2016) EcoFlex: a multifunctional MoClo kit for *E. coli* synthetic biology. ACS Synth Biol 5(10): 1059–1069. https://doi.org/10.1021/acssynbio.6b00031
15. Hernanz-Koers M, Gandia M, Garrigues S, Manzanares P, Yenush L, Orzaez D, Marcos JF (2018) FungalBraid: a GoldenBraid-based modular cloning platform for the assembly and exchange of DNA elements tailored to fungal synthetic biology. Fungal Genet Biol 116:51–61. https://doi.org/10.1016/j.fgb.2018.04.010
16. Kowarschik K, Hoehenwarter W, Marillonnet S, Trujillo M (2018) UbiGate: a synthetic biology toolbox to analyse ubiquitination. New Phytol 217(4):1749–1763. https://doi.org/10.1111/nph.14900
17. Occhialini A, Piatek AA, Pfotenhauer AC, Frazier TP, Stewart CN Jr, Lenaghan SC (2019) MoChlo: a versatile, modular cloning toolbox for chloroplast biotechnology. Plant Physiol 179(3):943–957. https://doi.org/10.1104/pp.18.01220
18. Vasudevan R, Gale GAR, Schiavon AA, Puzorjov A, Malin J, Gillespie MD, Vavitsas K, Zulkower V, Wang B, Howe CJ, Lea-Smith DJ, McCormick AJ (2019) CyanoGate: a modular cloning suite for engineering cyanobacteria based on the plant MoClo syntax. Plant Physiol. https://doi.org/10.1104/pp.18.01401
19. Crozet P, Navarro FJ, Willmund F, Mehrshahi P, Bakowski K, Lauersen KJ, Perez-Perez ME, Auroy P, Gorchs Rovira A, Sauret-Gueto S, Niemeyer J, Spaniol B, Theis J, Trosch R, Westrich LD, Vavitsas K, Baier T, Hubner W, de Carpentier F, Cassarini M, Danon A, Henri J, Marchand CH, de Mia M, Sarkissian K, Baulcombe DC, Peltier G, Crespo JL, Kruse O, Jensen PE, Schroda M, Smith AG, Lemaire SD (2018) Birth of a photosynthetic chassis: a MoClo toolkit enabling synthetic biology in the microalga Chlamydomonas reinhardtii. ACS Synth Biol 7(9):2074–2086. https://doi.org/10.1021/acssynbio.8b00251
20. Prielhofer R, Barrero JJ, Steuer S, Gassler T, Zahrl R, Baumann K, Sauer M, Mattanovich D, Gasser B, Marx H (2017) GoldenPiCS: a Golden Gate-derived modular cloning system

for applied synthetic biology in the yeast Pichia pastoris. BMC Syst Biol 11(1):123. https://doi.org/10.1186/s12918-017-0492-3

21. Engler C, Youles M, Gruetzner R, Ehnert TM, Werner S, Jones JD, Patron NJ, Marillonnet S (2014) A golden gate modular cloning toolbox for plants. ACS Synth Biol 3(11):839–843. https://doi.org/10.1021/sb4001504

22. Gantner J, Ordon J, Ilse T, Kretschmer C, Gruetzner R, Lofke C, Dagdas Y, Burstenbinder K, Marillonnet S, Stuttmann J (2018) Peripheral infrastructure vectors and an extended set of plant parts for the modular cloning system. PLoS One 13(5):e0197185. https://doi.org/10.1371/journal.pone.0197185

Chapter 2

Golden Gate Cloning of Multigene Constructs Using the Modular Cloning System MoClo

Sylvestre Marillonnet and Stefan Werner

Abstract

Modular cloning systems that rely on type IIS enzymes for DNA assembly have many advantages for construct engineering for biological research and synthetic biology. These systems are simple to use, efficient, and allow users to assemble multigene constructs by performing a series of one-pot assembly steps, starting from libraries of cloned and sequenced parts. The efficiency of these systems also facilitates the generation of libraries of construct variants. We describe here a protocol for assembly of multigene constructs using the modular cloning system MoClo. Making constructs using the MoClo system requires to first define the structure of the final construct to identify all basic parts and vectors required for the construction strategy. The assembly strategy is then defined following a set of standard rules. Multigene constructs are then assembled using a series of one-pot assembly steps with the set of identified parts and vectors.

Key words Modular cloning, Synthetic biology, Biological parts, DNA assembly, Multigene constructs

1 Introduction

For many years, DNA constructs were made using protocols that required a succession of labor-intensive DNA manipulation steps that included, for example, DNA digestion using one or two restriction enzymes, partial or complete filling of DNA ends, dephosphorylation of vector ends, gel extraction, ligation, transformation of *Escherichia coli* cells, and screening of colonies. These protocols worked, and are still used today, but are relatively time-consuming and inefficient. Fortunately, they can now be replaced by more efficient strategies that include homology-based assembly methods and type-IIS-based assembly methods. Type-IIS-based methods rely on Golden Gate cloning, a method that allows assembly of multiple DNA fragments using a simple one-pot, one-step assembly reaction. Golden Gate cloning combines the use of a type IIS enzyme with ligase in a restriction-ligation reaction [3]. Type

Daniel Schindler (ed.), *Golden Gate Cloning: Methods and Protocols*, Methods in Molecular Biology, vol. 2850,
https://doi.org/10.1007/978-1-0716-4220-7_2,

IIS enzymes, such as BsaI or BpiI, are characterized by the fact that the restriction enzyme cleavage site lies outside of the enzyme DNA recognition site. This allows DNA digestion and ligation to be performed simultaneously, as recognition sites for the type IIS enzyme used for cloning are eliminated during the cloning reaction. Golden Gate has been used as the base of several related but not identical modular cloning systems for plants, yeast, mammalian cells, and fungi [1, 2, 7–10, 12–22]. These cloning systems differ by the specific type IIS enzymes used for cloning at the various assembly steps, by the sequences of the type IIS cleavage sites that are used for cleavage and assembly, or by the antibiotic selection markers present on the various vector backbones. We present here a protocol for the modular system MoClo that was first designed to be used in plants [20], but that can easily be adapted for use in other organisms such as yeast and bacteria. The MoClo system is a hierarchical assembly system by which transcription units are assembled from a set of level 0 modules that contain basic genetic elements such as promoters, coding sequences, and terminators. The resulting level 1 constructs are then assembled in multigene constructs using two alternative strategies: (i) assembly in level 2 constructs by successive addition of sets of up to six level 1 transcription units to a growing construct or (ii) assembly of several intermediate level M multigene constructs, which are then assembled in a level P construct (Fig. 1) (*see* **Note 1**). This chapter describes how to make multigene constructs using level M and P constructs.

2 Materials

2.1 Cloning

1. 10 U/μL *BsaI-HF®v2* (NEB, New England Biolabs Inc., Ipswich, MA, USA) (*see* **Note 2**).
2. 10 U/μL *Bpi*I (Thermo Fisher Scientific).
3. 3 U/μL T4 DNA Ligase or 20 U/μL T4 DNA Ligase (HC), both supplied with 10× ligation buffer: 300 mM Tris-HCl pH 7.8, 100 mM $MgCl_2$, 100 mM DTT, and 10 mM ATP.
4. A spectrophotometer for DNA quantification, for example the NanoDrop ND2000 (Thermo Fisher Scientific). For more precise DNA quantification, a fluorometer can be used such as the Qubit 4 Fluorometer (Thermo Fisher Scientific) (*see* **Note 3**).
5. Luria-Bertani (LB) medium: 1% (*w/v*) bacto-tryptone, 0.5% (*w/v*) yeast extract, 1% (*w/v*) NaCl in deionized water, adjusted to pH 7.0 with 5 N NaOH. For plates, 1.5% (*w/v*) agar is added.

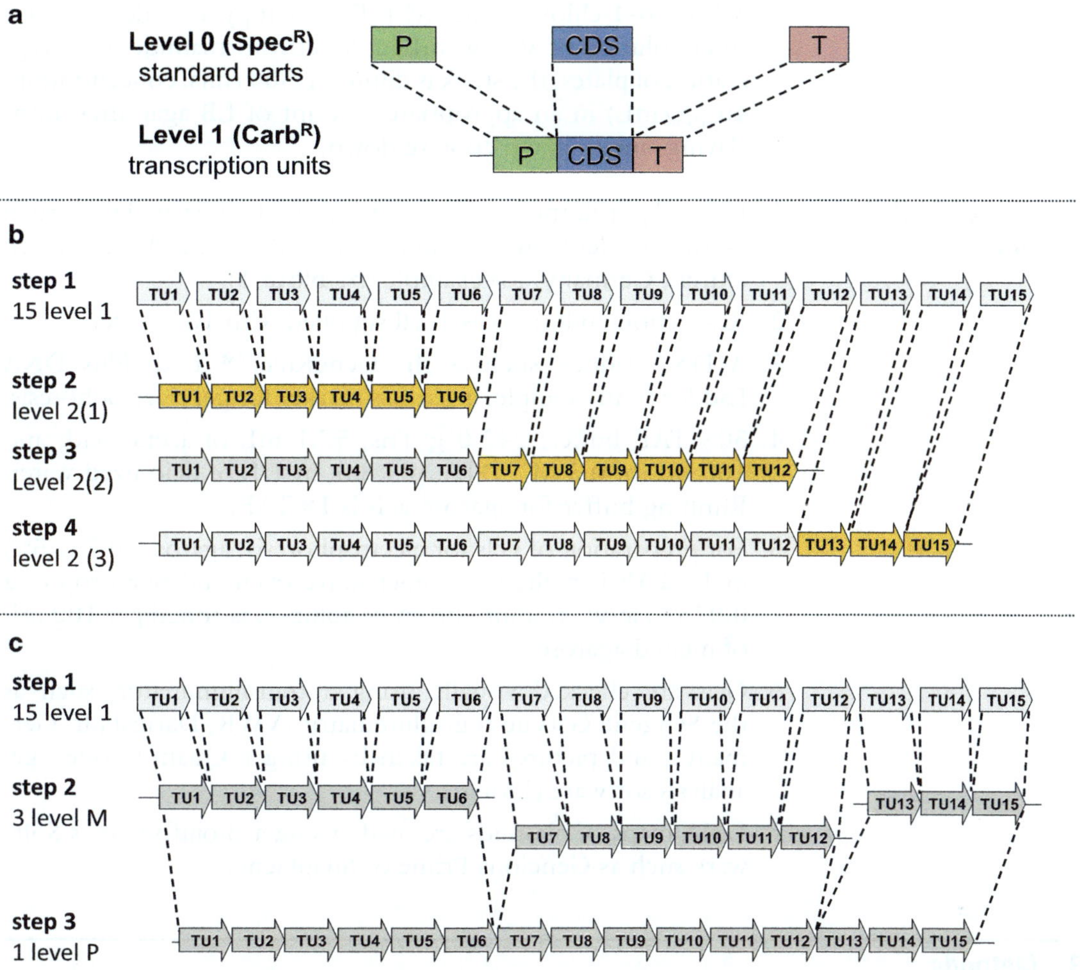

Fig. 1 Comparison of multigene assembly using level 2 versus levels M and P. (**a**) Level 1 constructs are assembled from standard parts (level 0 modules) such as promoters (P), coding sequences (CDS), and terminators (T) using a first Golden Gate cloning reaction, resulting in constructs containing a transcription unit (TU). To make a construct containing 15 TU (as an example), 15 level 1 constructs (represented as arrows in **b** and **c**) are assembled in parallel from selected level 0 modules. (**b**) Assembly of multigene constructs using level 2 constructs. Up to six level 1 constructs are assembled at each step into a level 2 construct that grows larger at each step. For example, steps 2–4 show the successive addition of 6, 6, and 3 transcription units to a level 2 construct (transcription units cloned directly from level 1 constructs are shown in yellow). (**c**) Assembly of multigene constructs using level M and P constructs. Groups of up to six transcription units are preassembled in level M constructs. The level M constructs are then assembled in a level P construct

6. Antibiotics carbenicillin (used instead of ampicillin) and kanamycin: filter-sterilized stocks of 50 mg/mL in H_2O (stored in aliquots at −20 °C) are diluted 1:1000 (final concentration: 50 μg/mL) in an appropriate amount of medium after the medium has been autoclaved and cooled down. For spectinomycin, a stock of 40 mg/mL is made and is used at a final concentration of 100 μg/mL (dilution 1:400).

7. 5-Bromo-4-chloro-3-indolyl-β-D-galactopyranoside (X-gal): stock solution of 20 mg/mL in dimethylformamide. For preparation of plates, the stock is diluted 1:500 (final concentration: 40 μg/mL) in an appropriate amount of LB agar after autoclaving/melting and cooling down.

2.2 Screening of Colonies

1. Plasmid purification kit for preparation of miniprep DNA, such as the NucleoSpin® Plasmid Quick Pure (Macherey-Nagel, Düren, Germany) for example (*see* **Note 4**).
2. Restriction endonucleases, all supplied with 10× buffer.
3. A DNA ladder, such as the GeneRuler™ 1 kb Plus DNA Ladder as an example, is used as marker for gel electrophoresis.
4. 50× TAE buffer: 242.0 g Tris, 57.1 mL of acetic acid, and 100 mL of 0.5 M EDTA, pH 8.0, in 1 L of deionized water. Running buffer for agarose gels is 1× TAE.
5. For preparation of gels for electrophoresis, agarose (0.7–1.5%) in 1× TAE is melted in a microwave oven and one drop of a 0.025% (*w/v*) ethidium bromide solution is added per 100 mL of melted agarose.
6. Gels are checked visually using a transilluminator, such as the Syngene GelVue transilluminator (VWR, Darmstadt, Germany), and pictures are taken by using a Quantity one® gel analysis software (Biorad).
7. DNA maps of plasmids are made using a Bioinformatics Software such as Geneious Prime or SnapGene.

3 Methods

The first and perhaps most important step in the assembly of multigene constructs is planning the structure of the final construct and determining the assembly strategy. For planning the construct, one needs to consider the composition of each transcription unit (or gene), the number of genes, the order and orientation of the genes in the assembled construct, and the potential use of a selection marker for stable plant transformation, which, in fact, simply consists of adding one more gene to the multigene construct. There is no theoretical limit to the number of genes that can be assembled; however, the limit is set by the size and the stability of the plasmid that has to be cloned and maintained in *E. coli*. The process of selecting standard or individualized parts for the transcription units, the choice of destination vectors, and their assembly into higher-order constructs are described below.

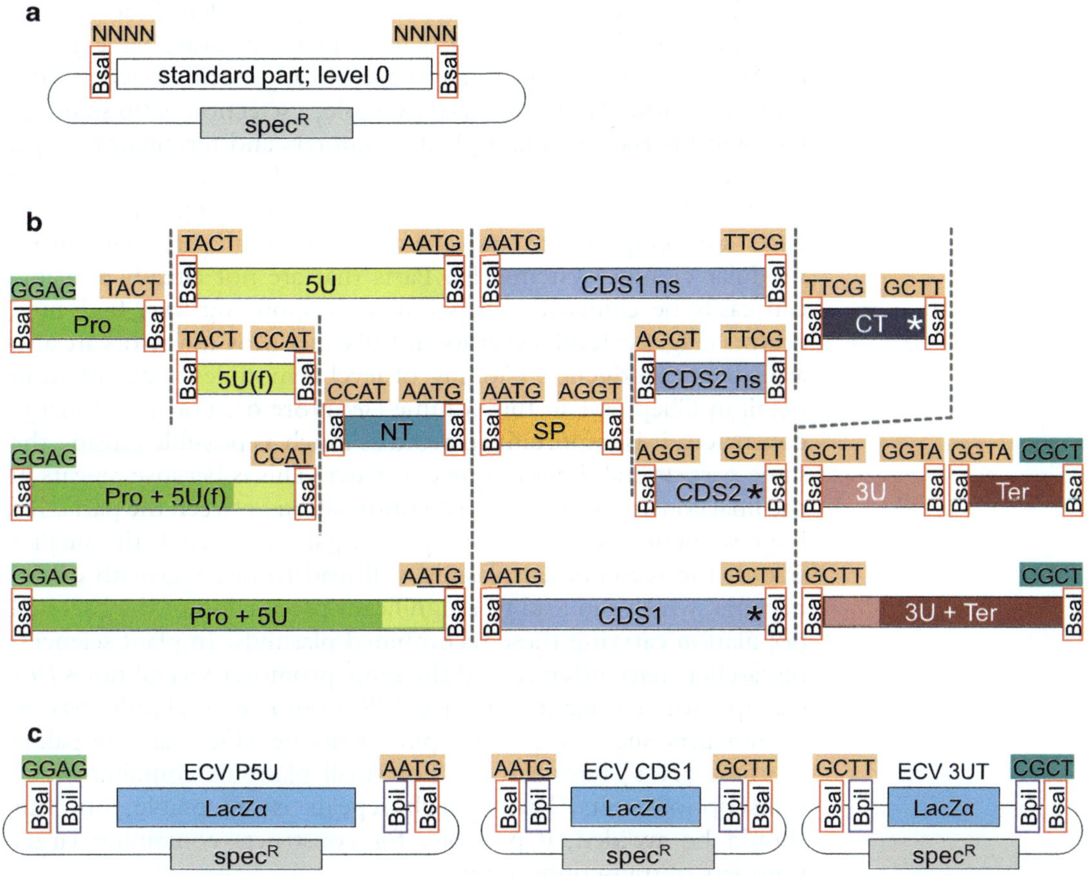

Fig. 2 Level 0 modules (standard parts) and cloning vectors. (**a**) General structure of a level 0 plasmid. The level 0 module is flanked by two inverted *BsaI* sites. (**b**) Different types of standard parts with their corresponding fusion sites. Promoter (Pro), 5′- and 3′-UTR (5U, 3U), N- and C-terminal tag (NT, CT), signal peptide (SP), coding sequence (CDS) with stop codon (*) or without (ns), terminator (Ter). (**c**) Level 0 cloning vectors. Only three vectors are shown as example, but vectors for each module type are available

3.1 Selection of Standard Parts and Vectors

3.1.1 Selection of Standard Parts

Assembly of multigene constructs starts with the assembly of transcription units (genes) from basic parts (Fig. 1a). MoClo basic parts, also called level 0 modules or standard parts, consist of a DNA fragment of interest flanked by two *Bsa*I restriction sites in opposite orientations cloned in a plasmid with a spectinomycin resistance cassette (Fig. 2a). The various types of level 0 modules comprise, but are not limited to, promoters, 5′ untranslated regions (5′UTRs), coding sequences, 3′UTRs, and terminators (*see* **Note 5**). The 4 bp sequences of the *Bsa*I cleavage sites (the "fusion sites") are specific for each type of module and define which parts can be fused together. For example, a promoter module (with TACT as 3′ fusion site) can be fused only to an untranslated leader sequence with the same fusion site at the 5′ end. The different types of standard parts and their fusion sites are shown in Fig. 2b.

Making a multigene construct first requires identification of all basic parts needed. Several parts may already be available, e.g., the promoters or terminators may have already been cloned by the same or another laboratory. For example, a selection of 96 standard biological parts (including plant promoters and terminators, tags, reporter genes, selectable markers) can be obtained from Addgene (https://www.addgene.org/cloning/MoClo/) [4]. In addition, parts are being cloned and characterized by research labs using modular cloning systems [5]. Parts that are not already available can easily be cloned by users into corresponding level 0 cloning vectors (Fig. 2c; level 0 vectors and all other MoClo vectors are also available at Addgene). Cloning of level 0 modules is described in detail in Chapter 1 in this volume (*see* **Note 6**). For making large constructs, it is important to avoid as much as possible reusing the same part several times in one construct. This is because repeats in the final construct can lead to recombination between the parts. If a large sequence between two repeated parts is deleted, the smaller size of the recombined plasmid will lead to faster growth of the bacteria, which can lead to a significant proportion of the bacterial population carrying these recombined plasmids. In plant sciences, researchers have often reused the same promoter several times (for example, the strong constitutive 35S promoter, or *Agrobacterium* terminators such as the octopine synthase [Ocs] and nopaline synthase [Nos] terminators). For small plasmids containing two transcription units, having such repeats is acceptable, but this should be avoided, if possible, for constructs containing larger numbers of transcription units.

3.1.2 Selection of Level 1 Destination Vectors

The position and orientation of each gene in a final construct determine which level 1 destination vector must be chosen for assembly of a transcription unit. A total of 14 vectors are available (pL1F1–7, pL1R1–7) for cloning of genes in either forward or reverse orientation at each of 7 possible positions (Fig. 3). These 14 level 1 vectors differ only by the sequence of the fusion sites. The two external fusion sites (*Bpi*I cleavage sites) of each vector are designed to be compatible with the fusion sites of the vectors from the positions before and after. Furthermore, the downstream fusion site of the vector at position 7 is compatible to the upstream fusion site of the vector at position 1 (TGCC). Thus, multigene constructs with more than seven genes can be assembled without the need for additional cloning vectors. In practice, a gene that is planned to be cloned at position 8 in a multigene construct is simply cloned in a position 1 destination vector. Similarly, a gene at position 9 has to be cloned in a position 2 vector and so on. An example for making a multigene construct containing 18 genes is shown in Fig. 4. All level 1 vectors have identical internal fusion sites (*Bsa*I cleavage sites) to allow assembly of the same level 0 parts in all vectors, i.e., the first level 0 module must have GGAG as

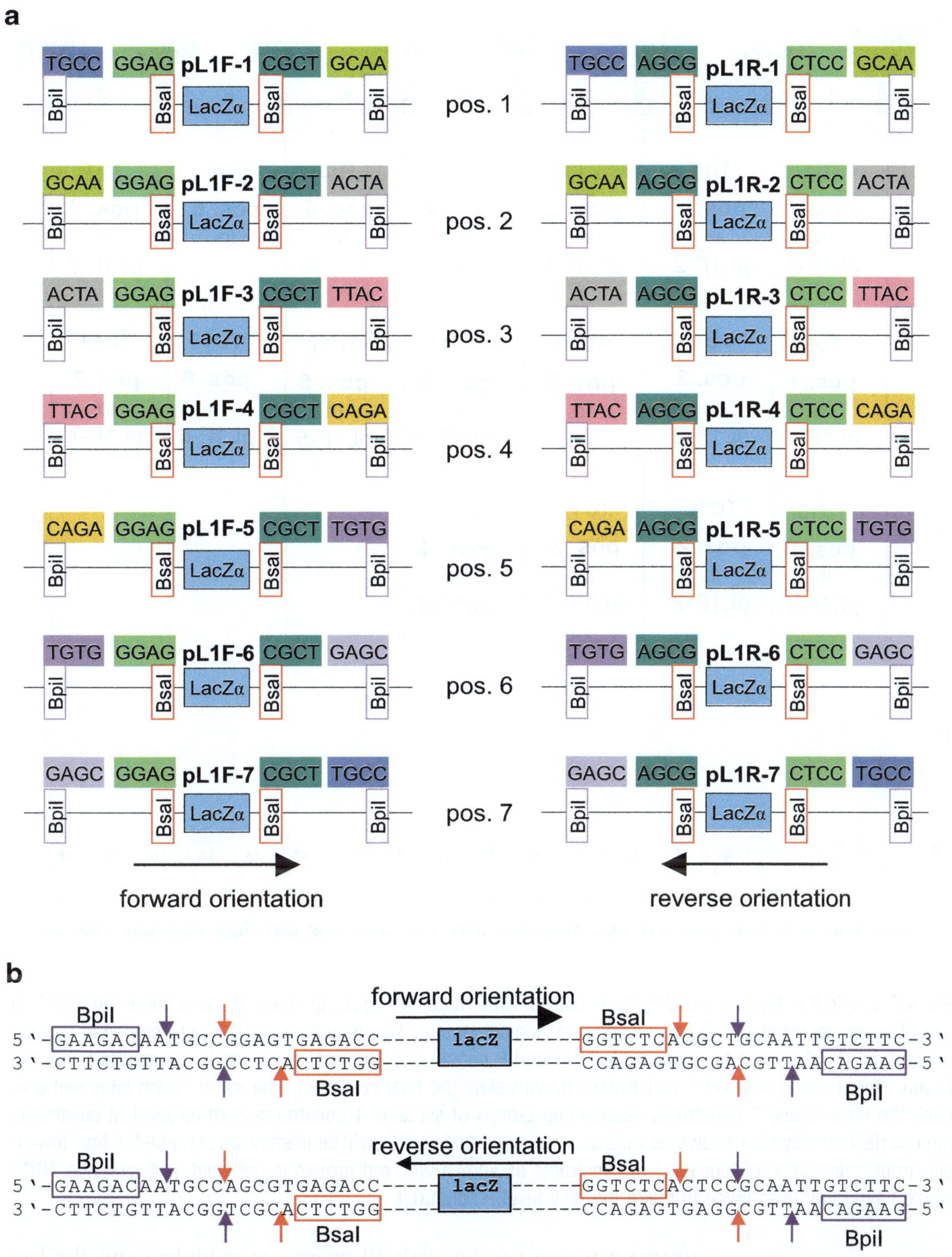

Fig. 3 Level 1 destination vectors. (**a**) Genes can be cloned at one of seven positions in forward (pL1F-vectors) or reverse (pL1R-vectors) orientation. (**b**) Sequence of the type IIS recognition (rectangles) and cleavage sites (arrows) of vectors pL1F-1 and pL1R-1

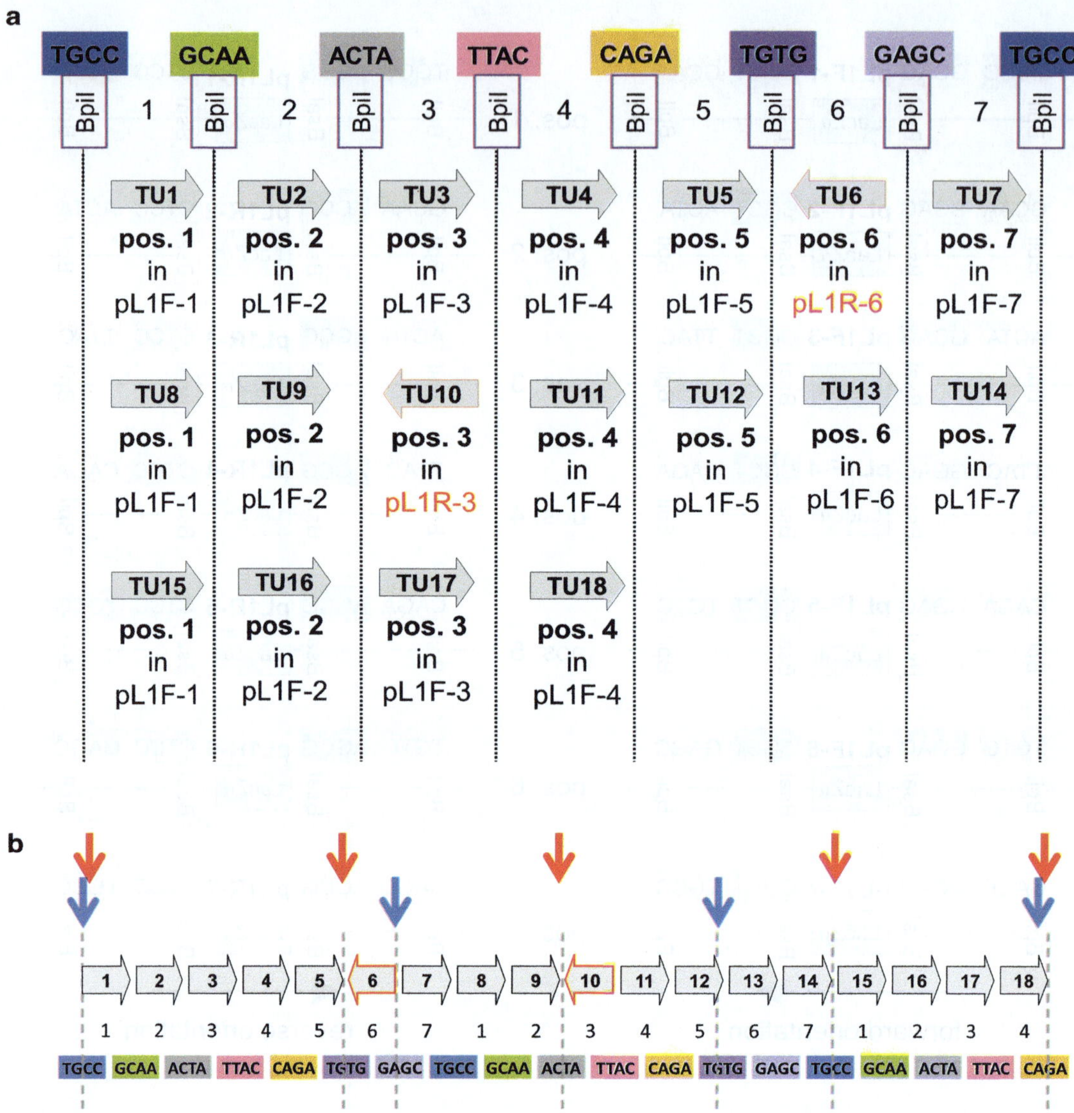

Fig. 4 Example for planning a multigene construct with 18 genes. (**a**) Assignment of transcription units (TU) to one of seven positions and the corresponding cloning vectors. Two genes (TU6, TU10) are cloned in reverse orientation. (**b**) Visual aid for planning level M and P constructs. The final construct is represented with the fusion sites flanking all level 1 constructs shown below the final construct. The same fusion sites will also flank the level M and P constructs. Subcloning groups of six level 1 constructs in three level M constructs (groups delimited by blue arrows) is valid as none of the fusion sites will be used twice to make the final level P construct. Subcloning of four level M constructs as shown with red arrows is not valid as fusion site TGCC would have to be used twice to make the final level P construct

upstream fusion site (as with all promoter modules) and the last module must have CGCT as downstream fusion site (as with all terminator modules). An example for cloning a gene with three basic parts is shown in Fig. 5. All level 1 vectors are binary plasmids

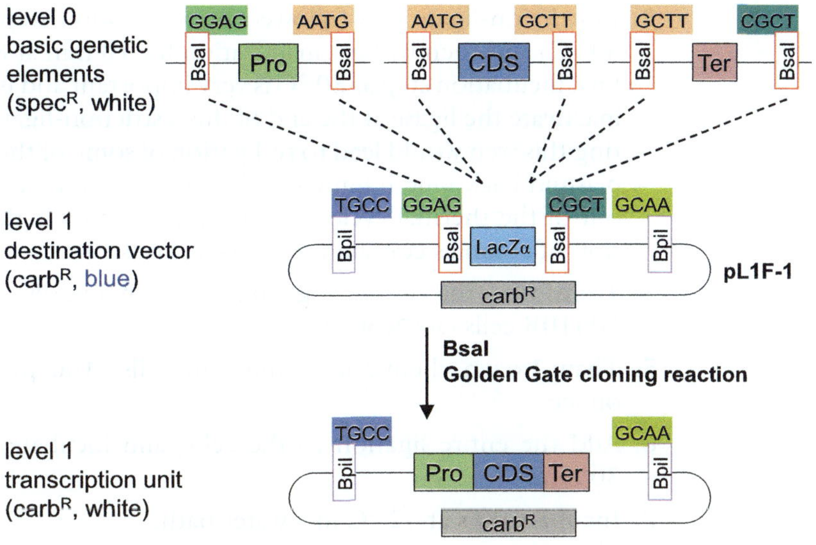

Fig. 5 Example of cloning a transcription unit consisting of three basic parts. Correct clones are identified by the antibiotic resistance marker (carb^R) and the color of the colonies on plates with X-gal (white)

that can be used for *Agrobacterium*-mediated transient expression in plants. These vectors are nevertheless suitable for making constructs for other organisms, as level 1 constructs are usually used only as intermediate constructs, and the only feature that is required for this purpose is replication and selection in *E. coli*. Other vectors (for example, non-binary vectors) could also be used as long as they contain the features outlined in Fig. 3 and an appropriate selectable marker (an ampicillin/carbenicillin resistance gene).

3.2 Assembly of Transcription Units

Once the level 0 modules and the destination vector are made, or are already available from previously made constructs, assembling the transcription unit only requires pipetting all components into a reaction mix and incubating the reaction vessel in a thermal cycler.

1. Set up a restriction-ligation by pipetting into a tube 20 fmol (approximately 50 ng; *see* **Note 7**) of each level 0 module (*see* **Note 8**) and of the vector, 2 μL 10× ligation buffer, 10 U (1 μL) of *Bsa*I, and either 3 U (1 μL) of ligase for assembly of two to four modules or 20 U (1 μL) HC ligase for assembly of more than four modules (final volume of 20 μL; *see* **Note 9**).
2. Incubate the restriction-ligation mix in a thermal cycler. For assembly of two to four level 0 modules, incubation for 60–120 min at 37 °C is sufficient. If more modules are ligated together, the incubation time is increased to 6 h, or cycling is used as following: 2 min at 37 °C followed by 3 min at 16 °C, both repeated 50 times (*see* **Note 10**).

3. Restriction-ligation is followed by a digestion step (5 min at 50 °C) and then by heat inactivation for 10 min at 80 °C. The final incubation step at 80 °C is very important and is needed to inactivate the ligase at the end of the restriction-ligation. Omitting this step would lead to re-ligation of some of the insert and plasmid backbone fragments when the reaction vessel is taken out of the thermal cycler, and would lead to a higher proportion of colonies containing incorrect constructs.
4. Transform the entire ligation into chemically competent DH10B cells (*see* **Note 11**).
5. Thaw frozen chemically competent cells (100 μL per tube) on ice.
6. Add the entire ligation to the cells, and incubate on ice for 30 min.
7. Incubate 90 s at 42 °C in a water bath.
8. Let the cells recover on ice for 5 min.
9. Add 1 mL of LB medium to the cells, and incubate the tube at 37 °C in a shaker-incubator (150 rpm) for 45 min to 1 h.
10. After incubation, plate 25–100 μL of the transformation on LB agar plates containing carbenicillin and X-gal.
11. Incubate the plates overnight at 37 °C. Many white and very few blue colonies should be obtained.
12. Pick a few white colonies from the plate and transfer to liquid LB medium containing carbenicillin (*see* **Notes 12** and **13**).
13. Grow bacterial culture overnight at 37 °C and use for isolation of plasmid DNA.
14. Check plasmids by restriction digest using *Bpi*I (for MoClo level 1 modules) or any other suitable enzyme.

3.3 Assembly of Level M Constructs

Level 1 constructs are preassembled in groups of up to six into level M vectors. The resulting level M constructs will later be further assembled to make a level P construct. Level M vectors contain two *Bpi*I recognition sites in opposite orientation for insertion of level 1 modules. The upstream fusion site defined by the first *Bpi*I site (for example, TGCC for pLM-1; Fig. 6) is compatible with fusion sites flanking upstream fusion sites of the level 1 vectors, whereas the downstream fusion site consists of a universal sequence (GGGA). This design allows cloning of two to six genes in the same vector (using more genes in this step would lead to incorrect clones because the same fusion site would be present on different modules). The last gene is fused to the vector using a compatible linker. We use here an example for the cloning of four transcription units in a final level P vector, with cloning of transcription units by groups of two in level M constructs (this small number of genes

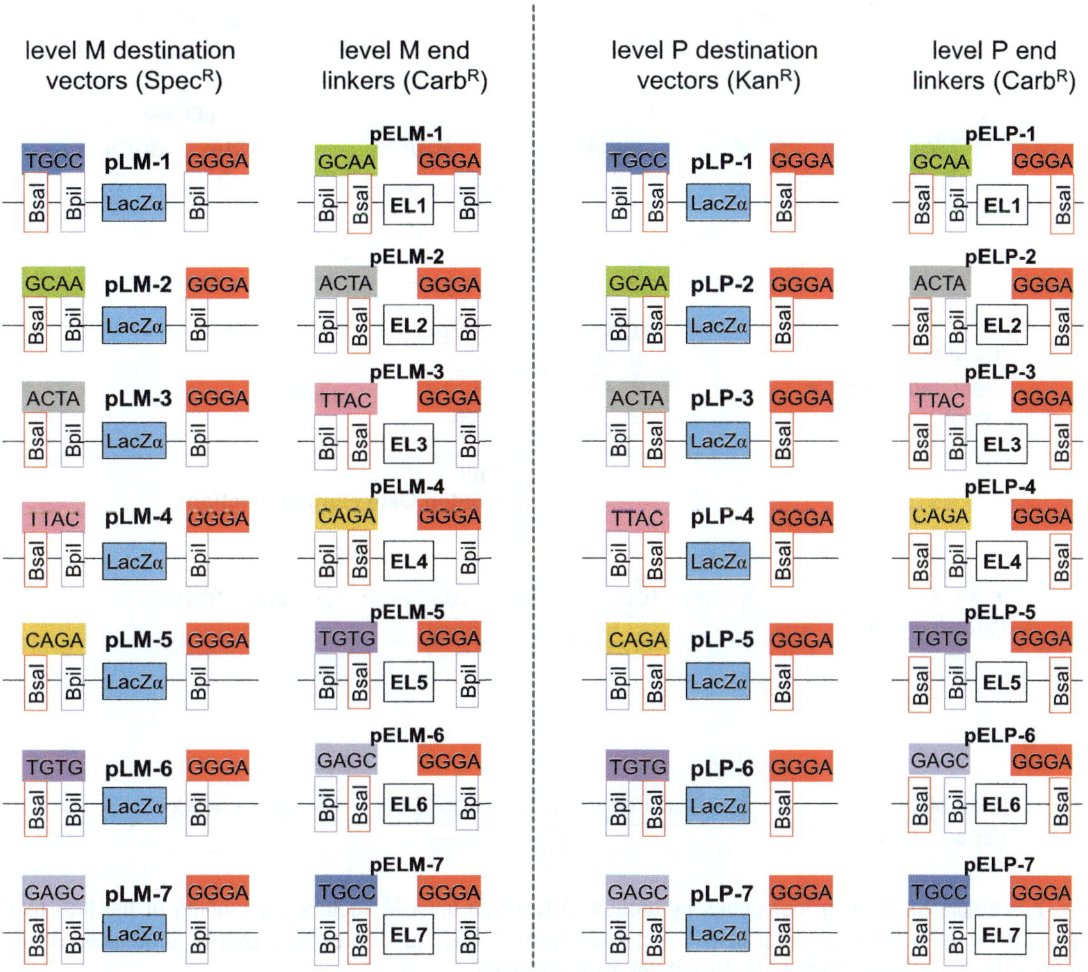

Fig. 6 Level M and P destination vectors and end-linkers

could be cloned directly in a final level M construct, but is chosen here for the sake of space in the figures). Cloning of two transcription units in the first level M construct will require using vector pLM1 and end-linker pELM-2 (see compatibility of the fusion sites in Figs. 6 and 7). Cloning of the next two transcription units will require vector pLM3 and end-linker pELM-4.

After cloning, level M constructs become flanked by two BsaI sites, one coming from the level M vector, and the other coming from the level M end-linker (Figs. 6 and 7). The fusion sites flanking these level M vectors are derived from those used previously in level 1 constructs. It is important to check, before starting cloning of level M constructs, that all level M constructs that are made will be compatible for assembly in level P vectors. We can take an example for defining a strategy for making a construct containing 18 genes (see Fig. 4b). One possibility is to clone the transcription units in level M vectors in three groups of six genes. This strategy is

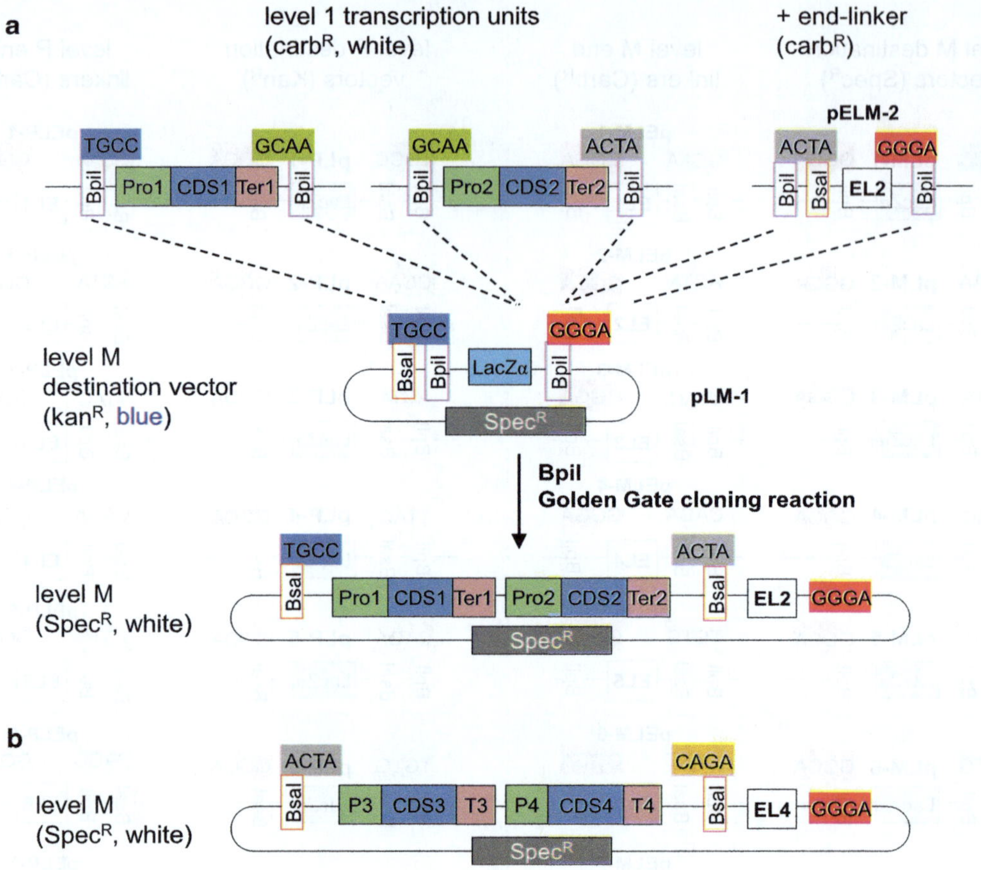

Fig. 7 Example for cloning four genes by groups of two into level M vectors. (**a**) Cloning of the first two transcription units requires vector pLM-1 and end-linker pELM-2. (**b**) The second level M is obtained using vector pLM-3 and end-linker pELM-4 (part details not shown)

valid as the fusion sites that will flank the level M constructs (TGCC, GAGC, TGTG, and CAGA) are all different (Fig. 4b, blue arrows). Alternatively, one could decide to clone the level 1 constructs in groups of no more than 5 transcription units for higher cloning efficiency, for example in 4 groups of 5, 4, 5, and 4 transcription units. However, this strategy is not valid as the fusion sites that would flank the resulting level M constructs are TGCC, TGTG, ACTA, TGCC, and CAGA, with TGCC present twice (Fig. 4b, red arrows). An alternative would be to clone groups of 5, 4, 4, and 5 transcription units, with fusion sites TGCC, TGTG, ACTA, GAGC, and CAGA flanking the resulting level M constructs.

The Golden Gate reaction for making level M constructs is virtually the same as described in Subheading 3.2 except that the enzyme *Bpi*I is used instead of *Bsa*I.

1. Set up a restriction-ligation by pipetting into a tube 20 fmol of each level 1 module, of the end-linker and of the level M vector, 2 μL 10× ligation buffer, 10 U (1 μL) of *Bpi*I, and either 3 U (1 μL) of ligase for assembly of two to four modules or 20 U (1 μL) HC ligase for assembly of more than four modules (final volume of 20 μL).
2. Incubate the restriction-ligation mix in a thermal cycler. For assembly of two to four level 1 modules, incubation for 60–120 min at 37 °C is sufficient. If more modules are ligated together, the incubation time is increased to 6 h, or cycling is used as following: 2 min at 37 °C followed by 3 min at 16 °C, both repeated 50 times (*see* **Note 10**).
3. Restriction-ligation is followed by a digestion step (5 min at 37 °C; *see* **Note 14**) and then by heat inactivation for 10 min at 80 °C.
4. Transformation of the reaction mix into *E. coli* is performed as described in Subheading 3.2, and cells are spread on LB plates with spectinomycin and X-gal. Colonies containing correct constructs should be white whereas colonies containing uncut or re-ligated vector should be blue. Plasmids should be isolated from a few colonies and should be analyzed by restriction digest with a suitable enzyme.

3.4 Assembly of Level P Constructs

Level P vectors and end-linkers are similar to level M, but all BpiI sites and BsaI sites are reciprocally exchanged. In addition, level P vectors have a different antibiotic selectable marker: Kan^R instead of $Spec^R$ (Fig. 6). As for level M constructs, cloning in level P vectors requires selecting a suitable level P vector and an end-linker. In the example where two level M constructs consisting of two transcription units are cloned, vector pLP-1 and end-linker pELP-4 are used (Fig. 8). The Golden Gate reaction is the same as described in Subheading 3.2. Level P constructs are flanked by BpiI restriction sites and have therefore a similar structure as level 1 constructs (except for the selectable marker). As such, several level P constructs can be assembled into level M vectors. As discussed with level M constructs, the user needs to make sure that all level P vectors that need to be assembled are compatible with respect to their fusion sites. The suitability of the fusion sites has to be determined at the beginning of the project, before starting making level M constructs.

An important parameter to consider when choosing a level P vector is the nature of its backbone, particularly in regard to the type of replication origin (ori) that it contains. The ori controls the copy number at which a plasmid will replicate in a cell. Replicating large plasmids represents a large metabolic load for the host bacteria, resulting in a reduced growth rate of cells carrying these large plasmids. This effect is amplified when the plasmid replicates at a

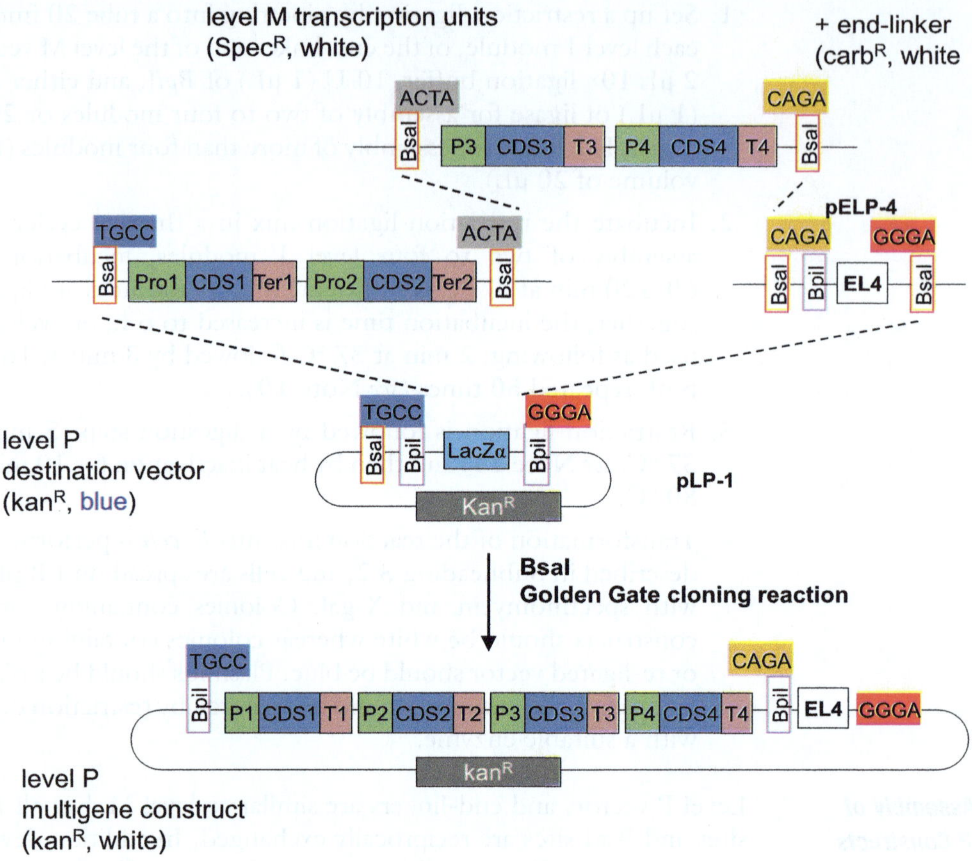

Fig. 8 Example of level P cloning from two level M constructs

high copy number. In that case, any recombination event leading to deletion of some sequences will lead to an increase in the growth rate of host bacteria, up to the point where mostly bacteria containing recombined plasmids will be growing. As a result, predominantly recombined plasmid DNA may be recovered in miniprep DNA. We have encountered cases in our research where plasmids with several large repeats were unstable in high copy plasmids but dramatically more stable in vectors with a lower copy ori [6]. We have also encountered cases where large plasmids with nine transcription units were unstable, even though they did not contain any noticeable repeat (unpublished data). But the same constructs cloned in a backbone with a lower copy ori such as p15a (level P vector pAGM62636, Addgene Plasmid #169095) or pBR322 (level P vector pAGM82783) were completely stable, as evidenced by restriction analysis of miniprep DNA. Therefore, for large constructs containing many transcription units, using vectors with a medium copy or low copy ori is recommended.

It is important to note that using plasmids with a low copy number ori may not always be advantageous for all experiments.

Indeed, plasmids replicating at a low copy number will be harvested with a lower yield. Consequently, larger volumes of the DNA preparation may be required for cloning purposes. However, plasmid DNA is sometimes contaminated with some amount of chromosomal DNA from the bacterial host. Therefore, with plasmid DNA isolated from low copy plasmids, the ratio of plasmid DNA to chromosomal DNA contaminations may be lower. This is a problem with Golden Gate cloning as contaminating chromosomal DNA will contain many BsaI and BpiI sites, which will give rise to a multitude of DNA fragments after digestion. Some of these fragments will ligate inappropriately with the DNA fragments of the cloning reaction, lowering the efficiency of the assembly reaction. Therefore, for small and stable constructs (level 0 and level 1 plasmids), it is generally preferable to use standard high copy vectors.

4 Notes

1. Several level P constructs can be assembled in level M vectors to make larger constructs. This process can be repeated indefinitely by using level M and level P vectors alternately. The limitation will, of course, be the size of the final construct and its ability to be transformed and to replicate in *E. coli*.
2. Type IIS restriction enzymes and ligase can be purchased from several commercial suppliers. Users should nevertheless be aware that various isoschizomers of type IIS restriction enzymes may not necessarily all cleave DNA with the same efficiency when using the reaction conditions described in this protocol (especially the restriction-ligation buffer). We have obtained good results with the enzymes and reaction conditions described here, but use of enzymes from other sources might require adaptation of the reaction conditions to obtain optimal results. For example, with the conditions that we are using, better cloning efficiency is obtained when using BsaI rather than Eco31I.
3. When working with plasmids that replicate at high copy and that do not differ widely in size, measuring plasmid DNA concentration using a device that relies on UV spectrometry, such as a Nanodrop, is generally adequate. However, when working with plasmids with a low copy backbone or harboring large inserts (such as multigene constructs comprising several transcription units), miniprep DNA yields can be quite low, and estimation of plasmid DNA concentration can become much less precise. This can lead to cloning reactions where some inserts or the vector can be either over- or under-estimated related to other fragments, which can lead to low assembly

efficiency. Therefore, when working with low copy or large plasmids, it is recommended to measure the DNA concentration of all plasmids needed for a cloning reaction using a fluorometer. This recommendation also applies for the preparation of construct libraries, even if all plasmids involved have high copy backbones, as precise determination of the concentration of all DNA components will serve to optimize the assembly reaction and maximize the library titer.

4. Plasmid purification can be done with any commercially available kit providing high-purity DNA. Cloning of multigene constructs can result in very large plasmids for which standard miniprep kits are not suited. In these cases, the use of kits designed for purification of large plasmids is recommended (e.g., NucleoBond® PC from Macherey-Nagel).

5. Some basic parts may be made for constructing something else than a transcription unit. For example, a recombination site sequence may be cloned as a level 0 part between fusion sites GGAG CGCT. In that case, the level 1 construct will be assembled from this part alone in the appropriate level 1 vector.

6. Level 0 modules should not contain recognition sites for the type IIS restriction enzymes used during the different cloning steps (BsaI, BpiI, and optionally BsmBI for use of the modules with level 2 vectors [11] or for other applications). If recognition sites for these enzymes are present, they will have to be removed using a process called domestication. Domestication is usually done by PCR amplification of the sequences of interest using primers designed to introduce point mutations (silent mutations for sites in coding sequences) in the restriction site sequences. Another possibility is to have the sequence lacking these restriction sites synthesized by a commercial gene synthesis provider. This solution provides the added benefit of allowing codon-optimization (in case of coding sequences), which might be useful for expression in non-native host species.

7. In practice, if all module plasmids and the vector have approximately the same size (4–5 kb), simply adding 50 ng of DNA of each module and of the vector will work relatively well. However, when plasmids with widely different sizes are used, calculating an equimolar amount should provide a higher cloning efficiency. We have previously recommended using 40 fmol of each fragment and of the vector for cloning. It, however, appears that using less DNA leads to higher cloning efficiency. We therefore now recommend using 20 fmol for each insert and vector. The following formula (from the NEB catalog) can be used: 1 μg of a 1000 bp DNA fragment corresponds to 1.52 pmol. Therefore, the volume of DNA to pipet (in μL) to have 20 fmol is given by the equation: 20 (fmol) × size (bp) of

the DNA fragment/(concentration [ng/μL] × 1520). Frequently, this formula will lead to volumes <1 μL with standard miniprep DNA. In these cases, we recommend diluting the DNA in order to avoid pipetting errors.

8. The Golden Gate cloning reaction is quite efficient and leads to a large proportion of colonies containing the desired construct. This high efficiency makes it possible to construct libraries of construct variants rather than individual constructs. Construct libraries are made by using several modules of the same type (flanked by the same fusion sites) in the ligation mix rather than a single module of each type. In that case, if *n* variant modules replace a single module, the amount of each variant is divided by *n*, but the total amount of fragments for this module type remains the same, i.e., 20 fmol. Constructs libraries can be made in vectors containing a CCDB marker for counter-selection to avoid growing empty vectors, but can also be made in normal constructs containing a standard LacZ marker if the presence of a few empty plasmid vectors in the resulting library is acceptable for the project. Libraries can be made for level 1 constructs, but also for constructs of all successive levels M and P.

9. The total volume of enzymes in the reaction should not exceed 10% (*v/v*). Otherwise, glycerol from the storage buffer of the enzymes can lead to decreased efficiencies.

10. Both the temperature conditions and ligase concentration are complementary parameters. For simple constructs requiring ligation of only one or two inserts, any program should work efficiently, even with ligase at regular concentration, and in most cases continuous incubation at 37 °C should be sufficient. The cycling program may improve the cloning efficiency especially when more modules are used and when plasmids become larger. For some applications that require high cloning efficiencies, e.g., the cloning of promoter libraries, use of high-concentration ligase and cycling is recommended.

11. Any other *E. coli* strain can also be used as long as it provides complementation for the *lacZα* fragment to allow blue-white selection. However, different *E. coli* strains show different transformation efficiencies, which can be important for certain applications or when transforming large plasmids. In some cases and with some strains, electroporation can provide better results. In this case, DNA from the restriction-ligation mix should first be ethanol-precipitated and resuspended in 10 μL of water.

12. Two colonies should be sufficient in most cases, since the majority of colonies should contain the correct fragment. Clones with a correct restriction pattern do not necessarily

need to be sequenced since constructs are normally assembled only from sequenced level 0 modules.

13. When libraries are made rather than individual constructs, an aliquot of the transformation is plated to evaluate the efficiency of cloning. In this case, there is usually no need to screen these colonies. The rest of the transformation is grown as a single culture. A miniprep made from this culture can be used in the same way as a miniprep from a single construct for the next stage of cloning. It can also be digested to check the quality of the library; however, the size of the fragments will of course depend on what is present in the library.

14. The optimal temperature for restriction with BpiI is 37 °C. The final digestion is thus carried out at 37 °C.

References

1. Andreou AI, Nakayama N (2018) Mobius Assembly: a versatile Golden-Gate framework towards universal DNA assembly. PLoS One 13:e0189892
2. Binder A, Lambert J, Morbitzer R et al (2014) A modular plasmid assembly kit for multigene expression, gene silencing and silencing rescue in plants. PLoS One 9:e88218
3. Engler C, Kandzia R, Marillonnet S (2008) A one pot, one step, precision cloning method with high throughput capability. PLoS One 3: e3647
4. Engler C, Youles M, Gruetzner R et al (2014) A golden gate modular cloning toolbox for plants. ACS Synth Biol 3:839–843
5. Feike D, Korolev AV, Soumpourou E et al (2019) Characterizing standard genetic parts and establishing common principles for engineering legume and cereal roots. Plant Biotechnol J
6. Grutzner R, Schubert R, Horn C et al (2021) Engineering Betalain biosynthesis in tomato for high level Betanin production in fruits. Front Plant Sci 12:682443
7. Iverson SV, Haddock TL, Beal J et al (2016) CIDAR MoClo: improved MoClo assembly standard and new E. coli part library enable rapid combinatorial design for synthetic and traditional biology. ACS Synth Biol 5:99–103
8. Larroude M, Park YK, Soudier P et al (2019) A modular Golden Gate toolkit for Yarrowia lipolytica synthetic biology. Microbial Biotechnol
9. Lee ME, Deloache WC, Cervantes B et al (2015) A highly characterized yeast toolkit for modular, multipart assembly. ACS Synth Biol 4:975–986
10. Lin D, O'callaghan CA (2018) MetClo: methylase-assisted hierarchical DNA assembly using a single type IIS restriction enzyme. Nucleic Acids Res 46:e113
11. Marillonnet S, Werner S (2015) Assembly of multigene constructs using Golden Gate cloning. Methods Mol Biol 1321:269–284
12. Moore SJ, Lai HE, Kelwick RJ et al (2016) EcoFlex: a multifunctional MoClo kit for E. coli synthetic biology. ACS Synth Biol 5: 1059–1069
13. Occhialini A, Piatek AA, Pfotenhauer AC et al (2019) MoChlo: a versatile, modular cloning toolbox for chloroplast biotechnology. Plant Physiol 179:943–957
14. Pollak B, Cerda A, Delmans M et al (2019) Loop assembly: a simple and open system for recursive fabrication of DNA circuits. New Phytol 222:628–640
15. Prielhofer R, Barrero JJ, Steuer S et al (2017) GoldenPiCS: a Golden Gate-derived modular cloning system for applied synthetic biology in the yeast Pichia pastoris. BMC Syst Biol 11:123
16. Rajkumar AS, Varela JA, Juergens H et al (2019) Biological parts for Kluyveromyces marxianus synthetic biology. Front Bioeng Biotechnol 7:97
17. Sarrion-Perdigones A, Vazquez-Vilar M, Palaci J et al (2013) GoldenBraid 2.0: a comprehensive DNA assembly framework for plant synthetic biology. Plant Physiol 162:1618–1631
18. Taylor GM, Mordaka PM, Heap JT (2019) Start-Stop Assembly: a functionally scarless DNA assembly system optimized for metabolic engineering. Nucleic Acids Res 47:e17

19. Terfruchte M, Joehnk B, Fajardo-Somera R et al (2014) Establishing a versatile Golden Gate cloning system for genetic engineering in fungi. Fungal Genet Biol 62:1–10
20. Weber E, Engler C, Gruetzner R et al (2011) A modular cloning system for standardized assembly of multigene constructs. PLoS One 6:e16765
21. Werner S, Engler C, Weber E et al (2012) Fast track assembly of multigene constructs using Golden Gate cloning and the MoClo system. Bioeng Bugs 3:38–43
22. Wu D, Schandry N, Lahaye T (2018) A modular toolbox for Golden-gate-based plasmid assembly streamlines the generation of Ralstonia solanacearum species complex knockout strains and multi-cassette complementation constructs. Mol Plant Pathol 19:1511–1522

Chapter 3

Selection of Fusion-Site Overhang Sets for High-Fidelity and High-Complexity Golden Gate Assembly

Gregory J. S. Lohman

Abstract

Golden Gate Assembly depends on the accurate ligation of overhangs at fragment fusion sites to generate full-length products with all parts in the desired order. Traditionally, fusion-site sequences are selected by using validated sets of overhang sequences or by applying a handful of semi-empirical rules to guide overhang choice. While these approaches allow dependable assembly of 6–8 fragments in one pot, recent work has demonstrated that comprehensive measurement of ligase fidelity allows prediction of high-fidelity junction sets that enable much more complex assemblies of 12, 24, or even 36+ fragments in a single reaction that will join with high accuracy and efficiency. In this chapter, we outline the application of a set of online tools that apply these comprehensive datasets to the analysis of existing junction sets, the de novo selection of new high-fidelity overhang sets, the modification and expansion of existing sets, and the principles for dividing known sequences at an arbitrary number of high-fidelity breakpoints.

Key words Golden Gate Assembly, Fusion site selection, Ligation fidelity, Data-optimized assembly design, Online tools

1 Introduction

Golden Gate Assembly (GGA) utilizes type IIS restriction enzymes, which cut distal to their recognition site, generating short 5′-overhangs of arbitrary sequence independent of the recognition sequence [1]. As the cut site has no sequence restriction, overhangs of any sequence can be generated. GGA permits multiple fragments to be joined in a single reaction by the generation of overhangs in complementary pairs that are used to order the fragments, which are then joined by the ligase (Fig. 1). Fragments can be ligated back into their original holding vector, which regenerates the restriction site, or to the desired fusion-site partner, eliminating the restriction recognition site. Repeated rounds of cutting and ligation permit the generation of complete, full-length assemblies. By selection of unique, non-palindromic (self-complementary) overhangs for each junction, GGA can be used to join anywhere between 2 and upward

Daniel Schindler (ed.), *Golden Gate Cloning: Methods and Protocols*, Methods in Molecular Biology, vol. 2850,
https://doi.org/10.1007/978-1-0716-4220-7_3,

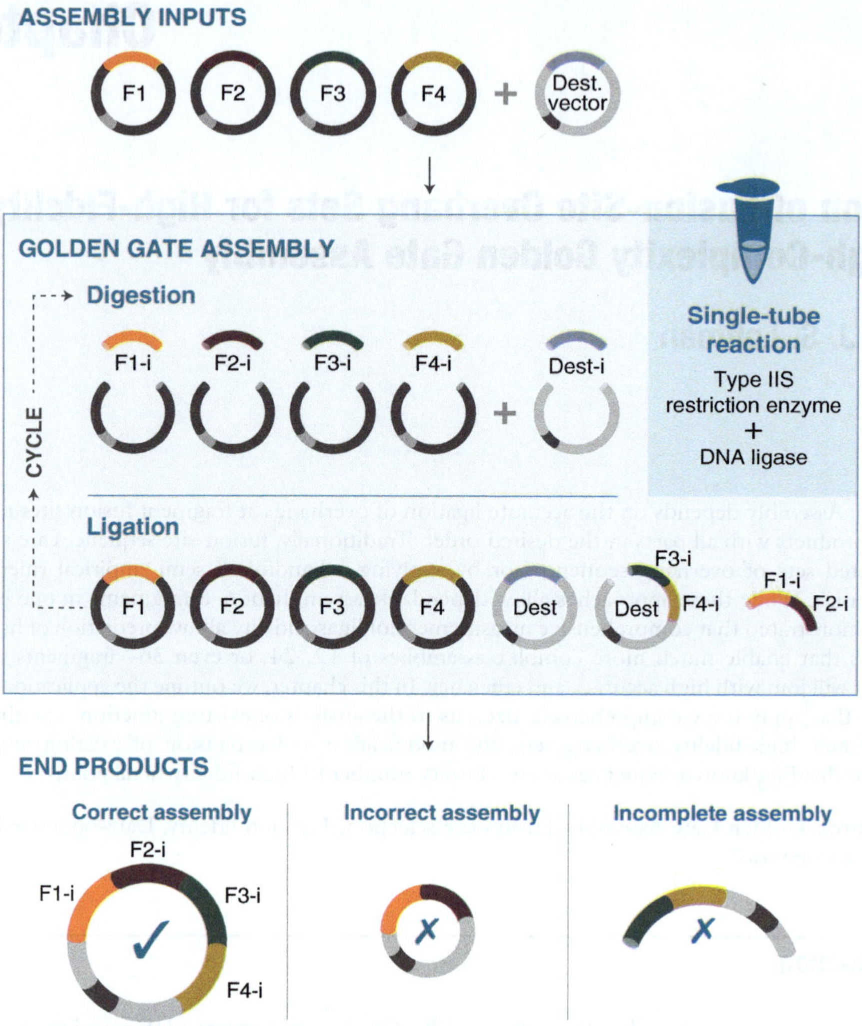

Fig. 1 Golden Gate Assembly depends on the use of type IIS restriction enzymes, which cut outside their recognition site. Thus, the same enzyme can be used to generate multiple overhangs of arbitrary sequence. Selection of complementary pairs allows for fragments to be joined in a specific order. Through repeated rounds of cutting and ligation, full-length products are generated with the fragments in the desired order in a destination construct that no longer contains the type IIS enzyme recognition site. Errors in ligation will join non-complementary overhangs, resulting in misordering of fragments producing products containing deletions or insertions. Incomplete reaction will lead to products that have fragments in the correct order but are not full length and circularized

of 50 fragments in one reaction [2–5]. Historically, constructs assembled from more than six to eight fragments have most often been accomplished through use of a modular hierarchical system such as MoClo to join sub-assemblies together over multiple rounds of reaction [6, 7]. Until recently, few assemblies were reported in the range of more than 12 fragments in one pot, with rapidly decreasing yields of correct transformants as fragment number increases.

If ligation occurs with perfect fidelity, there are 120 possible unique 4-base overhang pairs (excluding palindromes) that could be used as fusion sites. However, ligases tolerate mismatched base pairs to some extent, with the most common ligase for GGA, T4 DNA ligase, known to be quite promiscuous in a nick ligation context [8]. In GGA, ligation of an overhang with any sequence other than its fully Watson–Crick (WC) base-paired partner will lead to misassembly by joining two fragments out of order (Fig. 1). As these fragments will no longer contain the type IIS enzyme recognition site, this fusion is irreversible, and the product will at best consume DNA fragments non-productively and, at worst, generate inserts with misordered or missing fragments that will produce a colony with an incorrect insert. Thus, misligation leads to both reduced yields and increased screening requirements to locate correctly assembled product. Further, the likelihood of erroneous products increases exponentially with fragment number as more and more possible mismatched pairings exist in solution [4].

In prior work, the question of assembly fidelity was dealt with practically by using validated assembly sets such as those used in GGA standards [7]. Beyond direct validation of subsets, numerous rules and guidelines have been proposed for selecting fusion-site overhangs. Some rules, i.e., the need to avoid palindromic overhangs that can self-ligate, are unambiguously needed. A common recommendation to maximize fidelity is for every overhang to have at least two base pairs different from every other overhang within the same assembly. Following this rule ensures that any pairings other than the intended partner will have two mismatched positions and has been shown to be quite reliable at producing high-fidelity assemblies [9–11]. However, this rule limits the user to a maximum of eight possible fusion sites for four-base overhangs and five fusion sites for three-base overhangs, and in cases where junctions are limited by a fixed target sequence or other limitations in the placement of breakpoints, it may be quite difficult to select sets that fully keep to this rule. Beyond avoiding palindromes and keeping two-base differences across the set, other rules have been proposed such as avoiding extremes of GC content and avoiding having any two overhangs have the three bases in a row identical, but these rules have not been systematically validated. Largely, in the absence of using previously validated sets, users were often left to guesswork in selecting fusion sites.

Recently, we have used a single-molecule sequencing assay to comprehensively profile the ligation fidelity (discrimination against mismatched sequences) and bias (preference for some sequences over others) in a variety of end-joining contexts for several DNA ligases [4, 12–14]. Most relevant to GGA, we have carried out mock assemblies that include substrates that generate all possible four-base overhangs when cut with a type IIS enzyme. When treated with GGA enzyme mixtures under assembly cycling conditions, these substrates generate sequenceable libraries that allow quantitation of all possible ligation products, allowing assessment of the likelihood of any given mismatch pairing occurring relative to its WC-paired product [4]. Comprehensive ligation fidelity data have been used to select overhang sets that minimize predicted mismatch ligation products between all members of the set. Use of these sets in actual assemblies has shown that 12-fragment assemblies, rather than the extremes of what is possible, are quite robust with near quantitative assembly yields, when junction fidelity is maximized using this Data-optimized Assembly Design (DAD). Indeed, assemblies up to 24 fragments can be quite routine, with assemblies of 36 and even 50+ fragments in one pot now demonstrated as the practical limits of GGA [5]. Further, the flexibility of these design rules permits this high fragment number to be used within coding sequences to generate assemblies that require no changes to native sequence (aside from removal of native type IIS enzyme sites), yet provide high flexibility in both fragment number and fusion-site positioning [5, 15]. DAD has permitted applications such as the rapid assembly of T7 bacteriophage genomes [5], highly modular protein mutagenesis libraries from up to 20 fragments [16], and highly repetitive CRISPR arrays containing 12 fragments [17, 18].

This chapter discusses how to apply comprehensive ligation fidelity data to analyze the fidelity of existing junction sets; generate new sets ab initio that are predicted to have high assembly fidelity; modify existing sets to replace error-prone pairings or add new junctions without impacting overall fidelity; and to design high-complexity fusion-site sets chosen within an existing coding sequence to generate new assemblies for combinatorial synthesis or the rapid assembly of small genomes. The designs are compatible with standard methods of generating fragments and carrying out GGA discussed at length in other chapters and publications [19].

2 Materials

These methods require a personal computer running Windows, MacOS, or Linux with an Internet connection. All tools for applying ligase fidelity data to assembly design can be found at https://ligasefidelity.neb.com/ (*see* **Note 1**). It is also suggested to install software for the visualization and browsing of sequences such as Snap Gene, Geneious, or an online tool such as ApE (Chapter 5) [20] or the NEBridge™ Golden Gate Tool (https://goldengate.neb.com/#!/) for assembly visualization.

For Subheading 3.4, you will need the full sequence you desire to assemble in plain text format.

3 Methods

3.1 Evaluating the Fidelity of an Existing Junction Set Using the NEBridge Ligase Fidelity Viewer™

This method describes the use of the Ligase Fidelity Viewer to assess the fidelity of an existing set of overhangs, using the MoPET assembly standard junction overhangs (CTTG, CCAT, GGCT, GGAT, CGGG, GGTG, AGGC, TAAT) as an example [21]. This tool is essentially a browser of ligase and assembly fidelity datasets, permitting ease of viewing of a subset of overhangs. The tool accepts a list of overhangs, automatically adding their complements, and reports the qualitative overall expected fidelity of the set (*see* **Note 2**). In addition to the overall score, the tool will also display pairwise ligation predictions, flagging any pairs observed to lead to elevated levels of mismatch ligation (*see* **Note 3**). Subheading 3.3 will discuss how to use this information to change fusion-site overhangs to improve overall set fidelity.

1. Navigate to https://ligasefidelity.neb.com/viewset/run.cgi.
2. Select the overhang length appropriate for the set being used. For the example MoPET junction set, select 4-base.
3. Select the cycling conditions under which the fidelity of the set will be evaluated. For the MoPET example, select BsaI-HFv2 37 static (*see* **Note 4**).
4. Enter the overhangs in the set to be evaluated in a comma-delineated list, written in the 5′→3′ direction. For this example, enter "CTTG, CCAT, GGCT, GGAT, CGGG, GGTG, AGGC, TAAT" (*see* **Note 5**).
5. Click "Submit." The output generated will give an assessment of the overall predicted fidelity (*see* **Note 6**) and a matrix showing all overhangs plus their complements, marking any predicted interactions (Fig. 2).
6. To view predictions using the same set under different assembly conditions, select a new set of conditions from the "Ligation conditions" dropdown and again click "Submit." For this example, select "BsaI-HFv2, 37-16 cycling." The new results will be displayed (Fig. 2, *see* **Note 7**).

3.2 Generating a High-Fidelity Overhang Set Using the NEBridge GetSet™ Tool

This method describes the use of the NEBridge GetSet™ Tool for generating a high-fidelity overhang set for use in GGA. This tool is best used when overhangs can be independent of any required sequence context. For example, this tool could be used to generate a new set of fusion sites to be used in developing an assembly standard. The sample inputs included are chosen with this potential application in mind.

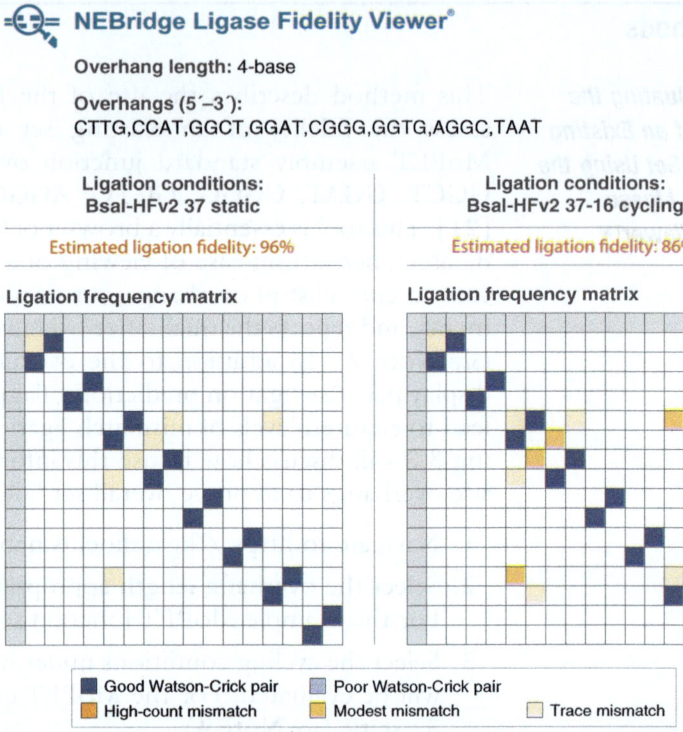

Fig. 2 Output of the NEBridge Ligase Fidelity Viewer for the example inputs used in Subheading 3.1. The grid shows each overhang and its Watson-Crick complement on each axis and marks any predicted ligation events. *Left panel*: results using "BsaI-HFv2 37 static" incubation conditions. All show strong expected ligation between complements (dark blue squares) with an overall fidelity score of 96%. "Trace" mismatch events are predicted for self-ligation of CTTG, ligation of GGCT with ATCC (reverse complement of GGAT) and AGGC, and ligation of AGCC (reverse complement of GGCT) with GGAT. *Right panel*: results predicted when using "BsaI-HFv2 37-16 cycling" conditions for assembly. In this case a lower overall fidelity of 86% is predicted. Mismatch ligation events between GGCT/AGGC and AGCC/GATT are now predicted to be "modest" frequency mismatches, and in addition to the trace mismatch ligation seen under "BsaI 37-HFv2 static" incubation conditions of CTTG/CTTG and AGCC/GCCT, additional trace mismatch ligation is predicted for GGCT self-ligation and AGGC self-ligation, and between AGCC and GCCT (AGGC reverse complement). *See* **Notes 4**, **6**, and **7** for more information on interpreting these fidelity estimates

1. Navigate to https://ligasefidelity.neb.com/getset/run.cgi.
2. Select the overhang length to be used in the standard from the dropdown menu. For this example, select "4-base."
3. Select the desired conditions to be used for the assembly with the set. For this example, select "BsaI-HFv2 37-16 cycling" (*see* **Note 4**).
4. Enter the number of fusion sites needed in the box following "Number of overhangs." For this example, enter "15" (*see* **Note 8**).

5. Enter any overhangs that must be included in the set in the box following "Required overhangs (5′→3′)." For this example, enter "GATG" (*see* **Note 9**).
6. Enter any overhangs to be excluded from consideration in the box following "Excluded overhangs (5′→3′)." For this example, enter "AAAA, CCCC, TAAA" (*see* **Note 10**).
7. Click "Submit." The output will be generated with the overhang set shown as a list that is followed by the estimated fidelity score, and a matrix showing predicted ligation events (Fig. 3, *see* **Note 11**).

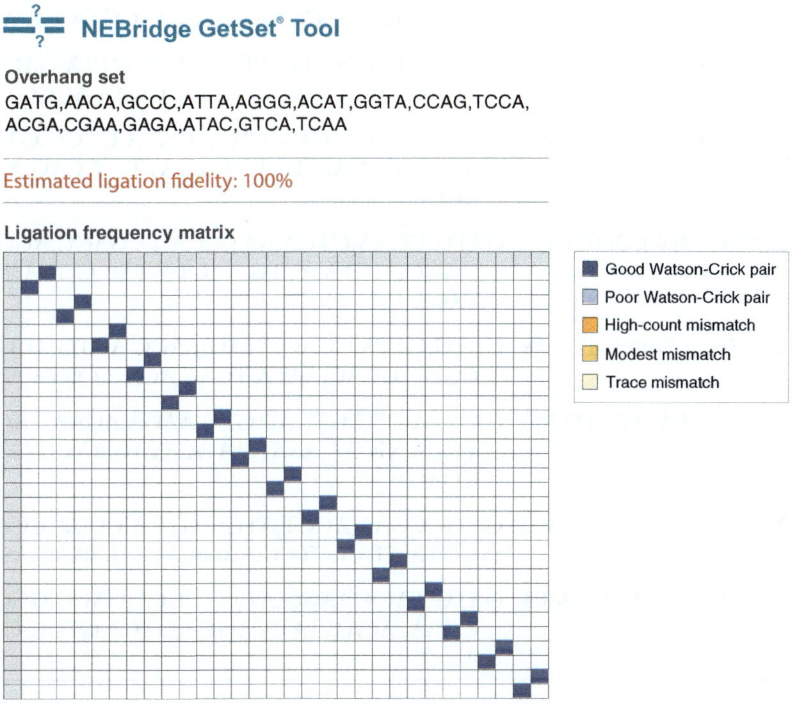

Fig. 3 The output generated by a run of the NEBridge GetSet™ Tool using the sample inputs outlined in Subheading 3.2. The "Request ID" 20230518185519020063 can be used to retrieve this exact output; as the results of the search are not deterministic, resubmission with the same inputs will typically return different sets matching the requirements (*see* **Note 11** and Table 1). The "Overhang set" is the list of overhangs determined to be a high-fidelity set matching any other input requirements. The "Estimated ligation fidelity" shows the fidelity score, in this case 100% (no mismatch ligation events predicted). Finally, the "Ligation frequency matrix" shows, in the same manner as the Ligation Fidelity Viewer (Fig. 2), each overhang and its complement in a grid form with all predicted ligation interactions highlighted. Note that figure layout differs from the full tool output; use of the Request ID on the webtool will retrieve the full results

Table 1
Example results from running Subheading 3.2 with the provided example parameters

NEBridge GetSet Request ID	Overhang set	Estimated ligation fidelity
20230518185519020063	GATG,AACA,GCCC,ATTA,AGGG,ACAT,GGTA, CCAG, TCCA,ACGA,CGAA,GAGA,ATAC,GTCA, TCAA	100%
20230518190225020970	GATG,AACC,AAGA,GGGA,GTTA,ACCG,CTCA, GCCA, CCAC,GAGC,TACA,CAGG,ATAA,AATA, CGAA	100%
20230518190308021207	GATG,AAGG,ACTG,AAAT,CCCA,CTCC,AGTA,ATCA, CTGA,GCCC,CGAA,TAGA,AACC,GCGA,ACCT	100%
20230518190411021458	GATG,CTCC,ACTG,ATAA,ACAT,CTTA,GACC,TCCA, CCAG,AGGG CGGC,AAAC,GCAA,ACCC,GTCA	100%
20230518190501021690	GATG,GACC,GGAA,AAGC,CGTA,CCGC,GTTA, ATAA, CAGA,CTCA,TCAA,ATTG,CCAG,AACT, ACCC	100%
20230518190530021906	GATG,CGAA,CTCA,AATA,CAGA,GACC,AGGG, CGGC, GCTA,ATTG,AACT,GTAA,GGGA,TCAA, TACA	100%
20230518190607022128	GATG,GTTA,CCAG,ATAG,AACC,ACTG,ACGA,CTCA, CGGC,GCAC,AATA,TCAA,AGAC,CGTA,AGCG	100%
20230518190656022366	GATG,AGCG,GTCA,ACTG,AAGA,AGAA,TACA, TCCA, GCAA,AAAT,AATA,ATAC,AACC,GCCC, CCAC	100%
20230518190727022580	GATG,CGAA,ACCT,ATGC,GTCA,CCCA,CCAG,AATA, GGTA,ATAA,CTCC,AAGA,ACTG,AAAC,AACG	100%
20230518190804022804	GATG,TACA,AAGG,ATTC,CCCA,GACC,GCGA, CGAA, ACAT,AATA,AGGA,GTCA,CTCC,CTTA, ACCC	100%

3.3 Modifying or Expanding an Existing Junction Overhang Set Using the NEBridge GetSet™ Tool

This method describes the use of the NEBridge GetSet™ Tool to modify existing assemblies. In the provided example, we revisit the MoPET overhang set from Subheading 3.1. In the first implementation, we remove the overhang GGCT from the set (*see* **Note 7**) and use the tool to replace it with a predicted compatible overhang that results in a higher fidelity set. Finally, we show that by increasing the desired number of overhangs, a much larger compatible set can be generated that does not decrease the predicted overall fidelity score.

1. Navigate to https://ligasefidelity.neb.com/getset/run.cgi.
2. Select the overhang length to be used in the standard from the dropdown menu. For this example, select "4-base."

3. Select the desired conditions to be used for the assembly with the set. For this example, select "BsaI-HFv2 37-16 cycling" (*see* **Note 4**).
4. Enter the number of fusion sites needed in the box following "Number of overhangs." For this example, enter "8" to generate a set the same size as the original MoPET standard, or "15" to generate a larger list of compatible overhangs (*see* **Note 12**).
5. Enter any overhangs that must be included in the set in the box following "Required overhangs (5′→3′)." For this example, enter "CTTG, CCAT, GGAT, CGGG, GGTG, AGGC, TAAT" (*see* **Note 13**).
6. Enter any overhangs to be excluded from consideration in the box following "Excluded overhangs (5′→3′)." For this example, leave this field blank.
7. Click "Submit." The output will be generated with the overhang set shown as a list that is followed by the estimated fidelity score, and a matrix showing predicted ligation events (Fig. 4, *see* **Note 14**).

3.4 Use of the NEBridge SplitSet™ Tool to Divide a Target Sequence into Fragments with High-Fidelity Junction Overhangs

This method describes how to use the NEBridge SplitSet™ Tool to divide a known sequence at optimal breakpoints. To use this tool, you will need the sequence you desire to assemble. The sequence may be entered as a linear or circular construct and permits the selection of an arbitrary number of breakpoints. By default, the tool will space the breakpoints approximately equidistant through the sequence, but the user may instead specify the regions where breakpoints are desired. The tool searches within the specified windows to find an optimized set of breakpoints, then returns a list of fragments that end at these fusion sites.

As an example, we will use a *lacI* and *lacZ* containing cassette presented in our previous publications [3–5, 15]. The sequence will be used as a linear insert, including the fusion sites to the pGGA-select vector. SplitSet will be used to divide this insert into 12 fragments with high-fidelity overhangs used that will be compatible with the vector fusion sites.

1. Navigate to https://ligasefidelity.neb.com/splitset/run.cgi.
2. Optionally, enter a name for your assembly in the field "Assembly name." For this example, enter the name "Lac_test1."
3. Select the appropriate choice "linear" or "circular" in the dropdown menu "Assembly type." For this example, select "linear" (*see* **Note 15**).
4. From the dropdown menu "Overhang length," select the overhang length that corresponds to the overhangs generated by the restriction enzyme to be used. For this example, select "4-base."

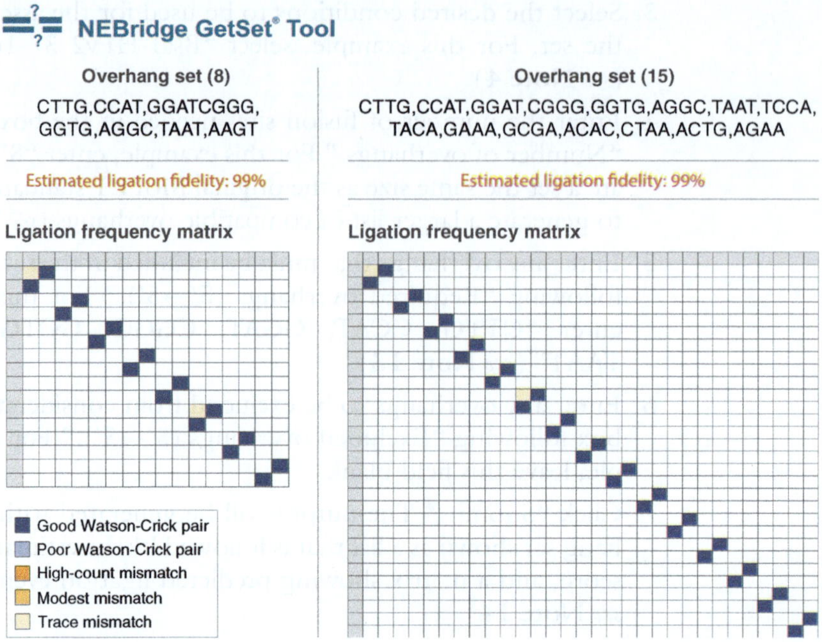

Fig. 4 The output generated by a run of the NEBridge GetSet™ Tool using the sample inputs outlined in Subheading 3.3, with the output matching the format described in Fig. 3. The input seven overhangs match those of the MoPET set analyzed in Subheading 3.1 (Fig. 1), with the overhang GCCT removed. *Left panel*: the results of a request for eight total overhangs (Request ID 20230616190557015815), removing only the GGCT overhang that is involved in the majority of predicted mismatch interactions. The resultant replacement of the GGCT overhang with AAGT leads to a predicted overall fidelity of 99%, compared to the 86% predicted in the original set. *Right panel*: the results of requesting 15 total overhangs (Request 20230616191248016496). Here we see additional overhangs can be added while maintaining 99% predicted fidelity despite the inclusion of seven new fusion sites. As in Subheading 3.2, the results of this search are not deterministic, and multiple runs with the same parameters will result in different 15 member sets. Examples of running the search ten times can be found in Table 2. Note that figure layout differs from the full tool output; use of the Request ID on the webtool will retrieve the full results

5. Select the assembly conditions to be used for the fidelity analysis from the dropdown menu "Ligation conditions." For this example, use "BsaI-HFv2 37-16 cycling" (*see* **Note 4**).
6. Paste the full nucleotide sequence to be analyzed in the box "Nucleotide sequence." The full sequence to be divided into fragments at high-fidelity junction sites must be included here, in plaintext format (*see* **Note 16**).
7. If the sequence is linear and represents an insert sequence to be cloned into a destination vector, check the box "Use terminal overhangs as fusion sites (for linear assemblies only)." For the provided example, select this check box (*see* **Note 15**).
8. "Define split regions based on" allows a choice of how to divide the assembly: desired number of fragments, maximum allowable fragment size, minimum allowable fragment size, or via

manual entry of the windows where junctions may be placed. For this example, select "Number of fragments" from the dropdown and enter "11" in the box. *See* **Note 17** for more information on these options.

9. Click "Define Split Regions." This opens a list showing the windows where the tool will search for overhangs. For this example, leave the default numbers in place. *See* **Note 17** for more information on how these fields may be used.
10. Optionally, enter any overhangs that should be excluded from the list of possible choices in the field "Excluded overhangs (5′→3′)." For this example, leave this field blank.
11. Click "Submit." The output will be generated with the overhang set shown as a list that is followed by the estimated fidelity score, and a matrix showing predicted ligation events (Fig. 5, *see* **Note 14**). The SplitSet Tool further lists the coordinates of all fusion sites within the sequence, and finally lists the full sequences of all fragments in a format compatible with FASTA files. The fragments are named MyAssembly_F1. MyAssembly_FN where the name entered in step 2 will appear in place of MyAssembly and N = the fragment number (*see* **Note 18**).

4 Notes

1. As an alternative to use of the existing tools, the published assembly fidelity data tables could be used to develop a new tool integrated into the user's typical workflow [3, 4, 13]. Additionally, the underlying Overhang Optimizer Code used by the NEBridge Ligase Fidelity tools to search for high-fidelity junction sets is available for download under a non-commercial use license at https://www.neb.com/forms/overhang-optimizer-code.
2. It is important to note that the predicted fidelity is a qualitative assessment of expected fidelity based on the original multiplexed assay. Actual observed fidelity can be impacted by many factors; for example, for assemblies that generate circular final products in a plasmid backbone, only misligation events that also lead to circular final products containing the antibiotic resistance marker will produce colonies with incorrect inserts, while misassemblies that do not circularize will lead only to reduced overall yield, not colonies with incorrect constructs. Further, incorrect assemblies that are unrelated to ligation errors, such as reactions with off-target PCR amplification products, contaminating genomic DNA, and uncut empty vectors, can also produce colonies lacking the desired insert that

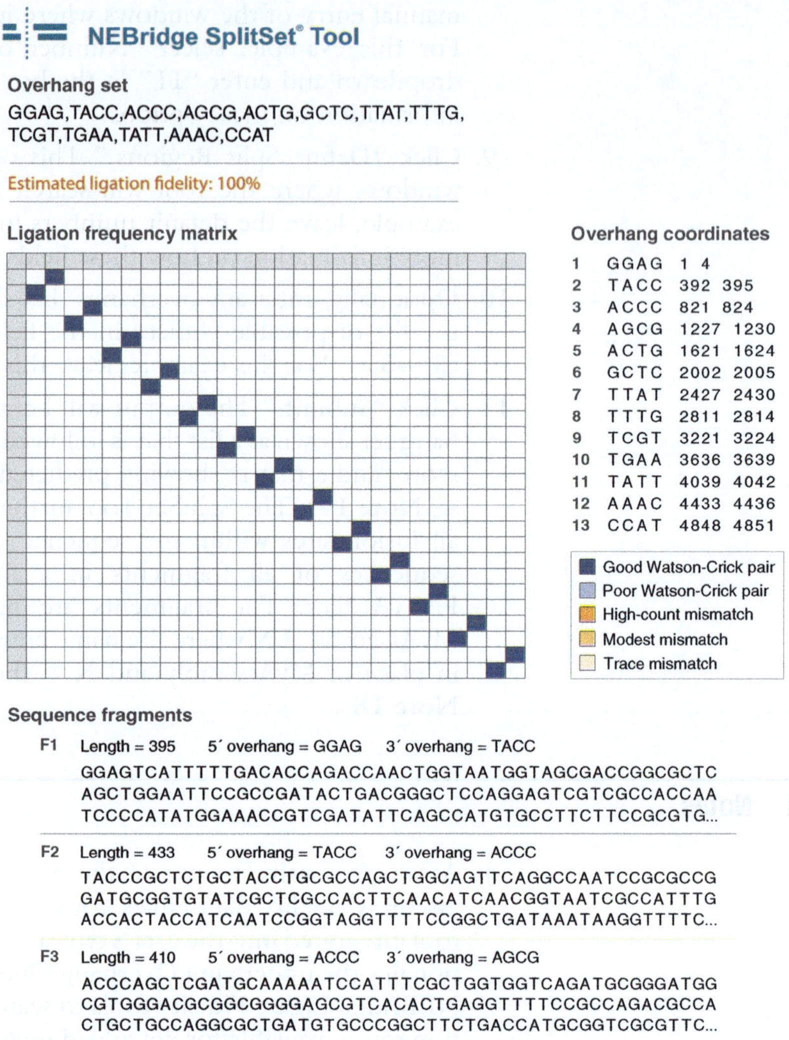

Fig. 5 The output generated by a run of the NEBridge SplitSet™ Tool using the sample inputs outlined in Subheading 3.4. The output (Request ID: 20230623165029004805) shows the "Overhang set" as a list that includes all overhangs selected to be the fusion sites, the estimated ligation fidelity of this set (100%), followed by a matrix table in the format described for the Ligase Fidelity Viewer (*see* Subheading 3.1 and Fig. 2). The SplitSet tool additionally provides a table that notes the coordinates of each fusion-site overhang within the provided sequence input, and a list of "Sequence fragments" that define the name, fragment length, coordinates within the original sequence, flanking fusion-site overhangs, and full sequence of each fragment. These outputs are abbreviated here for space, but the full output can be viewed at ligasefidelity.neb.com/splitset/run.cgi by entering the Request ID "20230623165029004805" in the "Retrieve an existing job" field at the bottom of the page. Note that figure layout differs from the full tool output; use of the Request ID on the webtool will retrieve the full results

will appear as assembly errors indistinguishable from those resulting from mismatch ligation events unless sequenced. See [15] for guidance for optimizing non-design features that impact assembly success.

3. Fidelity in the context of an assembly is a feature of a set of overhangs (and their complements), not a property of any single overhang in isolation. When looking at the full datasets showing all possible interactions, some sequences certainly ligate with non-perfectly matched partners more often than others. However, if all sequences that an overhang can form mismatch products with are excluded from a given set of overhangs, then the assembly will still be predicted to be high fidelity. That is, it does not matter what the fidelity of an overhang is when all possible ligation partners are present, only the fidelity of an overhang within the set of overhangs is considered. Thus, when a specific mismatch pairing is predicted, changing or eliminating only one member of the pair will be required to improve the fidelity of the overhang set.

4. Unless otherwise stated, all assembly conditions in the Ligase Fidelity applications are assumed to use T4 DNA ligase for assembly. The most important factors for accurate predictions are, in order: use of three- versus four-base overhangs, ligase identity, cycling conditions, buffer conditions, and type IIS enzyme identity [15]. Generally, the specific kinds of mismatch ligation events observed are affected by the overhang length and ligase identity, with other factors influencing the relative amount of each mismatch product observed. For a given overhang length, the specific type IIS enzyme used has only a small effect, and a set predicted to assemble with high fidelity for one type IIS enzyme specificity can be used with any commonly used GGA enzyme with equal levels of expected success [4]. In fact, more flexible high-fidelity sets can be reliably generated by using overall lower fidelity conditions. A set that is high fidelity under 37-16 cycling conditions will also be high fidelity under 37 static incubation, but the reverse is not true (Fig. 2).

5. When entering overhangs in this list, the tools will return errors if overhangs are of the wrong length or include non-canonical bases. The tool will discard these unexpected inputs and return an answer for the overhang set minus these entries. Similarly, duplicate overhangs will be removed and only appear once in the output. Inputs can be entered upper or lowercase, and all spaces are removed before calculating the fidelity matrix. Critically, the complement of each entered overhang sequence will also be added to the matrix, as mismatch events between all overhang sequences and their reverse complements must be considered. Consequently, any overhang set where the input

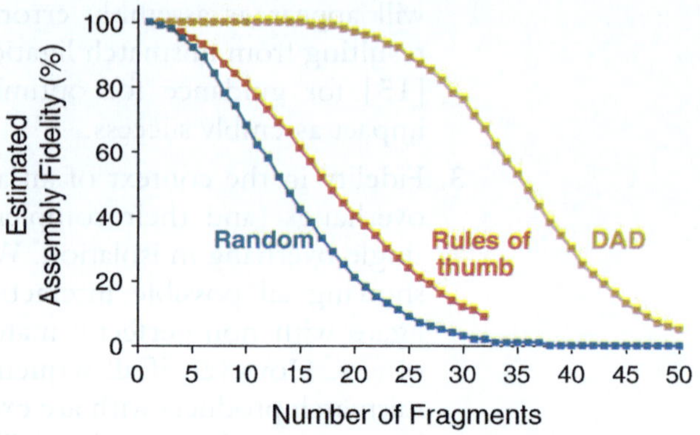

Fig. 6 Predicted fidelity by number of fragments under the BsmBI-v2 42–16 °C cycled assembly conditions using T4 DNA ligase. Predictions are based on the previously reported dataset under these conditions [4]. Average fidelity of randomly selected sets (blue line, 1000 sets per point) and of sets chosen with the classic rule of thumb to avoid having three bases in a row identical between any two overhangs (red line) show rapid decreases from perfect fidelity. The highest fidelity sets that can be selected using DAD (orange line) show that perfect fidelity can be maintained up to ~20 overhang sets, and fidelity scores above 80% are possible up to ~30 overhang sets

string contains a given overhang and its reverse complement will return the complement as a duplicate sequence.

6. While there is no sharp cutoff in estimated fidelity that separates a successful assembly from an unsuccessful one, in general, a higher estimate will result in both a higher yield of final assembly and a higher ratio of correct to incorrect inserts. Figure 6 shows maximum predicted assembly by number of junctions for a typical set of assembly conditions, which allows assessment of how far off ideal a given junction set is. In general, sets with lower predicted fidelity will require more screening to locate a colony with a correct insert. With predicted fidelity over 80%, we typically find good yields and need to pick only one or, at most, two colonies to find a construct containing the desired assembly.
7. For the MoPet Example (Fig. 2), the cycled conditions reveal more potential mismatch interactions than the 37 static conditions. While the latter conditions should allow this set to assemble with high fidelity, often the lower fidelity conditions can make debugging easier. In Fig. 2b, it is straightforward to see that nearly all mismatch interactions involve the GGCT/AGCC pair and that changing this junction could improve performance across all assembly conditions. In addition, it has been shown that while 37 static incubations are generally higher fidelity, much higher yields of transformants are generated using cycled assembly conditions, especially for high

fragment number assemblies [5]. We recommend using an overhang set with high predicted fidelity under cycled conditions, rather than a suboptimal set with reaction conditions that promote higher fidelity at the expense of assembly yield.

8. We find that in the absence of sequence restrictions, sets up to 20 fusion sites with essentially no mismatch ligation predicted can be found, and up to 30 fusion sites with a >90% fidelity score (Fig. 6) [4].
9. Required overhangs can be used to include any fusion sites that must be included in your assembly. For example, predetermined vector fusion sites can be entered in the "Required overhangs (5′→3′)" box, ensuring that all other selected overhangs will be compatible with the pre-existing sites. In the presented example, we include GATC as a fusion site that contains a start codon, a frequently employed strategy in GGA standards that will be used for assembling genes. See Note 5 for information on acceptable formats for entries in this field.
10. For excluded overhangs, any overhangs that should not be considered to be used as fusion sites can be entered. All self-complementary overhangs are excluded by default, as are the reverse complements of any overhangs you enter in "Excluded overhangs (5′→3′)." In the example used in Subheading 3.2, we exclude homopolymers AAAA and CCCC, as well as the TAAA/TTTA pair, which we have found to be notably low efficiency in some ligation experiments [3]. *See* **Note 5** for information on acceptable formats for entries in this field.
11. This search is not deterministic, and a new set of overhangs will be generated de novo each time the application is run. The smaller the number of overhangs in the set, the more possible sets with equivalent predicted fidelities can be found. Each set of outputs will be associated with a Request ID that can be used to retrieve a specific set of results. Table 1 shows the results of running the example parameters in Subheading 3.2 ten times. Each result includes the required GATG overhang, excludes AAAA, CCCC, and TAAA, and has equivalent predicted fidelity, but the specific suggested overhangs returned otherwise differ from run to run.
12. This input allows you to decide to simply replace the overhang you wish to remove or expand a set to include more overhangs. Typically, it is possible to expand a set to around 20 total overhangs without significant drop in predicted fidelity. Note that to change an existing set of parts to accommodate the new fusion sites, two parts will need to be changed to replace one fusion site (both flanking parts). Similarly, to add multiple new sites to the end of an assembly, the final existing fragment will

need to be changed such that the terminal overhang now matches the first new fusion site rather than the existing vector fusion site. Consequently, the last new part must end with an overhang that matches the vector fusion site. See other chapters in this book and references for examples on how to design and change parts [15].

13. This list includes all overhangs from the existing MoPET set, save the GGCT overhang that is the source of nearly all predicted mismatch ligation (Fig. 2, *see* **Note** 7).
14. As with Subheading 3.2, the results of this search are not deterministic, and rerunning the protocol will provide a different list. All versions should work, but specific results can be recalled by recording the Request ID. Table 2 shows the results of running the parameters in Subheading 3.3 ten times, which resulted in ten unique sets with nearly identical predicted fidelity.

Table 2
Example results from running Subheading 3.3 with the provided example parameters

NEBridge GetSet Request ID	Overhang set	Estimated ligation fidelity
20230616191248016496	CTTG,CCAT,GGAT,CGGG,GGTG,AGGC,TAAT,TCCA, TACA,GAAA,GCGA,ACAC,CTAA,ACTG,AGAA	99%
20230616191517016864	CTTG,CCAT,GGAT,CGGG,GGTG,AGGC,TAAT,CTGC, TAGA,ACAC,ACTG,GACA,AAAA,AACG,TGAA	99%
20230616191605017092	CTTG,CCAT,GGAT,CGGG,GGTG,AGGC,TAAT,AAGA, CGTA,TCCA,AATC,AGAA,CTAA,ACTG,AACT	99%
20230616191633017300	CTTG,CCAT,GGAT,CGGG,GGTG,AGGC,TAAT,ACTG, CTGC,GATA,ACGA,AACT,AAAA,ACAC,TACA	99%
20230616191706017514	CTTG,CCAT,GGAT,CGGG,GGTG,AGGC,TAAT,AGAA, ACTG,CTGC,GCTA,AAGA,CTAA,TACA,GTGA	99%
20230616191757017758	CTTG,CCAT,GGAT,CGGG,GGTG,AGGC,TAAT,AGAA, GAAA,AAGA,TCCA,ACAC,AACT,GCTA,CTAA	99%
20230616191828017981	TTG,CCAT,GGAT,CGGG,GGTG,AGGC,TAAT,GCTA, AACA,ACAC,AGAA,CTAA,GAAA,TCCA,ACTG	99%
20230616192013018312	CTTG,CCAT,GGAT,CGGG,GGTG,AGGC,TAAT,GAAA, TCCA,CGTA,GCTA,AAGA,CTGC,AACT,ACAC	99%
20230616192043018526	CTTG,CCAT,GGAT,CGGG,GGTG,AGGC,TAAT,GACA, ACGA,AACG,AATC,TGAA,GTAA,ACTG,ACAC	99%
20230616192115018746	CTTG,CCAT,GGAT,CGGG,GGTG,AGGC,TAAT,CTAA, CTGC,AAGA,AAAC,AGAA,GATA,AACG,TCCA	99%

15. The "linear" versus "circular" options allow flexibility in the way the sequence is entered, and the assembly is designed. When using a destination vector with fixed overhangs, as described in the presented example, it is often most convenient to use the linear option, and simply enter the insert sequence. In this case, the overhang sequences matching the vector fusion sites are included at the beginning and end of the insert sequence. Selecting the box "Use terminal overhangs as fusion sites (for linear assemblies only)," instructs the SplitSet to include the terminal sequences as overhangs in the fidelity calculation and ensures that the overhang set provided will be fully compatible with these fusion sites. Alternatively, one could enter the full circular sequence of vector and insert together and manually define the split regions to ensure no breaks are placed in the backbone and the fusion sites are predefined; see Note 17 for more on this option. The "circular" option is most effectively used when assembling a small genome or other circular sequence where there is no precise vector backbone or other destination construct. For an example of this sort of assembly, see our prior publications covering T7 bacteriophage genome assembly [5, 15].

16. Sequences entered here must contain only ACTG inputs. Make sure to trim any extraneous text from your entered sequence, i.e., copy paste only the sequence info from a FASTA file. Spaces and line breaks will be stripped automatically, but other characters will cause an error message to be returned until they are removed. The current version of this tool cannot support any ambiguous base positions in the assembly. Variable or unknown regions must be reduced to a single sequence. The split regions can then be defined to ensure that breakpoints are not permitted within the variable region.

17. The default way to use the tool to divide a sequence is to enter the number of fragments desired. Entering that number, clicking "Define split regions based on" will open a table automatically populated with 51 nt windows spread equally throughout the sequence. If a circular assembly is desired, a number of breakpoints will be shown equal to the number of requested fragments; for a linear assembly, the number of breakpoints will be one less than the number of desired fragments (plus the vector fusion sites on the sequence termini, if that option is selected as explained in **Note 15**). The options "maximal fragment size" and "minimal fragment size" are options that can be used when the number of fragments is not important, but there are limitations on size; e.g., if input fragments are to be ordered from a vendor with size limits on the provided DNA. In both cases, the number entered in the input box should be the desired size of the fragments in base pairs. Clicking on "Define split regions" will then populate a table with breakpoint windows of

51 nt spread throughout the sequence, such that all fragments will be approximately equally sized as close to the limit entered as possible staying under or over the limit depending on whether the "maximal" or "minimal" option was selected. Finally, by using "Split regions" (i.e., manual entry) you may enter the exact search regions desired as a comma-delineated list (e.g., "5-55, 600-650, 1200-1250, 1788-1791, 2600-2700"), which the program will use to place breakpoints at the listed sequence positions. In all cases, one may manually adjust the windows within the auto-populated table generated after clicking the "Define split regions" button. Both the start and end positions of each window may be freely adjusted. However, trying to define a window smaller than the requested overhang length or windows that overlap will return errors and prevent the tool from running.

18. The sequences provided are generally compatible with the FASTA format but include additional information on the positions covered within the target sequence and the identity of fusion-site overhangs noted in the header. Note that these fragments do not include the external type IIS sites that are required for performing the assembly. Before using these sequences in an assembly visualization tool, such as the NEBridge Golden Gate Tool, one must manually append the desired type IIS recognition site and required spacers to each end. The structure of the appended sequence should be at least 6 bp of outside spacer sequence (either as a primer extension or a holding vector), the desired type IIS enzyme recognition site, and an appropriate spacer for that type IIS enzyme (e.g., 1 bp for BsaI, BsmBI/Esp3I, and SapI/BspQI; 2 bp for BbsI/BpiI; and 4 bp for PaqCI/AarI). Standard sequences can easily be appended using the "concat" command in Excel.

Acknowledgments

I am grateful to Tasha José (NEB Scientific Illustrator) for the design and layout of Figs. 1, 2, 3, 4, 5, and 6 and to Andrew Sikkema, Sean Lund, and Karen Lohman for feedback on and editing of this manuscript.

References

1. Szybalski W, Kim SC, Hasan N, Podhajska AJ (1991) Class-IIS restriction enzymes--a review. Gene 100:13–26. https://doi.org/10.1016/0378-1119(91)90345-c
2. Engler C, Kandzia R, Marillonnet S (2008) A one pot, one step, precision cloning method with high throughput capability. PLoS One 3(11):e3647. https://doi.org/10.1371/journal.pone.0003647
3. Potapov V, Ong JL, Kucera RB, Langhorst BW, Bilotti K, Pryor JM, Cantor EJ, Canton B, Knight TF, Evans TC Jr, Lohman

GJS (2018) Comprehensive profiling of four base overhang ligation Fidelity by T4 DNA ligase and application to DNA assembly. ACS Synth Biol 7(11):2665–2674. https://doi.org/10.1021/acssynbio.8b00333

4. Pryor JM, Potapov V, Kucera RB, Bilotti K, Cantor EJ, Lohman GJS (2020) Enabling one-pot Golden Gate assemblies of unprecedented complexity using data-optimized assembly design. PLoS One 15(9):e0238592. https://doi.org/10.1371/journal.pone.0238592
5. Pryor JM, Potapov V, Bilotti K, Pokhrel N, Lohman GJS (2022) Rapid 40 kb genome construction from 52 parts through data-optimized assembly design. ACS Synth Biol 11(6):2036–2042. https://doi.org/10.1021/acssynbio.1c00525
6. Weber E, Engler C, Gruetzner R, Werner S, Marillonnet S (2011) A modular cloning system for standardized assembly of multigene constructs. PLoS One 6(2):e16765. https://doi.org/10.1371/journal.pone.0016765
7. Bird JE, Marles-Wright J, Giachino A (2022) A user's guide to Golden Gate cloning methods and standards. ACS Synth Biol 11(11): 3551–3563. https://doi.org/10.1021/acssynbio.2c00355
8. Lohman Gregory JS, Bauer RJ, Nichols NM, Mazzola L, Bybee J, Rivizzigno D, Cantin E, Evans TC Jr (2015) A high-throughput assay for the comprehensive profiling of DNA ligase fidelity. Nucleic Acids Res 44(2):e14–e14. https://doi.org/10.1093/nar/gkv898
9. Engler C, Gruetzner R, Kandzia R, Marillonnet S (2009) Golden gate shuffling: a one-pot DNA shuffling method based on type IIs restriction enzymes. PLoS One 4(5):e5553. https://doi.org/10.1371/journal.pone.0005553
10. Engler C, Marillonnet S (2011) Generation of families of construct variants using golden gate shuffling. Methods Mol Biol (Clifton, NJ) 729: 167–181. https://doi.org/10.1007/978-1-61779-065-2_11
11. Engler C, Marillonnet S (2014) Golden Gate cloning. Methods Mol Biol (Clifton, NJ) 1116: 119–131. https://doi.org/10.1007/978-1-62703-764-8_9
12. Potapov V, Ong JL, Langhorst BW, Bilotti K, Cahoon D, Canton B, Knight TF, Evans TC Jr, Lohman GJS (2018) A single-molecule sequencing assay for the comprehensive profiling of T4 DNA ligase fidelity and bias during DNA end-joining. Nucleic Acids Res 46(13):e79–e79. https://doi.org/10.1093/nar/gky303
13. Bilotti K, Potapov V, Pryor JM, Duckworth AT, Keck JL, Lohman GJS (2022) Mismatch discrimination and sequence bias during end-joining by DNA ligases. Nucleic Acids Res 50(8):4647–4658. https://doi.org/10.1093/nar/gkac241
14. Duckworth AT, Bilotti K, Potapov V, Lohman GJS (2023) Profiling DNA ligase substrate specificity with a Pacific biosciences single-molecule real-time sequencing assay. Curr Protoc 3(3):e690. https://doi.org/10.1002/cpz1.690
15. Sikkema AP, Tabatabaei SK, Lee YJ, Lund S, Lohman GJS (2023) High-complexity one-pot Golden Gate assembly. Curr Protoc 3(9):e882. https://doi.org/10.1002/cpz1.882
16. Oling D, Lan-Chow-Wing O, Martella A, Gilberto S, Chi J, Cooper E, Edstrom T, Peng B, Sumner D, Karlsson F, Volkov P, Webster CI, Roth R (2022) FRAGLER: a fragment recycler application enabling rapid and scalable modular DNA assembly. ACS Synth Biol 11(7):2229–2237. https://doi.org/10.1021/acssynbio.2c00106
17. Volke DC, Martino RA, Kozaeva E, Smania AM, Nikel PI (2022) Modular (de)-construction of complex bacterial phenotypes by CRISPR/nCas9-assisted, multiplex cytidine base-editing. Nat Commun 13(1):3026. https://doi.org/10.1038/s41467-022-30780-z
18. Yuan G, Martin S, Hassan MM, Tuskan GA, Yang X (2022) PARA: a new platform for the rapid assembly of gRNA arrays for multiplexed CRISPR technologies. Cells 11(16). https://doi.org/10.3390/cells11162467
19. Marillonnet S, Grutzner R (2020) Synthetic DNA assembly using Golden Gate cloning and the hierarchical modular cloning pipeline. Curr Protoc Mol Biol 130(1):e115. https://doi.org/10.1002/cpmb.115
20. Davis MW, Jorgensen EM (2022) ApE, A plasmid editor: a freely available DNA manipulation and visualization program. Front Bioinform 2:818619. https://doi.org/10.3389/fbinf.2022.818619
21. Weber E, Birkenfeld J, Franz J, Gritzan U, Linden L, Trautwein M (2017) Modular Protein Expression Toolbox (MoPET), a standardized assembly system for defined expression constructs and expression optimization libraries. PLoS One 12(5):e0176314. https://doi.org/10.1371/journal.pone.0176314

Chapter 4

Modular DNA Construct Design for High-Throughput Golden Gate Assembly

Peter Vegh, Elliott Chapman, Craig Gilmour, and Rennos Fragkoudis

Abstract

Golden Gate cloning enables the modular assembly of DNA parts into desired synthetic genetic constructs. The "one-pot" nature of Golden Gate reactions makes them particularly amenable to high-throughput automation, facilitating the generation of thousands of constructs in a massively parallel manner. One potential bottleneck in this process is the design of these constructs. There are multiple parameters that must be considered during the design of an assembly process, and the final design should also be checked and verified before implementation. Doing this by hand for large numbers of constructs is neither practical nor feasible and increases the likelihood of introducing potentially costly errors. In this chapter we describe a design workflow that utilizes bespoke computational tools to automate the key phases of the construct design process and perform sequence editing in batches.

Key words Computer-aided design, DNA assembly, Overhang, Plasmid, Golden Gate, Cloning, Restriction enzyme, Synthetic biology, Engineering biology

1 Introduction

The drive toward automation of laboratory processes, combined with the decreasing cost of DNA synthesis, has led to a recent expansion in the number of projects within the field of engineering biology. In order to generate the large numbers of DNA constructs required to equip cells with new or improved functionalities, Golden Gate cloning has been widely employed. This method utilizes type IIS restriction enzymes that cut outside of their recognition site; this allows the generation of single-stranded DNA overhangs (cohesive or "sticky" ends) that can be used to join modular DNA parts in a specific, user-determined, order into a suitable receiver plasmid. The advantage of this method is that the enzyme recognition site is destroyed upon successful construct formation; thus, assemblies can be set up as "one-pot" reactions,

Daniel Schindler (ed.), *Golden Gate Cloning: Methods and Protocols*, Methods in Molecular Biology, vol. 2850, https://doi.org/10.1007/978-1-0716-4220-7_4,

circumventing the requirement for labor-intensive DNA purification steps required for traditional cloning.

Performing sequence design and editing manually can be both error-prone and time consuming. The large number of constructs assembled with the use of automation requires not only parallelization of the wet-lab operations but also of the design process. This high-throughput design phase is facilitated by available software, such as the Edinburgh Genome Foundry's (EGF's) suite of free and open-source packages, and associated web apps. The EGF is a biofoundry [1], a high-throughput facility specializing in the design, construction, and testing of engineered synthetic DNA constructs, using high-level automation such as robotics and liquid handlers.

Scripting, that is, instructing DNA design software with a program code, is the most flexible and efficient approach to carry out the design work as it has all the associated advantages of laboratory-based automation: the script self-documents the process, enables batch processing, and greatly improves precision, reproducibility, and speed. Additionally, a script can be reused; therefore, a one-time writing effort can be applied to multiple projects, perhaps with small modifications.

As the first step in implementing the finalized design, the DNA parts are ordered from one of the commercial DNA synthesis companies; each supplier may put different constraints on the sequence that can be made, depending on their applied technology. The project-specific cloning procedure also puts constraints on the sequence, such as the selection of the overhangs, or avoiding homologies within the constructs, which all need to be considered at the design phase.

This chapter provides a walk-through of planning a Golden Gate assembly, with focus on common design considerations and parallelization of the process. The design process consists of four main steps:

1. Evaluation of the project to assess the best approach and to estimate costs.
2. Design of the assembly plan and the desired constructs.
3. Preparation of the DNA part sequence files. At this stage, the sequences are evaluated, and domesticated (i.e., optimization for synthesis and assembly, and addition of flanking restriction enzyme recognition sites).
4. In silico simulation of the cloning procedure and a review of the results.

2 Materials

Design of sequences for Golden Gate assembly requires specialized software tools that permit efficient handling of sequence data. Common sequence data file formats are FASTA and GenBank. FASTA files usually have the extension ".fa" or ".fasta" and have a very simple format with a header line followed by sequence lines. The GenBank [2] file format (".gb" or ".gbk" extension) can store annotations and, therefore, it is the preferred format. These text files can be opened in text editors (as opposed to a word processor) such as Geany (https://geany.org), and sequence editors.

A sequence editor visualizes the sequence map (an overview with annotations) and enables edits to be made, while also facilitating various tasks, such as cloning simulation or sequence alignment. Examples of sequence editors are SnapGene (from Dotmatics; available at https://www.snapgene.com), or the Benchling cloud-based platform and website (https://www.benchling.com).

Various open-access tools are provided via the EGF's Collection of Useful Biological Apps (https://cuba.genomefoundry.org/). The web apps are powered by free and open-source software packages, written in the Python programming language. Full documentation for each of these packages is available online (https://github.com/Edinburgh-Genome-Foundry).

Ideally, the sequence modifications are performed using scripts that instruct the various software packages. A programming-based approach enables full utilization of all built-in package functionalities, while also allowing the process to be automated. Perhaps the best approach is to use a web-based development environment such as Jupyter Notebooks (https://jupyter.org), in tandem with a Python environment manager such as Anaconda (https://www.anaconda.com) to separate various Python versions and installed package ecosystems.

Ubuntu (https://ubuntu.com) is the recommended operating system, as it was used to both develop and test all of EGF's packages; alternatively, packages can be installed and used on a virtual system, using Docker (https://www.docker.com). Further instructions on installing this are provided at EGF's GitHub page (https://github.com/Edinburgh-Genome-Foundry/EGF_Docker_Jupyter). Key considerations on selecting software for sequence design and editing work are listed in **Note 1**.

3 Methods

With respect to designing a Golden Gate assembly, the key components are the individual DNA parts to be assembled, the receiver plasmid ("backbone"), the enzyme used for digesting the parts, and the chosen overhang ("sticky end") DNA sequences that determine the order of the parts in the assembled construct. During the design process, the project is evaluated to find the best assembly approach. The two design approaches are: (i) decomposition of a desired sequence into parts (top-down), or (ii) building from parts (modular bottom-up). A modular design is cost efficient and best suits Golden Gate cloning; therefore, constructs should be designed with that principle in mind. In this protocol, we focus on assembling multiple plasmids in the same project that are variants of each other.

Firstly, the individual part sequences and an assembly plan are prepared. The sequences are then evaluated and adapted for Golden Gate cloning. As a verification step, an in silico simulation of the cloning reaction is performed and the plasmid constructs are reviewed (Fig. 1). This simulation requires an assembly plan file, and the DNA part sequences (including the backbone).

Each assembly project is different and starts with different requirements and constraints, meaning design is not an entirely linear process. Therefore, the steps described below may need to be adapted to the project by performing the steps in a different order or considering certain steps together. In any case, a typical project has more than one iteration of the design cycle, that is, a review of the results is followed by further modification and optimization of the sequences.

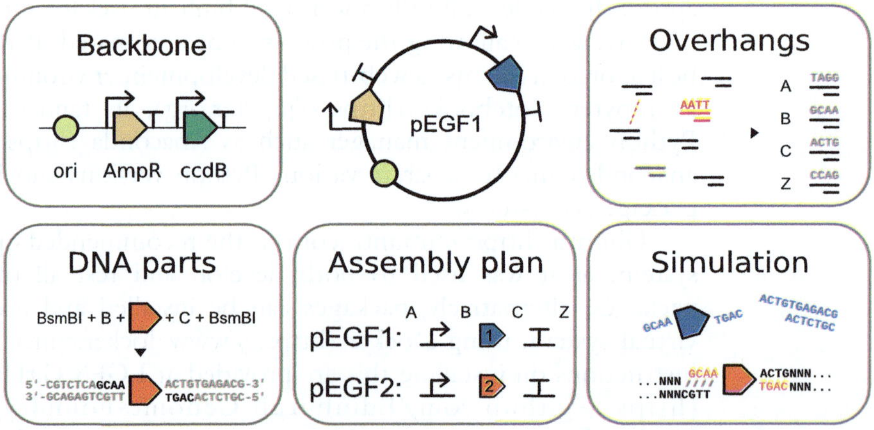

Fig. 1 Overview of the design workflow. Firstly, a backbone or an assembly standard is chosen for the project. In this example, the backbone has an *ori*, *psk* (AmpR), and a selection marker cassette (ccdB) that is removed upon assembly. Then the overhangs (DNA "sticky ends") are selected and evaluated. Undesired misannealing interactions (red dash) between overhangs, and weakly annealing overhangs (gray) are avoided. Palindromic overhangs (red) must not be used for assembly. The assembly plan specifies the DNA parts for each construct. The parts are then optimized for cloning, and flanking enzyme sites are added. Finally, the cloning reaction is simulated in silico, and the assembled construct sequences are reviewed

3.1 Assembly Plan

The steps below describe a *modular* (bottom-up) design approach, using the same receiver backbone for all constructs. *See* **Note 2** for additional comments that are specific to the *top-down* approach, which aims to find a decomposition of desired sequences into parts, for Golden Gate assembly. For the below steps, more details on the backbone, assembly standard or overhang selection, and part optimization are provided in the Subheadings 3.2, 3.3, and 3.4.

1. Determine the design approach, modular or top-down, based on the project constraints and goals. Choose an existing Golden Gate assembly standard or select a backbone for the plasmid constructs. This is usually predetermined by the experiment goals.
2. Create a spreadsheet for the assembly plan, and obtain or create a preliminary sequence for each DNA part. The first column lists the construct names, and the subsequent columns list in order which parts (and backbone) are assembled in each construct (one construct for each row).
3. Determine the sourcing of DNA parts: (i) de novo DNA synthesis or (ii) existing part repositories; examples for the latter are pre-made parts of an assembly standard or parts from a previous project. For new parts, consider DNA synthesis cost, quality, and speed. Check DNA synthesis pricing tiers and ensure length of the parts are within the desired range (e.g., maximum 3 kbp). For existing parts, calculate the sequencing verification costs. For resources on obtaining DNA parts (*see* **Note 3**).
4. Identify any sequence segments that are essential for the project and cannot be changed or replaced, and flexible elements that may be modified to meet constraints. Design the assembly in the context of all constructs so that part reuse is maximized.
5. Determine the type of fusion required for each overhang. The overhangs remain as fusion sites (seam or "scar") in the final sequence. An important design consideration is whether such scars are tolerated between functional units, or if a "scarless" assembly is necessary, where the overhangs form part of the original sequence.
6. Biosafety: ensure that the planned sequences are not under synthesis control (i.e., not a pathogen or a toxin), or obtain relevant permissions.
7. Select the overhangs (and the type IIS restriction enzyme) used for the assembly. This is predetermined by the backbone or the kit used; however, overhangs may still be added, removed, or replaced if required.
8. Avoid homologies *within* the constructs, as it may lead to recombination. For example, do not use the same promoter

or terminator multiple times in the same construct. Ensure you have as few parts as possible. Efficiency of the cloning reaction decreases with the number of parts, especially when using more than ten parts [3]. Coding sequences can be modified by synonymous codon replacement, but other sequences, such as promoters, cannot be modified without affecting their function. When more than one connector is used, check that they are different.

9. For modular designs, DNA parts should be domesticated (i.e., remove internal restriction enzyme recognition sites and add two flanking sites with overhangs, corresponding to their position in the assembly). See details in the following section.
10. Frequently, it is useful to draw an overview or sketch of a design template to facilitate discussion of the project. The EGF Sketch constructs web app is a quick way to prepare this template (https://cuba.genomefoundry.org/sketch_constructs). Alternatively, a DNA sequence map can be plotted with various editors or DNA Features Viewer [4] (https://cuba.genomefoundry.org/plot-sequence-features).
11. Consider the trade-off between part synthesis cost, use in future projects, and assembly efficiency (the cloning is more efficient with fewer parts).
12. Design the project in a redundant way if possible, because not all constructs might be successfully made. Ideally, the project should meet its goals even if some assemblies fail, especially in a high-throughput assembly approach.
13. Optional for validation step: if fragment analysis (FA) is used for validating the constructs, check and add restriction-digest sites if necessary. However, note that sequencing is a better validation approach as it detects point mutations, small recombinations, or indels.
14. Annotate the sequences: ideally all nucleotides in your sequences are accounted for. This records the biological or technical significance of each section and makes it easier to change them if needed. For example, some segments serve as spacers where only the length is important. In other cases the function is unknown, or a segment is not required, and is there only as a carryover from previous cloning work.
15. Review the constructs from a biological point of view. For example, ensure that expression cassettes are functional with a promoter and terminator; remove internal stop codons from fusion proteins; avoid bidirectional promoters; and check for presence of Kozak sequence, start and stop codons, where required. Keep in mind that there are alternative start codons for bacteria [5]. Check the presence of all functional elements required for the experiment, such as any markers, homology

arms, or recombination sites. If propagation is required in the target host, then a suitable replication origin (*ori*) must be added. This review step may be performed at the last stage as well, when reviewing simulated constructs.

16. Conform to any intellectual property limitations (e.g., patents) imposed on the sequences.
17. Optional for expressed proteins: include protein tags for easy detection or purification.
18. Optional: design and perform a trial (pilot) run for larger projects. This verifies feasibility of the assembly approach, and the suitability of the constructs in downstream experiments. Select the most important subset of constructs, or alternatively select assemblies that altogether use all parts. Ideally, build the project and the toolkit in an iterative manner. Verify that new parts work well, before incorporating more parts. Similarly, a scale-up approach also works well: start with fewer parts (and thus overhangs), previously used parts, and simpler or shorter new sequences, then verify that the assembly works. Then proceed to more complex parts and more overhangs.
19. Optional: for larger projects, select or design a control assembly that verifies that the cloning protocol was performed correctly. An example is a chromoprotein expression cassette construct [6]. This is especially useful for excluding problems with the protocol, during any troubleshooting.
20. Consider future use of the various DNA segments, and any further planned constructs in subsequent rounds of assemblies. This includes future work in different host strains.

3.2 Backbone (Receiver) Plasmid

Plasmids provide a convenient way to store and *propagate* DNA in bacteria; therefore, DNA parts are usually assembled into a receiver plasmid (backbone), a DNA part that contains the functional elements for propagation—most importantly a replication origin (*ori*) and a selection marker. Backbones also contain an element for counterselection of intact backbones, such as the *ccdB* toxin system, or a fluorescent marker. This element (dropout cassette) is removed upon assembly, ensuring propagation of the desired assembled product, rather than the intact backbone.

The constructed DNA may also be used and maintained in a *host* organism, relevant for the experiments and project goals. That may be different from the propagator organism (which is usually *Escherichia coli*), and may require a separate replication origin and other functional elements.

1. Choose a backbone (receiver plasmid) from an existing standard; alternatively onboard or synthesize a plasmid suitable for Golden Gate. Example existing backbones: the EMMA backbone for mammalian expression [7], the EcoFlex backbone for

bacterial expression [8], MoClo backbones [9], and JUMP backbones [10]. For comments on constructing a new backbone (*see* **Note 4**).

2. In all cases, a backbone plasmid for Golden Gate assembly must meet the following requirements: (i) two type IIS restriction enzyme sites flanking the location of insert. These two enzyme sites must have an orientation such that the recognition sites are also cleaved out from the plasmid during assembly. (ii) No other sites are present for these enzymes, otherwise the plasmid will be digested. While enzyme sites are theoretically allowed in the insert region (dropout cassette), it is better to have no unnecessary fragments (with "sticky ends") in the reaction. For multi-level assemblies, we must also ensure that the plasmid does not contain recognition sites for the restriction enzymes of the other levels. (iii) The overhangs of the two enzyme sites are suitable for the chosen overhangs set (standard) or the parts. (iv) Contains a selection marker for presence of the plasmid (i.e., post-segregational killing [*psk*]), for example, AmpR. (v) Additionally, it is good to ensure that it contains selection for proper assembly and against intact backbones. Common options for the excised segment are expression cassettes for the *ccdB* toxin, *lacZ* for blue-white screening, GFP/RFP (Green/Red Fluorescent Protein), or chromoproteins. (vi) Finally, the plasmid must be compatible with the strain used. Look up the strain genotype details in various databases and cross-check against the properties that the plasmid confers.
3. Research and review all required growing conditions specific to the plasmid and the propagator strain. Cross-check the compatibility with the genotype of the strain.
4. Note the resistance marker of the backbone, which will be retained in the assembled plasmids: choose an alternative marker for the DNA parts.
5. Optionally ensure presence of copy number control (*cop*), multimer resolution (*mrs*), and partitioning (*par*) systems. If multiple plasmids are transfected into the same host for propagation, ensure these use compatible systems.
6. Annotate potentially missing features in the plasmid, using pLannotate [11] (https://github.com/barricklab/pLannotate) or SnapGene (Features → Detect Common Features). Ensure target host species and genotype compatibility (resistance, toxicity, etc.).

3.3 Overhang Set and Assembly Standard

A crucial element of the plan is the overhang ("sticky end") set used for the assembly. In this context, an *overhang set* is a selection of mutually compatible overhangs that can be used together in an assembly by avoiding cross-complementarity between overhang pairs. An *assembly standard* is an ordered overhang set, combined with a defined function (e.g., "promoter") for each pair of consecutive overhangs; this allows each overhang pair to define a specific position within the standard. A group of positions may also be assigned a high-level function (e.g., "marker cassette"). Overhangs must fulfil a number of conditions and they must be mutually compatible, for efficient and proper assembly. They can be obtained from an existing assembly standard (MoClo, EMMA, etc.) or designed specifically for the project. In all cases, we must not use any overhang more than once within an assembly, and we must avoid self-complementary palindromic overhangs.

Overhangs may misanneal with non-complementing overhangs. While this happens at a low rate, it is important that the utilized overhangs are as high fidelity as possible [3, 12]. A quick way to achieve this is by ensuring at least two base-pair difference between each overhang pair in the set. A better approach is predicting misannealing with an experimental dataset, as explained below.

Individual overhang-related tasks are well covered by the many software packages, but addressing multiple objectives at the same time is more difficult and may need iterative work by hand. An example of such a task is decomposition of a protein coding sequence into parts, with the freedom to synonymously replace codons.

1. Determine the type IIS restriction enzyme used for the assembly. An assembly standard, existing backbone, or DNA parts determine the choice of the enzyme. For new projects, the presence of recognition sites in the desired sequence can inform of the better choice. Commonly used enzymes are BsmBI (Esp3I) and BsaI. For higher fidelity, it is recommended to use enzymes that create 4 bp overhangs, but 3 bp enzymes may enable simpler fusion protein designs.
2. If an assembly standard is used, that determines the overhangs in the project. Otherwise, select new overhangs with the Design overhangs web app [13] (cuba.genomefoundry.org/design-overhangs) or the NEBridge GetSet Tool (https://ligasefidelity.neb.com). Additional considerations are mentioned in **Note 5**.
3. Check misannealing and ligation fidelity using the NEBridge Ligase Fidelity Viewer (https://ligasefidelity.neb.com). Alternatively, use the EGF View Overhangs Crosstalk web app (https://cuba.genomefoundry.org/view-overhangs-crosstalk) or SnapGene. Exclude self-annealing, misannealing, and weakly annealing overhangs.

4. If the backbone's dropout cassette contains a recognition site, ensure that its digestion does not create one of the selected overhangs, and also does not misanneal with them.
5. Create the ***assembly standard*** file for future reference: this file describes which overhang pair is used for each assembly position. For example, each overhang can be assigned a letter from the alphabet, and thus a position—or any connection—can be denoted by two letters (i.e., the flanking overhangs).
6. Ensure that there is no frameshift in protein coding sequences that are built from joining parts. (This does not apply to decomposition designs.) Add two extra nucleotides to four-nucleotide overhangs at the joining sites [14]. These two nucleotides together with the four-nucleotide overhangs will be translated into two amino acids, thus keeping the sequence downstream in frame. Create a list of possible translations in all six frames, then choose as appropriate for the protein and the project, e.g., small amino acids or linker sequences. Avoid creating translation pause and arrest sites. Assign the extra nucleotides to the upstream or downstream positions as the translation requires. The report made by the "Overhang" package provides a list of the possible amino acid translations for each overhang (https://github.com/Edinburgh-Genome-Foundry/Overhang).

3.4 DNA Parts

In this section we assume that we start from undomesticated ("raw") part sequences, for which the position in the assembly, and thus overhangs, have already been selected. These parts are evaluated for patterns interfering with synthesis or cloning, then modified if necessary. The final step is the addition of flanking restriction enzyme recognition sites and overhangs at the 5′ and 3′ ends.

1. Sequences should be stored in a sequence file format. GenBank is preferred to FASTA as it can store annotation. Convert any sequences stored in text into GenBank format. The EGF Convert Sequence Files web app (https://cuba.genomefoundry.org/convert_sequence_files) can perform this conversion. Alternatively, SnapGene can create (File → New DNA File) or convert files (File → Save as → GenBank—Standard; or File → Batch Convert File Format).
2. Use the ".gb" file extension for GenBank files. An alternative is ".gbk." Check that the software tools can handle the chosen extension.
3. Store each sequence in a separate GenBank file for easy identification and handling. Additionally, working with the sequences is often easier by using the filenames as sequence names (without extension) instead of the internally stored sequence name.

4. Evaluate sequences for manufacturability. The sequences must be synthesizable and free from patterns interfering with cloning. Check for presence of these patterns: homopolymers and multimer repeats [15] (such as 9 × 2mer, 5 × 3mer, etc.), hairpins, extreme GC content, K-mers (homologies). This can be checked with the Evaluate Manufacturability web app (https://cuba.genomefoundry.org/evaluate_manufacturability).
5. Ensure sequence names are maximum 15 characters long. The original GenBank format has a name limit of 15 characters [2], which results in some software tools and DNA synthesis companies having the same limit. A practical way to conform to any naming requirements is using alphanumeric and underscore characters. Avoid using dash (hyphen) or space in part names.
6. One convenient way to note the position of a part is to prefix its name with the position ID (e.g., p7_GFP, pAD_CMVp). The EGF GeneDom package can use this prefix to perform part-domestication in a batch manner. A detailed description for using this tool and its web app (https://cuba.genomefoundry.org/domesticate_part_batches) is provided in Zulkower (2021) [16].
7. Optionally, create a lookup list in a spreadsheet, which matches short sequence names to any longer names and comments.
8. For protein coding sequences, check that the length is a multiple of 3 (accounting for introns); failure suggests editing errors.
9. Optimize sequences by removing any recognition sites for the enzyme used in the assembly. Optionally, remove any unwanted patterns identified at the evaluation stage. An unwanted enzyme site or pattern can be removed by (i) simply deleting one or more nucleotides if it is not in a functional region or the deletion does not affect function, (ii) replacing a codon with a synonymous codon in a coding sequence, using SnapGene or DNA Chisel [17] or its web app, and (iii) replacing the functional element or DNA part with an alternative that does not contain the pattern when the above is not possible (for example in a promoter).
10. At this stage also remove enzyme sites for any higher-level assemblies if applicable.
11. Optional: codon-optimize coding sequences, i.e., adapting codon-usage to the target species. This is dependent on the aims of the project.
12. Consider removing recognition sites for enzymes used by common assembly methods: BbsI, BsaI, and BsmBI [14]. This change makes the DNA part and the construct product more flexible for future use.

13. Domesticate the optimized parts by adding flanking enzyme sites and overhangs. Determine the desired positions in the designed assembly standard and add the corresponding overhangs/sites. Remember to add any additional nucleotides for protein fusion sites if necessary. Alternatively, perform the domestication in batches, using GeneDom as explained above.
14. For clonal parts, the requirements for the carrier plasmid containing the part are the same as for the backbone, with one important difference: the selection marker for the plasmid (i.e., post-segregational killing [*psk*] system, e.g., KanR) must be different from the backbone's selection marker.
15. For clonal parts, ensure that restriction enzyme sites do not overlap with methylation enzyme sites. Methylation can block restriction and cause failure of the assembly; crucially this can affect the BsaI enzyme. For synthesized clonal parts, add protective flanking nucleotides to ensure that the enzyme site cannot be a part of a methylation recognition site with the nucleotides adjacent in the carrier backbone. For more details (*see* **Note 6**).
16. For synthesized linear parts, add five flanking nucleotides at the 5′ and 3′ ends to protect the enzyme recognition sites from small terminal truncations.

3.5 Cloning Simulation

The final step in the Golden Gate assembly design process is to computationally validate the plan by simulating the cloning reaction. Several tools exist that can simulate cloning, including DNA Cauldron, SnapGene, or Benchling. The success of the simulation verifies that the technical aspect of the plan was designed correctly, after which the simulated constructs can be reviewed for their suitability. This last step requires the part sequences and the assembly plan, as prepared previously.

1. Collect the DNA part sequence files, including the backbone, and the assembly plan file.
2. Cross-check the assembly plan and part names to ensure they match, including any possible modifications made during the work. Attend to common errors: typos such as space character after the end of part names, mismatching letter case, or inconsistencies in using hyphen (dash), underscore, or space between names in the plan and sequence names. Other common errors include duplicate entries or missing parts.
3. Simulate the assembly reaction and ensuing constructs with SnapGene, Benchling, or the EGF Simulate Golden Gate Assemblies web app [16]. Ideally, simulate with the backbone (or other main part) in the same (first) position. This will make it easier to do multiple sequence alignments or other comparisons across the constructs.

4. Confirm that there are no errors raised during simulation.
5. Optional: evaluate manufacturability of simulated plasmids, to detect homologies between parts in the same construct (see previous sections).
6. Review the technical aspect of the assembly. Cross-check constructs against original sequences to ensure no unwanted changes and mistakes have been introduced. Ensure that a new restriction enzyme site has not been assembled at the joining regions (EGF's DNA Cauldron automatically flags this). Check that multi-part protein coding sequences stay in frame.
7. Review the sequences to confirm their suitability for the project's purpose, especially at regions where the parts join. Ensure presence of all required functionalities—resistance marker, suitable plasmid replication origin for the host, homology arms, etc.—as discussed previously.
8. Prepare the sequences in the required format for ordering DNA for synthesis (usually FASTA). If GeneDom (or its web app) was used for the assembly, then the domesticated files are available in FASTA and GenBank formats. Count the number of parts and the total number of nucleotides. Estimate cost of DNA synthesis.

4 Notes

1. *Materials.* The efficiency and the speed of the sequence design and editing work is largely determined by the quality of the utilized software tools. A key consideration when selecting software is *reliability*; can the package be trusted to reproducibly perform tasks correctly? A good indicator of reliability is the size of the user (and developer) base, as well as the maturity of the software. Additional assurances of quality are if an organization depends on the software, either as a product, or for the day-to-day running of its operations. Conformity to good software engineering practices is also a good sign; this includes good documentation, a test suite, and continuous integration. These characteristics also suggest the future and long-term availability of support and development.

   Finally, using software that is distributed under a *free and open-source* license can save on costs, but, more importantly, its source code can be reviewed to verify its workings, and it can also be adapted and extended to suit specific user needs. The EGF software suite recommended throughout the text fulfils all of these requirements: it is an integral part of the EGF operations in DNA design, assembly, and quality control, and has been continuously developed and expanded since EGF's foundation in 2016. It has also been extensively validated by thousands of successful DNA assemblies. Python scripting notebook templates for performing the design workflow using EGF software packages are provided in the EGF Examples repository (https://github.com/Edinburgh-Genome-Foundry/Examples).

2. *Sequence decomposition*. In this approach, we want to produce a *decomposition* of the desired construct sequences into smaller parts, which can be used to assemble the constructs. If we have multiple intended constructs, the first step is to identify *common blocks* that could be reused. This may be evident from the design of the project or the origin of the constructs, otherwise the EGF Find Common DNA Blocks web app can assist in this task (https://cuba.genomefoundry.org/find-common-blocks).

   The EGF Design Golden Gate Overhangs web app assists in finding sets of overhangs for multi-part DNA assemblies. It has two main use cases: (i) generate a collection of overhangs, or (ii) choose a set of overhangs from a final desired sequence, for a scarless assembly [13]. It provides a simple interface for specifying the sequence and the overhang set. The user can also set and annotate "cut" regions in the uploaded sequence, and the overhangs will be chosen from these. These regions usually do not have to be long, as any 100+ bp segment should contain at least one suitable overhang. If large common blocks exist between the planned constructs, then ideally these "cut" regions are set to flank these regions, so that they have to be created only once. The full workflow is described in Stracquadanio and Zulkower (2021) [13]. The web app is powered by the GoldenHinges Python package that can also utilize experimental misannealing data, as explained in its documentation (https://github.com/Edinburgh-Genome-Foundry/GoldenHinges).

3. *Part repositories*. Addgene is a global repository that was created to help scientists share plasmids [18]. The Standard European Vector Architecture (SEVA, http://seva-plasmids.com) platform is a repository of standard genetic parts [19]. SynBioHub is a design repository for Synthetic Biology [20]. Part sequences can also be found at the iGEM (International Genetically Engineered Machine) Registry of Standard Biological Parts (http://igem.org/Registry).

4. *Backbone*. The benefits of creating a new plasmid backbone are that we can make a bespoke sequence that is fully adapted to the project goals, while increasing cloning efficiency and decreasing costs by including regions that are common across

all constructs. There are multiple options for constructing a new backbone:

(a) Order a plasmid from a DNA provider. This is the most expensive option, but can be cost-effective if using an automated platform that is not suitable for single assemblies. An entirely new synthesized plasmid gives the most flexibility and can be more suitable for the project or for less commonly used organisms.

(b) Assemble a plasmid from existing or synthesized parts. This approach can save on costs but requires designing the assembly and a preparatory cloning work. First, decompose the designed backbone sequence into compatible fragments, using the EGF web app (or the EGF GoldenHinges Python package) as described in the main text. Then, using a sequence editor, join the first and last fragments into one sequence file, so that you obtain a circular sequence upon assembly. It is good practice to set the "cut" regions relatively evenly spaced and inside functional elements of the final plasmid, so that these elements are not expressed nor interfere with the selection process, when the part-plasmids are cloned or amplified.

(c) Modify an existing plasmid. If the basis of our plasmid design is an existing plasmid, then common DNA cloning and editing techniques can be used to modify it to suit Golden Gate. For example, an existing multiple cloning site (MCS) can be used to insert two correctly oriented restriction enzyme sites and a dropout cassette.

In all cases, it is recommended to sequence-verify the plasmid before using it. EGF provides free and open-source software tools for the design and interpretation of restriction-digest fragment analysis, Sanger sequencing [21], and Nanopore sequencing experiments [22].

5. *Overhang fidelity.* Many Golden Gate standards were developed without considering experimental overhang (misannealing or fidelity) data; therefore, an evaluation is recommended before use. Further optional considerations for overhangs: (i) Avoid overhang sequences that are part of the restriction enzyme recognition sequence. This avoids the possibility of assembling an enzyme site with the neighboring nucleotides, which leads to assembly failure. (ii) For joining two protein coding parts, avoid using an overhang with a stop codon in its sequence. This allows for more amino acid-pair translations to choose from. (iii) Similarly, avoid having a start codon in the overhang sequence; conversely, the inclusion of a start codon may allow for a scarless assembly between a promoter- (or Kozak-) and a coding-sequence.

6. *Methylation.* Consider the following sequence: "CCAGGTCTC"; this contains overlapping recognition sites for the Dcm methylase enzyme (CCAGG) and the BsaI restriction enzyme (GGTCTC). Growing this sequence in a Dcm + strain will result in methylation of the recognition site, which blocks restriction, leading to a failed assembly. Note that SnapGene flags potentially blocked enzyme sites on the sequence map with an asterisk. Add these nucleotides upstream and downstream to the BsaI recognition sites to avoid this problem: 5′ end—**TTC**GGTCTC … GAGACC**GAA**—3′ end. Otherwise, the BsaI site can potentially create a Dcm pattern with the adjacent nucleotides, when cloned into the part-carrier plasmid.

References

1. Holowko MB, Frow EK, Reid JC et al (2021) Building a biofoundry. Synth Biol 6:ysaa026. https://doi.org/10.1093/synbio/ysaa026
2. Benson DA, Cavanaugh M, Clark K et al (2013) GenBank. Nucleic Acids Res 41:D36–D42. https://doi.org/10.1093/nar/gks1195
3. Potapov V, Ong JL, Kucera RB et al (2018) Comprehensive profiling of four base overhang ligation fidelity by T4 DNA ligase and application to DNA assembly. ACS Synth Biol 7: 2665–2674. https://doi.org/10.1021/acssynbio.8b00333
4. Zulkower V, Rosser S (2020) DNA Features Viewer: a sequence annotation formatting and plotting library for Python. Bioinformatics 36: 4350–4352. https://doi.org/10.1093/bioinformatics/btaa213
5. Cao X, Slavoff SA (2020) Non-AUG start codons: expanding and regulating the small and alternative ORFeome. Exp Cell Res 391: 111973. https://doi.org/10.1016/j.yexcr.2020.111973
6. Liljeruhm J, Funk SK, Tietscher S et al (2018) Engineering a palette of eukaryotic chromoproteins for bacterial synthetic biology. J Biol Eng 12:8. https://doi.org/10.1186/s13036-018-0100-0
7. Martella A, Matjusaitis M, Auxillos J et al (2017) EMMA: an extensible mammalian modular assembly toolkit for the rapid design and production of diverse expression vectors. ACS Synth Biol 6:1380–1392. https://doi.org/10.1021/acssynbio.7b00016
8. Lai H-E, Moore S, Polizzi K, Freemont P (2018) EcoFlex: a multifunctional MoClo kit for E. coli synthetic biology. Methods Mol Biol 1772:429–444. https://doi.org/10.1007/978-1-4939-7795-6_25
9. Engler C, Youles M, Gruetzner R et al (2014) A Golden Gate modular cloning toolbox for plants. ACS Synth Biol 3:839–843. https://doi.org/10.1021/sb4001504
10. Valenzuela-Ortega M, French C (2021) Joint universal modular plasmids (JUMP): a flexible vector platform for synthetic biology. Synth Biol 6:ysab003. https://doi.org/10.1093/synbio/ysab003
11. McGuffie MJ, Barrick JE (2021) pLannotate: engineered plasmid annotation. Nucleic Acids Res 49:W516–W522. https://doi.org/10.1093/nar/gkab374
12. Pryor JM, Potapov V, Kucera RB et al (2020) Enabling one-pot Golden Gate assemblies of unprecedented complexity using data-optimized assembly design. PLoS One 15: e0238592. https://doi.org/10.1371/journal.pone.0238592
13. Stracquadanio G, Zulkower V (2021) Computer aided assembly and verification of synthetic chromosomes. In: Marchisio MA (ed) Computational methods in synthetic biology. Springer US, New York, pp 169–181
14. Bird JE, Marles-Wright J, Giachino A (2022) A user's guide to Golden Gate cloning methods and standards. ACS Synth Biol 11:3551–3563. https://doi.org/10.1021/acssynbio.2c00355
15. Hossain A, Lopez E, Halper SM et al (2020) Automated design of thousands of nonrepetitive parts for engineering stable genetic systems. Nat Biotechnol 38:1466–1475. https://doi.org/10.1038/s41587-020-0584-2
16. Zulkower V (2021) Computer-aided design and pre-validation of large batches of DNA assemblies. Methods Mol Biol 2229:157–166. https://doi.org/10.1007/978-1-0716-1032-9_6

17. Zulkower V, Rosser S (2020) DNA Chisel, a versatile sequence optimizer. Bioinformatics 36:4508–4509. https://doi.org/10.1093/bioinformatics/btaa558
18. Kamens J (2015) The Addgene repository: an international nonprofit plasmid and data resource. Nucleic Acids Res 43:D1152–D1157. https://doi.org/10.1093/nar/gku893
19. Martínez-García E, Fraile S, Algar E et al (2023) SEVA 4.0: an update of the Standard European Vector Architecture database for advanced analysis and programming of bacterial phenotypes. Nucleic Acids Res 51:D1558–D1567. https://doi.org/10.1093/nar/gkac1059
20. McLaughlin JA, Myers CJ, Zundel Z et al (2018) SynBioHub: a standards-enabled design repository for Synthetic Biology. ACS Synth Biol 7:682–688. https://doi.org/10.1021/acssynbio.7b00403
21. Zulkower V (2021) Computer-aided planning for the verification of large batches of DNA constructs. Methods Mol Biol 2229:167–174. https://doi.org/10.1007/978-1-0716-1032-9_7
22. Vegh P, Donovan S, Rosser S et al (2024) Biofoundry-scale DNA assembly validation using cost-effective high-throughput long-read sequencing. ACS Synth Biol 13:683–686. https://doi.org/10.1021/acssynbio.3c00589

Chapter 5

Using ApE for In Silico Golden Gate Cloning

M. Wayne Davis and Erik M. Jorgensen

Abstract

Golden Gate cloning allows rapid and reliable assembly of multiple DNA fragments in a defined orientation. Golden Gate cloning requires careful design of the restriction fragment overhangs to minimize undesired products and to generate the desired junctions. The ApE (A plasmid Editor) software package can assist in silico design of input fragments or to generate expected assembly products.

Key words ApE, In silico cloning, Golden Gate design, Golden Gate assembly

1 Introduction

Assembling different DNA fragments into plasmids in a reliable way is the foundation of molecular biology. Type II restriction enzymes are nucleases encoded by a single protein that cut at defined sequences, as opposed to type I enzymes that are complexes of multiple different subunits and can cut at undefined DNA sequences [1]. Type IIP ("Palindromic") restriction enzymes cut within a palindromic recognition sequence and have been the primary workhorse enzymes for generating DNA fragments to be assembled. Although cut-and-paste reactions using these robust enzymes laid the foundations of molecular biology, they are practically limited to assembling no more than three fragments for any single ligation. Thus, more complex constructions require a series of cloning reactions. Recombinase-based cloning, the Gateway system by Invitrogen for example, can be used to generate more complex four or five fragment assemblies, but these methods leave a relatively long fixed scar sequence at each junction from the recombinase recognition sequence. The development of Gibson assembly in 2009 [2] and Golden Gate cloning in 2014 [3] opened the floodgates for faster and more flexible single-step DNA assembly reactions. Gibson assembly can assemble sequences with scarless junctions, but requires generating matching sequences at each end

Daniel Schindler (ed.), *Golden Gate Cloning: Methods and Protocols*, Methods in Molecular Biology, vol. 2850,
https://doi.org/10.1007/978-1-0716-4220-7_5,

of fragments to be joined, usually by polymerase chain reaction (PCR), which may introduce mutations to the amplified product. The reaction uses a 5′ to 3′ exonuclease to generate 15- to 20-base single-stranded overhangs that hybridize to join the ends. The reaction then uses a polymerase and ligase to fill in any single-stranded gaps and seal the backbone. Gibson assembly can assemble more than ten fragments in a single tube, but near-identical sequences at the junctions can lead to mis-assembly and inverted repeats in the ends can lead to hairpin formation that can block assembly altogether.

Golden Gate assembly, the subject of this protocol, is perhaps the most robust and flexible of the modern assembly methods. Golden Gate uses type IIS ("shifted cleavage") restriction enzymes to generate three- or four-base single-stranded sticky ends that can be ligated. The key feature of type IIS restriction enzymes is that they cut any sequence adjacent to their recognition site. Because the recognition sequences are not incorporated into the junction fragments, ends that re-ligate to the starting configuration will simply be recut. Only fragments that do not include the recognition site will be incorporated into the assembly pathway (Fig. 1).

With careful design of the sticky ends, Golden Gate reactions can assemble more than 50 fragments in a single step [4]. Like Gibson assembly, the ends of each fragment need to be programmed with the correct compatible end and restriction recognition sequence, usually by donor plasmid, annealed synthetic DNA, or PCR. However, Golden Gate reaction is more robust than Gibson assembly to internal matching sequences. During the Gibson reaction, the exposure of internal duplications can lead to deletions or duplications. In Golden Gate reactions, the

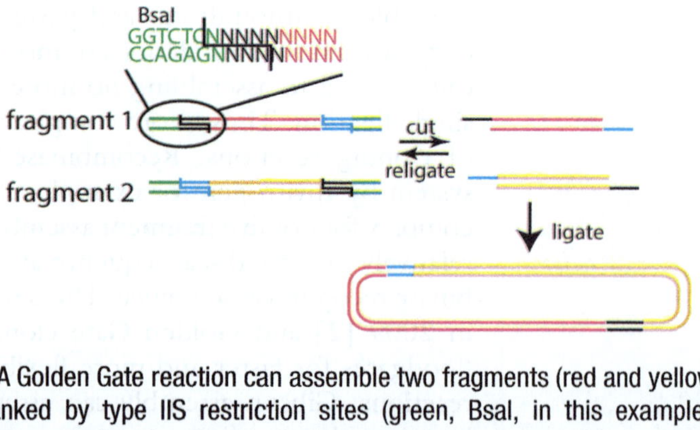

Fig. 1 A Golden Gate reaction can assemble two fragments (red and yellow) that are flanked by type IIS restriction sites (green, BsaI, in this example). The sequences of the four-base single-stranded overhangs (black and cyan) control the order of assembly. When a fragment containing the restriction recognition sequence is re-ligated onto an end, cycles of cleavage and ligation ensue. When the final product is eventually formed by ligation, it is refractory to further digestion

information for connecting fragments is encoded by three- or four-base single-stranded ends exposed by restriction digestion. However, Golden Gate cannot tolerate the presence of additional restriction recognition sequences within the assembled fragments. In most cases the reaction can be designed to remove internal sites by synonymous, single-base mutations during assembly.

An easy-to-use DNA analysis software is critical to any successful molecular biology campaign. For example, New England Biolabs provides a web-based Golden Gate assembler tool, NEBridge, available at https://goldengate.neb.com/. While there are several widely used application packages, such as Benchling, SnapGene, Mendelgen, or Geneious, this protocol will focus on the freely available program ApE ("A plasmid Editor"). ApE is available for all major desktop environments: MacOS, Windows, and Linux. ApE includes many analysis features such as automated and manual feature annotation, restriction site analysis, customizable graphic map creation, and DNA translation analysis. Further, ApE includes many more complex analysis tools such as in silico simulation and design of PCR reactions, Gibson assembly, restriction-ligation, and Golden Gate reactions. The use of the Golden Gate in silico designer and assembler tools will be discussed in this chapter.

2 Software

ApE is available for download at no cost from the author's website: https://jorgensen.biology.utah.edu/wayned/ape/.

The current version as of late 2023 is 2.1.4. The most recent version and current minimum hardware requirements can be obtained from the above website.

3 Methods

ApE has two separate tools for working with Golden Gate reactions: Golden Gate Designer and Golden Gate Assembler. The Designer tool is used to design new Golden Gate reactions out of individual DNA sequences. The Assembler tool simulates assembly from components provided by existing Golden Gate toolkit libraries.

3.1 Golden Gate Designer

The Golden Gate Designer is used to assemble DNA sequences into a final plasmid. The program relies on data from Pryor et al. [4] of the rate of ligation, correct and incorrect, of all possible three- or four-base single-stranded overhangs in a T4 ligase-based joining reaction. They found up to three-fold variation in rates of ligation of perfectly matched overhangs, showing that the efficiency of a Golden Gate reaction can be affected by the choice of junction sequences. More importantly, they found that some mismatched overhangs, many containing G-T pairings, have a low but non-zero rate of ligation. In fact, the rate of the strongest illegitimate junctions can be up to 25% the rate of the weaker legitimate ligations. These rates will compound and will lead to very low total efficiencies of correct assembly for a reaction with many fragments. Golden Gate Designer will inspect each DNA junction and move the seam by up to four bases to optimize the assembly overhangs to have the least propensity to ligate incorrect ends. The Golden Gate Designer tool produces a set of PCR primers that incorporates the required restriction recognition sequences, linker sequences, and hybridization site. In addition, it incorporates 5′ padding sequences required for the endonuclease to bind and cut the free end of the DNA fragment. Finally, it will generate a new sequence file representing the desired plasmid product. This sequence file will contain a comment section including the PCR primers, as well as feature annotations representing the PCR primers. Following are a detailed series of steps required to use the Golden Gate Designer tool.

1. The Golden Gate designer tool first asks the user to select the desired restriction enzyme to be used as the type IIS nuclease, such as BsaI, SapI, or PaqCI. The dropdown list will include all type IIS enzymes with three- or four-base overhang that are in the current user enzyme set (*see* **Note 1**). Once the enzyme has been added to the user set, the Designer tool can be reopened and the enzyme selected from the dropdown list.
2. Below the enzyme selection dropdown is an area for selecting and ordering DNA fragments. Each fragment contains a "+" and "-" button. Pressing the "+" button will add a new fragment immediately after the chosen fragment, whereas the "-" button will remove the chosen fragment.
3. At the top of each fragment area is a dropdown menu to select the DNA sequence to use as the source of the fragment. All DNA sequence files currently open in the program are available in this dropdown. The user then needs to use the second dropdown menu to choose the region of the sequence to include in the product (*see* **Note 2**). If the user wants the sequence to be added to the final product in the reverse orientation, they can select the reverse complement "Rev-Com" checkbox in the fragment area. Finally, if the user prefers to generate the fragment as a pair of complementary synthetic oligos rather than as a PCR product, they can select the "dsOligos" check button. In this case, both the top and bottom single-strand sequences must be ordered as single-stranded oligo DNA nucleotides (ssODNs). These oligos hybridize to generate the correct sticky-ended overhangs directly, without

restriction digestion. This option is not available for fragments that are selected as Golden Gate restriction fragments.

4. The program may issue several warnings for fragments that cannot produce a valid Golden Gate reaction. If the user selects a sequence that contains the restriction recognition site, there will be a warning text that appears, and the dialog will not permit completing the design. If two adjacent Golden Gate restriction fragments are selected that have incompatible overhangs, there will likewise be a warning that prevents completing the design.
5. Before completing the design, the user should set the default Golden Gate Designer values using the "Set Defaults" button. In this dialog, the user can set the following: (1) the default melting temperature for each PCR primer, excluding the 5′ additional spacer and restriction sequence; (2) the default search time in seconds—for all but extremely complex designs, one second should usually be long enough; (3) the default 5′ extension appended to the restriction recognition site—most restriction enzymes cannot cut at the extreme end of a DNA fragment and so require the addition of several extra padding bases to provide a comfortable binding surface; (4) the default linker sequence between the recognition site and the top strand cut site—if this sequence is longer than the enzyme requires, it will be truncated to the required length; (5) feature element generation in the final product sequence that delineate each of the input amplicons.
6. When all of the desired input fragments have been specified, the user can press the "Next" button to proceed to calculating the required DNA fragments. The tool will then display its calculated design.
7. The calculated design will first show the total efficiency of the design (*see* **Note 3**). In the calculated design, each fragment is presented with the algorithmically selected left and right single-stranded overhangs and the corresponding ligation efficiencies. For fragments that are designated as PCR products, the result shows the primer sequences including the 5′ extension sequence. It shows the calculated melting temperature for annealing to the initial template sequence as Tm1 and the melting temperature for annealing to the final PCR product (the Tm for the primer and its perfect complement) as Tm2. Double-stranded oligo fragments are shown as two single-stranded sequences, and restriction fragments are simply noted as such with their overhanging end.
8. When the user is satisfied with the design, they can select "Generate Product" to generate the final sequence in a new sequence window. All of the original feature annotations will be

carried over into the new sequence, as well as new annotations for each new PCR primer. The sequence comment area will contain a listing of all input sequences and the primer sequences or annealed oligo sequences needed to generate the input fragments. Finally, the comment area will contain a section called "Primer order" that can be copied as text and pasted into many DNA synthesis company's order pages.

3.2 GG Assembler

The ApE DNA Assembler tool simulates a Golden Gate reaction using pre-existing toolkits that have been designed to generate many constructs from libraries of compatible modular inputs. There are several such toolkits in wide use [5–8]. In addition, many laboratories have developed in-house toolkits of varying complexity for easy reuse of modular components. The following are a detailed series of steps required to use the Golden Gate Assembler tool.

1. The Golden Gate Assembler tool first asks the user to choose a type IIS restriction enzyme from a dropdown menu. As with the designer tool, the user can add enzymes to the current set using the Enzyme Selector Dialog (*see* **Note 4**).
2. Once an enzyme is selected, the program searches all open sequence files for restriction fragments that are flanked by two recognition sites in the correct orientation. The user can choose one of these fragments from a second dropdown menu button (*see* **Note 5**). Once this fragment is chosen, the program identifies all possible compatible Golden Gate fragments. If there is only one possible fragment, it is added to the assembly list and the program proceeds to identify the next compatible fragment. If fragments with compatible overhangs are available in the open sequence files, this process continues on until the final fragment is compatible with the free end of the first fragment. The calculation then stops and presents the user with the calculated efficiency of the reaction. The "OK" button is then activated, which will generate the final assembly product.
3. If during the process of identifying compatible fragments there are more than one possible fragments at a single position, the program will present the user with a dropdown menu button containing a list of the fragments that could be included at that position. If all of the possible fragments in one position lead to the same fragment in the next position, the program fills in the next compatible fragment in the assembly. On the other hand, if the alternative choices of fragment in this dropdown would lead to more than one alternative subsequent fragments, the program cannot select the next fragment in the assembly until the user selects one of the fragments from the dropdown.

4. If the final fragment in the assembly does not have a compatible end to circularize the product among the available Golden Gate fragments, the process stops with the message "No Product Possible" (*see* **Note 6**).
5. When all of the fragments of an assembly have been chosen, the user can select "OK" to generate the simulated product (*see* **Note 7**). The user can select the check box "Keep Assembler Dialog Open" to quickly generate a series of simulated products without having to reopen the tool. The comment text of the new sequence window will contain a list of all of the assembled fragments and their numeric endpoints.

4 Notes

1. If there are no type IIS enzymes in the list or the desired enzyme is not in the list, the user should add the desired enzyme to the user set. To do this, first close the Golden Gate Designer tool. Then open the enzyme selector dialog, available from the main menu item "Enzymes> Enzyme Selector." Once the dialog window is open, select "Enzymes> Edit Current Enzyme Set" from the menu bar. This new dialog can then be used to add the desired enzyme to the current user enzyme set by checking the box next to the enzyme name. Closing the Selection dialog will prompt the user to save the current set as the default set when the program is closed and reopened. It is probably most useful to add any frequently used type IIS enzymes to the default user set.
2. If the sequence is linear, the default fragment is the entire sequence; the other available choice is to use the currently selected region. If the user selects a different region while the tool is open, a new item appears in the dropdown providing that new sequence as an option. If the sequence window is a circular sequence, the dropdown provides the option of: all of the sequence (opened at the current origin); the currently selected region; or the complementary fragment of all of the sequence excluding the selection. If the sequence contains potential Golden Gate fragments, these fragments are also provided as an option in the dropdown menu.
3. This is calculated by multiplying the efficiency of each junction fragment. The efficiency of a junction is calculated by using the data in Pryor et al. [4] to find the frequency of the single-stranded overhang at that junction ligating to the correct complementary overhang divided by the sum of the frequency of ligating that overhang to each overhang present in the reaction. In most cases the efficiency should be better than 98%. In our experience, reactions below 95% efficiency, particularly

reactions with at least one large fragment, will require screening multiple candidate clones to find a correct assembly. Because the algorithm currently uses a Monte Carlo method to optimize the efficiency, pressing the "Recalculate" button will allow the algorithm to continue the calculation from the current best seed to try to find a better solution. Similarly, the "Calculate New" button will restart the calculation from a new seed.

4. Some toolkits are designed to use a mixture of more than one restriction enzyme, and an optional second or third enzyme can be added via an option at the bottom of the dropdown menu.
5. The program also searches for pairs of sequences less than 200 bases long that can hybridize to each other giving single-stranded overhangs of the correct length at both ends. This simulates hybridized single-stranded synthetic oligonucleotide inputs into a Golden Gate reaction. These pairs will be included in the assembly calculation as though they were the product of a restriction digest.
6. In some cases, this can be caused by the user having sequence files that represent circular DNAs but that are annotated as linear DNAs. This is most often the problem when users prefer to annotate all of their DNAs as linear. In this case, the user can select the check button "Treat All Sequences as Circular" to calculate the assembly as if all of the sequences represent circular plasmids.
7. The user can alternatively select the "With Calculator" button. This will generate the same new sequence window, and will also add all of the input plasmids to the Molar Ratio Reaction tool. Here the user can calculate the exact volume ratios of each plasmid to be added to a Golden Gate reaction to have the optimal molar ratios based on each plasmid molecular weight and concentration.

Acknowledgments

We would like to thank Adam Hefel and Brian Muller for comments on the manuscript. EMJ and MWD are supported by an HHMI Investigator Award and NIH GM146005.

References

1. Roberts RJ, Belfort M, Bestor T, Bhagwat AS, Bickle TA, Bitinaite J, Blumenthal RM, Degtyarev SK, Dryden DTF, Dybvig K, Firman K, Gromova ES, Gumport RI, Halford SE, Hattman S, Heitman J, Hornby DP, Janulaitis A, Jeltsch A, Josephsen J, Kiss A, Klaenhammer TR, Kobayashi I, Kong H, Krüger DH, Lacks S, Marinus MG, Miyahara M, Morgan RD, Murray NE, Nagaraja V, Piekarowicz A, Pingoud A, Raleigh E, Rao DN, Reich N, Repin

VE, Selker EU, Shaw P-C, Stein DC, Stoddard BL, Szybalski W, Trautner TA, Van Etten JL, Vitor JMB, Wilson GG, Xu S (2003) A nomenclature for restriction enzymes, DNA methyltransferases, homing endonucleases and their genes. Nucleic Acids Res 31:1805–1812. https://doi.org/10.1093/nar/gkg274

2. Gibson DG, Young L, Chuang R-Y, Venter JC, Hutchison CA, Smith HO (2009) Enzymatic assembly of DNA molecules up to several hundred kilobases. Nat Methods 6:343–345. https://doi.org/10.1038/nmeth.1318
3. Engler C, Kandzia R, Marillonnet S (2008) A one pot, one step, precision cloning method with high throughput capability. PLoS One 3:e3647. https://doi.org/10.1371/journal.pone.0003647
4. Pryor JM, Potapov V, Kucera RB, Bilotti K, Cantor EJ, Lohman GJS (2020) Enabling one-pot Golden Gate assemblies of unprecedented complexity using data-optimized assembly design. PLoS One 15:e0238592. https://doi.org/10.1371/journal.pone.0238592
5. Engler C, Youles M, Gruetzner R, Ehnert T-M, Werner S, Jones JDG, Patron NJ, Marillonnet S (2014) A Golden Gate modular cloning toolbox for plants. ACS Synth Biol 3:839–843. https://doi.org/10.1021/sb4001504
6. Nonet ML (2023) Rapid generation of C elegans single copy transgenes combining RMCE and drug selection. 2023.03.05.531207
7. Schwartz ML, Jorgensen EM (2016) SapTrap, a toolkit for high-throughput CRISPR/Cas9 gene modification in Caenorhabditis elegans. Genetics 202:1277–1288. https://doi.org/10.1534/genetics.115.184275
8. Chiasson D, Giménez-Oya V, Bircheneder M, Bachmaier S, Studtrucker T, Ryan J, Sollweck K, Leonhardt H, Boshart M, Dietrich P, Parniske M (2019) A unified multi-kingdom Golden Gate cloning platform. Sci Rep 9:10131. https://doi.org/10.1038/s41598-019-46171-2

Chapter 6

Standardized Golden Gate Assembly Metadata Representation Using SBOL

Gonzalo Vidal, Carolus Vitalis, and Johan Guillén

Abstract

Synthetic biology, also known as engineering biology, is an interdisciplinary field that applies engineering principles to biological systems. One way to engineer biological systems is by modifying their DNA. A common workflow involves creating new DNA parts through synthesis and then using them in combination with other parts through assembly. Assembly standards such as MoClo, Phytobricks, and Loop are based on Golden Gate, and provide a framework for combining parts. The Synthetic Biology Open Language (SBOL) has implemented a best practice for representing build plans to communicate them to other practitioners through whiteboard designs and in a machine-readable format for communication with lab automation tools. Here we present a software tool for creating SBOL representations of build plans to simulate type IIS-mediated assembly reactions and store relevant metadata.

Key words DNA Assembly, Golden Gate, Standard, SBOL, Synthetic Biology

1 Introduction

Synthetic Biology or Engineering Biology is an interdisciplinary field that applies engineering methods to biological systems. In its foundation, it relies on the idea that researchers can "write" arbitrary DNA sequences through DNA synthesis, rather than being constrained to "copy" and "paste" or random mutations from existing DNA sequences found in nature. Although the cost of DNA synthesis is decreasing over time, it is still expensive and remains out of reach for laboratories with limited resources.

DNA assembly techniques allow researchers to use DNA synthesis more efficiently, thereby reducing the cost of building DNA libraries. Two of the main DNA assembly methods are Gibson assembly [1] and type IIS-mediated assembly [2]. Gibson assembly can join two blunt DNA fragments with a ~ 20–40 base pairs homology, using an exonuclease to create complementary sticky ends, a DNA polymerase to close the gaps, and a DNA ligase to seal

Daniel Schindler (ed.), *Golden Gate Cloning: Methods and Protocols*, Methods in Molecular Biology, vol. 2850,
https://doi.org/10.1007/978-1-0716-4220-7_6,

the nicks [1]. Golden Gate is a type IIS-mediated DNA assembly method that employs restriction enzymes to generate DNA fragments with ~3–4 base pair overhangs. Complimentary overhangs anneal and are joined by a DNA ligase that seal the nicks creating a desired DNA fragment or plasmid [2]. Although Gibson assembly is a scarless method, it can produce sequence errors during steps that use a DNA polymerase such as polymerase chain reaction (PCR) fragment amplification or while filling the gaps in DNA. The community primarily use Golden Gate assembly because it is simple, less prone to mutations, and cost-efficient. To allow for the reuse of synthetic DNA fragments, synthetic biologists created a set of DNA parts with a defined biological function and assembly standards to join them. Some basic DNA parts are promoter, ribosome binding site, coding sequence, and terminator. Assembly standards like MoClo [3, 4], Loop [5], Golden Braid [6], and others [7] defined the syntax or the sequence of the sticky ends that flanked a DNA part. Using assembly standards synthetic biologists can share libraries of DNA parts with other laboratories improving collaboration and reproducibility. Combining standard parts from the literature with newly synthesized parts by another group allows them to create combinatorial transcriptional units with fewer resources than by pure DNA synthesis. Having standard parts also benefits the laboratory producing them as they can reuse their DNA parts in future projects. The Golden Gate assembly protocol is part of the workflow of most synthetic biologists and is the way how iGEM has decided to distribute DNA [7]. Furthermore, due to its simple one-pot one-step protocol, it is easy to implement in automated workflows using liquid handling robots [8].

A digital representation of parts and assemblies would help researchers handle larger combinations of parts, design an assembly strategy, and simulate it. This digital representation has to be human-readable to inform researchers the sequences that will be assembled, as well as machine-readable, to provide an interface with lab automation. The Synthetic Biology Open Language (SBOL) can represent these biological designs in a machine-readable format [9, 10]. SBOL visual allows for the visual representation of meta data encoded in SBOL using glyphs to abstract standard objects and relations [11, 12]. SBOL 3 has implemented a best practice for the representation of build plans that can represent Golden Gate assemblies [13].

In the standard representation of parts and assembly for build planning, a terminology that has an SBOL representation where the sequence information and metadata are stored is described. A *Part* is defined as a single contiguous linear DNA construct with a completely specified sequence. There are many types of parts, the simplest being the Part Core, which is a part that has no interfaces nor relation to an assembly, such as the sequence of GFP. Then to assemble it through Golden Gate following the CIDAR MoClo [3] standard for example, we need to design a *Part Insert* that can be the engineered fluorescent protein *Part Core* plus any 5′ and 3′ flanking sequences, in this case C (AATG) prefix and D (AGGT) suffix, to be placed into a designated insertion site of a *Backbone*. The *Backbone* is a DNA construct into which *Parts* are intended to be inserted at one or more insertion sites, for example, the Plasmid DVA_CD from the CIDAR MoClo kit. When the *Part Insert* is inserted into a *Backbone* insertion site, it creates a *Part in Backbone*, following the example it can be the engineered fluorescent protein in DVA_CD. Now this *Part in Backbone* can be used with other *Parts in Backbone* like a promoter, ribosome binding site (RBS), and terminator, and a *Backbone* in a Golden Gate *Assembly Plan*. When *Parts in Backbones* are digested by BsaI restriction enzyme, they create *Part Extracts*, which are parts plus any 5′ and 3′ flanking sequences that have been extracted from a *Part in Backbone*. On the other hand, the digestion of the *Backbone* will produce an *Open Backbone* and a *Drop-Out Sequence*. The *Part Extracts* and the Open Backbone will align by base complementarity and ligated producing a *Composite Part in Backbone* which is the final result of the assembly, a constitutive transcription unit that expresses the engineered fluorescent protein. All these terms store sequence information and metadata about functions and relations inside the Assembly Plan. For a more detailed description of the terminology and how metadata is stored, please refer to the publication [13].

Representing assembly plans in SBOL provides other benefits, such as making your data findable, accessible, interoperable, and reusable (FAIR), as well as the integration within the broader SBOL ecosystem. This ecosystem includes: SBOL Visual for hierarchical design visualization, the SynBioHub repository [14], and design tools such as SynBioSuite [15] and LOICA [16]. Here we present a software tool for the creation of build plan SBOL representations to simulate assembly reactions and store relevant metadata.

2 Materials

2.1 Dependencies

1. Installing SBOL-utilities via PIP will automatically install the required packages. These prerequisites include Python 3.7 or a later version. We recommend the use of an environment manager, such as Anaconda (https://www.anaconda.com/).
2. Additionally, users must install non-Python dependencies, Graphviz (https://graphviz.org/) to render graph-SBOL diagrams and Node.js (https://nodejs.org/en) to use the SBOL-converter.

2.2 Installation

1. The SBOL-utilities Python package is distributed using the Python Package Index (PyPI), which utilizes the Pip Installs Package (PIP) for installation and update management. The latest stable release of SBOL-utilities can be installed using the following commands (*see* **Note 1**):

```
pip3 install SBOL-utilities
```

2. To verify that the installation was successful, users should be able to run the following command with no errors

```
import SBOL-utilities
```

3 Methods

In the following sections, we detail how SBOL-utilities can be used to generate a Golden Gate representation. We exemplify its use by creating an assembly plan for one transcriptional unit (Fig. 1) and a collection with multiple fluorescent proteins (Fig. 2). To learn more about the advanced features, we recommend exploring the source repository https://github.com/SynBioDex/SBOL-utilities with more examples and tutorials, as well as the documentation https://SBOL-utilities.readthedocs.io

3.1 Creating an Assembly Plan for One Transcription Unit

To create an *Assembly Plan* using SBOL, we instantiate the restriction enzyme, the DNA *Parts in Backbone,* and *Backbone* involved in the reaction (*see* **Note 2**). Here we represent the Golde Gate assembly of the following *Parts in Backbone*: a constitutive promoter (J23100), an RBS (B0034), a green fluorescent protein (GFP), and a double terminator (B0015) in the *Backbone* odd receiver 1 (pOdd) from Loop assembly (Fig. 1). This process can be described step by step, as follows:

1. Create a SBOL document

```
doc = SBOL3.Document()
```

2. Set a namespace (*see* **Note 3**)

```
SBOL3.set_namespace('http://SBOLstandard.org')
```

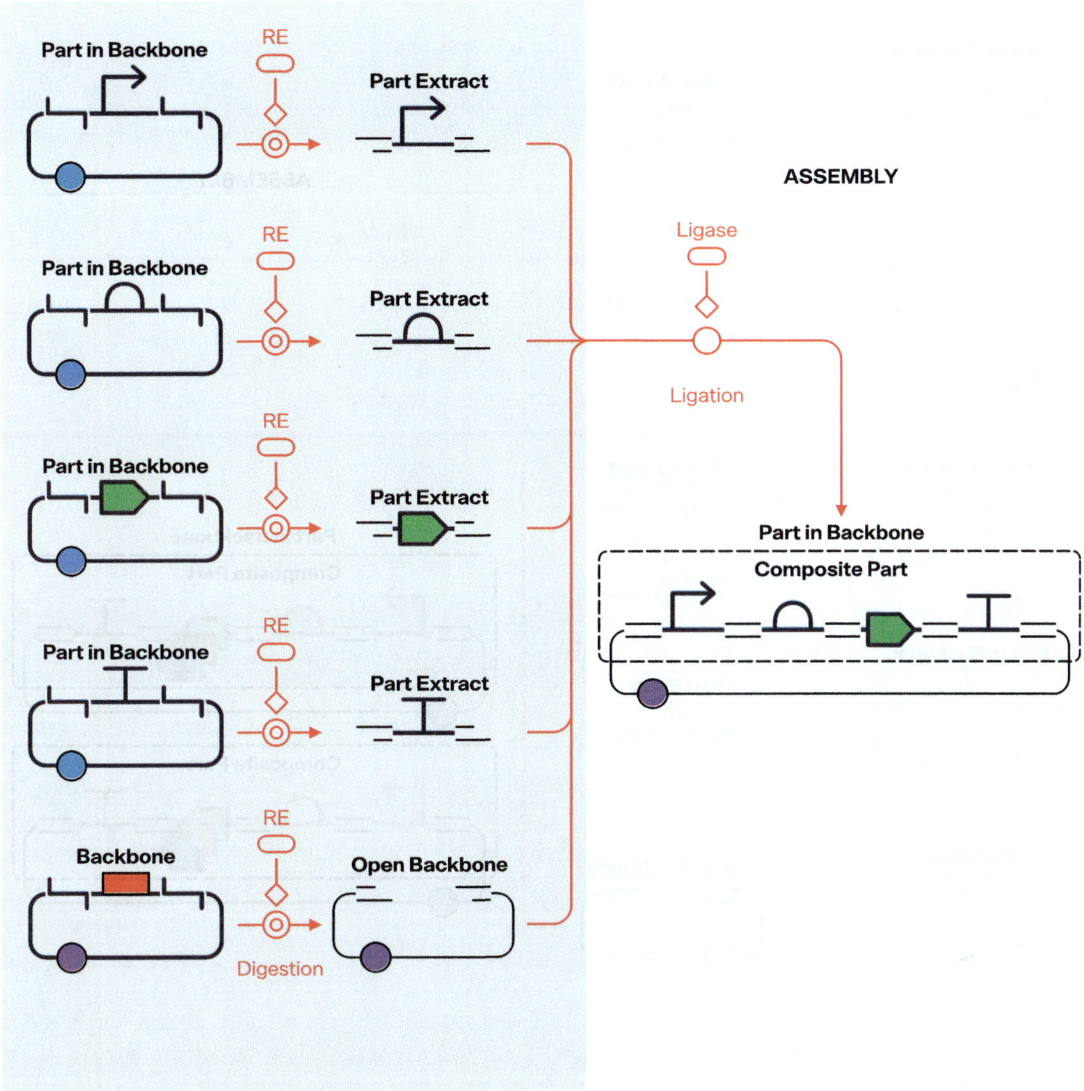

Fig. 1 Simple assembly SBOL visual representation. Using standardized build planning. In this Golden Gate assembly four *Parts in Backbone*: a promoter, an RBS, a CDSs and a Terminator, with an acceptor Backbone are mixed in a single reaction producing one transcription unit. Reagents and reactions are explicit in the diagram representing how the restriction enzyme (RE) regulates the digestion producing Part Extracts as an intermediate state; and how the ligase regulates the ligation producing a *Part in Backbone* with a *Composite Part* as the final product

3. Create a restriction enzyme (Table 1)

```
bsai = ed_restriction_enzyme('BsaI')
```

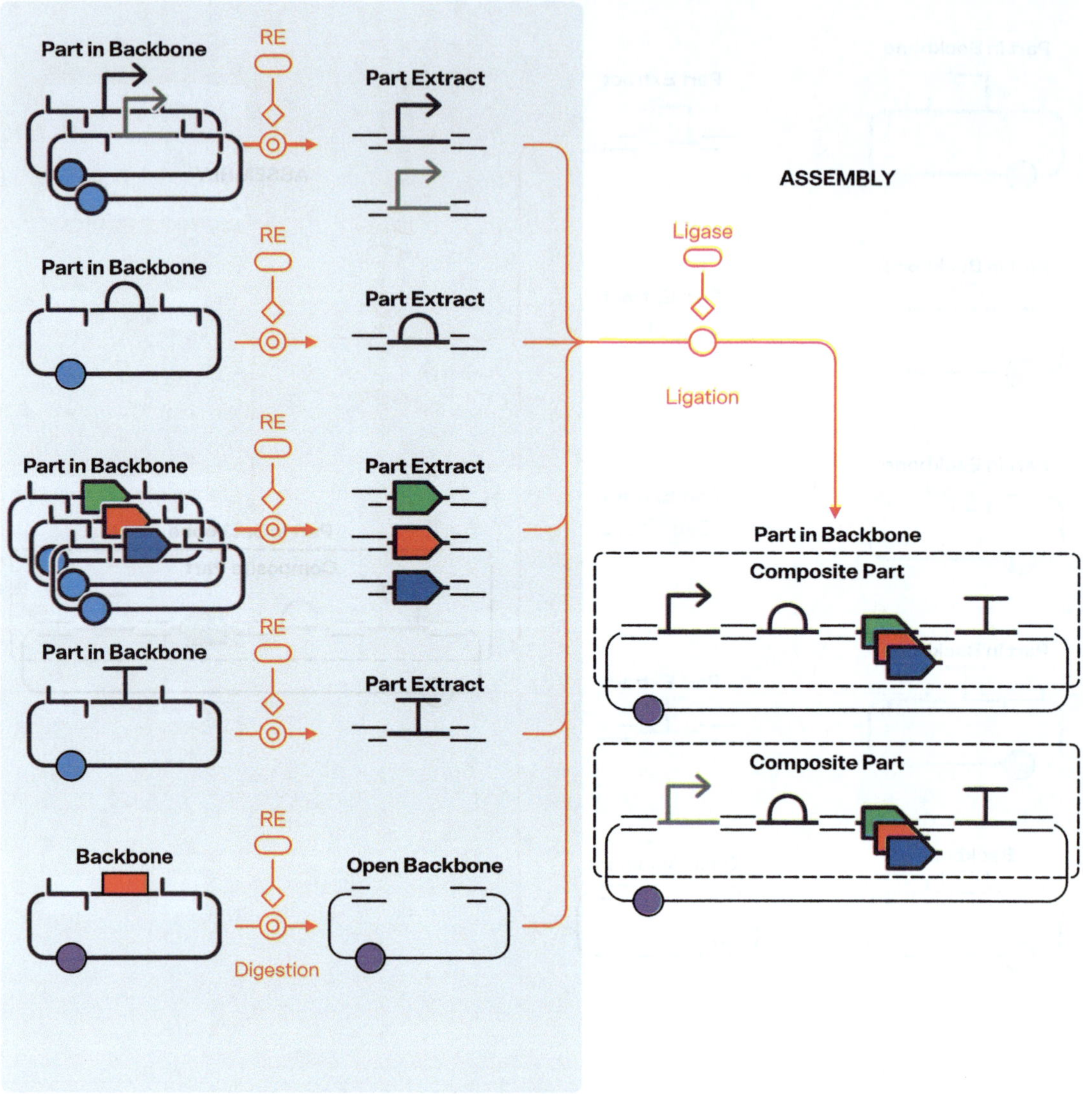

Fig. 2 Combinatorial assembly SBOL visual representation. Using standardized build planning. In this Golden Gate assembly seven *Parts in Backbone*, two promoters, one RBS, three CDSs and one Terminator, with an acceptor Backbone are mixed in six reactions producing a different transcription unit in each one in a combinatorial fashion. Reagents and reactions are explicit in the diagram representing how the restriction enzyme (RE) regulates the digestion producing Part Extracts; and how the ligase regulates the ligation producing *Parts in Backbone* with *Composite Parts*

4. Create an acceptor *Backbone* (Table 2)

```
poddl_dir = os.path.join('poddl.gb')      podd_doc = con-
vert_from_genbank(poddl_dir,
    'https://github.com/Gonza10V')
    podd_af = [top_level for top_level in podd_doc
```

Table 1
ed_restriction_enzyme

Argument	Description	Format/ values	Default
name	Name of the SBOL ExternallyDefined, used by PyDNA. Case sensitive, follow standard restriction enzyme nomenclature, i.e., "BsaI"	String	n/a
kwargs	Keyword arguments of any other ExternallyDefined attribute	n/a	n/a

Table 2
Backbone from SBOL

Argument	Description	Format/ Values	Default
identity	The identity of the Component. The identity of Sequence is also identity with the suffix '_seq'	String	n/a
SBOL_comp	The SBOL Component containing the DNA sequence to use	String	n/a
dropout_location	List of 2 integers that indicates the start and the end of the dropout sequence including overhangs. Note that the index of the first location is 1, as is typical practice in biology, rather than 0, as is typical practice in computer science	List [Integer]	n/a
fusion_site_length	Integer of the length of the fusion sites (eg. BsaI fusion site length is 4, SapI fusion site length is 3)	Integer	n/a
linear	Boolean that indicates if the backbone is linear, by default it is set to Flase which means that it has a circular topology	Boolean	False
kwargs	Keyword arguments of any other Component attribute	n/a	n/a

```
if type(top_level)==SBOL3.Component][0]
podd_backbone, podd_backbone_seq =
backbone_from_SBOL('pOdd_bb', podd_af,
[680,1770], 4, False, name='pOdd_bb')
doc.add([podd_backbone,podd_backbone_seq])
```

5. Obtain *Parts in Backbone* sequences from GeneBank files (*see* **Note 4**)

```
j23100_dir = os.path.join('ab_j23100.gb')     b0034_dir =
os.path.join('bc_b0034.gb')
    gfp_dir = os.path.join('ce_gfp.gb')
    b0015_dir = os.path.join('ef_b0015.gb')
```

6. Convert sequences to SBOL (*see* **Note 3**)

```
j23100_doc = convert_from_genbank
(j23100_dir,    'https://github.com/Gonza10V')
    j23100_ab = [top_level for top_level in j23100_doc
    if type(top_level)==SBOL3.Component][0]
    b0034_doc = convert_from_genbank(b0034_dir,
    'https://github.com/Gonza10V')
    b0034_bc = [top_level for top_level in b0034_doc
    if type(top_level)==SBOL3.Component][0]
    gfp_doc = convert_from_genbank(gfp_dir,
    'https://github.com/Gonza10V')
    gfp_ce = [top_level for top_level in gfp_doc
    if type(top_level)==SBOL3.Component][0]
    b0015_doc = convert_from_genbank(b0015_dir,
    'https://github.com/Gonza10V')
    b0015_ef = [top_level for top_level in b0015_doc
    if type(top_level)==SBOL3.Component][0]
```

7. Create standard *Parts in Backbone* (Table 3)

```
j23100_ab_in_bb, j23100_ab_in_bb_seq =    part_in_back-
bone_from_SBOL('j23100_ab_in_bb',
    j23100_ab, [479,513], [SBOL3.SO_PROMOTER], 4,
    False, name='j23100_ab_in_bb')
    doc.add([j23100_ab_in_bb, j23100_ab_in_bb_seq])
    b0034_bc_in_bb, b0034_bc_in_bb_seq =
    part_in_backbone_from_SBOL('b0034_bc_in_bb',
    b0034_bc, [479,499], [SBOL3.SO_RBS], 4, False,
    name='b0034_bc_in_bb')
    doc.add([b0034_bc_in_bb, b0034_bc_in_bb_seq])
    gfp_ce_in_bb, gfp_ce_in_bb_seq =
    part_in_backbone_from_SBOL('gfp_ce_in_bb',
    gfp_ce, [479,1195], [SBOL3.SO_CDS], 4, False,
    name='gfp_ce_in_bb')
    doc.add([gfp_ce_in_bb, gfp_ce_in_bb_seq])
    b0015_ef_in_bb, b0015_ef_in_bb_seq =
    part_in_backbone_from_SBOL('b0015_ef_in_bb',
    b0015_ef, [518,646], [SBOL3.SO_TERMINATOR], 4,
    False, name='b0015_ef_in_bb')
    doc.add([b0015_ef_in_bb, b0015_ef_in_bb_seq])
```

Table 3
Part in backbone from SBOL

Argument	Description	Format/Values	Default
identity	The identity of the Component. If it is a String it builds a new SBOL Component, and if it is None it adds on top of the input. The identity of Sequence is also identity with the suffix '_seq'	String—None	n/a
SBOL_comp	The SBOL3 Component that will be used to create the *part in backbone* Component and Sequence	SBOL3. Component	n/a
part_location	List of 2 integers that indicates the start and the end of the unitary part. Note that the index of the first location is 1, as is typical practice in biology, rather than 0, as is typical practice in computer science	List[Integer]	n/a
part_roles	List of strings that indicates the roles to add on the part	List[String]	n/a
fusion_site_length	Integer of the length of the fusion sites (eg. BsaI fusion site length is 4, SapI fusion site length is 3)	Integer	n/a
linear	Boolean that indicates if the backbone is linear, by default it is set to False which means that it has a circular topology	Boolean	False
kwargs	Keyword arguments of any other Component attribute	n/a	n/a

Table 4
Assembly plan composite in backbone single enzyme

Parameter	Description	Format/Values	Default
name	Name of the assembly plan Component	String	n/a
parts_in_backbone	Parts in backbone to be assembled	List[SBOL3. Component]	n/a
acceptor_backbone	Backbone in which parts are inserted into the assembly	SBOL3.Component	n/a
restriction_enzyme	Restriction enzyme with the correct name from Bio.Restriction as Externally Defined	String—SBOL3. ExternallyDefined	n/a
document	SBOL Document where the assembly plan will be created	SBOL3.Document	n/a

8. Create *Assembly Plan* (Table 4)

```
simple_assembly_plan =     Assembly_plan_composite_in_-
backbone_single_enzyme(
    name='simple_green_transcriptional_unit',
```

```
    parts_in_backbone=[j23100_ab_in_bb,
    b0034_bc_in_bb, gfp_ce_in_bb, b0015_ef_in_bb],
    acceptor_backbone=podd_backbone,
    restriction_enzyme=bsai,
    document=doc)
    simple_assembly_plan.run()
```

9. Inspect assembly products, its length should be 1 as the only product is a constitutive transcription unit expressing GFP, the Composite Part in Backbone J23100-B0034-GFP-B0015 in pOdd (Fig. 1).

```
len(simple_assembly_plan.products)
```

10. Inspect objects in assembly documents

```
for obj in     simple_assembly_plan.document.objects:
    print(obj.identity)
```

11. Inspect the sequence of an object. Select an object identity from the printed list and use it as check_obj. The user can swap object identities in check_obj to inspect the sequence of other objects. In this example, we selected the promoter J23100 Part Extract.

```
check_obj = 'http://SBOLstandard.org/testfiles/j23100_ab_
in_bb_part_extract'    for obj in simple_assembly_plan.doc-
ument.objects:
    if obj.identity == check_obj:
    print(obj.sequences[0].lookup().elements)
```

The output of this code is the string containing the sequence of the promoter J23100 and corresponds to the expected 43 bp long DNA sequence: "GGAGttgacggctagctcagtcctaggtacagtgctagcTACT".

This process can be iterated for all *Part Extract*, *Backbone*, and *Composite* to inspect its arrangement (Fig. 3).

3.2 Creating an Assembly Plan for a Combinatorial Design

Combinatorial designs allow researchers to simultaneously explore many conditions or variables systematically and efficiently. These approaches are used in synthetic biology to systematically generate diverse genetic constructs or pathways. This facilitates the screening of multiple variants to identify those with desired properties, such as enhanced productivity or improved function.

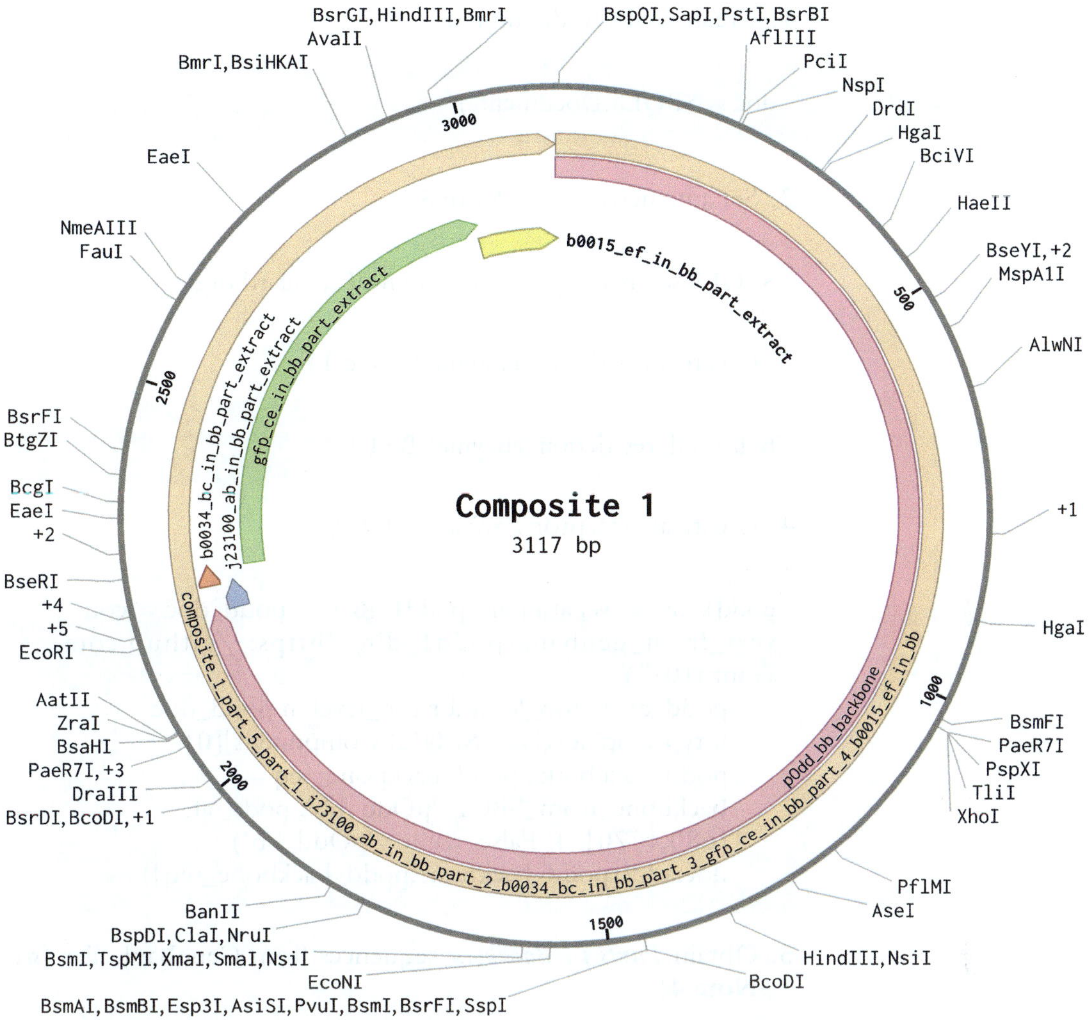

Fig. 3 Composite sequence diagram. DNA sequence of a *Composite* created using *Assembly plan composite in backbone single enzyme* was obtained from step 11 and uploaded to Benchling sequence manager. DNA sequences from *Composite*, *Part Extracts* and *Backbone* sequences were aligned and added as features using their identity as name

To create an *Assembly Plan* using SBOL, we instantiate the restriction enzyme, the DNA *Parts in Backbone*, and *Backbone* involved in the reaction (*see* **Note 2**). Here we represent the Golde Gate assembly of the following *Parts in Backbone*: two constitutive promoter variants (J23100 and J23101), an RBS (B0034), a green fluorescent protein (GFP), red fluorescent protein (RFP), cyan fluorescent protein (CFP), and a double terminator (B0015) in the *Backbone* odd receiver 1 (pOdd) from Loop assembly (Fig. 2). This process can be described step by step, as follows:

1. Create a SBOL document

```
doc = SBOL3.Document()
```

2. Set a namespace (*see* **Note 3**)

```
SBOL3.set_namespace('http://SBOLstandard.org')
```

3. Create a restriction enzyme (Table 1)

```
bsai = ed_restriction_enzyme('BsaI')
```

4. Create an acceptor *Backbone* (Table 2)

```
poddl_dir = os.path.join('poddl.gb')    podd_doc = con-
vert_from_genbank(poddl_dir, 'https://github.com/
Gonza10V')
    podd_af = [top_level for top_level in podd_doc
    if type(top_level)==SBOL3.Component][0]
    podd_backbone, podd_backbone_seq =
    backbone_from_SBOL('pOdd_bb', podd_af,
    [680,1770], 4, False, name='pOdd_bb')
    doc.add([podd_backbone,podd_backbone_seq])
```

5. Obtain *Parts in Backbone* sequences from GeneBank files (*see* **Note 4**)

```
j23100_dir = os.path.join('ab_j23100.gb')    j23101_dir =
os.path.join('ab_j23101.gb')
    b0034_dir = os.path.join('bc_b0034.gb')
    gfp_dir = os.path.join('ce_gfp.gb')
    rfp_dir = os.path.join('ce_mrfp1.gb')
    cfp_dir = os.path.join('ce_ecfp.gb')
    b0015_dir = os.path.join('ef_b0015.gb')
```

6. Convert sequences to SBOL (*see* **Note 3**)

```
j23100_doc = convert_from_genbank
(j23100_dir,    'https://github.com/Gonza10V')
    j23100_ab = [top_level for top_level in j23100_doc
    if type(top_level)==SBOL3.Component][0]
    j23101_doc = convert_from_genbank(j23101_dir,
```

```
'https://github.com/Gonza10V')
j23101_ab = [top_level for top_level in j23101_doc
if type(top_level)==SBOL3.Component][0]
b0034_doc = convert_from_genbank(b0034_dir,
'https://github.com/Gonza10V')
b0034_bc = [top_level for top_level in b0034_doc
if type(top_level)==SBOL3.Component][0]
gfp_doc = convert_from_genbank(gfp_dir,
'https://github.com/Gonza10V')
gfp_ce = [top_level for top_level in gfp_doc
if type(top_level)==SBOL3.Component][0]
rfp_doc = convert_from_genbank(rfp_dir,
'https://github.com/Gonza10V')
rfp_ce = [top_level for top_level in rfp_doc
if type(top_level)==SBOL3.Component][0]
cfp_doc = convert_from_genbank(cfp_dir,
'https://github.com/Gonza10V')
cfp_ce = [top_level for top_level in cfp_doc
if type(top_level)==SBOL3.Component][0]
b0015_doc = convert_from_genbank(b0015_dir,
'https://github.com/Gonza10V')
b0015_ef = [top_level for top_level in b0015_doc
if type(top_level)==SBOL3.Component][0]
```

7. Create standard *Parts in Backbone* (Table 3)

```
j23100_ab_in_bb, j23100_ab_in_bb_seq =     part_in_back-
bone_from_SBOL('j23100_ab_in_bb',
    j23100_ab, [479,513], [SBOL3.SO_PROMOTER], 4,
    False, name='j23100_ab_in_bb')
    doc.add([j23100_ab_in_bb, j23100_ab_in_bb_seq])
    j23101_ab_in_bb, j23101_ab_in_bb_seq =
    part_in_backbone_from_SBOL('j23101_ab_in_bb',
    j23101_ab, [479,513], [SBOL3.SO_PROMOTER], 4,
    False, name='j23101_ab_in_bb')
    doc.add([j23101_ab_in_bb, j23101_ab_in_bb_seq])
    b0034_bc_in_bb, b0034_bc_in_bb_seq =
    part_in_backbone_from_SBOL('b0034_bc_in_bb',
    b0034_bc, [479,499], [SBOL3.SO_RBS], 4, False,
    name='b0034_bc_in_bb')
    doc.add([b0034_bc_in_bb, b0034_bc_in_bb_seq])
    gfp_ce_in_bb, gfp_ce_in_bb_seq =
    part_in_backbone_from_SBOL('gfp_ce_in_bb',
    gfp_ce, [479,1195], [SBOL3.SO_CDS], 4, False,
    name='gfp_ce_in_bb')
    doc.add([gfp_ce_in_bb, gfp_ce_in_bb_seq])
```

```
rfp_ce_in_bb, rfp_ce_in_bb_seq =
part_in_backbone_from_SBOL('rfp_ce_in_bb',
rfp_ce, [479,1156], [SBOL3.SO_CDS], 4, False,
name='rfp_ce_in_bb')
doc.add([rfp_ce_in_bb, rfp_ce_in_bb_seq])
cfp_ce_in_bb, cfp_ce_in_bb_seq =
part_in_backbone_from_SBOL('cfp_ce_in_bb',
cfp_ce, [479,1198], [SBOL3.SO_CDS], 4, False,
name='cfp_ce_in_bb')
doc.add([cfp_ce_in_bb, cfp_ce_in_bb_seq])
b0015_ef_in_bb, b0015_ef_in_bb_seq =
part_in_backbone_from_SBOL('b0015_ef_in_bb',
b0015_ef, [518,646], [SBOL3.SO_TERMINATOR], 4,
False, name='b0015_ef_in_bb')
doc.add([b0015_ef_in_bb, b0015_ef_in_bb_seq])
```

8. Create *Assembly Plan* (Table 4)

```
combinatorial_assembly_plan =
Assembly_plan_composite_in_backbone_single_enzyme
(    name='combinatorial_rgb_transcriptional_units',
    parts_in_backbone=[j23100_ab_in_bb,
    j23101_ab_in_bb, b0034_bc_in_bb, gfp_ce_in_bb,
    rfp_ce_in_bb, cfp_ce_in_bb, b0015_ef_in_bb],
    acceptor_backbone=podd_backbone,
    restriction_enzyme=bsai,
    document=doc)
    combinatorial_assembly_plan.run()
```

9. Inspect assembly products; its length should be 6 as the products are constitutive transcription units expressing GFP, RFP, and CFP the *Composite Parts in Backbone* J23100-B0034-GFP-B0015 in pOdd, J23100-B0034-RFP-B0015 in pOdd, J23100-B0034-CFP-B0015 in pOdd, J23101-B0034-GFP-B0015 in pOdd, J23101-B0034-RFP-B0015 in pOdd, and J23101-B0034-CFP-B0015 in pOdd (Fig. 2).

```
len(combinatorial_assembly_plan.products)
```

10. Inspect objects in *Assembly Plan* documents

```
for obj in    combinatorial_assembly_plan.document.
objects:
    print(obj.identity)
```

11. Inspect the sequence of an object. Select an object identity from the printed list and use it as check_obj. The user can swap object identities in check_obj to inspect the sequence of other objects. In this example, we selected the promoter J23100 Part Extract.

```
check_obj = 'http://SBOLstandard.org/testfiles/j23100_ab_
in_bb_part_extract'    for obj in simple_assembly_plan.doc-
ument.objects:
    if obj.identity == check_obj:
    print(obj.sequences[0].lookup().elements)
```

The output of this code is the string containing the sequence of the promoter J23100 and corresponds to the expected 43 bp long DNA sequence: "GGAGttgacggctagctcagtcctaggtacagtgctagcTACT".

This process can be iterated for all *Part Extract*, *Backbone* and *Composite* to inspect its arrangement (Fig. 3).

4 Notes

1. The version of SBOL-utilities used in this example is 1.0a17 in this branch (https://github.com/Gonza10V/SBOL-utilities/tree/gonzalo-internship3).
2. The DNA parts can be a linear or circular vectors.
3. The namespace can be any valid internationalized resource identifier (IRI).
4. The paths used in the example assume that the file is in the same folder as the Python file. Be mindful of the file location and format to use it here. When running the Jupyter Notebook if you get the error "SSLCertVerificationError," please add the following two lines of code at the top of the script:

   import ssl ssl._create_default_https_context = ssl._create_-unverified_context If you don't have the package ssl, you have to install it by running the following command in the terminal: pip install ssl.
5. The script uses the Python library Biopyhton, if you don't have it in your environment, please install it running the following command in the terminal: pip install Bio.

Acknowledgements

The authors would like to acknowledge Jacob Beal, for all the guidance and feedback provided during the creation of this work

References

1. Gibson DG, Young L, Chuang RY, Venter JC, Hutchison CA, Smith HO (2009) Enzymatic assembly of DNA molecules up to several hundred kilobases. Nat Methods 6(5):343–345. https://doi.org/10.1038/nmeth.1318
2. Engler C, Kandzia R, Marillonnet S (2008) A one pot, one step, precision cloning method with high throughput capability. PLoS One 3(11). https://doi.org/10.1371/journal.pone.0003647
3. Iverson SV, Haddock TL, Beal J, Densmore DM (2016) CIDAR MoClo: Improved MoClo assembly standard and new E. coli part library enable rapid combinatorial design for synthetic and traditional biology. ACS Synth Biol 5(1):99. https://doi.org/10.1021/acssynbio.5b00124
4. Weber E, Engler C, Gruetzner R, Werner S, Marillonnet S (2011) A modular cloning system for standardized assembly of multigene constructs. PLoS One 6(2):e16765. https://doi.org/10.1371/journal.pone.0016765
5. Pollak B et al (2019) Universal loop assembly: Open, efficient and cross-kingdom DNA fabrication. Synth Biol 5(1):1–13. https://doi.org/10.1093/synbio/ysaa001
6. Sarrion-Perdigones A et al (2011) GoldenBraid: An iterative cloning system for standardized assembly of reusable genetic modules. PLoS One 6(7):e21622. https://doi.org/10.1371/journal.pone.0021622
7. Bird JE, Marles-Wright J, Giachino A (2022) A user's guide to Golden Gate cloning methods and standards. ACS Synth Biol 11(11):3551–3563. https://doi.org/10.1021/acssynbio.2c00355
8. Bryant JA, Kellinger M, Longmire C, Miller R, Wright RC (2023) AssemblyTron: flexible automation of DNA assembly with Opentrons OT-2 lab robots. Synth Biol 8(1). https://doi.org/10.1093/synbio/ysac032
9. McLaughlin JA et al (2020) The Synthetic Biology Open Language (SBOL) Version 3: simplified data exchange for bioengineering. Front Bioeng Biotechnol 8(September):1–15. https://doi.org/10.3389/fbioe.2020.01009
10. Buecherl L et al (2023) Synthetic biology open language (SBOL) *version 3.1.0.* J Integr Bioinform 20(1):20220058. https://doi.org/10.1515/jib-2022-0058
11. Quinn JY et al (2015) SBOL visual: a graphical language for genetic designs. PLoS Biol 13(12):e1002310. https://doi.org/10.1371/journal.pbio.1002310
12. Baig H et al (2021) Synthetic biology open language visual (SBOL Visual) version 2.3. J Integr Bioinform 18(3). https://doi.org/10.1515/jib-2020-0045
13. Beal J et al (2023) Standardized representation of parts and assembly for build planning. ACS Synth Biol 12:3655. https://doi.org/10.1021/ACSSYNBIO.3C00418
14. McLaughlin JA et al (2018) SynBioHub: a standards-enabled design repository for synthetic biology. ACS Synth Biol 7(2):682–688. https://doi.org/10.1021/ACSSYNBIO.7B00403/ASSET/IMAGES/LARGE/SB-2017-004037_0006.JPEG
15. Sents Z, Stoughton TE, Buecherl L, Thomas PJ, Fontanarrosa P, Myers CJ (2023) SynBioSuite: a tool for improving the workflow for genetic design and modeling. ACS Synth Biol 12(3):892–897. https://doi.org/10.1021/ACSSYNBIO.2C00597/ASSET/IMAGES/LARGE/SB2C00597_0002.JPEG
16. Vidal G, Vitalis C, Rudge TJ (2021) LOICA: integrating models with data for genetic network design automation. ACS Synth Biol 2022. https://doi.org/10.1021/ACSSYNBIO.1C00603/ASSET/IMAGES/LARGE/SB1C00603_0003.JPEG

Chapter 7

Use of a Golden Gate Plasmid Set Enabling Scarless MoClo-Compatible Transcription Unit Assembly

Stijn T. de Vries, Laura Kley, and Daniel Schindler

Abstract

Golden Gate cloning has become a powerful and widely used DNA assembly method. Its modular nature and the reusability of standardized parts allow rapid construction of transcription units and multi-gene constructs. Importantly, its modular structure makes it compatible with laboratory automation, allowing for systematic and highly complex DNA assembly. Golden Gate cloning relies on type IIS enzymes that cleave an adjacent undefined sequence motif at a defined distance from the directed enzyme recognition motif. This feature has been used to define hierarchical Golden Gate assembly standards with defined overhangs ("fusion sites") for defined part libraries. The simplest Golden Gate standard would consist of three-part libraries, namely promoter, coding and terminator sequences, respectively. Each library would have defined fusion sites, allowing a hierarchical Golden Gate assembly to generate transcription units. Typically, type IIS enzymes are used, which generate four nucleotide overhangs. This results in small scar sequences in hierarchical DNA assemblies, which can affect the functionality of transcription units. However, there are enzymes that generate three nucleotide overhangs, such as SapI. Here we provide a step-by-step protocol on how to use SapI to assemble transcription units using the start and stop codon for scarless transcription unit assembly. The protocol also provides guidance on how to perform multi-gene Golden Gate assemblies with the resulting transcription units using the Modular Cloning standard. The transcription units expressing fluorophores are used as an example.

Key words Golden Gate assembly, Start-stop cloning, SapI, Blunt-end cloning, Basic parts, Scarless cloning, Modular cloning

1 Introduction

Golden Gate cloning and the resulting Modular Cloning (MoClo) has become a popular DNA assembly strategy because of its modular and hierarchical structure, which allows for the economical construction of complex DNA assemblies [1, 2]. Golden Gate cloning is based on the use of type IIS restriction enzymes, which recognize a specific, directed DNA sequence but cut any adjacent DNA sequence at a defined distance (Fig. 1a). The type IIS enzymes used for Golden Gate assembly generate sticky ends.

Daniel Schindler (ed.), *Golden Gate Cloning: Methods and Protocols*, Methods in Molecular Biology, vol. 2850,
https://doi.org/10.1007/978-1-0716-4220-7_7,

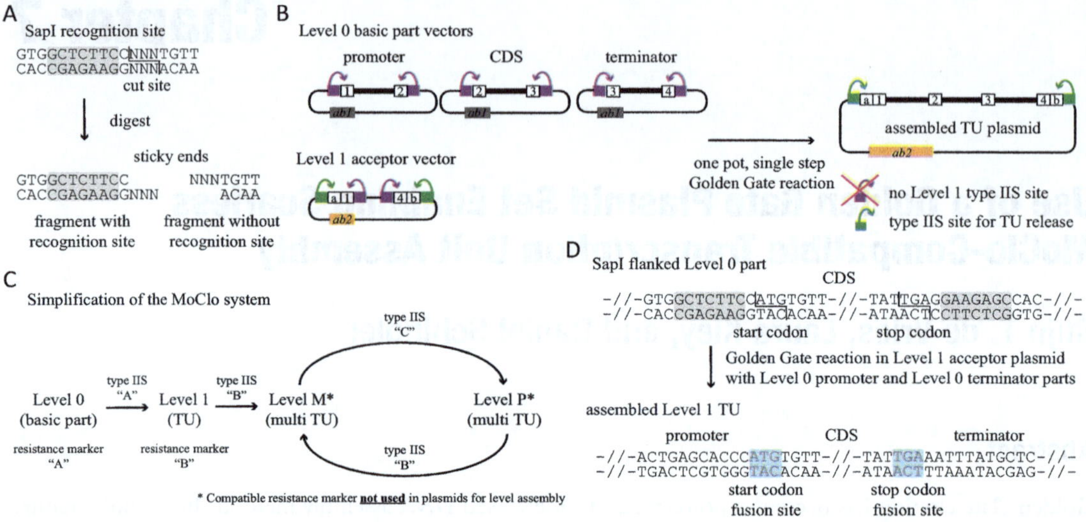

Fig. 1 Basic principles of Golden Gate cloning and MoClo. (**a**) Type IIS enzymes recognize a directed DNA sequence and cut any sequence at a defined distance from the recognition site. This results in two DNA fragments, one with and one without the recognition site. SapI is used as an example and its recognition site is shaded gray. N indicates that any DNA sequence can be at the cut site (note: enzyme activities can be inhibited by DNA methylation). (**b**) Golden Gate cloning relies on the orchestration of type IIS enzyme recognition and cleavage sites. The use of defined fusion sites (here represented by 1 to 4) allows the assembly of multiple parts into an acceptor plasmid. The basic part plasmids and the acceptor plasmid differ at least in the selection marker and the orientation of the type IIS enzyme recognition sites. The assembled plasmid does not contain the type IIS recognition site used to generate the plasmid. However, a different type IIS enzyme recognition site may be present in the acceptor plasmid to release the assembled DNA fragment with new fusion sites (here indicated by a and b) for higher order assembly. (**c**) The Modular Cloning (MoClo) system allows in theory for an infinite assembly rounds of DNA fragments [2]. The generated basic parts (Level 0) can be assembled in transcription units (TU). Multiple TUs can be assembled into higher order DNA assemblies. By swapping the type IIS enzyme and the antibiotic selection marker between each step, in particular between Level M and Level P allows in theory for unlimited cycles to combine DNA assemblies of growing complexities. The only requirement is careful planning to ensure that the fragments have the appropriate fusion sites. Important: In this protocol Level 1 plasmids are used that are compatible with Level P. The procedure is identical except for the type IIS enzyme and the antibiotic resistance markers. For simplicity, the standard MoClo system is visualized and the corresponding Level M compatible plasmids are also available. (**d**) Assembly of transcription units based on three nucleotide overhangs using the start codon (ATG) and stop codon (TGA) as fusion sites to allow scarless TU construction

These sticky ends can be defined by the user, allowing the creation of part libraries, for example, promoter, coding (CDS) and terminator sequences, each defined position with a specific type IIS-generated overhang ("fusion site") (Fig. 1b). The basic parts are usually cloned into defined Level 0 plasmids, sequence validated and can be released with a type IIS enzyme to allow the assembly of individual parts into transcription units (TU).

This modular structure allows for standardized workflows, making this DNA assembly strategy particularly amenable to automation. Another advantage of this standardized method is that once the basic parts have been sequence validated, they are no longer amplified, eliminating the risk of polymerase chain reaction (PCR)-based introduction of sequence errors. Therefore, only basic parts need to be sequence validated, TUs and higher order assemblies can be validated by simple, rapid and economic diagnostic methods such as colony PCR (cPCR) or restriction pattern analysis. We have recently developed protocols for semi-automated Golden Gate DNA assembly and long-read sequencing-based construct validation that allow economic DNA assembly reactions in 1 μL total volume and low-cost high-throughput sequence validation [3–6]. Establishing a systematic Golden Gate assembly standard with corresponding part libraries can therefore be highly rewarding, as it allows rapid combinatorial DNA assembly to answer biological questions or test the application potential of the gene(s) of interest.

The most interesting part of MoClo, however, is that it theoretically allows the assembly of DNA constructs of infinite size (Fig. 1c). The trick of the MoClo system is that TUs can be assembled into multiple hierarchical Level 1 plasmids, each corresponding to a defined position in higher order multi-gene DNA constructs. The Level 1 TU plasmids differ only in the fusion sites created when the fragments are released with the appropriate type IIS endonuclease. This allows for the assembly of multiple Level 1 plasmids into a higher order DNA assembly. With the original MoClo strategy, up to 6 TUs can be assembled into a Level M plasmid in combination with the appropriate end linker (EL) [2]. Again, there are multiple Level M plasmids, similar to Level 1 plasmids. Level M multi-gene constructs can be released with another type IIS enzyme and allow the assembly of multiple Level M constructs into a Level P plasmids. There are multiple Level P plasmids, similar to Level 1 and Level M. The clue is that at each Level, the enzyme and antibiotic selection marker are changed and the Level P DNA assemblies are again compatible with Level M-based Golden Gate cloning, allowing for an infinite cloning cycle. This strategy has been used to generate highly complex DNA constructs for various applications such as the generation of complex repetitive sequence arrays from sequence libraries or the construction of synthetic chromosomes [7, 8].

However, Golden Gate cloning has a small disadvantage. When using defined part libraries, the fusion sites are fixed. Mostly type IIS enzymes are used, which result in four nucleotide fusion sites. This results in small scars at the fusion sites, which can affect TU functionality. Especially when using experimental systems that are sensitive to small changes, such as the construction and application of small synthetic regulatory RNAs (sRNAs), scar sequences introduced by DNA assembly should be avoided [3, 9]. To minimize scar sequences introduced at the Level of TUs, the use of SapI seems to be promising. In contrast to the commonly used type

IIS enzymes, SapI has a three nucleotide fusion site, which allows the use of codons as fusion sites [10]. Recently, cloning fidelities have been determined for several type IIS enzymes, allowing for data-driven, optimized Golden Gate assembly strategies, including the SapI enzyme [11].

The protocol outlined describes the use of SapI for the construction of TUs, resulting in no assembly scars based on the three nucleotide sticky ends, in contrast to the commonly used four nucleotide sticky end-generating type IIS enzymes (Fig. 1d). The use of SapI allows for codon-based cloning, using the start (ATG) and stop (TGA) codons to link the CDS to the appropriate promoter and terminator sequences. The protocol describes the design and generation of basic parts (*see* Subheading 3.1) for their SapI-mediated assembly into TUs (*see* Subheading 3.2) and subsequent higher order assemblies (*see* Subheading 3.3). The assembled TUs (*see* Subheading 3.2) are compatible with the standard MoClo system developed by Sylvestre Marillonnet's group. The protocol concludes with an exemplary evaluation of the assembled multi-gene construct using spotting assays to test microbial growth and to measure the resulting fluorescence of the assembled multi-gene constructs (*see* Subheading 3.4).

2 Materials

The conduction of the described protocol requires the following standard laboratory equipment and consumables.

1. Standard reagents, consumables, and instrumentation for PCR reactions.
2. Standard reagents and equipment for gel electrophoresis.
3. Standard reagents, consumables, and instrumentation for microbial culturing.
4. Standard reagents, consumables, and instrumentation for transformation of *Escherichia coli*.
5. Standard reagents, consumables, and instrumentation for plasmid extraction and DNA clean up (*see* **Note 1**).
6. Standard micropipettes and consumables, 12 channel pipettes are advised.
7. Equipment to visualize distinct fluorescence signals on colony level (e.g., imaging system or binocular). Alternatively, a plate reader may be used.

2.1 Plasmids

All plasmids relevant for this protocol are listed in Table 1. Plasmids can be requested from the corresponding author.

Table 1
Plasmids used in this study

Name	Relevant features	Reference
pSL099	Level 0 acceptor plasmid for promoter position[a], spectinomycin resistance marker, Type IIS enzyme for Level 0 assembly: BpiI Type IIS enzyme for fragment release: SapI	Unpublished
pSL102	Level 0 acceptor plasmid for CDS position[a], spectinomycin resistance marker, Type IIS enzyme for Level 0 assembly: BpiI Type IIS enzyme for fragment release: SapI	Unpublished
pSL106	Level 0 acceptor plasmid for terminator position[a], spectinomycin resistance marker, Type IIS enzyme for Level 0 assembly: BpiI Type IIS enzyme for fragment release: SapI	Unpublished
pSL108	Level 1 TU acceptor plasmid position 1, ampicillin resistance marker, Type IIS enzyme for Level 1 assembly: SapI Type IIS enzyme for fragment release: BsaI	Unpublished
pSL109	Level 1 TU acceptor plasmid position 2, ampicillin resistance marker, Type IIS enzyme for Level 1 assembly: SapI Type IIS enzyme for fragment release: BsaI	Unpublished
pSL110	Level 1 TU acceptor plasmid position 3, ampicillin resistance marker, Type IIS enzyme for Level 1 assembly: SapI Type IIS enzyme for fragment release: BsaI	Unpublished
pMA67	Level P multi-gene acceptor plasmid, kanamycin resistance marker, Type IIS enzyme for Level P assembly: BsaI Type IIS enzyme for fragment release: BpiI	[7]
pMA676	Level P Endlinker plasmid for multi-gene assembly, chloramphenicol resistance marker, Type IIS enzyme for Level P assembly: BsaI	[12]
pSL842	Level 0 basic CDS part *mTurquoise*, spectinomycin resistance marker, Type IIS enzyme for fragment release: SapI	Unpublished
pSL843	Level 0 basic CDS part *gfp*, spectinomycin resistance marker, Type IIS enzyme for fragment release: SapI	Unpublished
pSL206	Level 0 basic CDS part *mCherry*, spectinomycin resistance marker, Type IIS enzyme for fragment release: SapI	Unpublished
pSLcol_16.03	Level 0 basic promoter part medium strength, spectinomycin resistance marker, Type IIS enzyme for fragment release: SapI	Unpublished
pSL209	Level 0 basic terminator part T1-T7Te, spectinomycin resistance marker, Type IIS enzyme for fragment release: SapI	Unpublished
pSL844	Level 1 position 1 plasmid containing neutral linker, ampicillin resistance marker, Type IIS enzyme for fragment release: BsaI	Unpublished

(continued)

Table 1
(continued)

Name	Relevant features	Reference
pSL845	Level 1 pos. 1 *mTurquoise* TU, ampicillin resistance marker, Type IIS enzyme for fragment release: BsaI	Unpublished
pSL846	Level 1 position 2 plasmid containing neutral, ampicillin resistance marker, Type IIS enzyme for fragment release: BsaI	Unpublished
pSL847	Level 1 pos. 2 *gfp* TU, ampicillin resistance marker, Type IIS enzyme for fragment release: BsaI	Unpublished
pSL848	Level 1 position 3 plasmid containing neutral, ampicillin resistance marker, Type IIS enzyme for fragment release: BsaI	Unpublished
pSL849	Level 1 pos. 3 *mCherry* TU, ampicillin resistance marker, Type IIS enzyme for fragment release: BsaI	Unpublished
pSL765	Level P multi-gene assembly with m*Turquoise*, *gfp*, and *mCherry* TU, kanamycin resistance marker, Type IIS enzyme for fragment release: BpiI	Unpublished
pSL766	Level P multi-gene assembly with m*Turquoise* TU, *gfp* TU, and neutral linker, kanamycin resistance marker, Type IIS enzyme for fragment release: BpiI	Unpublished
pSL767	Level P multi-gene assembly with m*Turquoise* TU, neutral linker, and *mCherry* TU, kanamycin resistance marker, Type IIS enzyme for fragment release: BpiI	Unpublished
pSL768	Level P multi-gene assembly with neutral linker, *gfp* TU, and *mCherry* TU, kanamycin resistance marker, Type IIS enzyme for fragment release: BpiI	Unpublished
pSL769	Level P multi-gene assembly with neutral linker at all three positions, kanamycin resistance marker, Type IIS enzyme for fragment release: BpiI	Unpublished

[a]Any of these plasmids would be suitable if used as template for blunt-end cloning of basic Level 0 parts (*see* Subheading 3.1.2) because the overhangs are attached with the oligonucleotides/synthesized DNA

2.2 Enzymes

Any enzyme with corresponding properties can be used with their supplied buffers (*see* **Note 3**).

1. BpiI (20,000 U/mL).
2. BsaI (20,000 U/mL).
3. SapI (10,000 U/mL).
4. T4 polynucleotide kinase (10,000 U/mL).
5. T4 DNA ligase (400,000 U/mL).
6. Proofreading DNA polymerase (in this case in-house purified).
7. Taq DNA polymerase (5000 U/mL).

Table 2
Oligonucleotides for the conduction of the described protocol

Name	Sequence (5′-3′)	Information
SLo0765	TGAAGAGCAGGCACGAACCC	Forward primer to amplify acceptor plasmid for blunt-end cloning
SLo0766	AGAAGAGCGAGCACAGAGTGC	Reverse primer to amplify acceptor plasmid for blunt-end cloning
SLo4231	TGCTCTTCTTTCAATAGGTCTGAGA ACGCGCGTTCTCAGACCTATTCGGT GAAGAGCT	Neutral DNA linker sequence, annealed with SLo4232 to generate double stranded DNA.
SLo4232	AGCTCTTCACCGAATAGGTCTGAGA ACGCGCGTTCTCAGACCTATTGAAA GAAGAGCA	Neutral DNA linker sequence, annealed with SLo4232 to generate double stranded DNA.
SLo04365	GTCGATTTTTGTGATGCTC	Primer for Sanger sequencing of Level 0 parts (*see* **Note 2**).

2.3 DNA Oligonucleotides

Relevant oligonucleotides for the conduction of the protocol are provided in Table 2. 100 μM stocks are generated with ddH_2O and stored at −20 °C. For PCR and Sanger sequencing reactions 10 μM working stocks are generated with ddH_2O and stored at −20 °C.

2.4 Antibiotics

All antibiotics used in this study are solved in sterile H_2O and are stored in 1 mL aliquots at −20 °C. Stock solutions are used 1:1000 (e.g., 500 μL stock solution in 500 mL medium) (Table 2).

1. Spectinomycin: 120 mg/mL, stock solution.
2. Ampicillin: 100 mg/mL, stock solution.
3. Kanamycin: 50 mg/mL, stock solution.

2.5 Chemicals, Buffers, and Media Components

1. Culture dilution buffer: 0.8% (w/v) NaCl.
2. SMM: 0.3% (w/v) KH_2PO_4, 0.678% (w/v) Na_2HPO_4, 0.05% (w/v) NaCl, 0.1% (w/v) NH_4Cl, 1.6% (w/v) tryptone, 1.0% (w/v) yeast extract, 0.5% (v/v) glycerol.
3. LB: 1% (w/v) tryptone, 0.5% (w/v) yeast extract, 1% (w/v) NaCl.

2.6 Consumables

1. Standard petri dishes.
2. Single well society for biomolecular screening (SBS) plates (for spotting experiment).
3. 96-well microtiter plates.
4. 1.5 mL reaction tubes.

Table 3
Strains used in this study

Name	Relevant features	Reference
E. coli DB3.1 (*see* Note 5)	F⁻ *gyrA462 endA1 glnV44* Δ(*srl-recA*) *mcrB mrr hsdS20*(r_B^-, m_B^-) *ara14 galK2 lacY1 proA2 rpsL20*(Str^R) *xyl5* Δ*leu mtl1*	Invitrogen
E. coli Top10	F⁻ *mcrA* Δ(*mrr-hsdRMS-mcrBC*) φ80*lacZ*ΔM15 Δ*lacX74 nupG recA1 araD139* Δ(*ara-leu*)7697 *galE15 galK16 rpsL*(Str^R) *endA1* λ⁻	Invitrogen

2.7 Strains

All relevant laboratory *E. coli* strains for the outlined protocol are provided in Table 3. Cells were made competent using an in-house RbCl procedure [13] (*see* **Note 4**).

3 Methods

3.1 Construction and Validation of Basic Parts

The first step in Golden Gate cloning is the design and domestication of DNA sequences into the selected standardized Golden Gate cloning hierarchy. However, before cloning can begin, the sequences must be identified and domesticated for cloning into Level 0 acceptor plasmids (Fig. 2a). This procedure can be performed with different strategies as described in Subheading 3.1.1. It is important that type IIS recognition sites used later are omitted from the basic part DNA sequences. In this protocol, basic parts are stored in Level 0 plasmids. DNA sequences can be cloned into the Level 0 plasmids using two different strategies. First, the blunt-end cloning strategy (3.1.2), where the fusion sites are attached to the DNA fragment to be stored as a basic part. Using this protocol will reduce the use of type IIS enzymes, but is limited to cloning single DNA fragments. Due to the nature of blunt-end cloning, the DNA sequence can be oriented forward or reverse in the plasmid. However, release of the fragment with the appropriate type IIS enzyme (here SapI) will result in the correct fragment with the appropriate fusion sites because the fusion sites are already added to the initial fragment (Fig. 2b). The other strategy is to use Golden Gate cloning (*see* Subheading 3.1.3) of DNA fragments in Level 0, which allows the combination of multiple DNA sequences in the appropriate order, which may be necessary if type IIS recognition sites need to be eliminated from the DNA sequence (Fig. 2c).

3.1.1 Design of Sequences (Synthesis/Amplification)

1. For hierarchical Golden Gate cloning standards, all needed sequences are generated as validated basic part plasmids, hereafter termed Level 0 plasmids.
2. Familiarize with the chosen Golden Gate cloning system. In this protocol, a procedure based on the Modular Cloning

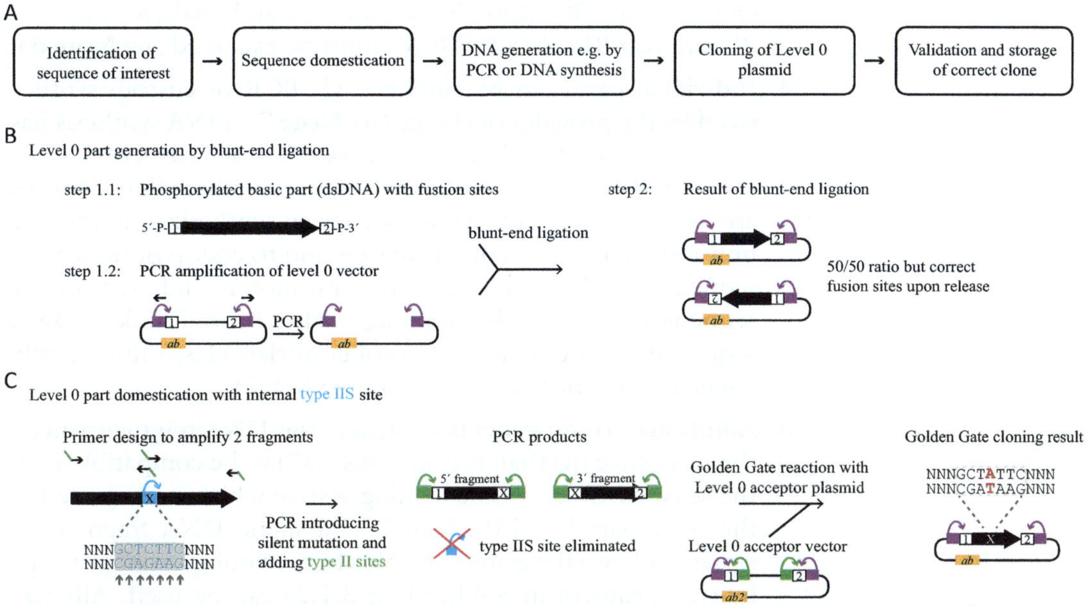

Fig. 2 Concept and requirements of sequences domestication and basic part generation. (**a**) Generalized and stepwise workflow to consider for Level 0 part design, generation and validation. (**b**) Simple generation of Level 0 parts by blunt-end ligation. If no type IIS recognition site needs to be eliminated, the DNA can be amplified as a single part. The oligonucleotides must have the appropriate fusion site attached (here, 1 and 2). The DNA fragment is generated either by PCR amplification or by annealing of complementary oligonucleotides. It is important that the oligonucleotides or the PCR amplicon are phosphorylated. Each acceptor plasmid is PCR amplified without the fusion sites. Importantly the plasmid will not be phosphorylated. The DNA is mixed in a simple ligation reaction. After transformation, the fragment will be in two orientations in the plasmid. However, because the fusion sites are attached to the DNA fragment and not originate from the plasmid, both orientations are fine. (**c**) In case the DNA sequence contains one or more unwanted type IIS recognition sites (visualized in blue) that need to be removed, amplification of multiple fragments with the appropriate type IIS enzyme for Level 0 basic part cloning (visualized in green) allows to eliminate the recognition site of any type IIS enzyme or other small DNA sequence motif. It is important to ensure that the mutation is not deleterious, for example, an amino acid change in the CDS. The fragments will have the appropriate fusion sites for Level 0 Golden Gate cloning as well as unique fusion sites to fuse the fragments (indicated by X). The fusion sites between the fragments eliminate type IIS recognition sites by introducing one or more sequence changes. The resulting plasmid has the recognition site eliminated (indicated by red letter in final plasmid). This strategy allows the modification of one to four consecutive nucleotides based on the typically used type IIS enzymes with four nucleotide fusion sites

system of the group of Sylvestre Marillonnet is used with alterations on Level 0. SapI is used to release Level 0 parts to assemble transcription units avoiding assembly scars. The general steps are compatible with any other system but may require alterations of type IIS enzymes and antibiotic selection markers throughout the steps.

3. *Important during the design process:* The part sequences must not contain any type IIS enzyme recognition sites used in the selected Golden Gate cloning system. In the outlined case, SapI

(5′-GCTCTTC-3′), BpiI (5′-GAAGAC-3′) and BsaI (5′-GGTCTC-3′) and BsmBI (5′-CGTCTC-3′) must be excluded (*see* **Note 6**).

4. Individual parts can be amplified via PCR or custom synthesized by the provider of choice (*see* **Note** 7). DNA synthesis has the advantage of having full control over each nucleotide, but this may not be economic. In this protocol, three fluorophores are used as an example which have been synthesized to remove unwanted enzyme recognition sites and to codon optimize the sequences. The used example promoter and terminator sequences were PCR amplified. The neutral linker DNA sequence was obtained as two oligonucleotides, subsequently annealed to generate double stranded DNA.
5. Additional to the general sequence, the DNA fragments need the respective overhangs ("fusion sites") to be compatible with the selected Golden Gate cloning system which is integrated at the basic part Level 0 plasmid. If a single DNA fragment is cloned in Level 0 plasmid, the blunt-end cloning strategy (step-by-step protocol in Subheading 3.1.2) can be used. Alternatively, or if basic parts consist of multiple fragments Golden Gate cloning for basic parts (step-by-step protocol in Subheading 3.1.3) must be used. Both strategies have different requirements for the design of the primer pairs for amplification which are outlined in **step 6** and 7, respectively.
6. Amplifying basic parts for blunt ligation needs to ensure the corresponding 3 assembly position specific nucleotides are attached to the primers (depicted by 20 Ns) to ensure proper assembly of the parts for transcription units (*see* **Note 8**):
   (i) Promoter sequences.
   forward oligo: 5′-**TTC**NNNNNNNNNNNNNNNNNNNN-3′.
   reverse oligo: 5′-**CAT**NNNNNNNNNNNNNNNNNNNN-3′.
   (ii) CDS sequences (*see* **Note 9**).
   forward oligo: 5′-**ATG**NNNNNNNNNNNNNNNNNNNN-3′.
   reverse oligo: 5′-**TCA**NNNNNNNNNNNNNNNNNNNN-3′.
   (iii) Terminator sequences.
   forward oligo: 5′-**TGA**NNNNNNNNNNNNNNNNNNNN-3′.
   reverse oligo: 5′-**CCG**NNNNNNNNNNNNNNNNNNNN-3′.
7. Cloning basic parts by Golden Gate assembly needs to ensure the recognition sites for the used type IIS enzyme and the respective fusion sites are added appropriately to the amplifying primers (depicted by 20 Ns) (*see* **Note 10**). Both elements are visualized in bold and underlined, respectively, in the examples.

(i) Promoter sequences.

forward oligo: 5′-T**GAAGAC**CATTTCNNNNNNNNNNNNNNNNNNN-3′.

reverse oligo: 5′-T**GAAGAC**CAACATNNNNNNNNNNNNNNNNNNN-3′.

(ii) CDS sequences (*see* **Note 9**).

forward oligo: 5′-T**GAAGAC**CATATGNNNNNNNNNNNNNNNNNNN-3′.

reverse oligo: 5′-T**GAAGAC**CAATCANNNNNNNNNNNNNNNNNNN-3′.

(iii) Terminator sequences.

forward oligo: 5′-T**GAAGAC**CATTGANNNNNNNNNNNNNNNNNNN-3′.

reverse oligo: 5′-T**GAAGAC**CAACCGNNNNNNNNNNNNNNNNNNN-3′.

8. Order the designed oligonucleotides from a respective vendor and amplify the sequences of interest or directly use the synthesized fragments (*see* **Note 11**).

3.1.2 Construction of Level 0 Parts by Blunt-End Cloning

1. Generate and purify the basic part DNA sequence by the method of choice (e.g., PCR, DNA synthesis) according to design guidelines in Subheading 3.1.1.

   Important: The 3 bp fusion sites must be added to the basic part as outlined in Subheading 3.1.1, **step 6**.

   Important: Phosphorylated and subsequently annealed oligonucleotides with blunt ends can be used directly in **step 10** (*see* **Note 12**).

2. Determine DNA quality and concentration of basic DNA parts with appropriate method(s) (e.g., gel electrophoresis, spectrophotometric, fluorescence dye assay).
3. PCR amplify the Level 0 plasmid ensuring the whole plasmid is amplified, including the recognition site of the type IIS enzyme for fragment release but excluding the fusion sites (Fig. 2b). By PCR amplification and subsequent DpnI digest high efficiency of cloning can be ensured.
4. PCR reaction mix using a proof reading DNA polymerase and reaction conditions are provided below to amplify example Level 0 plasmid pSL099 using primer pair SLo765/SLo766 (*see* **Note 13**).

Reagent	Stock concentration	Volume
Reaction buffer	5 X	10.0 μL
dNTPs	10 mM	1.0 μL
Forward primer	10 μM	2.5 μL
Reverse primer	10 μM	2.5 μL
DNA template	1–10 ng	–
Polymerase	2 U/μL	0.5 μL
ddH_2O	–	to 50 μL

Step	Settings	Cycles
Initial denaturation	98 °C for 20 s	–
PCR	98 °C for 20 s 67 °C for 20 s 72 °C for 2 min	35
Final extension	72 °C for 1 min	–
Hold	12 °C	–

5. Validate PCR reaction by performing gel electrophoresis with an appropriate size standard as reference.
6. Perform DpnI digest (*see* **Note 14**) by adding the appropriate volume of DpnI reaction buffer and 0.5 μL enzyme per 50 μL PCR reaction. Digest at 37 °C for at least 1 h up (*see* **Note 15**).
7. Purify the DpnI digested sample with the method of choice and quantify the DNA concentration. It is recommended to use molarity instead of mass.
8. The fragments to be cloned into the amplified Level 0 plasmid need to be phosphorylated (*see* **Note 16**). Reminder: Based on the nature of blunt-end cloning, the fragments can be cloned in forward or reverse orientation. However, because the fusion sites are attached to the DNA sequence, this does not matter.
9. Perform the phosphorylation according to the reaction mixture and reaction conditions outlined below (*see* **Note 17**):

Reagent	Stock concentration	Volume
Double stranded DNA (or single oligonucleotides prior annealing)	62.5 fmol/μL (maximum of 6.25 pmol)	4 μL
T4 DNA Ligase Buffer	10 X	0.5 μL
T4 Polynucleotide Kinase (PNK)	10 U/μL	0.5 μL
ddH_2O		to 5 μL

Step	Settings	Time
Incubation	37 °C	30 min
Heat inactivation	80 °C	20 min
Hold	12 °C	–

10. Because T4 ligase is used in the phosphorylation step, the DNA can just be added to the subsequent ligation step. Important: This needs to be included in the calculation for the reaction mix and the added volume is not counted towards the use of fresh T4 Ligase Buffer. The reaction mix is as following:

Reagent	Stock concentration	Volume
PCR amplified Level 0 plasmid	20 fmol/μL	1 μL
Phosphorylated fragment[a]	50 fmol/μL	1 μL
T4 DNA Ligase Buffer	10 X	1 μL
T4 DNA Ligase	400 U/μL	1 μL
ddH_2O		to 10 μL

[a]In case the fragment is phosphorylated as described in step 8, the used volume does not count to the total volume

11. Perform the ligation reaction using the following settings:

Step	Settings	Time
Incubation	16 °C	Overnight
Hold	4 °C	–

12. Transform 2.5 μL of the reaction mix in *E. coli* strain of choice. In this example, *E. coli* Top10 cells are used made competent with an in house RbCl procedure for heat shock transformation. *E. coli* Top10 is highly sensitive to the *ccdB* toxin used as an additional selection marker to prevent background.

However, any *E. coli* cloning strain and protocol for competent *E. coli* cells should be appropriate.

13. Add 1 mL of LB media to cells after transformation and recover at 37 °C with agitation for 30 to 60 minutes.
14. Plate transformation mixture on two plates with appropriate selection conditions (here LB spectinomycin). 100 μL of suspension is plated on the first plate. Afterwards the remaining transformation mixture is spun down 1 min at 10,000× *g*, the supernatant is discarded, the pellet resolved in the remaining liquid and plated on the second plate.
15. Incubate plates inverted overnight at 37 °C in a standing incubator.
16. Pick at least three single candidates and validate the expected insert with colony PCR or by plasmid extraction and subsequent test digest using the enzyme to release the insert.
17. Validate selected candidates using the sequencing service of choice.
18. Cryo-preserve sequence validated strain containing the Level 0 basic part with a unique identifier and the relevant information for the strain (e.g., plasmid map, sequence validation raw data).
19. If enough plasmid material is left, store at −20 °C for later use. If material is limited isolate plasmid with method of choice by inoculation cryo-preserved strain, determine DNA quality and concentration (*see* **Note 18**) and store until further use at −20 °C.

3.1.3 Construction of Level 0 Parts by Golden Gate Cloning

1. Purify the necessary Level 0 plasmids with the method of choice, perform quality control and determine the DNA concentration. It is recommended to set the concentration of the Level 0 plasmids to 50 fmol/μL.
2. PCR-amplify the sequences of choice with the respective primers and their optimal reaction conditions or use synthesized DNA fragments. Keep in mind the sequence requirements outlined in Subheading 3.1.1.
3. Validate PCR amplification by gel electrophoresis with an appropriate size standard (*see* **Note 19**).
4. Purify PCR reactions after size validation with the method of choice and determine DNA concentration. It is recommended to set the DNA concentration to 50 fmol/μL to simplify the reaction conditions. This is in particular important in case many reactions are done in parallel and/or the basic part sequence consists of multiple fragments.

5. Perform Golden Gate reaction using the outlined reaction mix and reaction conditions. The example is based on the used Level 0 plasmids and PCR amplified fragments (in this case the fluorophores) using a different Golden Gate system may requires to alter the used type IIS enzyme (*see* **Note 20**).

Reagent	Stock concentration	Volume
Level 0 acceptor plasmid	50 fmol/μL	1 μL
Insert fragment (*see* **Note 21**)	50 fmol/μL	3 μL
T4 DNA Ligase Buffer	10 X	1 μL
T4 DNA Ligase	400 U/μL	1 μL
BpiI	20 U/μL	1 μL
ddH_2O		to 10 μL

Step	Settings	Cycles
Type IIS enzyme optimum (*see* **Note 22**)	37 °C for 3 min	25
Ligase optimum	16 °C for 4 min	
Final digestion step[a]	50 °C for 20 min	–
Heat inactivation	80 °C for 20 min	–
Hold	12 °C	–

[a]At this step the ligase is inactivated which ensures minimal background because all wrongly ligated plasmids are removed from the reaction mix

6. Transform 5 μL reaction mix of the reaction mix in *E. coli* strain of choice. In this example, *E. coli* Top10 cells are used made competent with an in house RbCl procedure for heat shock transformation. *E. coli* Top10 is highly sensitive to the *ccdB* toxin used as an additional selection marker to prevent background. However, any *E. coli* cloning strain and protocol for competent *E. coli* cells should be appropriate.
7. From here follow **steps 12–18** of Subheading 3.1.2, the steps are identical.

3.2 DNA Assembly of Single Biological Parts into Level 1

Once a selection of Level 0 plasmids containing the desired parts has been generated, they can be combined into TUs by Golden Gate assembly. The hierarchical Golden Gate assembly system allows for the rapid construction of modified TUs (Fig. 3). For example, if a protein is not well expressed with the initially selected promoter sequence, the user can swap to alternative sequences or even use a library-based approach to identify the optimal promoter sequence. In the outlined protocol, only one promoter is used as an example. However, the protocol could be adapted to library-based

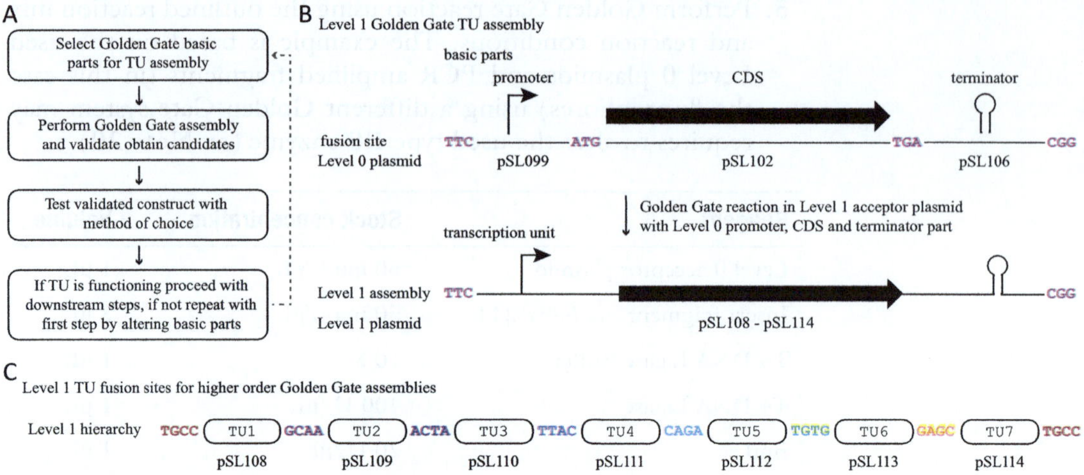

Fig. 3 Workflow and hierarchy of Level 1 Golden Gate TU assembly and subsequent assembly of multiple TUs. (**a**) Generalized workflow for constructing TUs from basic parts. Functional TUs can subsequently be used for their purpose. If a TU is not functional, a variation of the TU can be generated, for example, by changing the promoter and/or terminator (indicated by dotted line). (**b**) Golden Gate assembly using SapI-based TU cloning to use the ATG start codon and the TGA stop codon as fusion sites for scarless TU assembly. The system can be extended with additional fusion sites to allow for more complex TU assemblies, for example, using a C- and/or N-terminal tag for fusion proteins. Using SapI allows one to choose a specific codon as a fusion site. (**c**) The Golden Gate assembled TUs can be released for the assembly of multiple TUs. There are seven positions with hierarchical fusion sites based on the original MoClo system. The system allows the Golden Gate assembly of one to six TUs in the next assembly round. The fusion sites remain the same for later stages, for example, a multi-gene construct of TU1 to TU3 would result in the fusion sites `TGCC` and `TTAC` after release. Theoretically MoClo allows infinite rounds of DNA assembly (cf. Figure 1c)

cloning by simply changing the DNA sequence to a Level 0 plasmid pool. The remaining steps are the same.

In this protocol, the SapI enzyme, which generates three nucleotides fusion sites, is used to allow codon-based TU cloning (Fig. 3b). The use of SapI has the advantage of seamlessly assembling TUs that can subsequently be combined into higher order assemblies if appropriate Level 1 plasmids are used. For simplicity, a three-part cloning is outlined in this protocol. However, the user can extend the cloning system to combine more than three parts by selecting appropriate fusion sites. The resulting TUs can then be assembled into higher order constructs consisting of multiple TUs as described in Subheading 3.3. Importantly, the Level 1 plasmids used are compatible with Level P plasmids, unlike the standard MoClo procedure where Level 1 is compatible with Level M (cf. Figure 1c). However, Level 1 plasmids that are compatible with Level M are also available.

1. If not already available, extract the required Level 0 plasmids (here: pSLcol_16.03, pSL206, pSL209, pSL842, and pSL843) and the Level 1 acceptor plasmids (here: pSL108, pSL109 and

pSL110) with the method of choice and determine the DNA concentration. In this example, three basic parts, a promoter, a coding sequence, and a terminator are assembled into three different Level 1 acceptor plasmids. The reason for the three different acceptor plasmids is that the transcription units can be released with the respective type IIS enzyme and contain matching overhangs for a higher order assembly (Fig. 3c) (detailed step-by-step procedure in Subheading 3.3).

2. Set all individual plasmids to a final concentration of 50 fmol/μL.
3. In addition to the Level 0 plasmid assemblies, Level 1 plasmids with a "neutral spacer sequence" are generated directly in the three Level 1 positions, necessary for the assemblies in Subheading 3.3 (*see* **Note 23**).
4. Perform three Golden Gate reactions using the outlined reaction mix and reaction conditions. The example shown is based on three basic part positions in Level 0 plasmids (Promoter pSLcol_16.03, coding sequence pSL206 [*mcherry*], pSL842 [*mTurquoise*] or pSL843 [*gfp*], terminator pSL209) and Level 1 acceptor plasmids pSL108, pSL109 or pSL110.
5. The following six Golden Gate assemblies are performed:

   mTurquoise – TU: pSL108 with pSLcol_16.03, pSL842, and pSL209.

   GFP – TU: pSL109 with pSLcol_16.03, pSL843, and pSL209.

   mCherry – TU: pSL110 with pSLcol_16.03, pSL206, and pSL209.

   Important: Neutral linker were assembled previously in pSL108 to pSL110, resulting in pSL845, pSL847 and pSL849, respectively.

Reagent	Stock concentration	Volume
Level 1 acceptor plasmid	50 fmol/μL	1 μL
Each Level 0 part	50 fmol/μL	1 μL
T4 DNA Ligase Buffer	10 X	1 μL
T4 DNA Ligase	400 U/μL	1 μL
SapI	10 U/μL	1 μL
ddH_2O		to 10 μL

Step	Settings	Cycles
Type IIS enzyme optimum (*see* **Note 22**)	37 °C for 3 min	25
Ligase optimum	16 °C for 4 min	
Final digestion step[a]	50 °C for 20 min	–
Heat inactivation	80 °C for 20 min	–
Hold	12 °C	–

[a]At this step, the ligase is inactivated which ensures minimal background because all wrongly ligated plasmids are removed from the reaction mix

6. Transform 2.5 μL reaction mix of the reaction mix in *E. coli* strain of choice. In this example, *E. coli* Top10 cells are used made competent with an in house RbCl procedure for heat shock transformation. *E. coli* Top10 is highly sensitive to the *ccdB* toxin used as an additional selection marker to prevent background. However, any *E. coli* cloning strain and protocol for competent *E. coli* cells should be appropriate.
7. Add 1 mL of LB media to cells after transformation and recover at 37 °C with agitation for 30–60 min.
8. Plate transformation mixture on two plates with appropriate selection conditions (here LB ampicillin). 100 μL of suspension is plated on the first plate. Afterwards the remaining transformation mixture is spun down 1 min at 10,000× *g*, the supernatant is discarded, the pellet resolved in the remaining liquid and plated on the second plate.
9. Incubate plates inverted overnight at 37 °C in a standing incubator.
10. In the outlined experiment, the TU assembly can be visually validated by checking for fluorescent colonies (e.g., fluorescence binocular). In case no visible marker is used, it is recommended to perform colony PCR or in case of higher order assemblies perform plasmid extraction and restriction pattern digest of the DNA with the releasing type IIS enzyme (*see* **Note 24**).
11. Cryopreserve generated strains according to standard procedure and document all necessary information.
12. Extract plasmid DNA, perform quality control (e.g., test digest, $A_{260/280}$ ratio), and determine concentration with method of choice. Set concentration of plasmid to 50 fmol/μ L for subsequent higher order Golden Gate assembly. Store plasmid until further use at −20 °C.

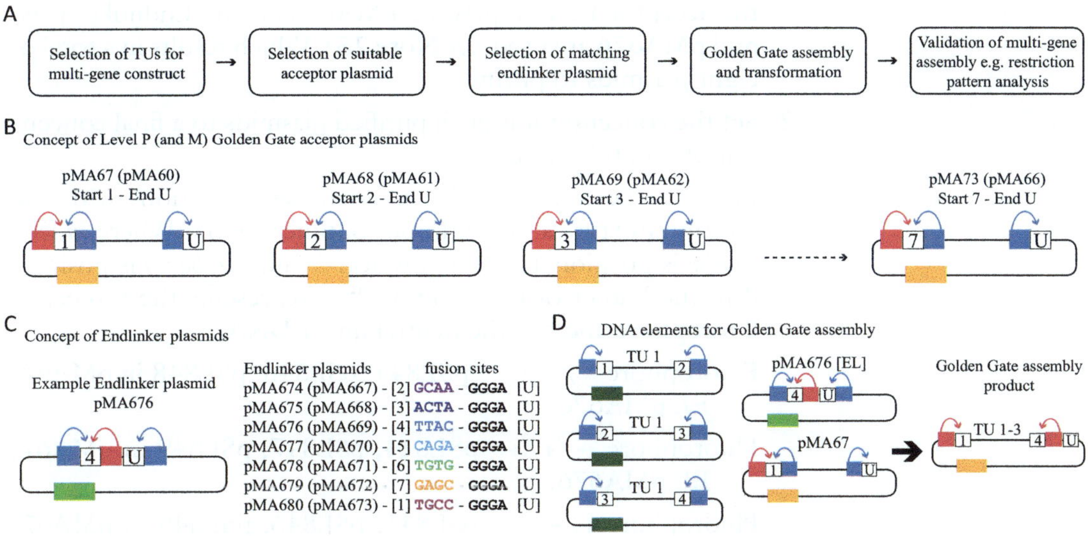

Fig. 4 Workflow and visualization of Golden Gate multi-gene assemblies based on the MoClo system. (**a**) Overview of the workflow for Golden Gate multi-gene assembly. (**b**) Concept of higher order acceptor plasmids. In contrast to Level 1 plasmids (cf. Fig. 3), these plasmids have a universal fusion site [U] with the sequence GGGA. As with Level 1 plasmids, there are 7 defined positions with the same fusion site sequences as in Level 1. The Endlinker (EL) plasmid provides the gap-filling fragment, allowing highly flexible DNA assemblies of up to 6 TUs in a single step. (**c**) The EL provides the assembly closure fragment by releasing it with the type IIS enzyme of the Golden Gate reaction. In addition, the EL provides the type IIS recognition sequence for a different type IIS enzyme for scarless release of the multi-gene construct for additional Golden Gate assembly rounds as needed. The combination of the 7 Level P (and M) plasmids with the corresponding EL plasmids provides maximum flexibility. (**d**) Example visualization of the Golden Gate assembly of three TUs (as done in this protocol) with the respective EL plasmid in the selected acceptor plasmid. The assembly results in a multi-gene construct of three TUs that can be released with a different type IIS enzyme. Type IIS recognition sites and cleavage sites are indicated by boxes with arrows, the color indicating a switch for a different type IIS enzyme. Colored boxes in plasmids represent different selection marker genes

3.3 DNA Assembly of Multiple TUs into Higher Order Levels

Using the Level 1 plasmids constructed in Subheading 3.2, a higher order assembly of multiple TUs can be performed (Fig. 4). The Level 1 plasmids are compatible with the MoClo Level M/P cloning system developed by the group of Sylvestre Marillonnet [2]. Here a previously generated Level M/P plasmid set with respective Endlinkers (EL) is used, theoretically allowing infinite rounds of DNA assembly [7, 8, 12]. The protocol outlined here only describes the assembly of three Level 1 plasmids into multi-gene constructs. The procedure for the next round of cloning would be similar, except that some reagents would be changed, including the acceptor plasmid, Endlinker plasmid, type IIS enzyme, and selection conditions.

1. Extract plasmids with method of choice and perform quality control and determine the concentration. For the outlined procedure the Level 1 plasmids generated in Subheading 3.2,

the acceptor Level P plasmid pMA67 and the Endlinker plasmid pMA676 are used (*see* **Note 25**). Which results in a total of eight plasmids to purify.

2. Set the concentration of all purified plasmids to a final concentration of 50 fmol/μL.
3. Perform the following five Golden Gate reactions with the reaction mixture and conditions outlined below. Where *mTurquoise* is position 1, *gfp* is position 2 and *mCherry* is position 3 in the higher order assembly. "+" represents the respective fluorophore and "-" the neutral linker DNA.

   Fluorophore −/−/− pSL844, pSL846, pSL848 in pMA67, EL pMA676.

   Fluorophore −/+/+ pSL844, pSL847, pSL849 in pMA67, EL pMA676.

   Fluorophore +/−/+ pSL845, pSL846, pSL849 in pMA67, EL pMA676.

   Fluorophore +/+/− pSL845, pSL847, pSL848 in pMA67, EL pMA676.

   Fluorophore +/+/+ pSL845, pSL847, pSL849 in pMA67, EL pMA676.

Reagent	Concentration	Volume
Level P acceptor plasmid pMA67	50 fmol/μL	1 μL
EL plasmid pMA676	50 fmol/μL	1 μL
Each Level 1 TU position	50 fmol/μL	1 μL
T4 DNA Ligase Buffer	10 X	1 μL
T4 DNA Ligase	400 U/μL	1 μL
BsaI	20 U/μL	1 μL
ddH_2O		to 10 μL

Step	Settings	Cycles
Type IIS enzyme optimum (*see* **Note 22**)	37 °C for 3 min	25
Ligase optimum	16 °C for 4 min	
Final digestion step[a]	50 °C for 20 min	–
Heat inactivation	80 °C for 20 min	–
Hold	12 °C	–

[a]At this step the ligase is inactivated which ensures minimal background because all wrongly ligated plasmids are removed from the reaction mix

4. Transform 5 μL of the Golden Gate reaction mix to the *E. coli* strain of choice. In this procedure in house prepared RbCl chemical competent *E. coli* Top10 cells are used.
5. Select after transformation on the appropriate medium, in this case LB kanamycin at 37 °C overnight.
6. Validate single colonies with the method of choice for the correct assembled plasmid. In case of larger sizes plasmid extraction and test restriction digest analysis with the assembly release type IIS enzyme is recommended. Colony PCR becomes inefficient with larger assemblies.
7. Validated strains should be cryo-preserved with unique identifiers and all necessary data needs to be stored accordingly.
8. Assembly can in theory have unlimited number of rounds using the Level M/P plasmids in consecutive steps (Fig. 1c). However, in this protocol the higher order assembly of three Level 1 TUs is the final step and the assemblies are compared in Subheading 3.4.

3.4 Macroscopic Validation of Assembled Multi-Gene Constructs

Once DNA constructs are generated, they need to be analyzed to study their biology or application potential. These experiments are very diverse and need to be adapted to the specific question. In this protocol, several multi-gene constructs containing up to three highly expressed fluorescent proteins were generated. To test their expression and potential effect on bacterial growth, a spotting experiment was performed to compare the different strains (Fig. 5). This procedure was used for visualization purposes and could be replaced by other methods such as plate reader or flow

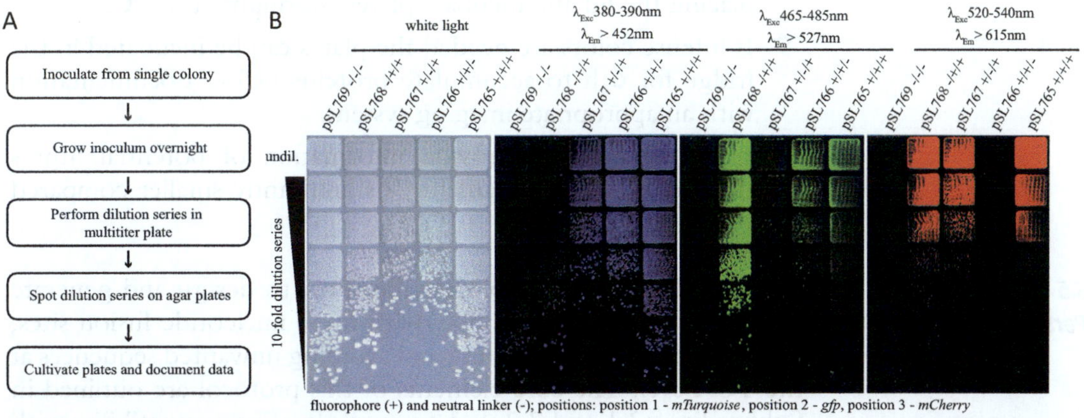

Fig. 5 Result of higher order assembly of multiple TUs. (**a**) Visualization of experimental workflow for growth assay in the form of a spotting assay. (**b**) Results of spotted cultures of strains with the assembled Level P multi-gene constructs. Shown are images of the same plate under white light and the respective fluorophore specific conditions. Fluorescence corresponds to the expected expression of fluorophores from the assembled multi-gene constructs

cytometry-based monitoring. However, this part is intended as an example and must be adapted to the assembled multi-gene construct of interest.

1. Streak the cryo-preserved strains on appropriate selective medium and incubate overnight at 37 °C.
2. Inoculate from a single colony for each strain into a glass tube with 3 mL of LB kanamycin. Cultivate shaking with 180 rpm at 37 °C overnight or until all strains reach the stationary phase.
3. Prepare single well plates with appropriate selective medium for spotting experiment (*see* **Note 26**).
4. Transfer 200 μL dense culture of the strains of interest into either column 1 or row A depending on if less or more than 8 strains are analyzed.
5. Add appropriate volume of culture dilution buffer to the remaining wells. The volume depends on the dilution series, for example, for tenfold dilutions dispense 180 μL, for fivefold dilutions dispense 160 μL.
6. Transfer the appropriate volume of the dense culture to the first dilution step using either an 8- or 12-channel pipette. Mix dilutions properly by pipetting up and down multiple times. Discard pipette tips and perform next dilution step by aspirating the just generated dilution with fresh sterile tips. Repeat for the appropriate number of dilution steps.
7. Spot the dilution series onto the single well plate using either a multi-channel pipette (*see* **Notes 27** and **28**) or if available a pin-pad based screening robot (Fig. 5) (*see* **Note 29**).
8. Let liquid of spots evaporate in a sterile environment before placing the lid and incubate plates overnight at 37 °C.
9. If intense colors are needed the plates can be incubated in the fridge for 6 h to accumulate proteins before documentation with an appropriate imaging system.
10. Colony sizes may provide information of potential fitness defects in case the diameter is significantly smaller compared to the control strain.

3.5 Summary & Perspectives

This step-by-step protocol describes how to design and generate Level 0 parts for TU assembly with three nucleotide fusion sites, allowing codon-based cloning and avoiding unwanted sequences at the TU Level. The core elements of this protocol are outlined in Subheadings 3.1 and 3.2, but an example of its compatibility with the Modular Cloning standard developed by Sylvestre Marillonnet is given in Subheading 3.3. A brief example of testing higher order DNA assemblies on solid media is given in Subheading 3.4. Fluorescent genes are used as an example, but the procedure could be adapted to any other type of multi-fragment assembly, such as the expression of heterologous pathways. In particular, the procedure outlines the onboarding of new Level 0 parts, which allows the creation of sequence libraries to test different combinations to identify the best performing DNA construct.

4 Notes

1. This protocol uses throughout an open-source procedure magnetic bead plasmid extraction and DNA purification procedure [14].
2. This primer is sufficient for sequencing smaller basic parts. If larger fragments are domesticated as basic parts, it may be necessary to use additional primer(s) to sequence the entire insert. For very large constructs, it may be more economical to use a whole plasmid sequencing service.
3. All enzymes used in this protocol were supplied by NEB. However, any other supplier of enzymes should be fine.
4. Any type of competent cells can be used. However, it is recommended to test their competence in advance to ensure high cloning efficiency.
5. This strain is relevant for the propagation of *ccdB* containing plasmids. However, any other *ccdB* resistant strain would be suitable. *ccdB* is a toxin that inhibits gyrase [15]. The strain must contain the *ccdA* antitoxin or the appropriate gyrase mutation(s).
6. Ensure that no type IIS recognition sites are formed in subsequent assembly steps and that the acceptor plasmids do not contain the type IIS recognition sites used, except where necessary for DNA assembly purposes.
7. The advantage of DNA synthesis is that the sequence is free in the design space. This allows, for example, to easily eliminate DNA motifs and/or to freely design the codon usage of a CDS. A good tool for automated, rational in silico DNA sequence optimization is DNA Chisel [16].
8. The nucleotides are specific for the plasmid set used here. If other Golden Gate systems are used, the fusion site sequences may differ. If custom fusion sites are used, it is recommended to use the appropriate tools to predict ideal fusion site combinations (e.g., the NEB tools driven by ligase fidelity data [11, 17, 18].
9. Ensure that the start and stop codon are not duplicated due to the ATG and TGA selected as the fusion site for CDS primers in this Golden Gate cloning system.

10. For the plasmid set used here, Golden Gate cloning in Level 0 uses BpiI. This may vary depending on the Golden Gate cloning plasmids used.
11. If gene synthesis is performed instead of fragment synthesis, vendors offer subcloning into custom plasmids. In this case, it is useful to deposit the Level 0 plasmid used with the vendor to directly obtain validated Level 0 parts. In this case, the protocol steps for Level 0 plasmid generation can be skipped and the user can proceed directly to **step 3.2**.
12. Due to the PNK heat inactivation step, the phosphorylated oligonucleotides are annealed after the PNK reaction by briefly boiling the mixture and slowly cooling the mixture to room temperature. A PCR cycler is used for this. The program heats the mixture to 95 °C for 5 min and then cools the samples to 12 °C at the minimum ramp speed of the PCR cycler (here 0.1 °C/sec). After several freeze-thaw cycles, it is recommended to repeat the annealing reaction.
13. This PCR amplification strategy and the attachment of the fusion site to the DNA fragments allow great flexibility with the fusion sites and allow modifications to be made as needed. For example, the use of more complex cloning schemes that require the addition of C- and/or N-terminal fusion tags to the gene of interest. When using alternative fusion sites, it is recommended to predict optimal fusion sites using appropriate computational tools, for example, the NEB tools which are driven by ligase fidelity data [11, 17, 18].
14. DpnI digests methylated GATC sequence motifs on the adenine base. GATC is methylated by the Dam methylase in standard *E. coli* strains. The motif randomly occurs every 256 bp. If an *E. coli* Δ*dam* strain is used, this reaction is not effective. However, this step is highly recommended to reduce potential background during subsequent cloning.
15. DpnI digestion can also be performed overnight and would reduce the amount of enzyme used to 0.1 μL per reaction.
16. To avoid this step, oligonucleotides and/or DNA fragments can be ordered already phosphorylated. Alternatively, oligonucleotides can be phosphorylated prior to PCR, similar to the steps described in **step 8**. However, it is recommended to check the reaction conditions and units of T4 polynucleotide kinase used.
17. If multiple fragments are phosphorylated, it is recommended to make a 2.5 x master mix containing water, T4 DNA Ligase Buffer and T4 PNK.
18. It is recommended to store the validated Level 0 parts with a defined molarity, so that they can be used directly in

subsequent Golden Gate assemblies without the need for frequent volume calculations to match the required mass.

19. If the sequence was amplified from a plasmid containing the same antibiotic resistance marker as the Level 0 plasmid, a careful performed DpnI digestion is recommended (*see* **step 5** in Subheading 3.1.2).
20. In contrast to Subheading 3.1.2, the fusion sites are already present in the Level 0 plasmid. The fusion sites of the plasmid and the fragment(s) must match to ensure an efficient reaction and to use the fragment(s) for transcription unit assembly.
21. In case of a single fragment Golden Gate cloning reaction, the fragment can be provided in approximately threefold access to the acceptor plasmid. However, for multiple fragments, equimolar amounts of all DNA fragments are recommended.
22. If alternative type IIS enzymes are used, it is imperative to check the compatibility of the enzyme with the protocol. For example, BsmBI has an optimum temperature of 55 °C. However, this temperature already inactivates the ligase. Therefore, the temperature would be increased to 42 °C to have the optimum for BsmBI activity but minimal inactivation of the ligase.
23. Neutral spacer sequences were generated using *SiteOut* [19]. The DNA was ordered as oligonucleotides, phosphorylated, annealed, and inserted into all PCR amplified Level 1 acceptor plasmids using the standard blunt-end cloning procedure.
24. Validation by sequencing is not necessary in most cases because the basic parts have not been amplified by PCR and have been sequence validated at the Level 0. However, if the construct causes high load or toxic products, it may mutate. An indication of this could be a low number of colonies obtained or very low efficiency during restriction pattern analysis of generated constructs. If the sequence cannot be altered in this case, it may be necessary to use an alternative host as a recipient for the Golden Gate assembly in order to propagate the DNA.
25. In the original MoClo system, Level 1 TUs are assembled in Level M. In this protocol, we use Level 1 plasmids that are compatible with Level P. However, Level 1 plasmids for SapI TU assembly are available for assembly in Level M.
26. Ensure that the plates are dry but not overdried for best quality results. The surface should be flat.
27. The volume to be dispensed depends on the medium and must be tested beforehand to avoid merging of individual spots.

28. To improve manual spotting coordination, it may be helpful to place a grid under the single well plate, such as the lid of a 96-well microplate or the inlet of a 96-tip tip box.
29. In this protocol, the Singer Instruments Rotor HDA+, a screening robot with reusable pin pads and a 7 × 7 spotting program, was used to reliably spot squares on agar plates.

Acknowledgments

This work was supported by the Max Planck Society within the framework of the MaxGENESYS project (DS) and the European Union (NextGenerationEU) via the European Regional Development Fund (ERDF) by the state Hesse within the project "biotechnological production of reactive peptides from waste streams as lead structures for drug development" (DS). We are grateful to all laboratory members for extensive discussions on Golden Gate cloning in particular Tania S. Köbel for her superior technical support. We thank Torsten Waldminghaus for providing plasmids. All material is available from the corresponding author upon request.

References

1. Engler C, Kandzia R, Marillonnet S (2008) A one pot, one step, precision cloning method with high throughput capability. PLoS One 3(11):e3647. https://doi.org/10.1371/journal.pone.0003647
2. Weber E, Engler C, Gruetzner R, Werner S, Marillonnet S (2011) A modular cloning system for standardized assembly of multigene constructs. PLoS One 6(2):e16765. https://doi.org/10.1371/journal.pone.0016765
3. Köbel TS, Melo Palhares R, Fromm C, Szymanski W, Angelidou G, Glatter T, Georg J, Berghoff BA, Schindler D (2022) An easy-to-use plasmid toolset for efficient generation and benchmarking of synthetic small RNAs in bacteria. ACS Synth Biol 11(9): 2989–3003. https://doi.org/10.1021/acssynbio.2c00164
4. Köbel T, Schindler D (2023) Automation and miniaturization of Golden Gate DNA assembly reactions using acoustic dispensers. arXiv. https://doi.org/10.48550/arXiv.2310.00325
5. Ramirez Rojas AA, Brinkmann CK, Kobel TS, Schindler D (2024) DuBA.flow—a low-cost, long-read amplicon sequencing workflow for the validation of synthetic DNA constructs. ACS Synth Biol. https://doi.org/10.1021/acssynbio.3c00522
6. Ramirez Rojas AA, Brinkmann CK, Schindler D (2024) Validation of Golden Gate assemblies using highly multiplexed Nanopore amplicon sequencing. arXiv. https://doi.org/10.48550/arXiv.2401.14191
7. Schindler D, Milbredt S, Sperlea T, Waldminghaus T (2016) Design and assembly of DNA sequence libraries for chromosomal insertion in bacteria based on a set of modified MoClo vectors. ACS Synth Biol 5(12):1362–1368. https://doi.org/10.1021/acssynbio.6b00089
8. Zumkeller C, Schindler D, Waldminghaus T (2018) Modular assembly of synthetic secondary chromosomes. Methods Mol Biol 1837: 71–94. https://doi.org/10.1007/978-1-4939-8675-0_5
9. Brück M, Berghoff BA, Schindler D (2023) In silico design, in vitro construction and in vivo application of synthetic small regulatory RNAs in bacteria. arXiv. https://doi.org/10.48550/arXiv.2304.14932
10. Taylor GM, Mordaka PM, Heap JT (2019) Start-stop assembly: a functionally scarless DNA assembly system optimized for metabolic engineering. Nucleic Acids Res 47(3):e17. https://doi.org/10.1093/nar/gky1182
11. Pryor JM, Potapov V, Kucera RB, Bilotti K, Cantor EJ, Lohman GJS (2020) Enabling

one-pot Golden Gate assemblies of unprecedented complexity using data-optimized assembly design. PLoS One 15(9):e0238592. https://doi.org/10.1371/journal.pone.0238592

12. Messerschmidt SJ, Schindler D, Zumkeller CM, Kemter FS, Schallopp N, Waldminghaus T (2016) Optimization and characterization of the synthetic secondary chromosome synVicII in Escherichia coli. Front Bioeng Biotechnol 4: 96. https://doi.org/10.3389/fbioe.2016.00096
13. Hanahan D (1983) Studies on transformation of Escherichia coli with plasmids. J Mol Biol 166(4):557–580. https://doi.org/10.1016/s0022-2836(83)80284-8
14. Oberacker P, Stepper P, Bond DM, Hohn S, Focken J, Meyer V, Schelle L, Sugrue VJ, Jeunen GJ, Moser T, Hore SR, von Meyenn F, Hipp K, Hore TA, Jurkowski TP (2019) Bio-On-Magnetic-Beads (BOMB): open platform for high-throughput nucleic acid extraction and manipulation. PLoS Biol 17(1): e3000107. https://doi.org/10.1371/journal.pbio.3000107
15. Bahassi EM, O'Dea MH, Allali N, Messens J, Gellert M, Couturier M (1999) Interactions of CcdB with DNA gyrase. Inactivation of Gyra, poisoning of the gyrase-DNA complex, and the antidote action of CcdA. J Biol Chem 274(16): 10936–10944. https://doi.org/10.1074/jbc.274.16.10936
16. Zulkower V, Rosser S (2020) DNA Chisel, a versatile sequence optimizer. Bioinformatics 36(16):4508–4509. https://doi.org/10.1093/bioinformatics/btaa558
17. Potapov V, Ong JL, Kucera RB, Langhorst BW, Bilotti K, Pryor JM, Cantor EJ, Canton B, Knight TF, Evans TC Jr, Lohman GJS (2018) Comprehensive profiling of four base overhang ligation fidelity by T4 DNA ligase and application to DNA assembly. ACS Synth Biol 7(11):2665–2674. https://doi.org/10.1021/acssynbio.8b00333
18. Potapov V, Ong JL, Langhorst BW, Bilotti K, Cahoon D, Canton B, Knight TF, Evans TC Jr, Lohman GJS (2018) A single-molecule sequencing assay for the comprehensive profiling of T4 DNA ligase fidelity and bias during DNA end-joining. Nucleic Acids Res 46(13):e79. https://doi.org/10.1093/nar/gky303
19. Estrada J, Ruiz-Herrero T, Scholes C, Wunderlich Z, DePace AH (2016) SiteOut: an nline tool to design binding site-free DNA sequences. PLoS One 11(3):e0151740. https://doi.org/10.1371/journal.pone.0151740

Chapter 8

Biofoundry-Assisted Golden Gate Cloning with AssemblyTron

John A. Bryant Jr and R. Clay Wright

Abstract

Golden Gate assembly is a requisite method in synthetic biology that facilitates critical conventions such as genetic part abstraction and rapid prototyping. However, compared to robotic implementation, manual Golden Gate implementation is cumbersome, error-prone, and inconsistent for complex assembly designs. AssemblyTron is an open-source python package that provides an affordable automation solution using open-source OpenTrons OT-2 lab robots. Automating Golden Gate assembly with AssemblyTron can reduce failure-rate, resource consumption, and training requirements for building complex DNA constructs, as well as indexed and combinatorial libraries. Here, we dissect a panel of upgrades to AssemblyTron's Golden Gate assembly capabilities, which include Golden Gate assembly into modular cloning part vectors, error-prone polymerase chain reaction (PCR) combinatorial mutant library assembly, and modular cloning indexed plasmid library assembly. These upgrades enable a broad pool of users with varying levels of experience to readily implement advanced Golden Gate applications using low-cost, open-source lab robotics.

Key words DNA assembly, Mutant library, Modular toolkit, Automation, Error-prone PCR, Directed evolution

1 Introduction

Golden Gate assembly is a principal enabling technology in synthetic biology, and automated implementation has transformed it into a high-throughput genetic engineering tool. Golden Gate assembly is a scarless DNA assembly strategy capable of assembling up to 40 fragments [1–4]. Plasmid libraries of standardized modular parts are often used to facilitate rapid Golden Gate assembly, circumventing polymerase chain reaction (PCR) amplification. Originally, the Modular Cloning system (MoClo) implemented this strategy and has inspired a wide array of plasmid toolkits for different applications [5–9], which have been reviewed by Casini et al. [10]. Modular cloning toolkits define a standard of unique and seamless overhangs for predefined part types, which are then

Daniel Schindler (ed.), *Golden Gate Cloning: Methods and Protocols*, Methods in Molecular Biology, vol. 2850,
https://doi.org/10.1007/978-1-0716-4220-7_8,

used to build libraries of genetic parts such as promoters, terminators, markers, and coding sequences that can be assembled in combinatorial fashion. In the Sky biofoundry, RoboMoClo combines lab automation with the MoClo system to create a biofoundry-assisted Golden Gate platform for metabolic pathway-level DNA assembly [11]. However, RoboMoClo does not provide openly available code, requires expensive robotics, and is intended as a service platform rather than a transferable tool.

Prior to AssemblyTron, resources for automating customized, in-lab Golden Gate workflows were scarce. Golden Gate assembly is often performed with commercial kits, where just one or two fragments are inserted into a predefined entry vector—a gross underutilization of the method. Customized Golden Gate greatly expands the flexibility and potential construct complexity; however, manual Golden Gate implementation requires careful design and extensive molecular biology background knowledge [3]. Fortunately, many automation tools exist for designing custom Golden Gate assemblies [12–15], but manually interpreting and implementing their output files is labor intensive and error-prone. Modular cloning toolkits, such as the MoClo system, can simplify Golden Gate workflows; however, PCR is unavoidable for integrating a gene of interest into a part plasmid. Moreover, the complex, multi-level nature of toolkits limits their accessibility for inexperienced researchers. We designed AssemblyTron to help resolve these issues with automation.

AssemblyTron is an open-source python package that, when paired with an OpenTrons OT-2 lab robot, can provide biofoundry capabilities with minimal startup investment [16]. AssemblyTron 1.0 uses j5 DNA design files as input and generates fully customized OpenTrons Application Program Interface (API) protocol scripts for assembling plasmids via homology mediated assembly or Golden Gate. We previously validated that AssemblyTron is adept at automating multi-part Golden Gate assembly and can simultaneously increase throughput, eliminate human errors, and reduce training demands [16]. AssemblyTron adds to a very sparse field of open-source build tools for automating Golden Gate assembly. Since our initial publication, we have expanded AssemblyTron's toolkit of capabilities.

Here, we will review AssemblyTron 2.0 upgrades for automating biofoundry-enabling Golden Gate protocols. First, we describe a modified version of AssemblyTron's standard Golden Gate assembly protocol, which accepts any type IIS restriction-enzyme-digestible part entry vector as the construct backbone. Next, we introduce a protocol for assembling mutant libraries with Golden Gate. In this strategy, AssemblyTron uses error-prone PCR with tunable error-rates to mutate user-specified DNA fragments that are assembled into plasmid libraries via Golden Gate. Finally, we introduce the modular cloning toolkit builder, which is the most recent and developmental installment of AssemblyTron. This feature provides an interactive graphical user interface (GUI), inspired by Fig. 2 of the Yeast MoClo Toolkit publication [7]. The current toolkit builder offered by AssemblyTron is for the auxin modular cloning toolkit, which is a custom, in-house toolkit in the Wright Lab (*see* **Note 1**). However, our automation strategy for toolkit building can be adapted for any toolkit system as a fork or pull-request of the AssemblyTron github repository https://github.com/PlantSynBioLab/AssemblyTron.

2 Materials

2.1 Hardware

1. OpenTrons OT-2 liquid handling robot, equipped with temperature module and thermocycler module (*see* **Note 2**).
2. Any gradient thermocycler.
3. Any Microvolume Spectrophotometer.
4. Agarose DNA gel electrophoresis setup.
5. Any imaging system for visualizing gel electrophoresis DNA bands.

2.2 Software

1. AssemblyTron (https://github.com/PlantSynBioLab/AssemblyTron) [16].
2. Python (https://www.python.org/downloads/).
3. J5 and DIVA (https://j5.jbei.org/), for plasmid design [12].
4. Anaconda (https://www.anaconda.com/download), for managing AssemblyTron software dependencies (Pandas, Shutil, Numpy, and Tkinter).
5. OpenTrons run app (https://OpenTrons.com/ot-app/).

2.3 PCR

1. High-fidelity DNA polymerase for amplification of backbone fragments.
2. Taq polymerase supplied with 10x magnesium-free buffer and 25 mM $MgSO_4$, for error-prone fragment amplification.
3. Custom primers can be ordered from any commercial vendor.
4. Plasmid MiniPrep kit for isolating template plasmid and assembled plasmid from Escherichia coli (*E. coli*).
5. DNA gel extraction kit for purifying PCR products prior to assembly (*see* **Note 3**).
6. DNA clean and concentrate kit for purifying PCR products prior to assembly (*see* **Note 3**).
7. DNA ladder: 1 kb DNA ladder for gel electrophoresis.

8. 50x Tris-acetate-EDTA (TAE) buffer: 2 M Tris base, 1 M acetic acid, and 20 mM Ethylenediamine tetraacetic acid (EDTA).
9. Nucleic acid gel stain.
10. For preparing gels for electrophoresis: 0.5% agarose in 1x TAE is melted in a microwave oven, and 3 μL of nucleic acid gel stain is added per 50 mL of agarose.
11. Running buffer for agarose gels is 1xTAE.

2.4 Cloning

1. Genbank files for template plasmids and the plasmid design. These should be referenced when linearizing template plasmids, during downstream plasmid use, etc.
2. Physical template plasmids for PCR amplification or restriction enzyme digestion.
3. Restriction endonuclease BsaI (10 U/μL).
4. T4 DNA ligase (2000 U/μL).
5. BSA, molecular biology grade or recombinant albumin, molecular biology grade, diluted to 10 mg/mL.
6. Lysogeny broth (LB) medium: 1% bacto-tryptone, 0.5% yeast extract, 1% NaCl in deionized water, adjusted to pH 7.0 with 5 N NaOH. For plates, 1.5% (w/v) bactoagar was added.
7. Antibiotics: ampicillin and kanamycin. Filter-sterilized stocks of 100 mg/mL ampicillin and 50 mg/mL kanamycin (stored in aliquots at −20 °C) are diluted 1:1000 (final concentration: 100 μg/mL ampicillin and 50 μg/mL kanamycin) in autoclaved and cooled liquid medium.
8. High efficiency competent *E. coli* cells (top10 chemically competent *E. coli* for standard plasmid assembly and NEB10B electrocompetent *E. coli* for library assembly).

2.5 Sequencing and Screening

1. Plasmid assembly can be assessed by chromoprotein expression, Sanger sequencing aligned with A Plasmid Editor [17], and whole plasmid sequencing.
2. Custom sequencing primers can be design for any construct using A Plasmid Editor [17].

3 Methods

AssemblyTron was developed and debugged using Windows OS, so we recommend implementing it with Windows OS (*see* **Note 4**). The most up to date version and documentation of AssemblyTron are available at https://github.com/PlantSynBioLab/AssemblyTron. Prior to installing AssemblyTron, users need the latest versions of Python and R installed on their OT-2 connected computers. The package can also be downloaded from pypi. We recommend using Anaconda to easily install and manage AssemblyTron dependencies, which are listed in Subheading 2.2. The OpenTrons run app should also be installed for uploading AssemblyTron scripts to the OT-2.

3.1 Plasmid Design and Preliminary Setup

A plasmid design must be prepared with j5 and DIVA prior to building it with AssemblyTron. For entry vector backbone parts, the forced assembly strategy should be set to "digest." Eugene rules should be added to specify which parts belong together, unless a complete combinatorial design is desired. When running the j5 algorithm, the design strategy should be set as "Combinatorial Golden Gate" (*see* **Note 5**). After running j5, download results, unzip the compressed file, and store the design folder on your local hard drive. Primers can be found in masteroligolist.csv or in the pmas00001_combinatorial.csv design file and must be ordered prior to moving forward. To ensure successful PCR amplification, digest any circular template plasmids with a single-cutter restriction endonuclease prior to PCR amplification (*see* **Note 6**). Optionally, entry vectors can be digested with type IIS restriction endonuclease and gel purified to reduce template vector carry-through. This step can be skipped if the entry vector contains a toxicity gene (e.g., *ccdB*) or a fluorescent marker that distinguishes it from positive transformants.

3.2 Entry Vector Golden Gate Assembly

AssemblyTron enables flexible Golden Gate assembly into any entry vector, which is a critical capability for assembling level 0 part plasmids for modular toolkits as well as reducing dependency on commercial Golden Gate kits.

1. Navigate to the ~/AssemblyTron/src/AssemblyTron directory on your local drive. Go to the Golden_Gate/ folder, and select the Setup_digests_gradient.py script (*see* **Note 7**).
2. Use the dialog box to navigate to your j5 design folder, which was previously saved to the local hard drive (*see* Subheading 3.1). Select confirm, and AssemblyTron automatically parses the j5 design file by importing a parsing module (*see* **Note 8**).
3. Screen the reagent_setup.txt file, which will be used to stage primers, templates, reagents, labware, and tubes on the OT-2 deck. Ensure that all templates and primers are assigned to a deck position. Close it until needed.
4. A parameter selection dialog box with default values for PCR and assembly is displayed. Most defaults do not need to be modified; however, the following parameters will require consideration or adjustment:
   (a) Check the Gradient PCR or touchdown PCR option. The default is touchdown PCR, but this may need to be

changed depending on the desired DNA yield (*see* **Note 9**) or how prone the PCR is to failure (e.g., if the range of annealing temperatures for fragments is large, it is safer to use the gradient amplification option.)

(b) Check the "PaqCI?" option. This specifies whether PaqCI or a different type IIS restriction enzyme is being used for Golden Gate assembly. PaqCI requires an activator additive, which alters reagent volumes in the final Golden Gate assembly mixes.

(c) There are two blank parameter entry windows available so that users can add additional parameters to protocols without modifying the dialog box generator scripts. These do not have to be entered.

(d) Entry windows for template concentrations (ng/μL) must be entered to proceed.

5. Select confirm once parameters are updated.
6. A dialog box is displayed for entering concentrations of the entry vector backbones used in this assembly. Enter concentrations in ng/μL and select confirm (*see* **Note 10**).
7. In the next dialog box, select which parts of the protocol to run. Users can choose from Primer and template dilution, PCR preparation, DpnI digest (template removal), Golden Gate assembly setup, and Golden Gate assembly run. This option makes it easy to skip certain parts of the protocol without modifying python code. For example, primers may already be diluted, so that part of the protocol would not need to be selected.
8. Select confirm. The gradient optimization algorithm runs automatically, unless the touchdown PCR option was specified.
9. Screen the reaction_setup.txt file, which specifies where to place tubes for PCRs and final assemblies. Ensure that each PCR and assembly is assigned to a tube and close the file until needed.
10. The setup script imports the primer dilution and Golden Gate protocol writer modules (dilution_24_digests_writer.py and GoldenGate_digests_separatepcrruns_gradient_writer.py) and executes them automatically. The final customized protocol scripts then open automatically. Close them to allow the setup script to finish running.
11. Navigate to the new directory, labeled with the current date, that was just generated.
12. Protocol scripts are ready to be uploaded directly to the OT-2 (*see* **Note 11**).

13. *See* Subheading 3.5 for instructions on uploading protocols to the OT-2.
14. Amplify fragments, run Golden Gate assembly, clean and concentrate assembled plasmids, and transform into high efficiency chemically competent *E. coli*.

3.3 Mutant Library Assembly

Mutant library screening is a common strategy for protein engineering and directed evolution (reviewed by Wang et al.) [18]. AssemblyTron provides protocols for facilitating mutant library assembly with error-prone PCR and Golden Gate assembly. Targeted mutagenesis of gene regions followed by assembly with unaltered backbone fragments provides high-quality plasmid libraries for functional screening [19, 20]. Using AssemblyTron to automate library assembly ensures consistent library quality, which saves time and resources. AssemblyTron allows users to tune error rate (Fig. 1a), which accommodates various target fragment sizes and libraries of varying degeneracy. We assembled libraries of pink chromoprotein expression plasmids with different error rates and screened for functional variation by colony color (Fig. 1b, c). Our data indicates that increasing error-rate settings with AssemblyTron correlate with increasing loss-of-function colorless colonies and diminishing pink colonies.

Here, we provide a detailed protocol for using AssemblyTron's library assembly feature. Our workflow consists of three stages. In stage one, an initial PCR is performed to amplify backbone fragments and a template amplicon for the error-prone fragment with high-fidelity polymerase (*see* **Note 12**). In stage two, error-prone PCR is used to mutate the target fragment with a user-specified error rate (Fig. 1a). In stage three, Golden Gate is used to assemble the mutant fragment pool with backbone fragments.

Golden Gate Primers with 5′ Golden Gate overhangs (*see* **Note 13**) should be ordered prior to implementing this protocol. Additionally, shorter primers lacking 5′ Golden Gate overhangs must be ordered to generate the initial template amplicon in stage one. This template will be used in stage two error-prone PCR. Primers lacking Golden Gate overhangs prevent the initial template amplicon from being assembled into the library in stage three, which would decrease the overall error-rate of the library.

Stage 1

1. Navigate to the ~/AssemblyTron/src/AssemblyTron/ directory and go to the Error_prone_PCR_Golden_Gate/ folder.
2. Select a stage one setup script. 1_Setup_STD_GoldenGate_digests_PCR.py is for libraries that will be assembled into entry vector backbones, and 1_Setup_STD_GoldenGate_PCR.py is for libraries that will be assembled strictly from PCR amplicons (*see* **Note 14**). Click on the appropriate script.

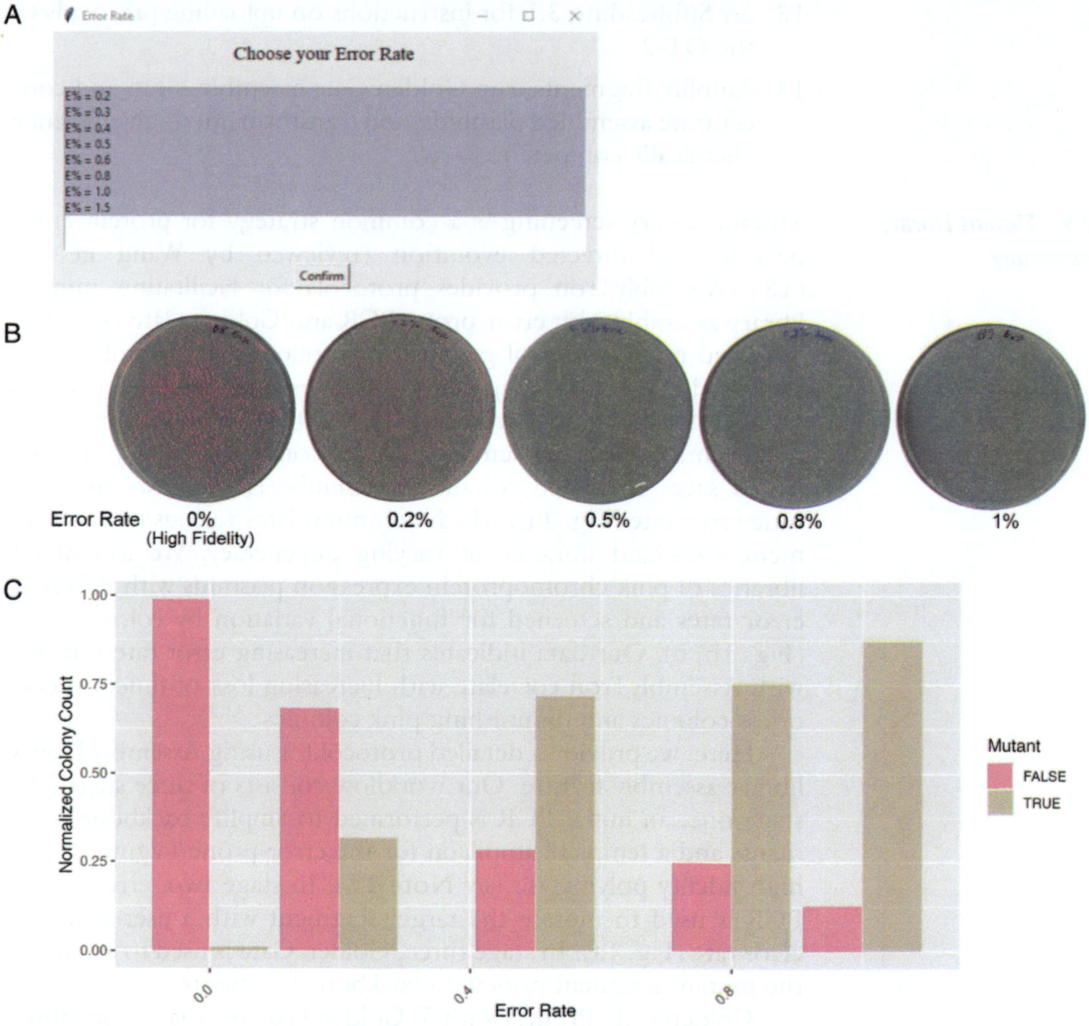

Fig. 1 Tuning mutagenesis with AssemblyTron error prone PCR. (**a**) A dialog box provided by AssemblyTron is used to tune the error rate. (**b**) Qualitative mutant accumulation increases as higher mutation rates are specified with AssemblyTron. (**c**) Quantitative mutant accumulation increases as higher mutation rates are specified based on a plot of normalized colony count for 10X dilutions of plates in B

3. Use the dialog box to navigate to your j5 design folder, which was previously saved to the local hard drive (*see* Subheading 3.1). Select confirm, and AssemblyTron automatically parses the j5 design file by importing a parsing module (*see* **Note 8**).
4. Screen the reagent_setup.txt file, which will be used to stage primers, templates, reagents, labware, and tubes on the OT-2 deck. Ensure that all templates and primers are assigned to a deck position. Close it until needed (*see* **Note 15**).
5. A parameter selection dialog box with default values for PCR and assembly is displayed. Most defaults do not need to be

modified; however, the following parameters will require consideration or adjustment:

(a) Check the Gradient PCR or touchdown PCR option. Stage one PCR is set to the gradient option by default. This is to ensure clean bands with high DNA concentration for high library quality and successful yields after gel extraction (*see* **Note 16**).

(b) Template concentrations (ng/μL) must be entered to proceed.

6. Select confirm once parameters are updated.
7. In the next dialog box, select which parts of the protocol to run. Users can choose from Dilution and PCR mix in this stage, since assembly will not be performed until stage three (*see* **Note 17**).
8. Screen the reaction_setup.txt file, which specifies where to place PCR tubes on the on-deck thermocycler. Ensure that each PCR is assigned to a tube and close the file until needed.
9. The setup script imports the primer dilution and Golden Gate protocol writer modules (dilution_24_digests_writer.py and GoldenGate_digests_separatepcrruns_gradient_writer.py) and executes them automatically. The final customized protocol scripts open automatically. Close them to allow the setup script to finish running.
10. Navigate to the new dated folder (e.g., 202310011038_PCR/).
11. The final customized protocol scripts are ready to be uploaded directly to the OT-2 (*see* **Note 11**). *See* Subheading 3.5 for instructions on uploading protocols.
12. After initial dilution and PCR protocols run, isolate fragments with gel electrophoresis, extract fragments, and measure DNA concentration before proceeding to stage two (*see* **Note 18**).

Stage 2

13. Navigate to the ~/AssemblyTron/src/AssemblyTron/Error_-prone_PCR_Golden_Gate/ directory and go to the dated PCR folder generated in stage one (e.g., 202301001 1038_PCR/).
14. Click on the stage two setup file (e.g., 2_2023010011038_ Setup_Error_prone_PCR.py).
15. Use the dialog box to navigate to your j5 design folder, which was previously saved to the local hard drive (*see* Subheading 3.1). Select confirm, and AssemblyTron automatically parses the j5 design file by importing a parsing module (*see* **Note 8**).

16. Use the next dialog box to select which fragments in the design are targets for mutagenesis by placing a 1 in the entry box. This specifies which fragments will be amplified with error-prone PCR.
17. Screen the reagent_setup.txt file, which will be used to stage primers, templates, reagents, labware, and tubes on the OT-2 deck. Ensure that all templates and primers are assigned to a deck position. Close it until needed.
18. A parameter selection dialog box with default values for PCR is displayed. Template concentrations are the only entries necessary, and defaults do not need to be modified (*see* **Note 19**).
19. In the next dialog box, select which parts of the protocol to run. Users can choose from Dilution and PCR mix in this stage, since assembly will not be performed until stage three (*see* **Note 17**).
20. Choose an error rate in the next dialog box (Fig. 1a). Depending on the user-specified error rate, AssemblyTron will alter ratios of manganese to magnesium and dNTPs to tune error rate and minimize transition/transversion bias (Fig. 1b) (*see* **Note 20**).
21. The setup script imports primer dilution and error-prone PCR protocol writer modules (dilution_Error_prone_PCR_writer.py and Error_prone_PCR_writer.py) and executes them automatically. The final customized protocol scripts open automatically. Close them to allow the setup script to finish running.
22. Navigate to the new stage two folder that was just created (e.g., 2_202310021038_EPPCR/).
23. The final customized protocol scripts that are ready to be uploaded directly to the OT-2. *See* Subheading 3.5 for details on uploading protocols to the OT-2 (*see* **Note 11**).
24. Run error-prone PCR on OT-2, clean and concentrate, and measure concentration of PCR products before proceeding to stage three.

Stage 3

25. Navigate to the ~/AssemblyTron/src/AssemblyTron/Error_-prone_PCR_Golden_Gate/ directory and go to the dated PCR folder generated in stage one (e.g., 2023010011038_PCR/).
26. Click on the stage three setup file (e.g., 3_2023010011038_Assembly.py).
27. Use the dialog box to navigate to your j5 design folder, which was previously saved to the local hard drive (*see* Subheading 3.1). Select confirm, and AssemblyTron automatically parses the j5 design file by importing a parsing module (*see* **Note 8**).

28. Screen the reagent_setup.txt file, which will be used to stage reagents and labware on the OT-2 deck. Close it until needed.
29. A parameter selection dialog box with default values for assembly is displayed. Most defaults do not need to be modified; however, if using PaqCI for assembly, ensure the "PaqCI?" option is checked. This specifies whether PaqCI or a different type IIS restriction enzyme is being used for Golden Gate assembly. PaqCI requires an activator additive, which alters reagent volumes in the final Golden Gate assembly mixes. Select confirm to confirm the assembly options.
30. In the next dialog box, select which parts of the protocol to run. Users can choose from Golden Gate setup and Golden Gate run in stage three, since PCR steps were performed in stages one and two.
31. Use the next dialog box for uploading concentrations of each assembly fragment in ng/μL. Each entry box is labeled by well number to avoid confusion. A new window will appear for each assembly (e.g., if there are four assemblies, four concentration input windows will appear consecutively).
32. Screen the reaction_setup.txt file, which specifies where to place fragment PCR tubes and final assembly PCR tubes on the on-deck thermocycler. Close the file until needed for setting up the OT-2 deck.
33. The setup script imports the Golden Gate protocol writer module (GoldenGate_digests_separatepcrruns_gradient_writer.py) and executes it automatically. The final customized protocol script opens automatically. Close it to allow the setup script to finish running.
34. Navigate to the new stage three folder that was just created (e.g., 3_202310021038_Assembly/).
35. The final customized protocol script is ready to be uploaded directly to the OT-2 (*see* **Note 11**). *See* Subheading 3.5 for details on uploading protocols to the OT-2.
36. Run the library assembly on the OT-2, clean and concentrate, and transform into high efficiency competent *E. coli* cells.

3.4 Modular Cloning Toolkit Builder

Automated modular cloning indexed plasmid library assembly has proved to be a successful platform strategy for biofoundries [11]. Combinatorial part assembly and shuffling will inevitably remain a useful cloning strategy even as DNA synthesis prices become trivial, so we have begun offering it as an AssemblyTron feature. Here we introduce a preliminary, open-source protocol for automating construct assembly from an auxin modular toolkit we have built in-house, based on part standards of the Yeast Modular Cloning Toolkit [7]. This toolkit-based build strategy will serve as a

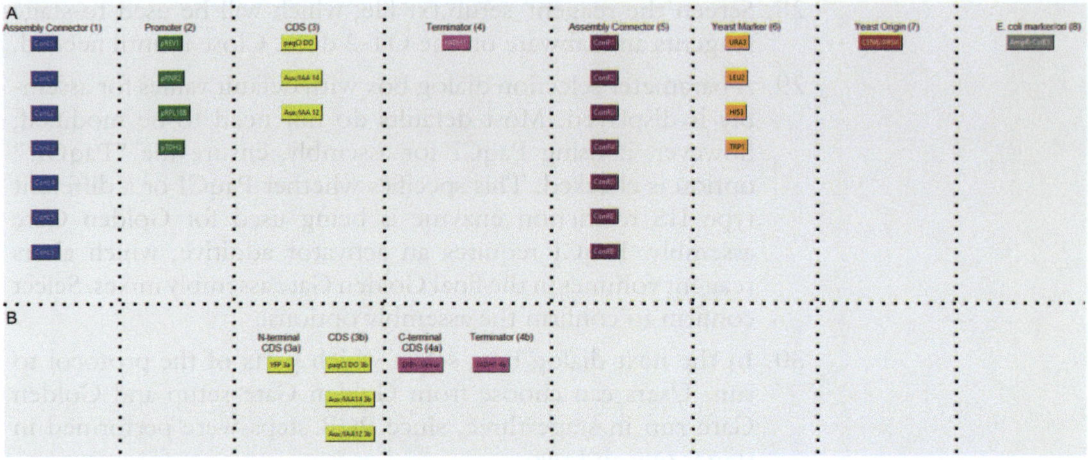

Fig. 2 AssemblyTron graphical user interface for modular part selection. (**a**) Any combination of eight color-coded part types can be specified for assembly. (**b**) A pair of type 3a and 3b parts can be selected in place of a standard type 3 part for N-terminal protein fusions. Similarly, a type 4a and 4b part can be selected in place of a standard type 4 part for a C-terminal protein fusion

model for expanding AssemblyTron platform to other modular cloning toolkits in the future.

1. Navigate to the ~/AssemblyTron/src/AssemblyTron/ directory and go to the MoClo_builder/ folder.
2. Select the Setup_MoClo.py script by clicking on it.
3. Use the dialog box to specify how many construct variants you are assembling from the toolkit.
4. A graphical user interface for selecting modular parts appears (Fig. 2). Choose the desired combination of parts. One part from each column must be selected for a total of eight (or more, *see* **Note 21**) parts (Fig. 2a, b).
5. Select confirm after selecting parts for the first construct. New selection windows will be consecutively displayed for specifying each additional construct.
6. A parameter selection dialog box with default values for assembly is displayed. Most defaults do not need to be modified. Select confirm.
7. Navigate to the new dated folder that was just created (e.g., 202310021254_MoClo/).
8. Run the MoClo_writer.py protocol writer script by clicking on it. This will generate the final customized protocol script that is ready to be uploaded directly to the OT-2.
9. *See* Subheading 3.5 for details on uploading protocols to the OT-2.

10. Run construct assembly on the OT-2, clean and concentrate, and transform into high efficiency chemically competent *E. coli* cells.

3.5 Uploading and Running Protocols on the OT-2

1. Open the OpenTrons run app, turn on the OT-2 and modules, and connect the app to the OT-2.
2. Navigate to the protocols tab on the left side of the app, select import, and upload protocol scripts in the correct order.
3. Open reagent_setup.txt and reaction_setup.txt. Use these files to guide physical setup of the OpenTrons deck (*see* **Notes 22** and **23**).
4. Click start run. Facilitate protocols by following any directions that are displayed on the OpenTrons run app during protocol pauses.

4 Notes

1. The auxin modular cloning toolkit should be published by the end of 2024.
2. Minor modifications could be made to the AssemblyTron source code to remove dependency on the modules and avoid purchasing them.
3. The protocol specifies when to use gel extraction or DNA clean and concentrate kit.
4. While running AssemblyTron on Mac OS may be possible, it will likely require additional debugging and programming knowledge.
5. Update "Golden Gate Recognition sequence" and "Golden Gate Term Extra Seq" fields within the "j5 parameters" if a different type IIS restriction enzyme besides the default BsaI is being used.
6. Digesting circular template plasmids with a single-cutter restriction endonuclease prior to PCR amplification is optional, but skipping this step often leads to PCR failure.
7. If the design does not use an entry vector, select an alternative setup script. For standard Golden Gate with 24 primers and templates, select Setup_nodigests_seppcr_gradient_24.py. For scaled standard Golden Gate with 96 primers and templates, select Setup_nodigests_seppcr_gradient_96.py. The workflow for each of these is nearly identical to Entry Vector Golden Gate.
8. The j5 parsing modules are located at ~/AssemblyTron/src/AssemblyTron/. Either j5_to_csvs.py or j5_to_csvs+digests.py is used depending on the assembly design.

9. Gradient amplification typically yields cleaner bands and higher DNA yields, while touchdown amplification has a higher likelihood of producing nonspecific bands and lower DNA yields.
10. In standard Golden Gate assembly, this dialog box does not appear.
11. A batch transfer script is no longer necessary for using AssemblyTron, thanks to the protocol writer modules.
12. We recommend gel purifying this template amplicon to reduce template DNA carry-through to the final library. This increases library quality.
13. We typically use j5/DIVA, https://public-diva.jbei.org/, to design these primers [12].
14. We recommend assembling libraries into entry vectors, since this typically increases assembly efficiency by reducing the number of fragments in the assembly. This also prevents unnecessary backbone PCR amplification.
15. Use the shortened 3′ only primers lacking Golden Gate overhangs for the initial template amplicon to be used in stage 2.
16. Touchdown PCR isn't a good option for error-prone PCR stage 1 since it is more likely to amplify off-target fragments and sometimes doesn't provide the highest possible concentration of amplified product.
17. There is no need for methylated template digestion with DpnI in the error-prone PCR workflow since all fragments are either gel extracted or amplified from a PCR amplicon with no Golden Gate overhangs.
18. Gel purification reduces template carry-through and ensures a high-quality library at the conclusion of the workflow.
19. Remember to use the amplicon from stage 1 as template for error-prone PCRs in stage 2.
20. PCR tubes for error-prone amplification should be placed in the same location on the thermocycler as their initial template amplicons from stage 1.
21. A combination of a type 3a and a type 3b part may be selected in place of a standard type 3 part to make an N-terminal protein fusion. Similarly, a type 4a and a 4b part combination may be selected in place of a standard type 4 part to make a C-terminal protein fusion.
22. The OpenTrons run app provides a deck map under labware setup>map view.
23. Ensure that shortened 3′ only primers lacking Golden Gate overhangs are used to generate the initial template amplicons in error-prone PCR stage one. Backbone fragments should be amplified from standard Golden Gate overhang primers in stage one.

Acknowledgments

We thank Cameron Longmire, Mason Kellinger, Sriya Sridhar, Ryan Miller, and Samuel Janousek for their contributions to lab automation projects in the Wright lab.

References

1. Engler C, Kandzia R, Marillonnet S (2008) A one pot, one step, precision cloning method with high throughput capability. PLoS One 3: e3647. https://doi.org/10.1371/journal.pone.0003647
2. Engler C, Gruetzner R, Kandzia R, Marillonnet S (2009) Golden gate shuffling: a one-pot DNA shuffling method based on type iis restriction enzymes. PLoS One 4:e5553. https://doi.org/10.1371/journal.pone.0005553
3. Engler C, Marillonnet S (2014) Golden gate cloning. In: Valla S, Lale R (eds) DNA cloning and assembly methods. Humana Press, Totowa, NJ, pp 119–131
4. Pryor JM, Potapov V, Bilotti K et al (2022) Rapid 40 kb genome construction from 52 parts through data-optimized assembly design. ACS Synth Biol 11:2036–2042. https://doi.org/10.1021/acssynbio.1c00525
5. Engler C, Youles M, Gruetzner R et al (2014) A golden gate modular cloning toolbox for plants. ACS Synth Biol 3:839–843. https://doi.org/10.1021/sb4001504
6. Iverson SV, Haddock TL, Beal J, Densmore DM (2016) CIDAR MoClo: improved MoClo assembly standard and new E. coli part library enable rapid combinatorial design for synthetic and traditional biology. ACS Synth Biol 5:99–103. https://doi.org/10.1021/acssynbio.5b00124
7. Lee ME, DeLoache WC, Cervantes B, Dueber JE (2015) A highly characterized yeast toolkit for modular, multipart assembly. ACS Synth Biol 4:975–986. https://doi.org/10.1021/sb500366v
8. Moore SJ, Lai H-E, Kelwick RJR et al (2016) EcoFlex: A multifunctional MoClo Kit for *E. coli* synthetic biology. ACS Synth Biol 5: 1059–1069. https://doi.org/10.1021/acssynbio.6b00031
9. Weber E, Engler C, Gruetzner R et al (2011) A modular cloning system for standardized assembly of multigene constructs. PLoS One 6:e16765. https://doi.org/10.1371/journal.pone.0016765
10. Casini A, Storch M, Baldwin GS, Ellis T (2015) Bricks and blueprints: methods and standards for DNA assembly. Nat Rev Mol Cell Biol 16: 568–576. https://doi.org/10.1038/nrm4014
11. Kang DH, Ko SC, Heo YB et al (2022) Robo-MoClo: a robotics-assisted modular cloning framework for multiple gene assembly in bio-foundry. ACS Synth Biol 11:1336–1348. https://doi.org/10.1021/acssynbio.1c00628
12. Hillson NJ, Rosengarten RD, Keasling JD (2012) j5 DNA assembly design automation software. ACS Synthetic Biology 1:14–21. https://doi.org/10.1021/sb2000116
13. Nielsen AAK, Der BS, Shin J et al (2016) Genetic circuit design automation. Science 352:aac7341. https://doi.org/10.1126/science.aac7341
14. Pryor JM, Potapov V, Kucera RB et al (2020) Enabling one-pot golden gate assemblies of unprecedented complexity using data-optimized assembly design. PLoS One 15:e0238592. https://doi.org/10.1371/journal.pone.0238592
15. Cloud-based Platform for Biotech R&D | Benchling. https://www.benchling.com/. Accessed 6 Nov 2023
16. Bryant JA Jr, Kellinger M, Longmire C et al (2023) AssemblyTron: flexible automation of DNA assembly with OpenTrons OT-2 lab robots. Synthetic Biology 8:ysac032. https://doi.org/10.1093/synbio/ysac032
17. Davis MW, Jorgensen EM (2022) ApE, a plasmid editor: a freely available DNA manipulation and visualization program. frontiers. Bioinformatics 2
18. Wang Y, Xue P, Cao M et al (2021) Directed evolution: methodologies and applications. Chem Rev 121:12384–12444. https://doi.org/10.1021/acs.chemrev.1c00260
19. Bosley AD, Ostermeier M (2005) Mathematical expressions useful in the construction, description and evaluation of protein libraries. Biomol Eng 22:57–61. https://doi.org/10.1016/j.bioeng.2004.11.002
20. Fromant M, Blanquet S, Plateau P (1995) Direct random mutagenesis of gene-sized DNA fragments using polymerase chain reaction. Anal Biochem 224:347–353. https://doi.org/10.1006/abio.1995.1050

Chapter 9

Automation and Miniaturization of Golden Gate DNA Assembly Reactions Using Acoustic Dispensers

Tania S. Köbel and Daniel Schindler

Abstract

Golden Gate cloning has become one of the most popular DNA assembly techniques. Its modular and hierarchical structure allows the construction of complex DNA fragments. Over time, Golden Gate cloning allows for the creation of a repository of reusable parts, reducing the cost of frequent sequence validation. However, as the number of reactions and fragments increases, so does the cost of consumables and the potential for human error. Typically, Golden Gate reactions are performed in volumes of 10–25 μL. Recent technological advances have led to the development of liquid handling robots that use sound to transfer liquids in the nL range from a source plate to a target plate. These acoustic dispensers have become particularly popular in the field of synthetic biology. The use of this technology allows miniaturization and parallelization of molecular reactions in a tip-free manner, making it sustainable by reducing plastic waste and reagent usage. Here, we provide a step-by-step protocol for performing and parallelizing Golden Gate cloning reactions in 1 μL total volume.

Key words Golden Gate cloning, DNA assembly, Acoustic liquid handling, In vitro, Synthetic biology, nL reactions, Sustainability

1 Introduction

Golden Gate cloning has become one of the most popular DNA assembly strategies. Initially developed as a one-step, one-pot reaction, it quickly demonstrated its modular and high-throughput capabilities [1, 2]. All Golden Gate technologies are based on the use of type IIS enzymes, which recognize a specific, directed motif and cleave any sequence at a defined distance. This principle allowed the development of Golden Gate cloning and its subsequent hierarchical modularization in the form of the Modular Cloning System (MoClo). Since its initial development, several organism-specific Golden Gate toolboxes have been developed for rapid and reliable engineering of the chosen host [3–6]. In addition, Golden Gate cloning has been used to standardize and/or modularize cloning steps for technologies, such as the rapid and

Daniel Schindler (ed.), *Golden Gate Cloning: Methods and Protocols*, Methods in Molecular Biology, vol. 2850,
https://doi.org/10.1007/978-1-0716-4220-7_9,

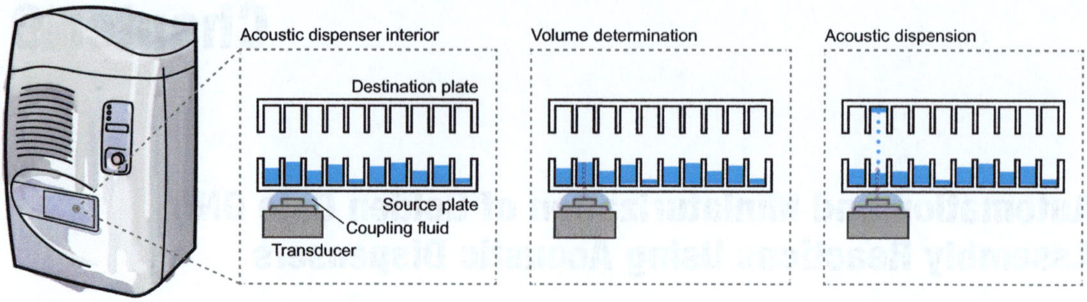

Fig. 1 Principle and application of acoustic dispensers for molecular reactions. Acoustic dispensers are systems controlled by an external computer. Within the acoustic dispenser there are three core elements: the transducer, the source plate, and the destination plate. The transducer is the sound source coupled to the source plate by the coupling fluid. The source plate contains the liquids to be dispensed into the target plate. The emitted sound waves are used to determine the volume of each individual well to confirm that an adequate volume is available at the start of acoustic dispensing and to confirm volume reduction after dispensing. For acoustic dispensing, the sound energy is adjusted to eject machine-defined nL-sized droplets (e.g., Echo650T = 2.5 nL and Echo525 = 25 nL) from the source plate to the target plate

modular construction of CRISPR/Cas gRNA arrays [7], genetic engineering toolboxes [8–10], post-transcriptional control toolkits [11, 12], and synthetic chromosome construction [13, 14].

A common feature of standard Golden Gate protocols is that the reactions are typically performed in 10–25 μL volumes. Modern scientific approaches often require an increased amount of DNA synthesis and assembly for the construction of large metabolic pathways or whole synthetic chromosomes [15, 16]. In addition to the increasing number of DNA assemblies, the complexity of Golden Gate reactions is also increasing, with >50 DNA parts being successfully assembled in a single step [17]. The standardization achieved by MoClo is further enabling the increased implementation of laboratory automation and is driving the establishment of biofoundries, dedicated sites to bundle technology for the automation of biological workflows. Acoustic dispensers have become very valuable in this context. Acoustic dispensers use sound to transfer liquid from a source plate to a target plate in fixed nanoliter-sized droplets (Fig. 1) [18]. This technology has recently been established for DNA assembly at the nL scale [19]. We use this technology on a daily basis for various applications, such as the Golden Gate-based construction of synthetic small regulatory RNAs (sRNAs) for post-transcriptional control in bacteria [11].

Here, we provide a detailed step-by-step protocol for the use of acoustic dispensers for Golden Gate assemblies, their subsequent high-throughput transformation, and hit identification for subsequent validation and application. As a proof of concept, we have assembled transcription units for the expression of fluorescent proteins with unknown promoter sequences to identify promoters of varying strength for future applications. The process can be completed within 3–4 days, allowing for rapid and sustainable DNA assembly.

2 Materials

The following non-standard and standard laboratory equipment and supplies are required to perform this protocol.

1. Acoustic dispenser with cherry pick software and appropriate consumables.
2. Standard polymerase chain reaction (PCR) reagents, consumables, and instrumentation.
3. Standard reagents and equipment for gel electrophoresis.
4. Standard reagents, consumables, and instrumentation for microbial cultivation.
5. Standard reagents, consumables, and instrumentation for *Escherichia coli* transformation.
6. Standard micropipettes and consumables, 12-channel pipettes are recommended.
7. A colony picking robot and a 96/384-pin-based screening robot are recommended but not required.

2.1 Plasmids

All plasmids relevant to this protocol are listed in Table 1. Plasmids can be obtained from the corresponding author. Any other Golden Gate cloning acceptor plasmid can be used. However, it is imperative to adapt the outlined protocol to the cloning standard with respect to the desired type IIS enzymes, corresponding overhangs and plasmid characteristics (e.g., antibiotic resistance) (*see* **Notes 1** and **2**).

2.2 DNA Oligonucleotides

The oligonucleotides used in the protocol are listed in Table 2. Oligonucleotides are dissolved in ddH_2O to make 100 μM stocks and stored at −20 °C. For PCR and sequencing reactions, 10 μM working stocks are generated with ddH_2O and stored at −20 °C.

2.3 Enzymes

For this protocol, any enzyme with the respective properties can be used (*see* **Note 3**). However, enzymes from any supplier should be suitable for the protocol outlined.

1. Necessary type IIS enzyme(s), in this protocol SapI.
2. T4 DNA ligase (400,000 units/mL).
3. Taq DNA polymerase (here, 2× ready-to-load master mix).

Table 1
Plasmids used in this study

Name	Relevant features	Parental plasmid	Reference
pSL009	pBAD-TOPO derivative serving as empty plasmid control; Kan^R		[11]
pSL099	Level 0 plasmid for subcloning of fragments to be released with SapI; $Spec^R$	pMA60 [9]	[22]
pSL135	Level 0 plasmid containing $P_L$lacO-1promoter to be released with SapI; $Spec^R$	pSL099	[22]
pSL137	pBAD-TOPO derivative allowing Golden Gate cloning of TUs with SapI; Kan^R	pSL009	[11]
pSL206	Level 0 plasmid containing *mCherry* to be released with SapI; $Spec^R$	pSL099	This study
pSL269	Level 0 plasmid containing a transcription terminator to be released with SapI; $Spec^R$	pSL099	This study
pSL541	Level 0 plasmid containing *gfp* to be released with SapI; $Spec^R$	pSL099	This study
pSLcol_16.01 to pSLcol_16.48	Level 0 plasmid collection containing distinct promoter sequences of a sequence library based on SLo5556 (cf. Fig. 2). Parts are released with SapI; $Spec^R$	pSL099	This study

Table 2
Oligonucleotides for the conduction of the described protocol

ID	Sequence (5′-3′)	Information
SLo1475	GCCTGACGAAGACCATTTC	Forward primer to amplify sequence library
SLo1476	ACCTAGCCGAAGACCAACAT	Reverse primer to amplify sequence library
SLo3961	CTGTCAAATGGACGAAGCAG	Colony PCR primer to validate TU assembly
SLo3962	CAGGCAAATTCTGTTTTATCAGACC	Colony PCR primer to validate TU assembly
SLo5556	CCTGACGAAGACCA TTTCNNNNNNNNNNNNNNNNN TTGACANNNNNNNNNNNNNNNNNNGATAC TNNNNNNNNAAAGAGGAGAAANNNNNNA TGTTGGTCTTCGGCTAG	Degenerated sequence of a promoter serving as PCR template with ribosome binding site

2.4 Antibiotics

All antibiotics used in this study are dissolved in ddH_2O and are stored in 1 mL aliquots at −20 °C. The stock concentration is 1000×.

Table 3
Strains used in this study

Name	Relevant features	Reference
E. coli MG1655	K-12 F^- λ^- (*see* **Note 8**)	[23]
E. coli DB3.1	F^- *gyrA462 endA1 glnV44* Δ(*srl-recA*) *mcrB mrr hsdS20*(r_B^-, m_B^-) *ara14 galK2 lacY1 proA2 rpsL20*(Str^R) *xyl5* Δ*leu mtl1* (*see* **Note 9**)	Invitrogen
E. coli Top10	F^- *mcrA* Δ(*mrr-hsdRMS-mcrBC*) φ80*lacZ*ΔM15 Δ*lacX74 nupG recA1 araD139* Δ(*ara-leu*)7697 *galE15 galK16 rpsL*(Str^R) *endA1* λ^- (*see* **Note 10**)	Invitrogen

1. Kanamycin (50 mg/mL stock).
2. Spectinomycin (120 mg/mL stock).

2.5 Chemicals, Buffers, and Media Components

1. LB medium: 1% (w/v) tryptone, 0.5% (w/v) yeast extract, 1% (w/v) sodium chloride, pH 7.0 ± 0.2.
2. 1 × tris-acetate-EDTA buffer (TAE): 1 mM ethylenediamine-tetraacetic acid (EDTA) Na_2, 20 mM acetate, 40 mM Tris base.
3. DNA dye: Thiazole orange dissolved in dimethyl sulfoxide (DMSO) (10,000× stock concentration: 13 mg/mL).
4. Agarose (standard).
5. DNA ladder.
6. Solid media is prepared with 2% Agar.

2.6 Consumables

1. Single-well microtiter plates (*see* **Note 4**).
2. 120 × 120 mm square petri dishes (*see* **Note 5**).
3. 384-well PCR plates (*see* **Note 6**).
4. Adhesive PCR plate seal.

2.7 Strains

All relevant laboratory *E. coli* strains for the outlined protocol are listed in Table 3. Cells were made competent using an in-house RbCl procedure [20] (*see* **Note 7**).

3 Methods

3.1 Preparation of DNA for Golden Gate Reaction

The first step in performing Golden Gate experiments is to generate and purify DNA fragments to be assembled into the selected acceptor vector (Fig. 2a). The number of parts and cloning steps depends on the user's experimental design and the selected Golden Gate assembly standard. In this example, three level 0 parts are assembled into one transcription unit (TU) in a pBAD-derived acceptor plasmid. The focus of this chapter is on experimental design and the

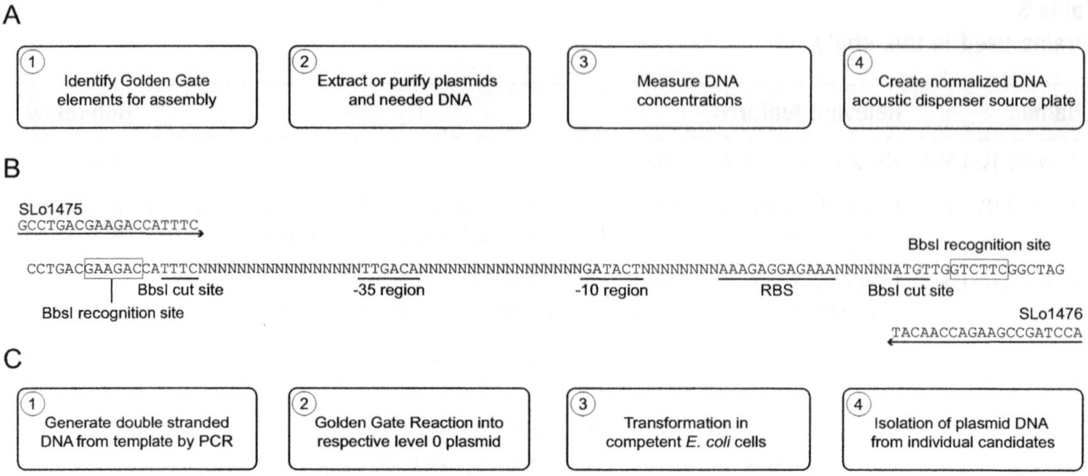

Fig. 2 Workflow for DNA preparation for Golden Gate reactions. (**a**) Outline of the steps required to perform acoustic dispenser-based Golden Gate assembly. The respective DNA parts to be assembled must be identified from the collection (or cloned and validated), purified, and the concentration must be measured before preparing the acoustic dispenser compatible source plate with normalized DNA concentrations. (**b**) Visualization of the promoter library concept. A degenerate oligonucleotide with defined −35, −10 region and ribosomal binding site (RBS) containing library amplification regions with BbsI restriction sites for Golden Gate cloning was designed and ordered. The indicated oligonucleotides (SLo1475/1476) were used to generate double-stranded DNA for Golden Gate cloning. The cloning of this element is not part of the protocol, but serves to clarify the promoter library used in this chapter. (**c**) Workflow for obtaining individual promoters from a random promoter sequence library

necessary steps to consider for highly parallelized, automated protocol execution. For this reason, we use a collection of promoter elements derived from a randomized sequence and fluorescent proteins as a direct readout of cloning success (Fig. 2b, c). The steps in Subheading 3.1 can be performed in about half a working day (excluding *E. coli* cultivation).

1. Design the appropriate Golden Gate assembly and ensure that all parts and the appropriate acceptor vector are available. In this example, different promoters of unknown sequence from isolates of a complex library (pSLcol_16.01 to pSLcol_16.48) are individually combined with one of the two fluorophores (pSL206 or pSL541) and a terminator (pSL269) into the expression plasmid pSL137 (Fig. 2b, c). The expression plasmid is a pBAD derivative for direct use in *E. coli* cells. The experimental setup results in 96 Golden Gate DNA assemblies by generating all combinations of the 48 promoter sequences with each of the two fluorophores.
2. Perform plasmid extraction of the selected level 0 parts and the acceptor plasmid(s) (*see* **Note 11**). If PCR fragments with appropriate Golden Gate compatible overhangs are used, perform the PCR purification of choice. Alternatively,

oligonucleotides can be annealed and used directly at the appropriate dilution (*see* **Note 12**).

3. Quantify the DNA by the method of choice (*see* **Note 13**).
4. To simplify subsequent DNA assembly steps, set the concentration of the DNA parts to 20 fmol/μL, DNA parts are used at equimolar concentrations (2 fmol each) in Golden Gate DNA assemblies (*see* **Note 14**). Normalization is ideally performed in acoustic dispenser compatible source plate(s) in combination with a bulk dispenser to quickly adjust the DNA concentration for each well. The plates are carefully sealed with an appropriate seal for storage at −20 °C (*see* **Note 15**).
5. Carefully document the location and information of each DNA for each well for future use and possible troubleshooting.
6. Store the normalized DNA at −20 °C until use (*see* **Note 16**).

3.2 Golden Gate Reaction Using an Acoustic Dispenser

Acoustic dispensers use sound to transfer nL droplets from a source plate to a destination plate. This technology allows reaction volumes to be drastically reduced and reactions to be parallelized. For molecular DNA assembly methods, a final volume of 1 μL in 384-well destination plates has been proven successful [11, 19]. Subheading 3.2 requires little hands-on time, but the Golden Gate reaction can be the rate-limiting step and is preferably performed overnight (Fig. 3). Subheadings 3.1 and 3.2 together can be performed during a standard working day.

1. Familiarize yourself with the acoustic dispenser available. Different dispensers have different specifications in terms of volume dispensed and liquid types. Consumables may vary from machine to machine, so it is imperative to be aware of all details before conducting experiments. Acoustic dispensers require regular maintenance. Important: Never use a machine without instruction from the local responsible operator.
2. Identify the appropriate source plate(s) for the Golden Gate assembly. For simplicity, this protocol requires one source plate containing all DNA parts as well as the reaction buffer, ddH_2O, and the enzymes (*see* **Note 17**). Prepare the pick list for the cherry pick software package (*see* **Note 18**). Below is an

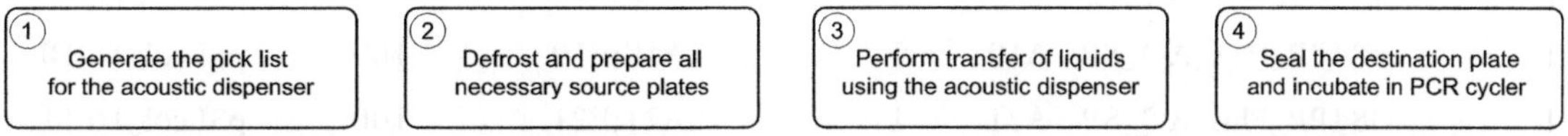

Fig. 3 Workflow for Golden Gate reactions using an acoustic dispenser. The first step is to create the pick list to define the appropriate acoustic dispenser liquid transfers. Ensure that all source plates are thawed and prepared to be used for the acoustic fluid transfer. After the transfer, seal the destination plate and perform Golden Gate DNA assembly in a PCR cycler

example pick list for assembling the first 12 TUs with *mCherry* and *gfp* for the outline example experiment. All DNA parts are used in equimolar concentrations:

Pick list (*see* **Note 19**):

Source plate name	Source plate type	Source well	Destination plate name	Destination well[a]	Transfer volume	Part name
1	384PP_Plus_AQ_SP	E1	1	A1;A3;A5; [. . .] E19;E21;E23	100	T4 ligase buffer
1	384PP_Plus_AQ_GP	E2	1	A1;A3;A5; [. . .] E19;E21;E23	100	T4 ligase
1	384PP_Plus_AQ_GP	E3	1	A1;A3;A5; [. . .] E19;E21;E23	100	SapI
1	384PP_Plus_AQ_SP	E4	1	A1;A3;A5; [. . .] E19;E21;E23	100	pSL137
1	384PP_Plus_AQ_SP	E5	1	A1;A3;A5; [. . .] E19;E21;E23	100	pSL269
1	384PP_Plus_AQ_SP	E6	1	A1;A3;A5; [. . .] A19;A21;A23	100	pSL206
1	384PP_Plus_AQ_SP	E7	1	E1;E3;E5; [. . .] E19;E21;E23	100	pSL541
1	384PP_Plus_AQ_SP	E8	1	A1;A3;A5; [. . .] E19;E21;E23	300	ddH_2O
1	384PP_Plus_AQ_SP	A1	1	A1;E1	100	pSLcol_16.01
1	384PP_Plus_AQ_SP	A3	1	A3;E3	100	pSLcol_16.02
1	384PP_Plus_AQ_SP	A5	1	A5;E5	100	pSLcol_16.03
1	384PP_Plus_AQ_SP	A7	1	A7;E7	100	pSLcol_16.04
1	384PP_Plus_AQ_SP	A9	1	A9;E9	100	pSLcol_16.05
1	384PP_Plus_AQ_SP	A11	1	A11;E11	100	pSLcol_16.06
1	384PP_Plus_AQ_SP	A13	1	A13;E13	100	pSLcol_16.07
1	384PP_Plus_AQ_SP	A15	1	A15;E15	100	pSLcol_16.08
1	384PP_Plus_AQ_SP	A17	1	A17;E17	100	pSLcol_16.09
1	384PP_Plus_AQ_SP	A19	1	A19;E19	100	pSLcol_16.10
1	384PP_Plus_AQ_SP	A21	1	A21;E21	100	pSLcol_16.11
1	384PP_Plus_AQ_SP	A23	1	A23;E23	100	pSLcol_16.12

[a][...] indicates that A7 through A23 and E1 through E17 are omitted for visualization purposes

3. Transfer the source plate from its storage to ice and ensure that all reagents are completely thawn (*see* **Note 20**). Allow the source plate equilibrate to room temperature prior to use to ensure proper transfer.
4. Centrifuge the source plate at max. 2000× g for ~10 s to remove any air bubbles and to prevent a meniscus at the air/-liquid interface (*see* **Note 21**).
5. Transfer the .csv file to the computer running the acoustic dispenser. Important: Follow your local IT best practice guidelines for data transfer.
6. Launch the cherry pick software to operate the acoustic dispenser. The instrument should automatically connect at this stage (*see* **Note 22**).
7. Open a new file in the cherry pick software to create a new protocol.
8. Adjust the settings in the category "protocol" according to the used equipment and consumables. In this chapter, the settings were as follows:

   Sample Plate Format: 384 PP.

   Sample Plate Type: 384 PP_AQ_SP (for Echo525).

   384 PP_AQ_SP2 (for Echo650T)

   Destination Plate Type: 384 PCR plate (*see* **Note 23**).
9. Navigate to the option "pick list" option and import the pick list created (cf. Subheading 3.2, **step 2**).
10. Double-check the selections, then click on the blue play button to save the experimental settings and to run the protocol.
11. Click on the start button and follow the instructions given by the operating system. Each step must to be confirmed individually by clicking the "ok" button. The steps are described below:
    (a) Insert source plate.
    (b) Insert destination plate.
12. The transfer is performed automatically based on the pick list provided. The system measures the volume of the source plate at the beginning and end of the experiment. The instrument provides a corresponding report file. It is recommended to always check the report file for "exception errors" (*see* **Note 24**).
13. After transfer, the source and destination plate are ejected and need to be handled accordingly.
14. After transfer, the source and destination plate are ejected and need to be handled accordingly.

15. Seal the source plate and return it to the appropriate storage conditions if not needed.
16. Seal the destination plate and transfer the reactions to a 384-well PCR cycler. Start the preferred Golden Gate assembly reaction protocol. This chapter uses a cycling program with the following settings for the assembly (*see* **Note 25**):

$$[4\ \text{min}\quad 37\,^\circ\text{C} - 4\ \text{min}\quad 16\,^\circ\text{C}]_{25\text{cycles}} - 10\ \text{min}\quad 50\,^\circ\text{C}$$
$$-\ 20\ \text{min}\quad 65\,^\circ\text{C}\ -\ \text{hold at}\ 8\,^\circ\text{C}.$$

17. Ideally, the Golden Gate reaction should be transformed to the preferred cells immediately after the reaction (see procedure in Subheading 3.3). However, Golden Gate reactions can be stored for a short time (<48 h) in the refrigerator or for a longer time (>48 h) at −20 °C.

3.3 High-Throughput Transformation

Golden Gate assemblies have proven to be very efficient. An advantage of using sequence verified level 0 parts is that there is usually no need to verify sequence integrity by sequencing in higher order assemblies. Depending on the complexity of the Golden Gate assembly, the user can usually estimate the percentage of successful assemblies. In most cases, picking two to four colonies is sufficient to obtain at least one successfully assembled construct (*see* **Note 26**). A limiting step is the high-throughput transformation of Golden Gate reactions into recipient cells. We use a workflow that allows high-throughput transformation into *E. coli* cells and subsequent selection of 12 or up to 96 transformations on a single agar plate. Subheading 3.3 requires little hands-on time and can be performed flexibly during a normal working day (Fig. 4). Selection plates are typically incubated overnight. The protocol uses chemically competent cells prepared in-house using the RbCl method [20]. However, any other method that produces sufficiently competent cells is applicable and is therefore not described in this protocol.

1. Prepare the appropriate agar medium with the required antibiotic. In this case, LB 2% agar supplemented with kanamycin as a selection marker is used (*see* **Note 27**).
2. Carefully pour the agar plates, avoiding the formation of bubbles. Do not move plates until completely set to ensure a flat and even surface. For drop plating, use 120 × 120 mm plates are used (Fig. 4b). For 96-well plating with a pin-based screening robot, use Society for Biomolecular Screening (SBS) format or other appropriate consumables (Fig. 4c) (*see* **Notes 28** and **29**).
3. Make sure the plates are properly dried but not overdried (*see* **Note 30**).

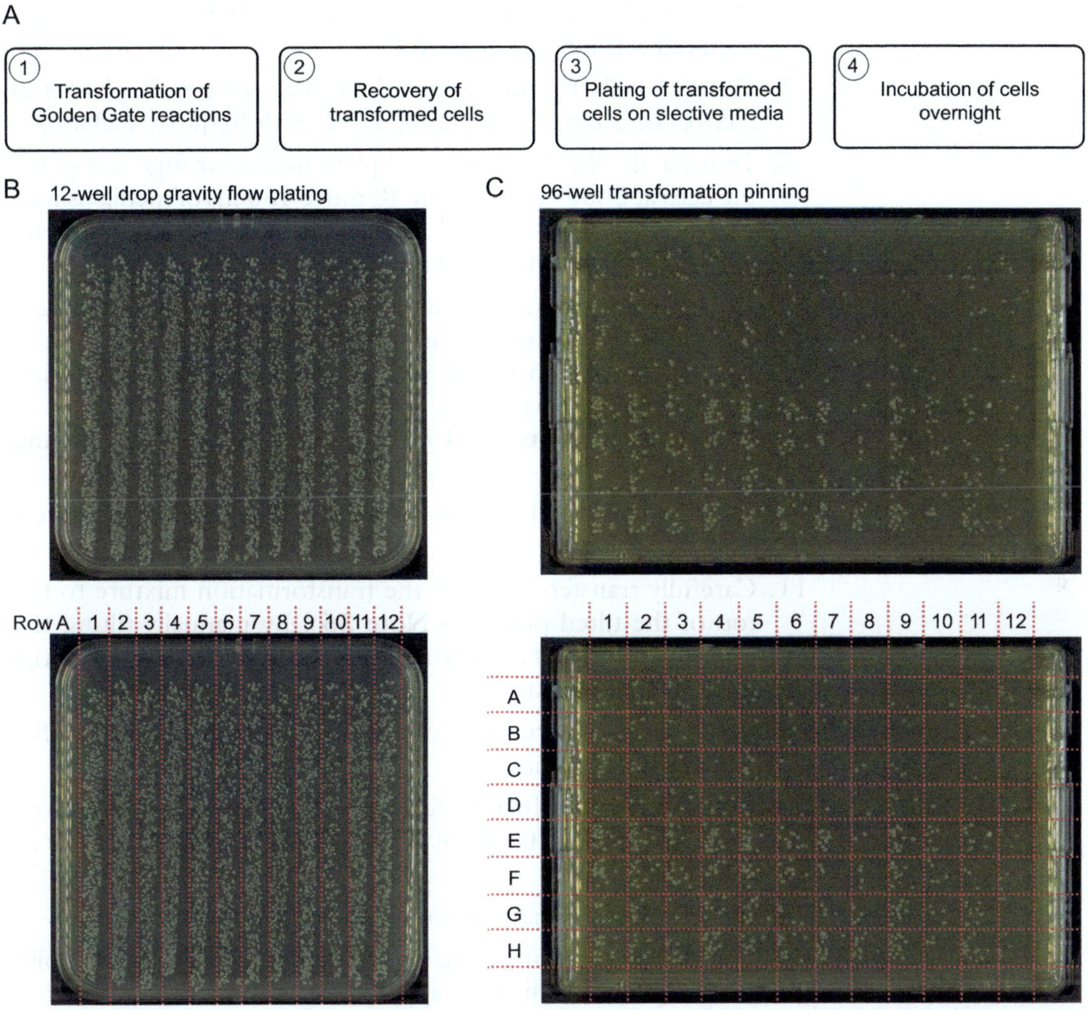

Fig. 4 Workflow for high-throughput transformation of Golden Gate reactions. (**a**) Workflow visualization of the individual steps described in 3.4. (**b**) Results of transformation plating by 12-well drop gravity flow using a 12-channel pipette. The top panel shows the plain transformation plate; the bottom panel highlights the individual lanes corresponding to the respective wells of the transformation plate. 20 μL of each transformation mixture is carefully transferred to a 120 × 120 mm petri dish, prearranged at an angle of ~30°. The drops create a gradient of the transformation mixture by gravity flow. A representative plate is shown after overnight incubation at 37 °C. (**c**) Results of 96-well transformation spotting using a 96-well pin pad-based robot. The top panel shows the plain transformation plate; the bottom panel highlights the individual lanes and rows corresponding to the respective wells of the transformation plate. The pin pad-based robot transfers droplets in a 7 × 7 grid and visits the transformation mixture only once per row to create a dilution series. A representative plate is shown after overnight incubation at 37 °C

4. Transformation procedure: Add 20 μL of competent cells to the Golden Gate reactions and incubate for 30 minutes on ice (*see* **Note 31**).
5. Prepare a 96-well deep-well plate with 500 μL recovery medium and preheat the plate(s) to 37 °C (*see* **Note 32**).

6. Heat shock cells in a preheated 384-well PCR cycler with the required settings (*see* **Note 33**).
7. Transfer cells to the prepared 96-well deep-well plate and recover cells for 45–60 min at 37 °C with frequent shaking.
8. Transfer the 96-well deep-well plate to a centrifuge and pellet the cells at 3000× *g* for 5 min. Remove 450 μL of media using a 12- or 96-channel multichannel pipette and gently resuspend the cells in the remaining liquid (*see* **Note 34**).
9. Two platting methods can be used to reliably plate high-throughput transformations at higher densities (Fig. 2b, c).

   Starting at step 10: 12-well drop gravity flow plating using multichannel pipettes (Fig. 4b).

   Starting at **step 14**: 96-well transformation pinning (Fig. 4c).
10. Position a 120 × 120 mm plate at an angle of approximately 30°.
11. Carefully transfer 20 μL of the transformation mixture to the top of the tilted plate (*see* **Note 35**). By manually tilting the plate to an angle of up to 90°, the speed and run distance of the drops can be adjusted.
12. When the drops are sufficiently spread, place the plate on a flat bench to dry the liquid.
13. Plate one row of the transformation mixture from the 96-well plate per 120 × 120 mm plate. After drying the plates move to **step 17**.
14. Transfer transformations to a 384-microtiter plate.
15. Use the 96-pin pin-pad to spot transformations on SBS solid media plates with the appropriate selection media in a 7 × 7 grid.
16. It is important to visit the transformation mixture only once for each row of the 7 × 7 grid to obtain a miniaturized dilution series.
17. Incubate the transformation plates inverted at 37 °C overnight (*see* **Note 36**).
18. Document your transformation plates using an imaging system. In this case, fluorescent TUs are assembled; they should be imaged under the appropriate fluorescence conditions and standard illumination settings.

3.4 cPCR Validation of Constructs for Subsequent Sequence Validation

Once DNA constructs have been made and transformed, they need to be validated. The method presented here has a direct visible readout based on the expressed fluorescent protein. However, for applications without direct readout, other validation procedures are required. Therefore, a method based on colony PCR is described (Fig. 5). The use of *E. coli* cells as the PCR template allows the

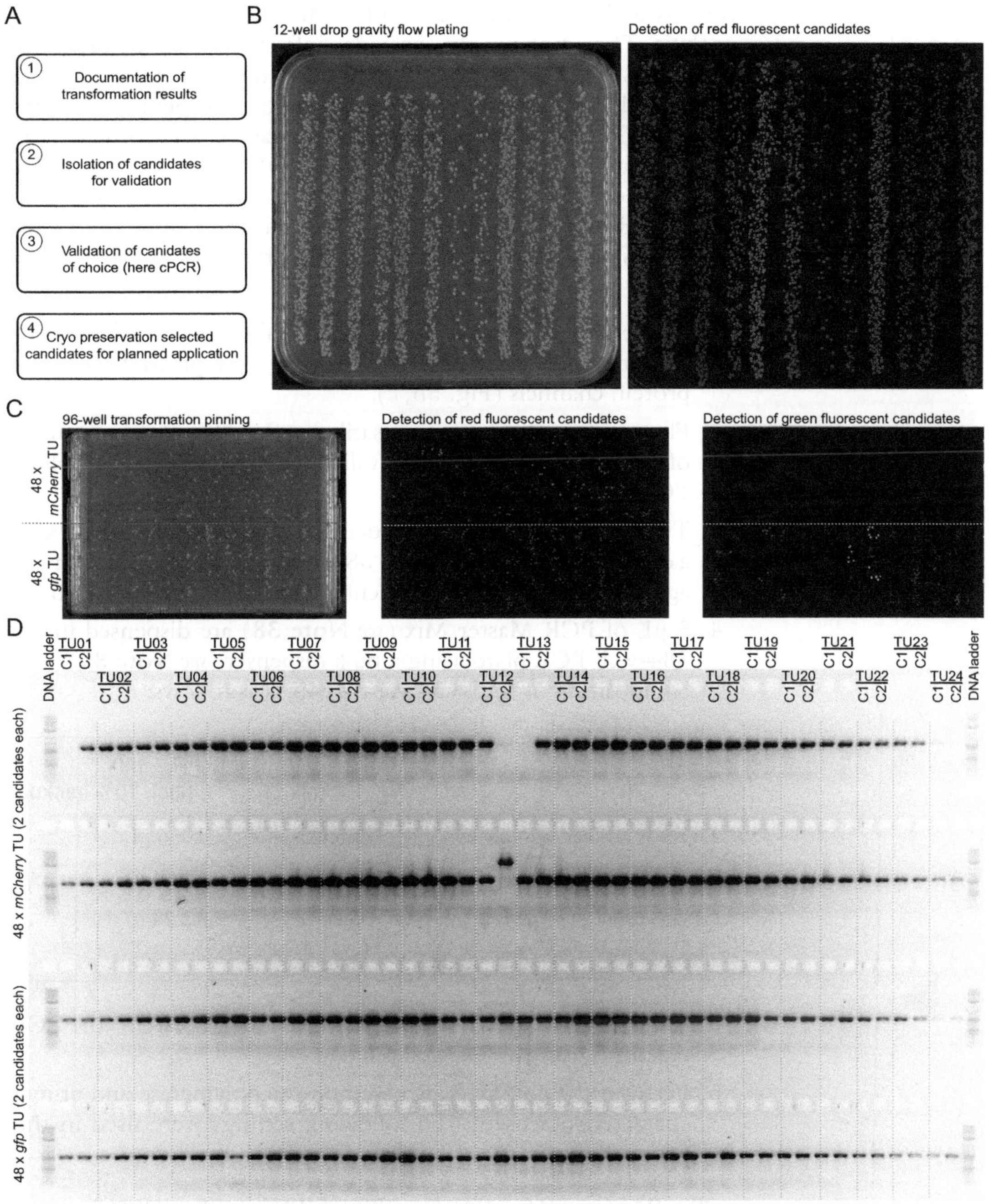

Fig. 5 Workflow and example of construct validation. (**a**) Workflow visualization of the individual steps described in Subheading 3.4. (**b**) Example image of 12-well drop gravity plating for the assembly of *mCherry* transcription units. Shown is a standard image (left) and an image under conditions that allow the detection of mCherry (right). Different fluorescence intensities can be observed at the colony level and are an indicator of the strength of promoters of unknown sequence. A representative result is shown. (**c**) Example images of 96-well plating using a pinning robot for the assembly of *mCherry* (top) and *gfp* (bottom) transcription units. Shown is a standard image (left) and images under conditions that allow detection of mCherry (middle) and GFP (right). Different fluorescence intensities can be observed at the colony level and are an indicator of the strength of promoters of unknown sequence. A representative result is shown. (**d**) Two candidates from each transformation for the 48 × *mCherry* and 48 × *gfp* TU constructs were tested by cPCR to validate the TU

process to be accelerated in contrast to restriction digest pattern analysis. The amplicon size provides information about the DNA assemblies and may allow identification of the problematic part(s). In general, a diagnostic method should be sufficient to confirm Golden Gate assemblies beyond the basic part construction. Sequencing of plasmids is only required for basic part construction or if methods have been used that potentially introduce mutations (e.g., PCR amplification steps). cPCR validation of Golden Gate assemblies takes approximately half a working day from start to finish.

1. In the case of this protocol, functionality can be validated by imaging plates in the respective red and green fluorescent protein channels (Fig. 5b, c).
2. Fluorescence does not necessarily confirm the correct assembly of candidates, so additional validation is performed by colony PCR (cPCR) (*see* **Note 37**).
3. Two colonies from each Golden Gate assembly are picked with a colony picking robot into a 384-well grid onto LB kanamycin agar plates. The plates are incubated at 37 °C for at least 6 h.
4. 5 μL of PCR Master Mix (*see* **Note 38**) are dispensed into a 384-well PCR plate using a bulk dispenser (*see* **Note 39**). The composition of the PCR Master Mix is as follows:

	1 × mix	384 × mix (incl. 10% backup)
2 × ready-to-load master mix	2.5 μL	1056 μL
Forward primer (10 μM)	0.1 μL	42.24 μL
Reverse primer (10 μM)	0.1 μL	42.24 μL
ddH_2O	2.3 μL	971.52 μL

5. Transfer cell material from the prepared agar plate to the PCR reactions using a 384-well pin pad.
6. Perform the colony PCR based on the polymerase and primer characteristics used. The following settings were used in this protocol:

$$30\ \text{s}\ 94\,^\circ\text{C} - [15\ \text{s}\ 94\,^\circ\text{C} - 30\ \text{s}\ 50\,^\circ\text{C} - 70\ \text{s}\ 68\,^\circ\text{C}]_{35\text{cycles}}$$
$$-5\ \text{min}\ 68\,^\circ\text{C} - \text{hold at}\ 8\,^\circ\text{C}$$

Fig. 5 (continued) constructs. Nearly all reactions show the expected band and at least one candidate for each TU construct appears to be correct. For visualization purposes, only the top row is labeled. The top half shows the 2 × 48 × *mCherry* assemblies and the bottom half shows the 2 × 48 × *gfp* assemblies. The respective *mCherry* and *gfp* positions correspond to the same promoter sequences. C1 = Candidate 1, C2 = Candidate 2, red dotted lines separate the different TU assemblies

7. Load 4 μL of the PCR reaction onto a 1% agarose gel and run at 100 V for 1 h (*see* **Note 40**). DNA loading dye is added directly to the gel.
8. Image and analyze the agarose gel using an appropriate documentation system (Fig. 5d).
9. Select suitable candidates from the arrayed agar plate used as cell material donor for the cPCR and inoculate overnight in appropriate medium for cryopreservation and downstream applications.

3.5 Summary and Perspectives

Acoustic dispensers have revolutionized molecular biology [19]. They allow highly parallelized reactions in nanoliter volumes. The transfer of liquids is free of consumables, making the use of acoustic dispensers an economical and sustainable strategy. Their versatility has been demonstrated in many different applications. We provide a detailed protocol for using acoustic dispensers for high-throughput Golden Gate assemblies. We also provide details on high-throughput transformation, selection and validation of DNA assemblies. We believe this technology will become the standard for molecular microbiology applications, and we have shown only one aspect of the use of acoustic dispensers.

4 Notes

1. It is mandatory to perform efficiency tests for new acceptor plasmids, altered numbers, or drastic fragment size changes prior to large-scale experiments to estimate DNA assembly efficiency.
2. It is recommended that each new batch of competent cells be tested for efficiency. Efficiency can be determined by transforming a standard plasmid of known concentration. Colony Forming Unit (CFU) can be used to estimate transformation efficiency in terms of CFUs per μg of plasmid DNA.
3. Efficiency testing is recommended when reagents are changed. This is especially true when new suppliers or in-house purified enzymes are used.
4. Consumables should be compatible with the screening robot used; it is recommended to use the supplier's single well plates or tested alternative consumables. In this chapter, the Rotor HDA+ from Singer Instruments has been used with the single well plates from the same manufacturer, which have been optimized with regard to their surface area.

5. Other formats will work as long as they are compatible in size with the multichannel pipettes used. If high-throughput plating is not used, standard petri dishes and glass bead-based spreading or alternative methods will also work.

6. 384-well plates are used as acceptor plates for the acoustic dispenser. Standard 96-well plates are too high for the acoustic dispensers used in this chapter. Low profile 96-well plates may be compatible with the acoustic dispenser. However, they must be skirted and a PCR cycler compatible with these plates must be available. For several reasons, we have adapted our protocol to use 384-well PCR plates as acceptor plates. Important: The dimensions of the 384-well PCR plate must be adapted to the operating system of the acoustic dispenser.

7. Any type of competent cells with sufficient transformation efficiency can be used. Transformation can also be performed by electroporation. However, this method may not be compatible with high throughput. It is strongly recommended to test the efficiency of batches of competent cells prior to large-scale experiments. Assembly efficiencies should be tested with varying numbers of fragments to ensure that the correct volume is plated after transformation.

8. Wild-type strain used in this study. Any other experimental model strain is also suitable.

9. This strain is required to propagate the *ccdB* containing acceptor plasmids used. CcdB is a gyrase inhibitor and causes cell death in strains without CcdA antitoxin or gyrase mutations (e.g., *gyrA462*).

10. Standard cloning strain that is highly sensitive to CcdB. However, any other cloning strain or wild-type strain will work. CcdB sensitivity is advantageous to minimize the risk of cells containing only the acceptor plasmid. New strains can be easily tested by transforming a *ccdB*-encoding plasmid; a positive control (e.g., pUC-based plasmid) must be included in this experiment to ensure that the cells are competent.

11. Any plasmid purification method is suitable. However, for large-scale purification of plasmid DNA protocols using magnetic beads are preferred. Suitable open source protocols are for example available at www.bomb.bio [21] and are used here and in previous studies in combination with the acoustic dispenser [11].

12. For annealing, 4.5 μL each of 100 μM primer and 1 μL 10X annealing buffer (10X annealing buffer: 10 mM Tris–HCl, pH 7.5–8.0, 50 mM NaCl and 1 mM EDTA) in a PCR reaction tube. The mixture is transferred to a PCR cycler. The sample is heated to 98 °C for 30 s and cooled to 8 °C at the lowest possible ramp speed of the instrument. The sample

is then diluted to 20 fmol/µL and used immediately or stored at −20 °C. After multiple freeze-thaw cycles and an observed decrease in efficiency, a new annealing reaction should be performed. Important: If multiple oligonucleotides are fused, the DNA must be phosphorylated with, for example, T4 Polynucleotide Kinase according to the manufacturer's instructions prior to annealing.

13. Determination of DNA concentration using the absorbance ratio of $A_{260/280}$ is usually sufficient. However, if DNA assembly is inefficient, determination of DNA concentration using fluorescent dyes may be helpful in troubleshooting.
14. As an estimate, 40 ng of a 3 kb plasmid is equivalent to 20 fmol.
15. Carefully handle the plates to avoid cross-contamination when removing the seal. Before removing the seal, spin the plate briefly to ensure that the liquid is at the bottom of the wells.
16. Even when carefully sealed evaporation may be observed during prolonged freezer storage.
17. It is recommended to use multiple source plates. A master plate for enzymes and buffers and the appropriate DNA collection plates. Importantly, some acoustic dispenser models (e.g., Echo650T used in this chapter) are capable of using tubes in racks. The tubes can be arranged according to the user's needs and are particularly useful for avoiding multiple freeze-thaw cycles of valuable reagents.
18. Several tools have been developed for the automated generation of pick lists, many of which are available as open source on GitHub (e.g., from Henrike Niederholtmeyer's group: https://github.com/HN-lab/PyEcho).
19. This pick list is based on the Echo525 acoustic dispenser. Depending on the machine type, the pick list may differ in terms of the source plate type, for example. However, the general concept of defining the source well, destination well, and transfer volume remains the same.
20. Carefully remove the seal, ideally while the reagents are still frozen, to minimize the risk of cross-contamination. It may be necessary to pulse spin the plate prior to seal removal.
21. Carefully inspect the source plate and repeat this step if necessary. Do not spin at too high x *g* to avoid damaging the source plate.
22. Operation may vary from device to device. In this protocol, the Echo525 or Echo650T (Labcyte) acoustic dispensers were used and the differences are indicated.
23. This plate definition is defined by the user based on the 384-well PCR plate consumables used. Any type of 384-well PCR plate can be used as long as the internal gripper can hold

the plate and the plate is rigid enough not to bend. Importantly, 384-well plates that deviate from the standard square format may not be properly gripped and may fall out of the gripper during inverting.

24. Possible causes of exception errors are too low volume (outside of 15–65 μL range) or volume cannot be calculated (e.g., air bubbles, plate too cold). Exception errors may indicate that a particular reaction will not work because one or more reagents have not been transferred.
25. It is recommended to optimize the settings for each type of Golden Gate assembly (e.g., in regard to number of fragments and/or acceptor vector) to optimize the efficiency for optimal large-scale experiments.
26. However, if the assembly complexity is changed, it is always recommended to perform an initial small-scale test to obtain information on assembly efficiency.
27. Antibiotics and supplements in the media may be heat sensitive. Do not add additional components until the media has cooled to at least 60 °C.
28. In this procedure, a Rotor HDA+ is used to inoculate the transformation mixture onto agar plates. It is recommended to use Singer PlusPlates in combination with this machine. However, any other pinning robot may be suitable for this process. However, it is important to test different consumables for optimal results.
29. SBS plates are ideally prepared with 35–50 mL of medium using either a 50 mL conical tube or an appropriate pipette. Carefully transfer the warm medium, avoiding air bubbles. Any air bubbles that are present can be either punctured or pushed to the edge with a sterile pipette tip as long as the agar has not solidified.
30. For best results, pour the plates the day before the assay and store the plates upside down at room temperature after the agar has solidified. The storage location should be away from direct sunlight, heat sources, or constant air flow (e.g., air conditioning). If plates are not to be used within 24 h, it is recommended that the stack of plates be carefully wrapped in cling film or placed in a plastic bag and stored in the refrigerator.
31. In this experiment, competent cells are transferred from a reservoir (stored on ice) using a 12-channel multichannel pipette or 96-well pipetting station. Ensure proper mixing of DNA and cells by gently pipetting the transformation mixture up and down.

32. LB medium is used here. However, richer media or specific recovery media (e.g., SOC) may be used if transformation results in low colony numbers.
33. This protocol uses in-house prepared RbCl competent *E. coli* cells with the following heat shock settings 45 s 42 °C.
34. If efficiencies are high, cell concentration may not be necessary.
35. Ensure that the plates have a flat surface without bubbles and are dry. Otherwise, lanes may merge and reactions may be cross-contaminated.
36. If no standard *E. coli* strain or temperature sensitive DNA constructs are used, the settings must be adjusted accordingly, for example, incubation temperature and/or incubation time.
37. It is highly recommended to aim for an amplicon size of 500–1500 bps and ideally, positive and negative candidates will result in amplicons of different sizes.
38. To prepare the correct amount of PCR Master Mix, it is recommended to include 10–15% excess.
39. In this case, a ready-to-load 2× Master Mix was used to speed up the sample analysis by gel electrophoresis. However, any DNA polymerase is suitable for this experiment and the most economical was used.
40. The settings must be adapted to the available gel electrophoresis equipment.

Acknowledgments

This work was supported by the Max Planck Society within the framework of the MaxGENESYS project (DS) and the European Union (NextGenerationEU) via the European Regional Development Fund (ERDF) by the state Hesse within the project "biotechnological production of reactive peptides from waste streams as lead structures for drug development" (DS). We are grateful to all laboratory members for extensive discussions on the cloning and validation steps of Golden Gate assemblies. We thank Ehmad Chehrghani Bozcheloe for the visualization of the acoustic dispenser. All materials are available on request from the corresponding author.

References

1. Engler C, Kandzia R, Marillonnet S (2008) A one pot, one step, precision cloning method with high throughput capability. PLoS One 3(11):e3647. https://doi.org/10.1371/journal.pone.0003647
2. Werner S, Engler C, Weber E, Gruetzner R, Marillonnet S (2012) Fast track assembly of multigene constructs using Golden Gate cloning and the MoClo system. Bioeng Bugs 3(1): 38–43. https://doi.org/10.4161/bbug.3.1.18223

3. Bird JE, Marles-Wright J, Giachino A (2022) A user's guide to Golden Gate cloning methods and standards. ACS Synth Biol 11(11): 3551–3563. https://doi.org/10.1021/acssynbio.2c00355
4. Lee ME, DeLoache WC, Cervantes B, Dueber JE (2015) A highly characterized yeast toolkit for modular, multipart assembly. ACS Synth Biol 4(9):975–986. https://doi.org/10.1021/sb500366v
5. Goosens VJ, Walker KT, Aragon SM, Singh A, Senthivel VR, Dekker L, Caro-Astorga J, Buat MLA, Song W, Lee KY, Ellis T (2021) *Komagataeibacter* tool kit (KTK): a modular cloning system for multigene constructs and programmed protein secretion from cellulose producing bacteria. ACS Synth Biol 10(12): 3422–3434. https://doi.org/10.1021/acssynbio.1c00358
6. Stukenberg D, Hensel T, Hoff J, Daniel B, Inckemann R, Tedeschi JN, Nousch F, Fritz G (2021) The Marburg collection: a Golden Gate DNA assembly framework for synthetic biology applications in *Vibrio natriegens*. ACS Synth Biol 10(8):1904–1919. https://doi.org/10.1021/acssynbio.1c00126
7. McCarty NS, Shaw WM, Ellis T, Ledesma-Amaro R (2019) Rapid assembly of gRNA arrays via Modular Cloning in yeast. ACS Synth Biol 8(4):906–910. https://doi.org/10.1021/acssynbio.9b00041
8. Stukenberg D, Hoff J, Faber A, Becker A (2022) NT-CRISPR, combining natural transformation and CRISPR-Cas9 counterselection for markerless and scarless genome editing in *Vibrio natriegens*. Commun Biol 5(1):265. https://doi.org/10.1038/s42003-022-03150-0
9. Schindler D, Milbredt S, Sperlea T, Waldminghaus T (2016) Design and assembly of DNA sequence libraries for chromosomal insertion in bacteria based on a set of modified MoClo vectors. ACS Synth Biol 5(12):1362–1368. https://doi.org/10.1021/acssynbio.6b00089
10. Guo Y, Dong J, Zhou T, Auxillos J, Li T, Zhang W, Wang L, Shen Y, Luo Y, Zheng Y, Lin J, Chen GQ, Wu Q, Cai Y, Dai J (2015) YeastFab: the design and construction of standard biological parts for metabolic engineering in *Saccharomyces cerevisiae*. Nucleic Acids Res 43(13):e88. https://doi.org/10.1093/nar/gkv464
11. Koebel TS, Melo Palhares R, Fromm C, Szymanski W, Angelidou G, Glatter T, Georg J, Berghoff BA, Schindler D (2022) An easy-to-use plasmid toolset for efficient generation and benchmarking of synthetic small RNAs in bacteria. ACS Synth Biol 11(9): 2989–3003. https://doi.org/10.1021/acssynbio.2c00164
12. Philipp N, Brinkmann CK, Georg J, Schindler D, Berghoff BA (2023) DIGGER-Bac: prediction of seed regions for high-fidelity construction of synthetic small RNAs in bacteria. Bioinformatics 39(5). https://doi.org/10.1093/bioinformatics/btad285
13. Zumkeller C, Schindler D, Waldminghaus T (2018) Modular assembly of synthetic secondary chromosomes. Methods Mol Biol 1837: 71–94. https://doi.org/10.1007/978-1-4939-8675-0_5
14. Messerschmidt SJ, Schindler D, Zumkeller CM, Kemter FS, Schallopp N, Waldminghaus T (2016) Optimization and characterization of the synthetic secondary chromosome synVicII in *Escherichia coli*. Front Bioeng Biotechnol 4: 96. https://doi.org/10.3389/fbioe.2016.00096
15. Hughes RA, Ellington AD (2017) Synthetic DNA synthesis and assembly: putting the synthetic in synthetic biology. Cold Spring Harb Perspect Biol 9(1). https://doi.org/10.1101/cshperspect.a023812
16. Schindler D, Dai J, Cai Y (2018) Synthetic genomics: a new venture to dissect genome fundamentals and engineer new functions. Curr Opin Chem Biol 46:56–62. https://doi.org/10.1016/j.cbpa.2018.04.002
17. Pryor JM, Potapov V, Bilotti K, Pokhrel N, Lohman GJS (2022) Rapid 40 kb genome construction from 52 parts through data-optimized assembly design. ACS Synth Biol 11(6):2036–2042. https://doi.org/10.1021/acssynbio.1c00525
18. Ellson R, Mutz M, Browning B, Lee L, Miller MF, Papen R (2003) Transfer of low nanoliter volumes between microplates using focused acoustics—automation considerations. JALA: J Assoc Lab Autom 8(5):29–34. https://doi.org/10.1016/s1535-5535-03-00011-x
19. Kanigowska P, Shen Y, Zheng Y, Rosser S, Cai Y (2016) Smart DNA fabrication using sound waves: applying acoustic dispensing technologies to synthetic biology. J Lab Autom 21(1): 49–56. https://doi.org/10.1177/2211068215593754
20. Hanahan D (1983) Studies on transformation of *Escherichia coli* with plasmids. J Mol Biol 166(4):557–580. https://doi.org/10.1016/s0022-2836(83)80284-8
21. Oberacker P, Stepper P, Bond DM, Hohn S, Focken J, Meyer V, Schelle L, Sugrue VJ, Jeunen GJ, Moser T, Hore SR, von Meyenn F, Hipp K, Hore TA, Jurkowski TP (2019) Bio-on-magnetic-beads (BOMB): open

platform for high-throughput nucleic acid extraction and manipulation. PLoS Biol 17(1):e3000107. https://doi.org/10.1371/journal.pbio.3000107

22. Brueck M, Berghoff BA, Schindler D (2023) *In silico* design, *in vitro* construction and *in vivo* application of synthetic small regulatory RNAs in bacteria. arXiv. https://doi.org/10.48550/arXiv.2304.14932

23. Blattner FR, Plunkett G 3rd, Bloch CA, Perna NT, Burland V, Riley M, Collado-Vides J, Glasner JD, Rode CK, Mayhew GF, Gregor J, Davis NW, Kirkpatrick HA, Goeden MA, Rose DJ, Mau B, Shao Y (1997) The complete genome sequence of *Escherichia coli* K-12. Science 277(5331):1453–1462. https://doi.org/10.1126/science.277.5331.1453

Chapter 10

Validation of Golden Gate Assemblies Using Highly Multiplexed Nanopore Amplicon Sequencing

Adán A. Ramírez Rojas, Cedric K. Brinkmann, and Daniel Schindler

Abstract

Golden Gate cloning has revolutionized synthetic biology. Its concept of modular, highly characterized libraries of parts that can be combined into higher order assemblies allows engineering principles to be applied to biological systems. The basic parts, typically stored in Level 0 plasmids, are sequence validated by the method of choice and can be combined into higher order assemblies on demand. Higher order assemblies are typically transcriptional units, and multiple transcriptional units can be assembled into multi-gene constructs. Higher order Golden Gate assembly based on defined and validated parts usually does not introduce sequence changes. Therefore, simple validation of the assemblies, e.g., by colony polymerase chain reaction (PCR) or restriction digest pattern analysis is sufficient. However, in many experimental setups, researchers do not use defined parts, but rather part libraries, resulting in assemblies of high combinatorial complexity where sequencing again becomes mandatory. Here, we present a detailed protocol for the use of a highly multiplexed dual barcode amplicon sequencing using the Nanopore sequencing platform for in-house sequence validation. The workflow, called DuBA.flow, is a start-to-finish procedure that provides all necessary steps from a single colony to the final easy-to-interpret sequencing report.

Key words Golden Gate assembly validation, DNA assembly validation, Nanopore sequencing, amplicon sequencing, Enzymatic fragmentation, Dual-barcode amplicon sequencing, Colony PCR, Laboratory automation

1 Introduction

Once *in vitro* DNA constructs have been produced using the method of choice, they are transformed into recipient cells. Recipients are usually special *Escherichia coli* cloning strains for the propagation and archiving of DNA assemblies in the form of plasmids. After transformation, each colony typically contains a unique plasmid. Several methods have been established to confirm the integrity of plasmids. One rapid method is diagnostic polymerase chain reaction (PCR) directly on cell material, called colony PCR (cPCR), which is suitable for selecting candidates for downstream

Daniel Schindler (ed.), *Golden Gate Cloning: Methods and Protocols*, Methods in Molecular Biology, vol. 2850,
https://doi.org/10.1007/978-1-0716-4220-7_10,

analysis or applications, but does not confirm DNA sequence integrity [1]. An alternative method suitable for identifying candidates with similar results is diagnostic restriction digest pattern analysis of purified plasmid DNA [2]. However, the final step is the determination of the DNA sequence. The method of choice for sequencing plasmid DNA or amplicons is Sanger sequencing. Sanger sequencing allows the sequencing of stretches of approximately 1000 base pairs (bp) [3]. Sanger sequencing provides reliable results in the laboratory routine but becomes impractical when longer sequences or large numbers of constructs need to be validated.

In recent years, alternative sequencing methods have emerged that allow the parallel sequencing of DNA, the so-called next generation sequencing (NGS) platforms [3]. This technology is a game changer, allowing the routine analysis of any type of DNA [4]. In addition to their primary application for whole genome sequencing and various other global sequence analysis methods, NGS platforms have been used to sequence plasmid DNA [5, 6]. This approach allows the sequencing of a large number of plasmids in parallel. However, it requires either large investments in infrastructure or long turnaround times if external services have to be used. Furthermore, the analysis of the rather short reads is not as trivial as the analysis of Sanger sequencing results. In particular, structural variations may not be detected.

Based on the successes and limitations of NGS strategies, long-read sequencing technologies have emerged. Long-read sequencing technologies combine the advantage of long sequences, similar to Sanger sequencing, with the parallelization of NGS technologies [3]. Therefore, long-read sequencing technologies are often referred to as third-generation sequencing (TGS) technologies. Initially, TGS provided lower read quality with relatively high error rates. However, continuous improvements have made TGS technologies an established methodology. One notable long-read sequencing technology is based on the alteration of a current as DNA moves through a membrane-inserted protein pore [7]. The current changes are specific to each base and allow a DNA strand to be decoded based on electrical signals. This technology is called Nanopore sequencing and has become a widely used technology. Its advantage is the rather small infrastructure investment and its easy operation. Several protocols have been developed to use Nanopore sequencing for the validation of DNA assemblies and whole plasmids [8, 9] and commercial services are becoming available. However, in-house validation of DNA assemblies using Nanopore sequencing is highly economical, especially when large numbers of DNA constructs need to be validated frequently.

We have recently developed a dual barcoding amplicon sequencing workflow called DuBA.flow which is available at GitHub (Fig. 1, https://github.com/RGSchindler/DuBA.flow) [10]. DuBA.flow is a start-to-finish workflow for generating and

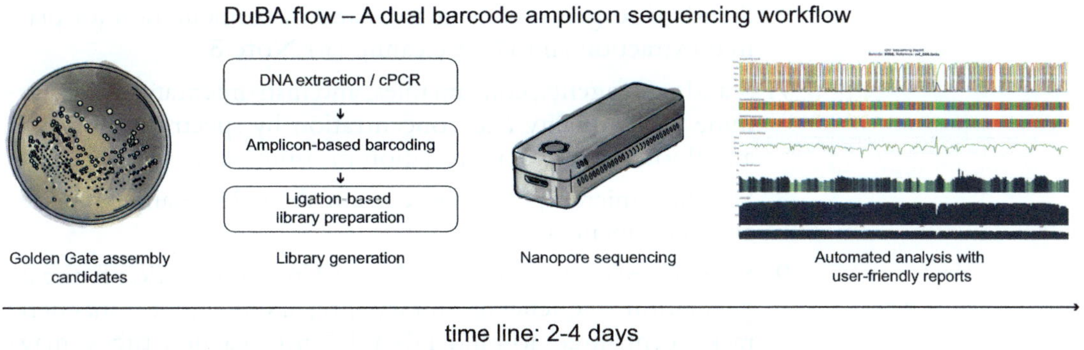

Fig. 1 DuBA.flow: a start-to-end long-read sequencing validation pipeline for assembled DNA constructs. The pipeline starts with candidates obtained from the Golden Gate reaction transformation. Within the pipeline, different strategies can be used to obtain barcoded amplicons, which are handed over to the Ligation Sequencing library preparation and subsequent Nanopore sequencing. Obtained data can be passed on to the automated computational analysis pipeline which returns user-friendly files for sequence validation

analyzing highly multiplexed amplicons, and returns easy-to-interpret result files, making the analysis of Nanopore sequencing results as easy as the analysis of Sanger sequencing data. Here we provide a detailed protocol for running our DuBA.flow pipeline from selected *E. coli* candidates to easy-to-interpret result files. The entire workflow can be performed within three days and is highly scalable. The protocol can be performed manually, but we also provide guidance on how to implement different types of laboratory automation equipment, making our DuBA.flow pipeline highly scalable. In particular, the use of acoustic dispensers allows for economic downscaling of reactions, making it highly economical.

2 Materials

The conduction of the described protocol requires the following nonstandard and standard laboratory equipment and consumables.

1. MinION Nanopore sequencer and Flongle adapter are used in this protocol (*see* **Note 1**).
2. Reagents and standard consumables to conduct the Oxford Nanopore Technologies Ligation Sequencing Kit protocol (*see* **Note 2**).
3. Standard reagents, consumables, and instrumentation for PCR reactions.
4. Standard reagents and equipment for gel electrophoresis.
5. Standard reagents, consumables, and instrumentation for microbial culturing.

6. Standard reagents, consumables and instrumentation for plasmid extraction and DNA cleanup (*see* **Note 3**).
7. Standard reagents, consumables, and instrumentation to determine DNA quality and concentration by spectrophotometric and fluorescent dye intercalation measurement.
8. Standard micropipettes and consumables, 12-channel pipettes are recommended.
9. Standard reagents, consumables, and instrumentation for next-generation sequencing library preparation (e.g., magnetic racks, centrifuge, lo-bind DNA 1.5 mL reaction tubes, magnetic beads).
10. For downscaled and high-throughput cPCR workflows respective equipment and consumables are necessary (*see* **Note 4**).
11. Standard thermomixer for 1.5 mL reaction tubes.

2.1 Computational Requirements (Hardware and Software)

Below are the settings of hardware and software listed used in this chapter—they do not represent the minimum specifications necessary. If new equipment is going to be purchased it is important to check for the most recent hardware requirements for the used software and to operate the MinION Nanopore sequencer (*see* **Notes 5** and **6**).

1. Nanopore sequencing: Intel Core i7-10610U CPU, SDD: 1 TB, RAM: 32 GB (*see* **Note** 7).

   Operating System: Microsoft Windows 10 Enterprise.

   Software: MinKNOW (up to version 23.11.4).
2. Basecalling: Intel Core i9-11900K CPU, SDDs: 1 TB, RAM: 64 GB, GPU: Nvidia.

   GeForce RTX3090, 24 GB RAM, Compute Capability 8.6 (*see* **Note 8**).

   Operating System: Ubuntu Linux 22.04.3.

   Software: CUDA (version 11.5), Guppy (up to version 6.5.7).
3. DuBA.flow analysis: Intel Core i7-10610U CPU, SDD: 1 TB, RAM: 32 GB (*see* **Note 9**).

   Operating System: Microsoft Windows 10 Enterprise.

   Software: Docker Desktop (version 4.12.0 (85629)), web browser (e.g., Mozilla Firefox, Google Chrome).

2.2 DNA Oligonucleotides

Relevant oligonucleotides for the conduction of the protocol are provided in Table 1. Example primers attaching the barcodes are provided in Table 2. 100 μM stocks are generated with ddH_2O and stored at −20 °C except indicated differently. For PCR and sequencing reactions 10 μM working stocks are generated with ddH_2O and stored at −20 °C. *Important:* Oligonucleotides for

Table 1
Oligonucleotides for the conduction of the described protocol

Name	Sequence (5′-3′)[a]	Information
SLo0100[b]	5′[phos] CTGTCTCTTATACACATCT	Tn5-ME reverse for transposase loading [11]
SLo0151	CCCAGTCACGACGTTGTAAAACG	M13 forward primer serving as control primer
SLo0152	AGCGGATAACAATTTCACACAGG	M13 reverse primer serving as control primer
SLo0673[b]	**CCCAGTCACGACGTTGTAAAACG**AGATGTG TATAAGAGACAG	Linker Tn5-ME with added M13 forward sequence for annealing with SLo0100
SLo0674[b]	**AGCGGATAACAATTTCACACAGG**AGATGTG TATAAGAGACAG	Linker Tn5-ME with added M13 reverse sequence for annealing with SLo0100
SLo1577	**CCCAGTCACGACGTTGTAAAACG**CGTCAA TTGTCTGATTCGTTACCA	Forward primer for cPCR amplification with added M13 forward sequence
SLo1578	**AGCGGATAACAATTTCACACAGG**CTTCTC TCATCCGCCAAAACA	Reverse primer for cPCR amplification with added M13 reverse sequence

Barcoding primers are listed in Table 2
[a] Bold letters indicate the overlaps for amplification (*see* **Note 10**) for the second PCR attaching the barcodes
[b] Indicated oligonucleotides for Tn5 loading are solved in annealing buffer (50 mM NaCl, 40 mM Tris-HCl pH 8.0) to a stock concentration of 140 μM

transposase loading are solved in annealing buffer (50 mM NaCl, 40 mM Tris-HCl pH 8.0) to a stock concentration of 140 μM.

2.3 Enzymes

Any enzyme with corresponding properties can be used. For the presented protocol, used enzymes were purchased from New England Biolabs (NEB) except for the Tn5 transposase and the HiFi polymerase.

KAPA HiFi HotStart ReadyMix (*see* **Note 11**).

1. Transposase; here in-house purified Tn5 transposase [11, 12].
2. Proof-reading polymerase.
3. Recommended enzymes for the Oxford Nanopore Technologies Ligation Sequencing kit enzymes (*see* **Note 12**).

2.4 Chemicals, Buffers, and Media Components

1. Ethanol (*see* **Note 13**).
2. Magnetic beads (*see* **Note 14**).
3. DNA intercalating dye for DNA visualization after gel electrophoresis (*see* **Note 15**).
4. Annealing buffer: 50 mM NaCl, 40 mM Tris–HCl pH 8.0.
5. Tn5 dilution buffer: 10 mM Tris–HCl pH 7.5, 150 mM NaCl.
6. 4 X tagmentation buffer: 40 mM Tris–HCl pH 7.5, 40 mM $MgCl_2$.
7. 100% dimethylformamide (DMF).

Table 2
Oligonucleotides for barcode attachment[a]

Name	Sequence (5′-3′)[b]	Information
SLo2420	**AAGAAAGTTGTCGGTGTCTTTGTG**CCCAG TCACGACGTTGTAAAACG	M13 forward primer with example barcode attached.
SLo2421	**TCGATTCCGTTTGTAGTCGTCTGT**CCCAG TCACGACGTTGTAAAACG	M13 forward primer with example barcode attached.
SLo2422	**GAGTCTTGTGTCCCAGTTACCAGG**CCCAG TCACGACGTTGTAAAACG	M13 forward primer with example barcode attached.
SLo2423	**TTCGGATTCTATCGTGTTTCCCTA**CCCAG TCACGACGTTGTAAAACG	M13 forward primer with example barcode attached.
SLo2424	**CTTGTCCAGGGTTTGTGTAACCTT**CCCAG TCACGACGTTGTAAAACG	M13 forward primer with example barcode attached.
SLo2516	**AAGAAAGTTGTCGGTGTCTTTGTG**AGCGGATAACAA TTTCACACAGG	M13 reverse primer with example barcode attached.
SLo2517	**TCGATTCCGTTTGTAGTCGTCTGT**AGCGGATAACAA TTTCACACAGG	M13 reverse primer with example barcode attached.
SLo2518	**GAGTCTTGTGTCCCAGTTACCAGG**AGCGGATAACAA TTTCACACAGG	M13 reverse primer with example barcode attached.
SLo2519	**TTCGGATTCTATCGTGTTTCCCTA**AGCGGATAACAA TTTCACACAGG	M13 reverse primer with example barcode attached.
SLo2520	**CTTGTCCAGGGTTTGTGTAACCTT**AGCGGATAACAA TTTCACACAGG	M13 reverse primer with example barcode attached.

[a] Five forward and reverse primer examples are provided. The barcodes can be freely defined by the user. Oxford Nanopore Technologies provides the sequences for their 96 established barcodes

[b] Bold letters indicate the barcode sequence

8. 100% dimethyl sulfoxide (DMSO).
9. Glycerol.

2.5 Consumables

1. PCR reaction vessels from single tubes to 384-well plates.
2. Appropriate seal for PCR plates.

3 Methods

3.1 Generation of Barcoded Amplicons

The key element of the described dual barcoding procedure is that the initial PCR or enzymatic fragmentation is performed with oligonucleotides containing a landing sequence for a subsequent PCR to add the user-defined barcodes (Fig. 2). This strategy has the advantage that the primers containing the barcodes are highly flexible. By simply changing the initial PCR primers to a new target sequence, the barcode primers can be reused. The protocol

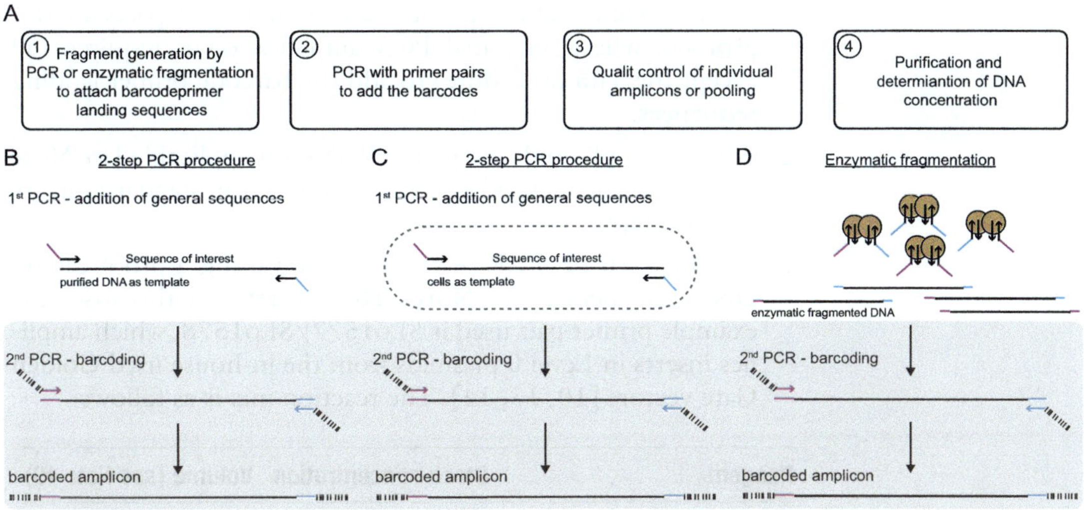

Fig. 2 Workflow and concept of the different types of dual barcoded DNA fragment generation. (**a**) Workflow for the generation of the dual barcoded amplicons. (**b**, **c**, and **d**) visualize the different strategies used to generate DNA fragments with the landing sequences for the primers attaching the barcode. The PCR attaching the barcodes is in all scenarios identical indicated by the gray background. The strategies differ in using purified DNA as a template for a distinct sequence of interest (**b**), using cells as a template (indicated by dotted rod-shaped bacterial cell) for a distinct sequence of interest (**c**) or enzymatically fragment purified DNA for whole plasmid analysis (**d**). The strategies were developed in parallel to allow the reuse of the barcoding primers for any type of application by only swapping the amplification primers with the landing sequences in the first step of (b) and (c)

describes three different approaches to generate the dual barcoded amplicons. The three different approaches differ only in the initial step (Fig. 2b, d). Enzymatic fragmentation has the advantage of providing dual barcoded amplicons for global analysis beyond a specific amplicon. However, this strategy results in a reduced sequencing output due to the generation of fragments with the same barcode at both ends (Fig. 2d). The generation of dual barcoded amplicons can be performed within half a day from purified DNA or single colonies.

3.1.1 Generation of Barcoded Amplicons from Purified Plasmids

1. Design primer sequences for target plasmid to generate PCR amplicons with M13 barcode primer landing sequences (Fig. 2B) (*see* **Note 16**). Below is an example design, in this case, “N” indicates the amplification-specific sequence to be added and the determined sequence is the M13 forward and reverse sequences:

   forward primer 5′-**CCCAGTCACGACGTTGTAAAACG**NNNNNNNNNNNNNNNNNNN-3′.

   reverse primer 5′-**AGCGGATAACAATTTCACACAGG**NNNNNNNNNNNNNNNNNNN-3Ç.

Important: These primers are used for all parts of this protocol where an initial PCR amplification is required (all except the Tn5 procedure) to add the barcode primer landing sequences.

2. Extract the plasmids (or genomic DNA if applicable [*see* **Note 17**]) using the method of choice and perform standard quality control steps.
3. Prepare a 10 μL PCR reaction mixture using a proofreading DNA polymerase (*see* **Notes 18** and **19**). In this case, the example primer pair used is SLo1577/SLo1578, which amplifies inserts in Level 0 plasmids from the in-house used Golden Gate vectors [10, 13, 14]. The reaction mix is as follows:

Reagent	Stock concentration	Volume (*see* Note 19)
Reaction buffer	5 X	2 μL
dNTPs	10 mM	0.2 μL
Forward primer (SLo1577)	2 μM	0.625 μL
Reverse primer (SLo1578)	2 μM	0.625 μL
DNA template	1–10 ng	–
Polymerase	2 U/μL	0.1 μL
ddH_2O	–	to 10 μL

4. Perform PCR reaction in a thermal cycler using the following settings (*see* **Note 20**):

Step	Settings	Cycles
Initial denaturation	98 °C for 20 s	–
PCR	98 °C for 20 s 66 °C for 20 s 72 °C for 30 s	10×
Final extension	72 °C for 1 min	–
Hold	12 °C	–

5. Prepare a 1:10 dilution of the initial PCR to be used as a template for the subsequent PCR.
6. Prepare 10 μL PCR reactions using unique combinations of barcode primer pairs (examples are provided in Table 2):

Reagent	Stock concentration	Volume (*see* Note 19)
Reaction buffer	5 X	2 μL
dNTPs	10 mM	0.2 μL
Forward primer	2 μM	1 μL
Reverse primer	2 μM	1 μL
DNA template	–	1 μL
Polymerase	2 U/μL	0.1 μL
ddH_2O	–	to 10 μL

7. Perform the PCR reaction in a thermal cycler using the following settings:

Step	Settings	Cycles
Initial denaturation	98 °C for 30 s	–
PCR	98 °C for 10 s 66 °C for 10 s 72 °C for 30 s	25×
Final extension	72 °C for 1 min	–
Hold	12 °C	–

8. Pool all PCR reactions and perform PCR purification using the method of choice (*see* **Notes 3** and **21**).
9. Determine DNA concentration using a spectrophotometer and a fluorescent dye-based method (*see* **Note 22**).
10. It is recommended to perform an electrophoretic analysis of the DNA to ensure that the size distribution of the DNA represents the expected sizes.
11. Store samples at −20 °C until further use or proceed directly to **step 1** in Subheading 3.3 (*see* **Note 23**).

3.1.2 Generation of Barcoded Amplicons from Microbial Colonies

1. The protocol has been tested for *E. coli* and *Saccharomyces cerevisiae* colonies (*see* **Note 24**). The cPCR procedure is described in the next two points, respectively.
2. For *E. coli* candidates: Transfer cell material to a PCR tube (e.g., using a sterile toothpick) containing 50 μL ddH_2O and vortex briefly mix to resuspend. Use 1 μL as a template in the PCR reaction (**step 4**). Ensure that that cell material is preserved, e.g., by storing the cell solution in the refrigerator or by cultivating it in appropriate media for subsequent cryopreservation.

3. For *S. cerevisiae* candidates: Transfer cell material to a PCR tube (e.g., using a sterile toothpick) containing 50 μL ddH_2O and vortex briefly to resuspend. Transfer 25 μL of the cell suspension to a PCR tube containing 25 μL of 40 mM NaOH. Store the remaining 25 μL cell suspension in the refrigerator. Transfer the PCR tube with 50 μL suspension to a PCR cycler or heat block and lyse cells at 95 °C for 15 min. Use 1 μL as a template in the PCR reaction.
4. Prepare PCR reactions with 10 μL total volume depending on the number of candidates according to the following reaction mix:

Reagent	Stock concentration	Volume (*see* Note 19)
Reaction buffer	5 X	2 μL
dNTPs	10 mM	0.2 μL
Forward primer (SLo1577)	2 μM	0.625 μL
Reverse primer (SLo1578)	2 μM	0.625 μL
DNA template	–	1 μL
Polymerase	2 U/μL	0.1 μL
ddH_2O	–	to 10 μL

5. Perform the PCR reaction in a thermal cycler using the following settings (*see* **Note 20**):

Step	Settings	Cycles
Initial denaturation	98 °C for 3 min	–
PCR	98 °C for 20 s 66 °C for 20 s 72 °C for 30 s	10×
Final extension	72 °C for 1 min	–
Hold	12 °C	–

6. Dilute the PCR reaction 100-fold to serve as a template for the next step to ensure a maximum of barcoded amplicons.
7. Prepare PCR reactions with 10 μL total volume depending on the number of candidates according to the following reaction mix. Ensure that each reaction has a unique combination of barcode primer pairs (see Table 2 for examples):

Reagent	Stock concentration	Volume (*see* Note 19)
Reaction buffer	5 X	2 μL
dNTPs	10 mM	0.2 μL
Forward primer	2 μM	1 μL
Reverse primer	2 μM	1 μL
DNA template	–	1 μL
Polymerase	2 U/μL	0.1 μL
ddH_2O	–	to 10 μL

8. Perform the PCR reaction in a thermal cycler using the following settings:

Step	Settings	Cycles
Initial denaturation	98 °C for 30 s	–
PCR	98 °C for 10 s 66 °C for 10 s 72 °C for 30 s	25×
Final extension	72 °C for 1 min	–
Hold	12 °C	–

9. Pool all PCR reactions and perform PCR purification using the method of choice (*see* **Notes 3** and **21**).
10. Determine DNA concentration using a spectrophotometer and a fluorescent dye-based method (*see* **Note 22**).
11. It is recommended to perform an electrophoretic analysis of the DNA to ensure that the size distribution of the DNA corresponds to the expected sizes.
12. Store samples at −20 °C until further use or proceed directly to **step 1** in Subheading 3.3 (*see* **Note 23**).

3.1.3 Generation of Barcoded Amplicons from Plasmids by Enzymatic Fragmentation

Prelude: The following protocol is based on the use of the $Tn5_{(R27S,\ E54K,\ L372P)}$ transposase developed by Hennig et al. [11]. This mutated version of Tn5, produced in-house according to the author's procedures, was selected because it generates larger fragments than the regular Tn5 used for library preparation for short-read sequencing technologies. The DNA fragmentation and adapter addition steps are based on the procedure described by Vonesch et al. [12]. Limitations of this procedure are that the transposase integrates adapter sequences in a random manner. This results in fragments with the same barcode at both ends of the sequence. In the theoretical ideal scenario, 50% of the fragments would have two different barcodes at both ends and 25% would have identical barcodes of either the forward or reverse landing pad sequence.

1. Extract the plasmids of interest using the method of choice and perform quality control by determining the $A_{260/280}$ using a spectrophotometer and determining the DNA concentration using a fluorescence assay of choice.
2. Dilute samples to a final concentration of 10–20 ng/μL and store at −20 °C until use.
3. Prepare annealed oligonucleotides for Tn5 loading. Resuspend lyophilized oligonucleotides SLo0100, SLo0673, and SLo0674 (Table 2) in annealing buffer to a stock concentration of 140 μM and mix independently: add one volume (50 μL) of SLo0673 or SLo0674 to one volume of SLo0100. The resulting working stock is 70 μM.
4. Perform the following program in a thermal cycler to anneal the oligonucleotides:

Step	Settings	Time
Denaturation	95 °C	5 min
Annealing[a]	95 to 65 °C	0.1 °C/s
Hold	65 °C	5 min
Slow cooling[a]	65 to 4 °C	0.1 °C/s
Hold	4 °C	–

[a]This can usually be programmed by setting the PCR ramp rate to the minimum value

5. Dispense the annealing reaction in 10 μL aliquots on ice into PCR tubes and store at −20 °C.
6. Prepare the following reaction mix to load the Tn5 with the adapters:

Reagent	Stock concentration	Volume
Reaction buffer	0.5 mg/mL	2 μL
Annealed adapter (M13 forward)	140 μM	1 μL
Annealed adapter (M13 reverse)	140 μM	1 μL
Tris-HCl pH 7.5	20 mM	8 μL
Total	–	11 μL

7. Incubate at 23 °C for 30–60 min at 300 rpm in a thermomixer (*see* **Note 25**).

8. Add 89 μL dilution buffer for immediate use. Alternatively, add 11 μL of 100% glycerol to store the adapter-loaded Tn5 mixture at −20 °C (50% final concentration). For the stored Tn5 mixture, dilute 1:5 with dilution buffer prior use.
9. Perform enzymatic fragmentation of each plasmid DNA separately by preparing the reaction mixture listed below. Keep all the components and final reactions on ice.

Component	Stock concentration	Volume
Tagmentation buffer	4 X	1.25 μL
DMF	100%	1.25 μL
Diluted Tn5 adapter complex	–	1.25 μL
Plasmid DNA (10–20 ng/μL)	–	1.25 μL
Total	–	5 μL

10. Incubate the reactions in a thermal cycler at 55 °C for 30 s and inactivate at 80 °C for 5 min.
11. Store samples on ice if barcoding PCR is to be performed subsequently. Alternatively, store at −20 °C until further use.
12. Prepare 7 μL PCR reactions using KAPA HiFi HotStart ReadyMix (Roche) to add barcodes to the enzymatically fragmented DNA. Add a unique barcode combination to each reaction using different primer pairs (see Table 2 for examples of primers) (*see* **Note 26**):

Reagent	Stock concentration	Volume (*see* Note 19)
KAPA HiFi Master Mix	2 X	3.5 μL
DMSO	100%	0.52 μL
Forward primer	10 μM	0.21 μL
Reverse primer	10 μM	0.21 μL
DNA template	–	2 μL
ddH_2O	–	to 7 μL

13. Perform the following PCR reaction in a thermal cycler which is optimized for the M13 landing sequences (*see* **Note 20**):

Step	Settings	Cycles
Gap filling (*see* **Note 27**)	72 °C for 3 min	
Initial denaturation	95 °C for 30 s	–

(continued)

Step	Settings	Cycles
PCR	98 °C for 20 s 66 °C for 15 s 72 °C for 5 min	20×
Final extension	72 °C for 10 min	–
Hold	12 °C	–

14. Pool all PCR reactions and perform PCR purification using the method of choice (*see* **Notes 3** and **21**).
15. Determine DNA concentration using a spectrophotometer and a fluorescent dye-based method (*see* **Note 22**).
16. Perform an electrophoretic analysis of the DNA to ensure the distribution of DNA sizes represents the expected sizes. In this case, a DNA smear and no distinct band is expected.
17. Store samples at −20 °C until further use or proceed directly to **step 1** in Subheading 3.3 (*see* **Note 23**).

3.2 Utilizing Automation for Downscaling of Reactions

High-throughput library preparation requires a large amount of reagents and is prone to human error. Recent technological developments allow for the use of laboratory automation to increase the number of samples while reducing the use of reagents. Acoustic dispensers have become a valuable tool for dispensing of complex reaction mixtures [15, 16]. Acoustic dispensers are used to automate and parallelize Golden Gate DNA assembly reactions [13]. Acoustic dispensers have previously been used to scale down the preparation of short-read sequencing libraries for plasmid validation [5, 6]. The acoustic dispenser is used here to accurately dispense the barcode primer pair combinations. The experimental procedure takes about half a day including PCR reactions and DNA purification. This procedure has been tested to sequence up to 1536 amplicons on a single Flongle flow cell [10].

3.2.1 Barcoded Amplicons from Purified Plasmids

1. Purify plasmids in a 96-well format using the method of choice. Alternatively, arrange plasmids in 96-well format.
2. Prepare a 1:100 plasmid dilution and transfer 1 μL per sample into a 384-well plate using 8-, 12-, or 96-channel pipettes (*see* **Note 28**).
3. Dispense 1 μL of the first PCR master mix into each well. The reaction mixture per sample is as follows:

Reagent	Stock concentration	Volume (*see* Note 19)
Reaction buffer	5 X	0.4 μL
dNTPs	10 mM	0.04 μL
Forward primer	2 μM	0.125 μL
Reverse primer	2 μM	0.125 μL
Polymerase	2 U/μL	0.02 μL
ddH_2O	–	0.29 μL

4. Seal the PCR plate with an appropriate seal.
5. Ensure that the reaction mix is at the bottom of all wells by gently spinning the plate in an appropriate centrifuge.
6. Perform the first PCR with optimized settings for the primer pairs used and the expected amplicon size. For the standard primer pair used in this study (SLo1577/SLo1578) and expected amplicon sizes the settings are as follows:

Step	Settings	Cycles
Initial denaturation	98 °C for 20 s	–
PCR	98 °C for 20 s 66 °C for 20 s 72 °C for 30 s	10×
Final extension	72 °C for 1 min	–
Hold	12 °C	–

7. Remove the seal from the plate and dispense unique barcoded primer combinations into each reaction mix. In this protocol, an acoustic dispenser is used to dispense 100 nL of each oligonucleotide (final concentration of 0.2 μM) into the reaction mix in 2.5 or 25 nL increments, depending on the machine type (*see* **Note 29**).
8. Prepare the master mix for the second PCR reaction as described below and dispense 3 μL into each well. The total reaction volume is 5 μL.

Reagent	Stock concentration	Volume (*see* Note 19)
Reaction buffer	5 X	1 μL
dNTPs	10 mM	0.1 μL
Polymerase	2 U/μL	0.05 μL
ddH_2O	–	1.85 μL

9. Seal the PCR plate with an appropriate seal and perform the second PCR with optimized settings for the primer pairs. In the case of the standard M13 barcode primer pair used in this study (Table 2), the settings are as follows:

Step	Settings	Cycles
Initial denaturation	98 °C for 30 s	–
PCR	98 °C for 10 s 66 °C for 10 s 72 °C for 30 s	25×
Final extension	72 °C for 1 min	–
Hold	12 °C	–

10. Combine all PCR reactions in a single tube and proceed with the preferred amplicon purification procedure (*see* **Note 30**).
11. After purification, determine DNA quality and concentration using a spectrophotometer and a fluorescence-based method.
12. Test the purified DNA by an electrophoresis method of choice.
13. Store samples at −20 °C until further use or proceed directly to **step 1** in Subheading 3.3 (*see* **Note 23**).

3.2.2 Barcoded Amplicons from Microbial Colonies

1. Bacterial cells are arrayed in a 384-grid format on solid agar using a colony-picking robot.
2. Solid plates are incubated at 37 °C for several hours to obtain sufficient but not excessive cell material. Alternatively, liquid cultures can be grown overnight at 37 °C with vigorous shaking. Here, a humidity-controlled incubator to incubate 384-well microtiter plates at 550 rpm, 3 mm shaking, 37 °C, and 75% relative humidity is used. With these settings, it is possible to work without a seal for microbial culture (*see* **Note 31**). Liquid culture may have an advantage in that the transfer of cell material is more standardized. At the same time, however, transferring media to low-volume cPCR reactions can cause inhibition of the PCR reaction, especially if cultures are incubated too long.
3. Dispense 2 μL cPCR master mix to the wells of a 384 PCR plate. The reaction mixture is as follows:

Reagent	Concentration	Volume (*see* Note 19)
Reaction buffer	5 X	0.4 μL
dNTPs	10 mM	0.04 μL
Forward primer	2 μM	0.125 μL
Reverse primer	2 μM	0.125 μL
Polymerase	2 U/μL	0.02 μL
ddH_2O	–	1.29 μL.

4. Transfer cell material using a 384-well pinpad either from the incubated solid media plate or the liquid culture to the cPCR PCR plate containing the dispensed cPCR master mix (*see* **Note 32**).
5. Ensure that the reaction mix is at the bottom of all wells by gently pulse-centrifugation of the plate. Too much centrifugation may cause cells to pellet at the bottom of the wells and prevent efficient cPCR (*see* **Note 33**).
6. From this step onward the protocol is identical to Subheading 3.2.1 from **step 6**. Perform the following steps according to **step 6** and following Subheading 3.2.1.

3.3 Nanopore Sequencing Library Preparation and Nanopore Sequencing

Nanopore sequencing library preparation relies on the library preparation kits provided by Oxford Nanopore Technologies. For the DuBA.flow workflow, the Ligation Sequencing kit series is used (*see* **Note 34**). The procedure of the library preparation takes around two hours and sequencing on a Flongle flow cell, which is sufficient for most applications in regard to DNA assembly validation takes 24 h.

1. If not already done, determine the concentration of the sample using a fluorescence-based method. It is also recommended to measure the sample with a spectrophotometer to determine the $A_{260/280}$ ratio, which should be 1.8.
2. The procedure outlined is based on the Oxford Nanopore Technologies Ligation Sequencing Kit (SQK-LSK109). Due to the constantly improving chemistry, it is recommended to frequently check for updated protocols and to follow the latest procedure for the available kit version (*see* **Note 35**).
3. Perform library preparation according to the most recent protocol (*see* **Note 36**).
4. Assemble the Nanopore sequencer with the Flongle adapter and connect the sequencer to the operating computer.

5. Perform the sequencing experiment with the preset settings for Flongle flow cells. No live basecalling is performed in this scenario. If the used hardware supports GPU-based basecalling, live basecalling is recommended.
6. After the sequencing run, disassemble and store all equipment appropriately. If Flongle cells are used, dispose according to the local regulations. If standard flow cells are used act according to the current return procedure of Oxford Nanopore Technologies (*see* **Note 37**).
7. Appropriately store and back up the primary, not basecalled data.
8. Perform basecalling of raw data according to the preferred settings. In this protocol, basecalling is performed using the Oxford Nanopore Technologies basecaller Guppy in the GPU mode (*see* **Note 38**). The following command is used in a Linux OS terminal:

```
guppy_basecaller -i /path_to_input_fast5_files/ -r -s /path_-to_output_folder/ --flowcell type_of_flow_cell --kit type_of_library_kit --device cuda:0 --num_callers 24 --disable_qscore_filtering --compress_fastq
```

Important: The elements in *italics* need to be adjusted by the operator.

Proceed to **step 3.4** to analyze the sequencing data using the automated DuBa.flow analysis pipeline.

3.4 Analysis Pipeline and Validation of Sequences

The most important part of sequence analysis is a rapid and throughout analysis pipeline and easy-to-interpret result files (Fig. 3). The DuBA.flow workflow provides a start-to-end solution for in-house sequence analysis. The computational analysis software is available at https://github.com/RGSchindler/DuBA.flow under a CC BY-NC-SA 4.0 license (*see* **Note 39**). The analysis pipeline can be executed on a standard computer and needs minimal computational resources and time. However, a bottleneck may be the basecalling of the raw data, which requires a dedicated GPU and the respective computational environment for rapid base calling. The current basecaller of Oxford Nanopore Technologies is Guppy and is available from the Nanopore Community (*see* **Note 40**). The DuBA.flow analysis takes only little hands on and computation depends on the complexity and amount of data but usually the needed time is neglectable.

1. If no live basecalling was performed during the sequencing experiment, the raw data have to be basecalled (*see* **step 8** in Subheading 3.3).

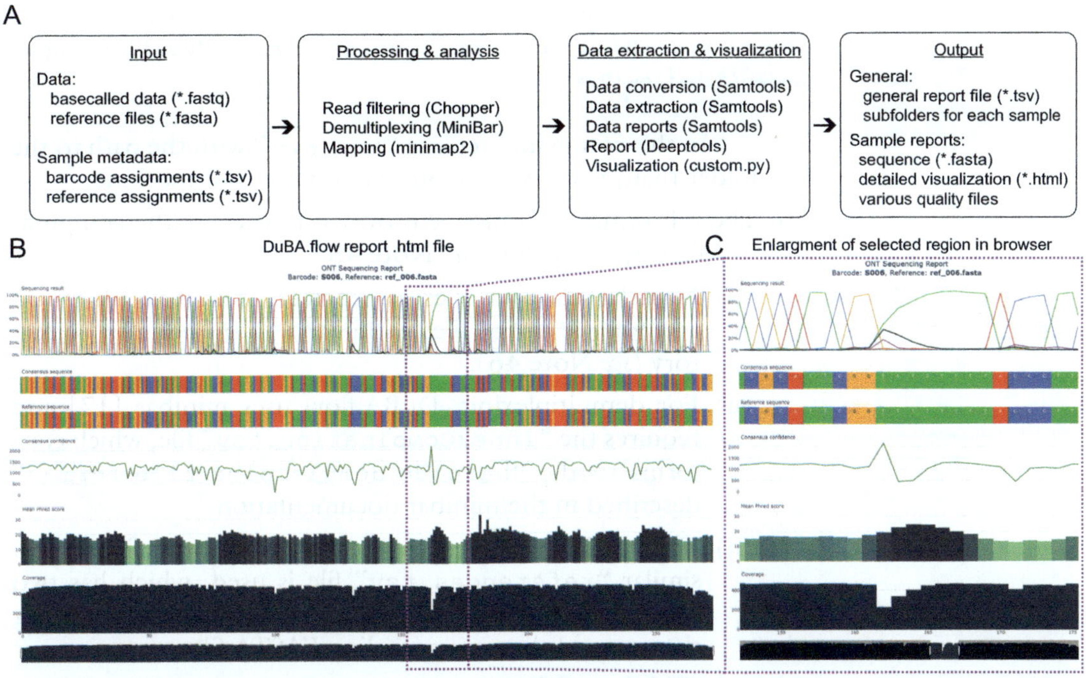

Fig. 3 DuBA.flow analysis pipeline and user-friendly output files. (**a**) Workflow of the automated analysis pipeline which relies on user provided sequencing data, reference data and two sample metadata files. The data are processed automatically and returns user-friendly analysis files in subfolders for each sample, as well as an overview report. (**b**) Example of interactive results file generated by the analysis workflow. The .html file can be opened with standard web browsers and allows the user to move and zoom along the navigation bar; the indicated purple dotted box represents the enlargement shown in (**c**). Zooming in allows to analyze the Nanopore sequencing results similar to Sanger sequencing traces. Figure and figure legend from DuBa.flow GitHub description (https:/github.com/RGSchindler/DuBA.flow)

2. The DuBA.flow computational pipeline is provided as Docker image (*see* **Note 41**). The most recent documentation can be found at https://github.com/RGSchindler/DuBA.flow.
3. Install Docker Desktop (https://www.docker.com/) on the machine used for analysis purposes if not already installed (*see* **Notes 42** and **43**).
4. After successfully installing and launching Docker, the analysis pipeline can be pulled using the command "`docker pull cedkb/dubaflow:0.4`" from the operating system terminal (*see* **Note 44**). This step is only necessary when (a) installing DuBA.flow for the first time or (b) after deleting the DuBA.flow image from Docker.
5. Before submitting the input files and running the analysis pipeline, it is essential to bind the volume of the Docker container to a specific folder on the host machine. To initiate the container and create the volume mount, type the following command in the system terminal:

```
"docker run --name dubaflow --workdir /home/dubaflow -it --
mount type=bind,src=home_path,dst=/home/dubaflow/input
cedkb/dubaflow:0.4"
```

In this command, replace "`home_path`" with the path to the folder that provides the input files on the host machine.

6. The following files must be provided to execute the computational analysis pipeline (*see* **Note 45**):
   (i) Provide reference files as separate, valid FASTA (`*.fa` and/or `*.fasta`) files within the "`references`" directory (*see* **Note 46**).
   (ii) For demultiplexing, DuBA.flow uses minibar [17] and requires the "`IndexCombination.tsv`" file, which conforms to the "`barcode demultiplex file format`" described in the minibar documentation.
   (iii) To associate samples with their respective references, a similar "`references.tsv`" file is used, which has two columns: SampleID and ReferenceFile, the latter being the name of the corresponding FASTA file.
   (iv) The base-called raw reads are provided in a single "`input.fastq.gz`" file.
7. The terminal, after executing the "`docker run`" command, should seamlessly transition to the terminal interface of the "`dubaflow`" docker container. After placing the input files in the designated folders, the pipeline can be easily initiated by running the command "`python3 DuBaFlow.py`" (*see* **Note 47**).
8. First DuBA.flow checks if all required files are available, if not an error message is returned, otherwise the calculation pipeline calculates everything automatically. Keep in mind the computation time needed depends on the specifications of the used computer and the complexity of the data.
9. Once the DuBa.flow analysis pipeline has been completed, an output directory is created in the specified home path, containing.
   (i) A comprehensive summary report file.
   (ii) Distinct subfolders for each sample encapsulating individual reports.

   A report includes the processing status of each sample and provides a statistical breakdown highlighting metrics such as the total number of sequence reads, the number of reads successfully mapped, the observed error rate, the average read length, and the average quality score. The individual sample reports provide a range of valuable reports and output files, including the interactive easy-to-use HTML report file (Fig. 3b) that conveys all essential information and a consensus file in FASTA.

3.5 Summary and Perspectives

Golden Gate DNA assembly is growing in popularity, complexity, and throughput. Long-read sequencing has become a widely applicable technology for fast and accurate in-house sequence validation and genome sequencing. The step-by-step protocol outlined here provides a detailed description of how to use the DuBA.flow workflow for rapid analysis of Golden Gate DNA assemblies from low to high throughput. The user-friendly result files of the analysis pipeline are valuable for sequence validation of basic parts, transcription units, and/or higher order Golden Gate assemblies.

4 Notes

1. Any other Oxford Nanopore Technologies Nanopore sequencer should be compatible with this procedure. If other manufacturers/sequencing technologies are used, the protocol must be adapted accordingly.
2. Other Nanopore sequencing library kits may be used. However, we experienced the best results with the ligation sequencing kit.
3. Any commercial kit or in-house procedure can be used. We used the open-source magnetic bead protocols available at www.bomb.bio for plasmid extraction and DNA purification [18].
4. This protocol uses the Echo525/650 T, the ARI Cobra, the Singer Instruments PIXL, and the Singer Instruments Rotor HDA+. Other instruments may be used. However, the procedure is optimized for the combination of instruments outlined.
5. Depending on the Nanopore sequencing platform used, a machine for operating the Nanopore sequencer may already be included (e.g., GridION).
6. The settings below are the systems used in this chapter; they are neither minimum nor maximum specifications. Other operating systems and software may be used.
7. It is recommended to check the latest hardware requirements for Nanopore sequencing before ordering a computer to operate a Nanopore sequencer.
8. GPU-based basecalling requires an Nvidia GPU with a minimum compute capability. It is recommended to check Guppy (or alternative basecaller) specifications for the latest requirements.

9. Hardware requirements are subject to change. For the latest DuBA.flow analysis information, visit https://github.com/RGSchindler/DuBA.flow.
10. The overhangs can be freely defined by the user. For historical reasons, the M13 sequences are used in this application. The only requirements for the overhangs are that the forward and reverse are not identical and do not bind elsewhere in the DNA samples.
11. Other products may be suitable, but this reagent worked best in the setup described.
12. All enzymes and consumables were used according to the Oxford Nanopore Technologies Ligation Sequencing Kit protocol, with the exception of the FFPE enzyme mix, which was not used, and the AMPure XP beads, which were purchased from an alternative supplier.
13. Always prepare ethanol dilutions fresh to avoid reduced alcohol content due to evaporation.
14. Protocols are available to generate in-house replacements for commercial supplies. A good reference is, e.g., https://doi.org/10.17504/protocols.io.bkppkvmn.
15. Thiazole orange has been established as a reliable and inexpensive dye for staining DNA [19]. Thiazole orange is dissolved in dimethyl sulfoxide (DMSO) (10,000× stock concentration: 13 mg/mL).
16. The M13 sequences are used for historical reasons but can be replaced by any other sequence. It can be an in silico designed sequence. These sequences should be unique to serve as ideal landing sequences for the barcoding primers.
17. This procedure can also be used to analyze amplicons from any other type of DNA, such as amplicons from genomes or metagenomic samples. Only the initial primers from **step 1** need to be redesigned to amplify, e.g., bacterial 16S or fungal ITS sequences.
18. We have obtained good results using the Q5 High-Fidelity DNA polymerase (NEB) with the standard buffer. However, each new primer pair is tested with a gradient PCR for optimal amplification. The optimal primer concentration may vary with other polymerases.
19. It is recommended to prepare a master mix for the PCR reaction for the number of samples to be analyzed. When preparing a master mix, it is recommended to include a 10% volume overage to account for volume loss during dispensing. The master mix contains all elements except the DNA template and barcode attachment primer, which are unique to each sample. These reagents are added to the individual reactions

after dispensing the Master Mix. When using small amounts of oligonucleotides, it is recommended to prepare 2 μM working solutions for more accurate pipetting.

20. If other primer sequences are used, it is recommended to test their optimal annealing temperature by performing a gradient PCR and gel electrophoresis to check for unspecific amplification.
21. When analyzing large numbers of amplicons, it does not make sense to purify and quality control each amplicon individually. It is more economical to repeat failed candidates in the next sequencing experiment than to perform quality control on individual amplicons. An appropriate number of candidates should be analyzed from the start, taking into account incorrect assemblies and amplification failures.
22. Spectrophotometric analysis is intended to provide information on DNA purity via the $A_{260/280}$ value, which should ideally be 1.8. The DNA concentration is determined using a fluorescence-based assay to be as accurate as possible.
23. Barcoded fragments from different fragment generation strategies can be combined as long as each barcode combination is unique. To obtain good coverage for each fragment, it is recommended to pool different samples based on their molarity (n) and not mass (m), as this may bias the resulting sequencing data toward shorter fragments.
24. This procedure can be adapted to any material, the only critical part is to perform an efficient initial PCR to ensure that there is enough template for the subsequent PCR with the primer pairs that attach the barcodes.
25. Ensure that the temperature is maintained at 23 °C. Do not exceed 60 min of loading, as the Tn5 enzyme will gradually lose activity.
26. To avoid pipetting very small volumes, it is recommended to mix the forward and reverse barcode primers 1:1, resulting in a 5 μM mixture.
27. The gap-filling step at 72 °C for 3 min is essential to fill the 5′ overhangs of the single-stranded linker oligonucleotides to allow for adapter primer binding and amplification of the enzymatically fragmented DNA.
28. Carefully track and document the location of each plasmid sample.
29. This procedure is optimized for acoustic dispensers. This has the advantage of using very small volumes and being tip-free. If an acoustic dispenser is not available, other types of equipment may be suitable. However, care must be taken to avoid cross-contamination of primers with attached barcodes.

30. Instead of pipetting each well into a tube, you can invert the PCR plate into a matching microplate lid and spin the plate briefly to collect all samples in the lid. From there, transfer to a reaction tube of the appropriate size for your amplicon cleanup procedure. *Important:* Test the settings before performing this step with valuable sample material.
31. Settings must be optimized for each incubator to avoid cross-contamination or evaporation, and to achieve optimal aeration and microbial growth.
32. This protocol uses a Singer Instruments Rotor HDA+. Different robots have different types of pinpads and may need to be optimized for this procedure. However, if the sample size is not too large, the procedure can be performed with a manual 384 replica plating stamp.
33. It is recommended to test the primer efficiency on a small scale with the same settings before performing a high throughput experiment.
34. Follow the most recent library preparation protocol for the chemistry provided in the Nanopore Community.
35. We do not use the recommended FFPE enzyme reagent in the outlined in the Oxford Nanopore Technologies protocol for the ligation sequencing protocol.
36. Carefully review the Oxford Nanopore Technologies protocol for the selected library sequencing kit. The kit does not contain all necessary reagents and materials, which must be purchased in advance, e.g., magnetic racks, enzymes, and magnetic beads.
37. Wash and store cells according to the most recent procedure provided in the Nanopore Community protocols and return to Oxford Nanopore Technologies as outlined by the company. We do not reuse MinION flow cells. While this option is available, it is recommended that different barcode combinations are used to avoid cross-contamination due to consecutive sequencing experiments.
38. It is recommended to check the Nanopore Community frequently for the latest basecalling software and documentation.
39. It is recommended to refer to the GitHub documentation for the latest user guides and software versions.
40. It is recommended to refer to the Oxford Nanopore Technologies documentation for the latest operating instructions and software versions of Guppy or alternative basecalling software.
41. For the development and deployment of our pipeline, Docker was used to address the complex and diverse dependencies and environmental configurations inherent in bioinformatics tools

and applications. This technology ensures consistent, reproducible, and reliable results across multiple operating systems and computing infrastructures.

42. Docker Engine is available on a variety of Linux distributions, MacOS, and Windows OS as a static binary installation. See the DuBA.flow documentation for compatible versions of Docker Desktop.
43. To run Docker on Windows machines, you must have a system running a recent version of Windows 10 or Windows 11 with the Windows Subsystem for Linux 2 (WSL2) installed and enabled. The pipeline was tested using a laptop running Windows 11 Home edition ver. 22H2 with 16 GB of RAM and an intel i5-11300H processor or better.
44. Example files are provided with the Docker image.
45. If a reference is not available or a sequence needs to be assembled de novo, the Ref.creator tool (https://github.com/RGSchindler/Ref.creator) can be used to generate a reference or de novo assembly based on the sequencing reads. The tool is available as a Docker image on the Docker Hub and its detailed documentation is maintained on GitHub and should be followed.
46. Depending on the Docker settings, you may need to type "`bash`" and press Enter before running the "`python3 DuBaFlow.py`" command.

Acknowledgments

This work was supported by the Max Planck Society within the framework of the MaxGENESYS project (DS) and the European Union (NextGenerationEU) via the European Regional Development Fund (ERDF) by the state Hesse within the project "biotechnological production of reactive peptides from waste streams as lead structures for drug development" (DS). We are grateful to all laboratory members for extensive discussions on the development of the DuBA.flow workflow and in particular Tania S. Köbel for her superior technical support. We thank Lars Steinmetz and Kim Remans at EMBL for sharing the transposase expression plasmids from Hennig et al., 2018. We thank Tobias Erb and the Erb Lab for constant support with protein purification and unrestricted access to equipment. We thank Ehmad Chehrghani Bozcheloe for the visualization of the petri dish and the MinION Nanopore sequencer in Fig. 1. All material is available from the corresponding author upon request; except transposase material originates from Hennig et al., 2018 and must be requested from the appropriate source.

References

1. Bergkessel M, Guthrie C (2013) Colony PCR. Methods Enzymol 529:299–309. https://doi.org/10.1016/B978-0-12-418687-3.00025-2
2. Thompson R, Hughes SG, Broda P (1974) Plasmid identification using specific endonucleases. Mol Gen Genet 133(2):141–149. https://doi.org/10.1007/BF00264835
3. Heather JM, Chain B (2016) The sequence of sequencers: the history of sequencing DNA. Genomics 107(1):1–8. https://doi.org/10.1016/j.ygeno.2015.11.003
4. van Heyningen V (2019) Genome sequencing-the dawn of a game-changing era. Heredity (Edinb) 123(1):58–66. https://doi.org/10.1038/s41437-019-0226-y
5. Suckling L, McFarlane C, Sawyer C, Chambers SP, Kitney RI, McClymont DW, Freemont PS (2019) Miniaturisation of high-throughput plasmid DNA library preparation for next-generation sequencing using multifactorial optimisation. Synth Syst Biotechnol 4(1):57–66. https://doi.org/10.1016/j.synbio.2019.01.002
6. Shapland EB, Holmes V, Reeves CD, Sorokin E, Durot M, Platt D, Allen C, Dean J, Serber Z, Newman J, Chandran S (2015) Low-cost, high-throughput sequencing of DNA assemblies using a highly multiplexed Nextera process. ACS Synth Biol 4(7):860–866. https://doi.org/10.1021/sb500362n
7. Wang Y, Zhao Y, Bollas A, Wang Y, Au KF (2021) Nanopore sequencing technology, bioinformatics and applications. Nat Biotechnol 39(11):1348–1365. https://doi.org/10.1038/s41587-021-01108-x
8. Currin A, Swainston N, Dunstan MS, Jervis AJ, Mulherin P, Robinson CJ, Taylor S, Carbonell P, Hollywood KA, Yan C, Takano E, Scrutton NS, Breitling R (2019) Highly multiplexed, fast and accurate nanopore sequencing for verification of synthetic DNA constructs and sequence libraries. Synth Biol (Oxf) 4(1):ysz025. https://doi.org/10.1093/synbio/ysz025
9. Emiliani FE, Hsu I, McKenna A (2022) Multiplexed assembly and annotation of synthetic biology constructs using long-read Nanopore sequencing. ACS Synth Biol 11(7):2238–2246. https://doi.org/10.1021/acssynbio.2c00126
10. Ramírez Rojas AA, Brinkmann CK, Köbel T, Schindler D (2024) DuBA.flow—a low-cost, long-read amplicon sequencing workflow for the validation of synthetic DNA constructs. ACS Synth Biol 13:457. https://doi.org/10.1021/acssynbio.3c00522
11. Hennig BP, Velten L, Racke I, Tu CS, Thoms M, Rybin V, Besir H, Remans K, Steinmetz LM (2018) Large-spcale low-cost NGS library reparation using a robust Tn5 purification and tagmentation protocol. G3 (Bethesda) 8(1):79–89. https://doi.org/10.1534/g3.117.300257
12. Vonesch SC, Li S, Szu Tu C, Hennig BP, Dobrev N, Steinmetz LM (2021) Fast and inexpensive whole-genome sequencing library preparation from intact yeast cells. G3 (Bethesda) 11(1). https://doi.org/10.1093/g3journal/jkaa009
13. Köbel T, Schindler D (2023) Automation and miniaturization of Golden Gate DNA assembly reactions using acoustic dispensers. arXiv. https://doi.org/10.48550/arXiv.2310.00325
14. Brück M, Berghoff BA, Schindler D (2023) *In silico* design, *in vitro* construction and *in vivo* application of synthetic small regulatory RNAs in bacteria. arXiv. https://doi.org/10.48550/arXiv.2304.14932
15. Kanigowska P, Shen Y, Zheng Y, Rosser S, Cai Y (2016) Smart DNA fabrication using sound waves: applying acoustic dispensing technologies to synthetic biology. J Lab Autom 21(1):49–56. https://doi.org/10.1177/2211068215593754
16. Ellson R, Mutz M, Browning B, Lee L, Miller MF, Papen R (2003) Transfer of low nanoliter volumes between microplates using focused acoustics—automation considerations. J Assoc Lab Autom 8(5):29. https://doi.org/10.1016/S1535-5535-03-00011-X
17. Krehenwinkel H, Pomerantz A, Henderson JB, Kennedy SR, Lim JY, Swamy V, Shoobridge JD, Graham N, Patel NH, Gillespie RG, Prost S (2019) Nanopore sequencing of long ribosomal DNA amplicons enables portable and simple biodiversity assessments with high phylogenetic resolution across broad taxonomic scale. Gigascience 8(5). https://doi.org/10.1093/gigascience/giz006
18. Oberacker P, Stepper P, Bond DM, Hohn S, Focken J, Meyer V, Schelle L, Sugrue VJ, Jeunen GJ, Moser T, Hore SR, von Meyenn F, Hipp K, Hore TA, Jurkowski TP (2019) Bio-on-magnetic-beads (BOMB): open platform for high-throughput nucleic acid extraction and manipulation. PLoS Biol 17(1):e3000107. https://doi.org/10.1371/journal.pbio.3000107
19. O'Neil CS, Beach JL, Gruber TD (2018) Thiazole orange as an everyday replacement for ethidium bromide and costly DNA dyes for electrophoresis. Electrophoresis 39(12):1474–1477. https://doi.org/10.1002/elps.201700489

Chapter 11

Modular Golden Gate Assembly of Linear DNA Templates for Cell-Free Prototyping

François-Xavier Lehr, Aukse Gaizauskaite, Katarzyna Elżbieta Lipińska, Sara Gilles, Arpita Sahoo, René Inckemann, and Henrike Niederholtmeyer

Abstract

Cell-free transcription and translation (TXTL) systems have emerged as a powerful tool for testing genetic regulatory elements and circuits. Cell-free prototyping can dramatically accelerate the design-build-test-learn cycle of new functions in synthetic biology, in particular when quick-to-assemble linear DNA templates are used. Here, we describe a Golden-Gate-assisted, cloning-free workflow to rapidly produce linear DNA templates for TXTL reactions by assembling transcription units from basic genetic parts of a modular cloning toolbox. Functional DNA templates composed of multiple parts such as promoter, ribosomal binding site (RBS), coding sequence, and terminator are produced in vitro in a one-pot Golden Gate assembly reaction followed by polymerase chain reaction (PCR) amplification. We demonstrate assembly, cell-free testing of promoter and RBS combinations, as well as characterization of a repressor–promoter pair. By eliminating time-consuming transformation and cloning steps in cells and by taking advantage of modular cloning toolboxes, our cell-free prototyping workflow can produce data for large numbers of new assembled constructs within a single day.

Key words *E. coli* cell-free systems, Golden Gate cloning, Linear DNA, Rapid prototyping, DNA assembly, In vitro transcription translation

1 Introduction

In cell-free expression, the molecular machinery for transcription and translation is isolated from cells and supplemented with substrates as well as DNA templates to drive the synthesis of mRNA and proteins [1]. Cell-free transcription and translation (TXTL) systems enable rapid prototyping of new genetic elements, such as promoters and ribosomal binding sites, as well as the engineering of transcription factors and riboswitches. TXTL reactions are typically performed in small volumes of a few microliters and fluorescent

François-Xavier Lehr and Aukse Gaizauskaite contributed equally.

Daniel Schindler (ed.), *Golden Gate Cloning: Methods and Protocols*, Methods in Molecular Biology, vol. 2850, https://doi.org/10.1007/978-1-0716-4220-7_11,

reporter outputs are measured in plate readers. This workflow allows the screening of libraries of genetic parts in high throughput [1]. Conveniently, it is possible to use linear DNA as expression templates to eliminate time-consuming cloning and transformation steps. To protect linear DNA from degradation by the RecBCD nuclease complex in *Escherichia coli* lysates, two approaches are commonly used. The first involves adding the phage protein GamS, which stabilizes linear DNA [2]. The second method includes short linear DNA strands containing repeats of the chi sequence, which act as competitive inhibitors, stalling the RecBCD nuclease [3].

Libraries of linear DNA templates to screen can be assembled by different methods. One option is to assemble and amplify them by polymerase chain reaction (PCR) using overlapping oligonucleotides and PCR products [4, 5]. However, this approach may require the purchase of long oligonucleotide libraries and may encounter issues with non-specific PCR byproducts. Another option is to assemble expression templates from multiple basic parts by Golden Gate assembly, followed by a PCR to amplify the transcription unit (TU) of interest [2].

Golden gate cloning relies on type IIS restriction enzymes that cleave outside their recognition sequence, generating short single-stranded overhangs that can be customized to assemble fragments in a defined order [6]. The modular cloning (MoClo) system [7] uses standardized fusion sites that allow the assembly of TUs or multigene constructs from multiple basic genetic parts like promoters, ribosomal binding sites (RBS), coding sequences (CDS), and terminators in the correct order in a single reaction. Several toolboxes, such as the Marburg collection [8], have been developed and contain collections of basic parts that are a useful starting point for many synthetic biology projects. In the Golden Gate cloning systems, basic sequence elements like promoters, RBSs, and CDSs, are called level 0 parts and are stored as level 0 plasmids. These are then assembled into TUs, so-called level 1 constructs. Golden Gate reactions are usually transformed into bacteria, where the assembled plasmid is propagated. However, an assembled level 1 TU can also simply be amplified by a PCR reaction with standard primers to produce linear DNA templates suitable for TXTL reactions. When starting with a collection of level 0 plasmids containing basic parts, large libraries of new constructs can be rapidly generated and characterized in a TXTL system.

The cell-free prototyping workflow we describe here (Fig. 1) allows TU assembly and measurements to be completed in a single day with minimal hands-on effort and the following incubation times: 2 h Golden Gate reaction, 3 h PCR and DNA purification, and 5 h TXTL plate reader reaction. We also describe the assembly of new level 0 plasmids because it is often necessary to expand existing collections of level 0 parts with custom-designed parts for

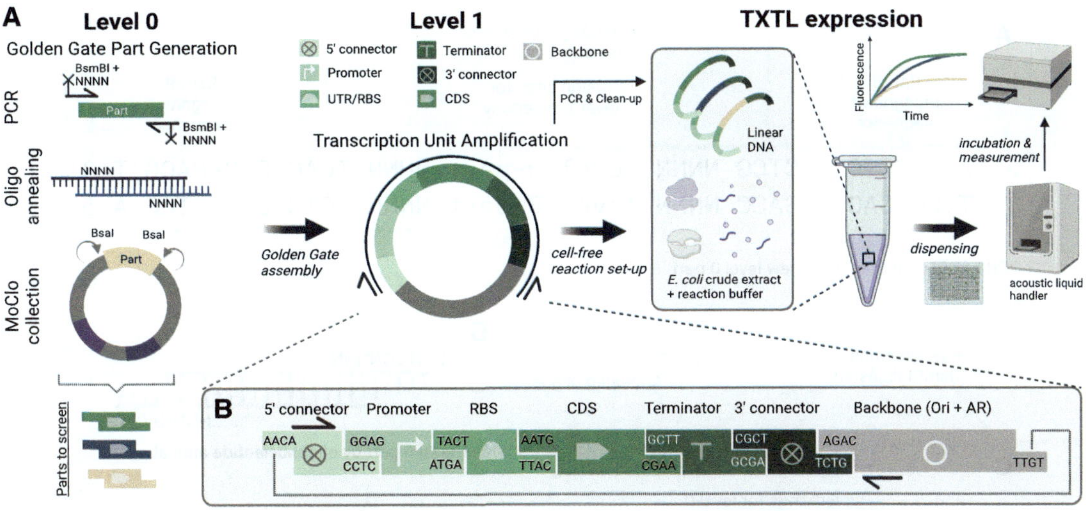

Fig. 1 Overview of the cell-free prototyping workflow. (**a**) Level 0 parts are assembled into level 1 constructs by Golden Gate assembly and subsequently amplified by PCR for functional testing of linear DNA in cell-free reactions. (**b**) Overview of assembly and fusion sites of the level 1 construct

specific projects. Finally, to facilitate the screening of larger libraries, we report the setup of reactions in multiwell plates assisted by an automatic robotic liquid handling system capable of dispensing low volumes.

2 Materials

2.1 Level 0 Plasmids

In the MoClo system, level 0 plasmids contain basic genetic parts (e.g., promoters, RBSs, CDSs, terminators) that can be assembled into TUs, or level 1 constructs via their specific fusion sites (Fig. 1b). Collections of level 0 parts for different organisms, for example, *E. coli* [9] or *Vibrio natriegens* [8] have been created and can be obtained from Addgene or other sources. Here, we used the Marburg collection [8] as a starting point and source for most of the level 0 parts. Additionally, to create new, project-specific level 0 parts, we give examples of their construction in Subheading 3.1. An overview of the level 0 plasmids used in this study is shown in Fig. 3a.

2.2 Oligonucleotides

Researchers may wish to add their parts of interest. Therefore, to illustrate the design, we show how to create new level 0 plasmids using two examples (*see* Subheading 3.1 for the method). Longer parts can be generated by PCR, as shown here for the TetR CDS. Short parts can be created by annealing two oligonucleotides, illustrated here by the promoter pTet. Primer design is illustrated for both cases in Fig. 2. All primer sequences are given in the $5' \rightarrow 3'$ direction.

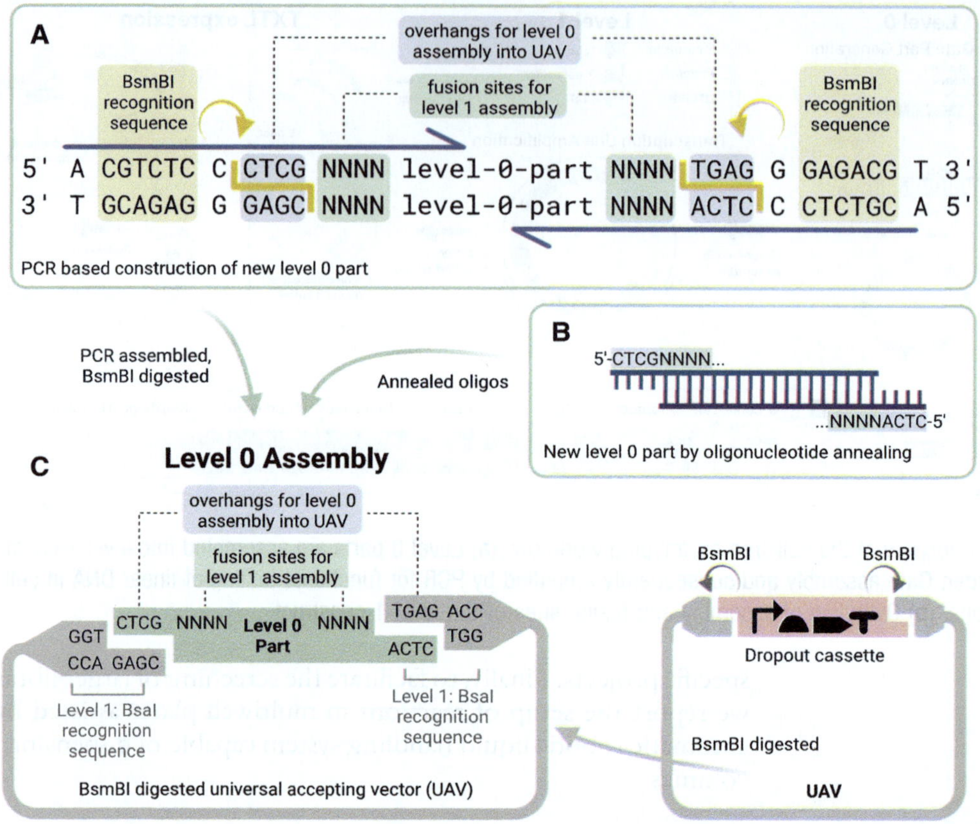

Fig. 2 Construction of new level 0 parts. (**a**) By PCR. (**b**) By annealing two single-stranded DNA oligonucleotides. (**c**) Parts are assembled into a universal accepting vector (UAV) using BsmBI restriction enzyme

1. Creating level 0 part by PCR:
   (a) Forward primer for level 0 part generated by PCR: ACGTCTCC*CTCG***NNNN**(n)$_{15–25}$.
   (b) Reverse primer for level 0 part generated by PCR: ACGTCTCC*CTCA***NNNN**(n)$_{15–25}$.
   (c) Example primers used for amplifying the TetR CDS level 0 part (forward/reverse):
   ACGTCTCC*CTCG***AATG**(tccagattagataaaagtaaagtg)
   ACGTCTCC*CTCA***AAGC**(agacccactttcacatttaagttg).

   where (n)$_{15–25}$ are nucleotides from the 5′ and 3′ end of the gene of interest. BsmBI recognition sites are underlined. The overhangs for level 0 part integration into the universal accepting vector (UAV) are italicized. Fusion sites (NNNN) of the Golden Gate level 1 assembly are denoted in bold (*see* **Note 1**).

2. Creating level 0 part by oligo annealing:
   (a) Forward oligonucleotide for level 0 part production by annealing:

   *CTCG***NNNN**$(n)_{15\text{-max}}$.

   (b) Reverse oligonucleotide for level 0 part production by annealing:

   *CTCA***NNNN**$(n)_{15\text{-max}}$.

   (c) Primers for generating the pTet part by oligo annealing (forward/reverse):

   *CTCG***GGAG(**ctgttctcagtgatagagattgacatccctatcagtg atagatataatgagcactactagagtact)

   *CTCA***AGTA**(ctctagtagtgctcattatatctatcactgataggg at gtcaatctctatcactgagaacagctcc).

where $(n)_{15\text{-max}}$ is the part sequence for annealing into a double-stranded product. The overhangs for level 0 integration into UAV are italicized and stay single-stranded after annealing of the oligonucleotides. Note that this level 0 part is not cut by BsmBI during level 0 plasmid assembly. The overhangs are already created by annealing. Phosphorylation of the oligonucleotides is not necessary. Fusion sites (**NNNN**) of the Golden Gate level 1 assembly are denoted in bold.

3. Amplifying level 1 assembly by PCR:
   (a) Forward primer for level 1 part generated by PCR:

   $(N)_{15\text{–}25}$.

   (b) Reverse primer for level 1 part generated by PCR:

   $(N)_{15\text{–}25}$.

   (c) Amplification primers used in this chapter (forward/reverse):

   A1: TGGAACCCAACTCGGAGCTC

   A2: ATTGCAGCACTGGGGCCAGATG.

where $(N)_{15\text{–}25}$ are sequences from the 5′ and 3′ ends of the TU. We recommend designing amplification primers with each of the primers binding to one distinct input plasmid of the Golden Gate assembly reaction. For example, designing two primers binding only in the destination plasmid backbone may result in unspecific amplification of the residual input plasmid sequence. For instance, the linear constructs tested in this chapter were obtained by a forward primer binding in the first basic part of the assembly (the 5′ connector) and the reverse primer binding on the backbone of the destination plasmid.

4. Oligonucleotides to generate chi sequence (Chi6) DNA by annealing:
   (a) Forward oligonucleotide Chi6: TCACTTCACTGCTGG TGGCCACTGCTGGTGGCCACTGCTGGTGGCCAC TGCTGGTGGCCACTGCTGGTGGCCACTGCTGGT GGCCA.
   (b) Reverse oligonucleotide Chi6: TGGCCACCAGCAGTG GCCACCAGCAGTGGCCACCAGCAGTGGCCACCA GCAGTGGCCACCAGCAGTGGCCACCAGCAGTGA AGTGA.

Forward and reverse Chi6 oligonucleotides are annealed by heating 100 μM of each oligonucleotide in 1x duplex buffer to 100 °C followed by slow cooling to room temperature (RT). Chi6 duplex DNA produced by annealing is stored at −20 °C.

2.3 Molecular Biology Reagents

1. 2× OneTaq DNA polymerase master mix—containing Taq DNA polymerase, dNTPs, $MgCl_2$, buffer components.
2. PCR purification Kit.
3. 3 M sodium acetate solution (pH 5.2).
4. BsaI—type IIS restriction enzyme.
5. BsmBI—type IIS restriction enzyme.
6. T4 DNA ligase enzyme.
7. T4/T7 DNA ligase buffer.
8. 1% agarose gel.
9. 1× TAE (Tris-acetate-EDTA) buffer.
10. 10× Duplex buffer: 1 M potassium acetate, 300 mM HEPES, pH 7.5.

2.4 Autolysate Preparation

1. 2xYTP Broth: 31 g/L pre-mixed 2xYT (containing 16 g/L tryptone, 10 g/L yeast extract, 5 g/L sodium chloride), 3 g/L monopotassium phosphate (KH_2PO_4), 6.97 g/L dipotassium phosphate (K_2HPO_4). Sterilized by autoclaving.
2. Starter culture: *E. coli* BL21-Gold (DE3) pAD-LyseR (Addgene #99244)—Autolysis *E. coli* strain with ampicillin resistance [10].
3. S30A buffer, pH 7.7: 14 mM Mg-glutamate (L-glutamic acid magnesium salt), 60 mM K-glutamate (L-glutamic acid potassium salt), 50 mM Tris (Tris-(hydroxymethyl)-aminomethane).
4. 1 M DTT (dithiothreitol) solution. Store at −20 °C.
5. S30A buffer w/ DTT: S30A buffer +2 mM DTT (always prepare fresh).

6. 2 M Tris base solution in water. Filter to sterilize, and store at RT.
7. 2 M IPTG (Isopropyl-ß-D-thiogalactopyranosid) solution. Filter to sterilize, and store at −20 °C.
8. 2.5 M D-Glucose solution. Use 40–60 °C heating with constant mixing to dissolve glucose. Filter to sterilize, store at RT.
9. 100 mg/mL ampicillin antibiotic solution. Filter to sterilize, and store at −20 °C.

2.5 Reaction Buffer Preparation

2.5.1 Reagents for Energy Solution Preparation

1. 2 M HEPES buffer (4-(2-hydroxyethyl)piperazine-1-ethanesulfonic acid). Adjust pH to 8.0 with KOH solution. Filter to sterilize, and store dark and at 4 °C.

   The following reagents need to be prepared freshly for every use, or to be frozen at −80 °C for extended storage:
2. 65 mM CoA (Coenzyme A) solution.
3. 175 mM NAD (β-nicotinamide adenine dinucleotide) solution. Adjust pH to 7.5–8 with 2 M Tris.
4. 650 mM cAMP (adenosine 3′,5′-cyclic monophosphate) solution. Adjust pH to 8.0 with 2 M Tris.
5. 34 mM folinic acid calcium salt solution.
6. 1 M spermidine (1,8-diamino-4-azaoctane) solution (warm-up briefly to 37 °C for better solubility).
7. 1 M putrescine (1,4-butanediamine dihydrochloride) solution.
8. 1.4 M 3-PGA (D-(−)-3-phosphoglyceric acid disodium salt) solution. Adjust pH to 7.5 with 2 M Tris.
9. Nucleotide mix stock solution: 156 mM ATP (adenosine 5′-triphosphate potassium salt), 156 mM GTP (guanosine 5′-triphosphate sodium salt), 94 mM CTP (cytidine 5′-triphosphate sodium salt), 94 mM UTP (uridine 5′-triphosphate sodium salt). Adjust pH to 7.5 with 3 M KOH.

2.5.2 Reagents for Reaction Buffer Preparation

1. 168 mM buffered individual amino acid solutions (except 140 mM leucine) (e.g., RTS amino acid sampler, Biotechrabbit).
2. 4× amino acid solution: 6 mM of all the amino acids, except leucine at 5 mM concentration. Preparation from individual amino acid solutions is described in Subheading 3.3.3.
3. 14× energy solution: 700 mM of HEPES (pH 8), 21 mM of ATP, 21 mM of GTP, 12.6 mM of CTP, 12.6 mM of UTP, 3.64 mM of CoA, 4.62 mM of NAD, 10.5 mM of cAMP, 0.95 mM of folinic acid, 14 mM of spermidine, 14 mM putrescine, and 420 mM of 3-PGA. Preparation from reagents listed in Subheading 2.5.1 is described in Subheading 3.3.4.

4. 40% PEG-8000 (polyethylene glycol, MW 7000–9000 g/mol) solution. Filter to sterilize.
5. 1 M ammonium glutamate (L-glutamic acid ammonium salt) solution. Filter to sterilize.
6. 1 M oxalic acid (potassium oxalate) solution. Filter to sterilize.
7. 1 M Mg-glutamate (L-glutamic acid magnesium salt) solution. Filter to sterilize.
8. 4 M K-glutamate (L-glutamic acid potassium salt) solution. Filter to sterilize.
9. 1 M DTT solution. *See* Subheading 2.4, **item 4**.

2.6 Echo® Dispensing and Plate Reader Measurement of Cell-Free Reactions

1. 2.3× reaction buffer: 16.24 mM Mg-glutamate, 162.4 mM K-glutamate, 3.48 mM amino acids solution, 2.3× energy solution, 2.32 mM DTT, 23.2 mM ammonium glutamate, 23.2 mM oxalic acid, 4.64% PEG-8000. Preparation is described in Subheading 3.3.5.
2. Autolysate. Preparation is described in Subheadings 3.3.1 and 3.3.2.
3. Amplified level 1 DNA constructs. Preparation is described in Subheading 3.2.
4. Echo® compatible source plate (e.g. 384PP 2.0, Labcyte).
5. 384-well target plate (e.g. Nunc 384-Shallow Well, black, 264705).

3 Methods

3.1 Construction of New Level 0 Plasmids

For level 0 parts that are not included in a preexisting MoClo collection, we advise generating and storing the desired level 0 part in a level 0 plasmid (*see* **Note 2**) (Fig. 2). For primer design see Subheading 2.2 and Fig. 2. We present two methods that can be chosen according to the size of the new level 0 part. For smaller parts (typically under 100 base pairs), it is convenient to directly anneal two oligonucleotides. Longer level 0 parts are generated by PCR amplification or DNA synthesis of the desired target sequence. If target DNA sequences contain internal BsmBI and/or BsaI restriction sites, these sites need to be removed in advance.

3.1.1 Amplification of a New Level 0 Part Sequence by PCR

In order to amplify a new DNA part of interest, a standard PCR reaction (Fig. 2a) can be run according to the manufacturer's recommendations. Fifty microliter PCR reaction containing the following components is given as an example:

1. 1 μL of 10 μM forward primer.
2. 1 μL of 10 μM reverse primer.
3. 25 μL of One*Taq* 2× master mix with standard buffer.
4. Template DNA and ddH_2O up to 50 μL.

Before further use, PCR products are checked by agarose gel electrophoresis, and then purified using a standard PCR purification kit according to the manufacturer's recommendations.

3.1.2 Generating a New Level 0 Part Sequence by Annealing Oligonucleotides

Short level 0 parts can be rapidly generated by annealing two single-stranded DNA oligonucleotides (Fig. 2b). An example of a 50 μL reaction mixture for oligonucleotide annealing is given below:

1. 1.5 μL of 10 μM forward oligonucleotide.
2. 1.5 μL of 10 μM reverse oligonucleotide.
3. 5 μL of T4 DNA ligase buffer.
4. 42 μL of ddH_2O.

Incubate the reaction at 95 °C in a heat block. After incubation, turn off the heat block, and allow the sample to slowly cool down to RT. Annealed oligonucleotides form a fully prepared DNA fragment, which has single-stranded overhangs suitable for Golden Gate assembly.

3.1.3 Assembly of New Level 0 Plasmids from Parts Created by PCR or Oligonucleotide Annealing

New DNA parts created by PCR or oligonucleotide annealing can be assembled into the level 0 universal accepting vector (UAV) (Fig. 2c) by using the following Golden Gate reaction mixture:

1. 100 ng of DNA insert.
2. 1 μL of UAV DNA (25 ng/μL).
3. 1 μL of T4 DNA ligase buffer.
4. 1 μL of T4 DNA ligase.
5. 0.5 μL of BsmBI restriction enzyme.
6. Up to 10 μL ddH2O.

Golden Gate assembly reaction conditions:

50 cycles (37 °C, 2 min; 16 °C, 5 min), followed by 1 cycle at 50 °C, 10 min, and finished with final cycle (80 °C, 10 min, 1 cycle). Finally, hold at 10 °C overnight.

New assemblies are transformed into a suitable cloning strain (such as *E. coli* DH5α) and plated on the appropriate antibiotic. If the accepting vector contains a dropout cassette that results in fluorescent colonies, colonies that likely contain the desired insert can be easily selected by loss of fluorescence (Fig. 2c). New level 0 plasmids are purified by miniprep and sequence verified.

3.2 Construction of Level 1 Cell-Free Expression Templates

3.2.1 Golden Gate Reaction for Level 1 Assembly

Level 1 constructs are assembled by combining several level 0 parts. To build a functional level 1 TU, the user must assemble at least a promoter, a RBS, a CDS, and a terminator. Although it is theoretically possible to use this minimum set of level 0 parts for the level 1 assembly and subsequent amplification, we recommend adding them to a level 1 universal destination vector (UDV) (along with the 5′ and 3′ connectors) for i) enhancing the assembly efficiency, and ii) enabling storage of the assembly as a plasmid, if desired. Figure 3 gives an overview over the constructs used in this chapter. To demonstrate rapid characterization of expression strengths, we combined five constitutive promoters with two different RBS to generate a small combinatorial library (Fig. 3b).

1. Dilute each level 0 part to 20 fmol/μL. This will help to ensure high assembly efficiency and facilitate further handling due to the standardization [6].
2. Thaw the level 0 parts and the T4 ligase buffer. If your ultimate goal is to create a linear template, you can leave out the UDV backbone (*see* **Note 3**).
3. Each level 1 assembly from level 0 plasmids is set up for a final reaction volume of 10 μL and follows this composition:
   (a) 0.5 μL of each level 0 part at 20 fmol/μL.
   (b) 1 μL of 10x T4 ligase buffer.

A **Level 0 Parts**

	5' Connector	Promoter						RBS		CDS		Terminator	3' Connector	Backbone
Description	5C1RLN	J23119	J23116	J23108	J23101	J23100	pTet	RBS0034	5' g10 UTR	sfGFP	TetR	B1006	3C1SN	Amp-ColE1
5' Fusion site	AACA	GGAG						TACT		AATG		GCTT	CGCT	AGAC
3' Fusion site	CCTC	ATGA						TTAC		CGAA		GCGA	TCTG	TTGT
Source	Marburg collection	Marburg collection					Oligo annealing	Marburg collection			PCR	Marburg collection		

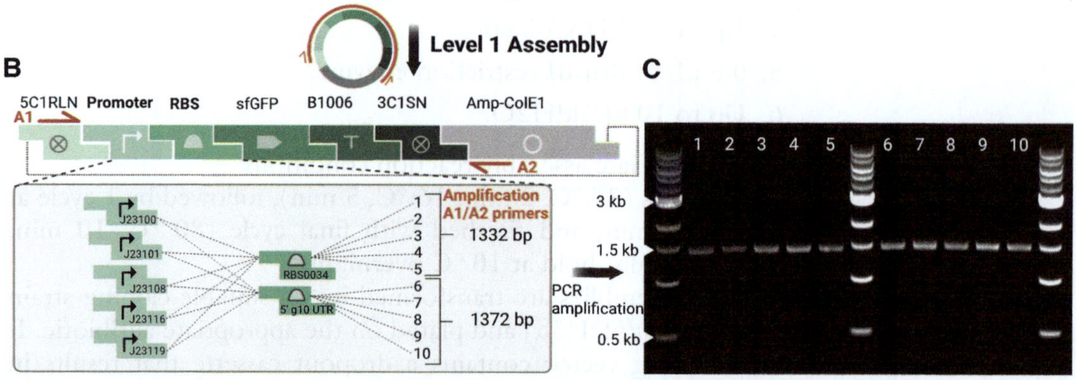

Fig. 3 Overview of the level 0 and 1 constructs used. (**a**) List of level 0 parts used in this protocol with their origin. Most parts were obtained from the Marburg collection, while the *tetR* and pTet level 0 parts were prepared by PCR and oligonucleotide annealing as described. (**b**) Examples of level 1 assemblies combining five different promoters with two RBS parts, resulting in ten different constructs. (**c**) Amplification of the resulting level 1 constructs with primers A1 and A2 produces DNA products that are suitable linear templates for TXTL reactions

(c) 1 μL of T4 Ligase.

(d) 0.5 μL of BsaI restriction enzyme.

(e) ddH_2O to 10 μL final reaction volume.

Start by pipetting the level 0 parts that vary throughout your assemblies (one tube per assembly). Don't pipet the common level 0 parts yet, they will be added in the master mix prepared in **step 4**.

4. Prepare the master mix assembly according to the final number of reactions as follows: Start by adding ddH_2O, then the level 0 parts common to all of your level 1 constructs, the T4 ligase buffer, and the enzymes T4 ligase and BsaI. Mix well and add the master mix to the tubes prepared in **step 3**.
5. Place your reactions in a thermocycler with the following protocol (*see* **Note 4**):

   15 cycles (37 °C, 1.5 min; 16 °C, 3 min), followed by a final cycle (50 °C, 5 min; 80 °C, 10 min). Finally, hold at 8 °C.
6. After the end of the thermal cycling reaction, dilute the assemblies to 1:10 with ddH_2O (*see* **Note 5**).
7. (*Optional*) Transform your assembly into a suitable cloning strain for plasmid propagation and purification. This is only possible if you add a UDV into the assembly reaction. Storing or testing the assembled plasmids in cells is optional. To generate linear template DNA for cell-free expression, transformation is not necessary.

3.2.2 Amplification and Purification of Level 1 Linear Templates

1. Set up a 100 μL PCR reaction containing the following components:

   (a) 2 μL of 10 μM forward primer A1.

   (b) 2 μL of 10 μM reverse primer A2.

   (c) 50 μL of one*Taq* 2× master mix with standard buffer.

   (d) 2 μL of Golden Gate level 1 assembly (diluted 1:10).

   (e) ddH_2O to 100 μL final reaction volume.

   PCR program: Initial denaturation (94 °C, 30 s), followed by 35 cycles of thermal cycling (94 °C, 30 s; 45–68 °C, 30 s; 68 °C, 1 min/kb), then final extension (68 °C, 5 min). Finally, hold at 8 °C.
2. Run the PCR product on a 1% agarose gel to confirm the correct size. The linear level 1 construct should be visible as a single band, without side products (Fig. 3c).
3. Proceed with the purification of your linear DNA using a PCR clean-up kit according to the manufacturer's instructions. We recommend using the Invitrogen PureLink PCR Purification Kit (ref. K310001) with the following adjustments for improved binding to the column:

(a) 100 μL of PCR reaction.
(b) 100 μL of ddH_2O.
(c) 800 μL of binding buffer B2.
(d) 10 μL of 3 M sodium acetate.

To improve DNA yield in the elution step, let the column incubate for up to 1 h at RT after adding the elution buffer, before proceeding to the centrifugation step.

4. Measure the concentration and purity of the PCR products by absorption at 260/280 nm.
5. (*Optional*) Send your samples for Sanger sequencing (*see* **Note 6**).

3.3 Preparation of TXTL Reagents

E. coli cell lysate for TXTL reactions is prepared from the *E. coli* BL21-Gold (DE3) pAD-LyseR strain. This autolysis strain expresses bacteriophage lambda gene R, causing efficient cell lysis after a freeze-thaw cycle, and produces a highly active lysate for cell-free expression [10]. For transcription and translation to occur in vitro, the autolysate needs to be supplemented with a reaction buffer to sustain the reaction. Our TXTL reaction buffer recipe, presented in Subheadings 3.3.3, 3.3.4 and 3.3.5, is adapted from [11].

3.3.1 Growing and Preparing Cell Biomass for Autolysate Preparation

1. Add ampicillin (Amp) to the 2xYTP medium to a final concentration of 100 μg/mL, and inoculate with the autolysis strain glycerol stock. Grow cells overnight at 37 °C, with 220 rpm shaking.
2. Fill two 1 L volume sterilized glass flasks with 400 mL 2xYTP medium each. Add 16 mL of 2.5 M glucose stock solution into each flask to prepare 2xYTPG medium.
3. Inoculate 400 mL of prepared 2xYTPG medium (from step no. 2) by adding 400 μL Amp stock solution, and 1 mL of autolysis strain overnight culture (from step no. 1). Grow cells at 37 °C, with 220 rpm shaking, until the optical density of the culture (600 nm wavelength, 1 cm path length) $OD_{600} = 0.5–0.6$.
4. (*Optional*) After reaching $OD_{600} = 0.5–0.6$, induce cells with 1 mM IPTG (add 200 μL of 2 M IPTG stock solution (2.4.6)). It is only required if you want to obtain a lysate with active T7 RNA polymerase for expression from T7 promoters.
5. Continue growing the cells at 37 °C, with 220 rpm shaking until $OD_{600} = 1.3–1.5$ (*see* **Note 7**).
6. Harvest cell biomass: centrifuge cells using a swinging-bucket rotor at 2000× *g* speed, for 15 min at RT.

7. Remove supernatant: Resuspend biomass from each batch (400 mL medium) in 45 mL of cold (4 °C) S30A buffer.
8. Preweigh two empty 50 mL volume falcon tubes and pour two portions of resuspended biomass into them.
9. Centrifuge using the same conditions (*see* **step 6**).
10. Remove the supernatant carefully, all obvious droplets and liquid leftovers must be cleared away. *Important:* Residual liquid will lead to an overestimation of the biomass amount and result in a lysate with lower expression activity (*see* **Note 8**).
11. Weigh the tubes with centrifuged biomass to determine the amount of biomass per batch. Don't forget to exclude the weight of the empty tubes, measured in **step 8**.
12. Add 2 volumes of cold (4–10 °C) S30A buffer freshly supplemented with 2 mM DTT according to the weight of cell biomass of each batch (for example, if the biomass weight is 1.235 mg, then 2.470 μL of S30A buffer with DTT must be added).
13. Thoroughly resuspend samples by vortexing (to fully resuspend biomass pellets in the buffer).
14. Freeze cells in buffer at −80 °C. Leave them in the freezer for at least 1 hour (*see* **Note 9**).

3.3.2 Extracting Autolysate from Cell Biomass

1. Thaw frozen cells from Subheading 3.3.1 by leaving them at RT for 10–15 min (*see* **Note 10**).
2. Vortex thawed cells vigorously for at least 5 min. Freeze-thawing and vortexing will lead to cell lysis and produce a very viscous mixture.
3. Incubate biomass at 37 °C, with 220 rpm shaking for 90 min. Briefly vortex tubes after 45 min of incubation.
4. Combine lysed biomass in small centrifugation tubes suitable for high-speed ultracentrifugation. This step may be difficult due to the viscous nature of the mixture (*see* **Note 11**).
5. Centrifuge samples for 45–60 min at 4 °C temperature in an ultracentrifuge with a fixed-angle rotor using 45,000–50,000× *g* speed (*see* **Note 12**).
6. Very gently move the supernatant to new 1.5 mL tubes (this is your autolysate extract), avoid transferring any pellet or other solid materials (*see* **Note 13**).
7. Centrifuge autolysate extract at RT for 3 min at the maximum speed of a table centrifuge (~21,000× *g*), to remove leftovers of larger particles.

8. Combine the cleared autolysate in one fresh tube and mix well. Then aliquot the autolysate extract into small tubes. Depending on the planned experiments, we recommend aliquot sizes of about 50 μL.
9. Freeze aliquots at −80 °C for storage.

3.3.3 Amino Acid Solution Preparation (6 mM)

Each individual amino acid is usually supplied at 1.5 mL, 168 mM concentration, except for leucine, which is supplied as a 140 mM solution. The final composition of the amino acid solution contains 6 mM of all the amino acids, except leucine (which is at 5 mM concentration).

1. Take all 20 amino acids from −20 °C and thaw them at RT. Vortex until amino acids dissolve (*see* **Note 14**).
2. After dissolving amino acids, put them on ice, *except for Asn, Phe,* and *Cys*: these three amino acids should be kept at RT.
3. Add sterile water to a sterile tube, and place it on ice.
4. To the pre-cooled sterile water, add the necessary amount of each amino acid in the *following order*: Ala, Arg, Asn, Asp, Gln, Glu, Gly, His, Ile, Lys, Met, Phe, Pro, Ser, Thr, Val, Trp, Tyr, Leu, Cys (*see* **Note 15**).
5. Aliquot amino acid solution on ice into sterile tubes. Vortex frequently to avoid unequal distribution of suspension.
6. Store at −80 °C.

3.3.4 Energy Solution Preparation (14×)

The final composition of the energy solution described in this section is 700 mM of HEPES (pH 8), 21 mM of ATP, 21 mM of GTP, 12.6 mM of CTP, 12.6 mM of UTP, 3.64 mM of CoA, 4.62 mM of NAD, 10.5 mM of cAMP, 0.95 mM of folinic acid, 14 mM of spermidine, 14 mM putrescine, and 420 mM of 3-PGA. The prepared energy solution is at 14× concentration as compared to the final TXTL reaction.

1. Thaw all necessary components on ice and keep them on ice while preparing the energy solution (*see* **Note 16**).
2. Place a sterile tube on ice, and mix stock solutions in the *following order* to reach the concentrations stated above: HEPES, water, nucleotide mix, CoA, NAD, cAMP, folinic acid, spermidine, putrescine, and 3-PGA. Shortly vortex the tube after each added component.
3. Keep the prepared energy solution on ice, and aliquot into sterile tubes. Vortex the main stock solution frequently to avoid unequal distribution.
4. Freeze aliquots in liquid nitrogen and store at −80 °C.

3.3.5 Reaction Buffer Preparation

The final composition of the reaction buffer is 16.24 mM Mg-glutamate, 162.4 mM K-glutamate, 3.48 mM amino acids solution, 2.3× energy solution, 2.32 mM DTT, 23.2 mM ammonium glutamate, 23.2 mM oxalic acid, 4.64% PEG-8000. The prepared reaction buffer will be further mixed with autolysate and DNA samples for cell-free TXTL reactions (*see* Subheading 3.4), and, therefore, will be additionally diluted 2.3 times.

1. Place the sterile tube on ice. To prepare a working solution of the reaction buffer with the concentrations stated above, mix together: Mg-glutamate, K-glutamate, amino acids solution, energy solution, DTT, ammonium glutamate, oxalic acid, PEG-8000. Shortly vortex solution after the addition of each item (*see* **Note 17**).
2. Keep the prepared reaction buffer on ice, and aliquot into sterile tubes. Vortex the main stock frequently to avoid unequal distribution.
3. Freeze aliquots in liquid nitrogen and store at −80 °C.

3.4 Cell-Free Expression Reactions

Cell lysate and reaction buffer for TXTL reactions from Subheading 3.3 can be prepared in large batches, aliquoted, and conveniently stored at −80 °C until the time of the experiment. If stored TXTL reagents and a collection of level 0 parts are available, rapid production of linear level 1 DNA templates by Golden Gate assembly and PCR (*see* Subheading 3.2) makes it possible to go from construct design to measurement within one day. In this section, we describe the setup of TXTL reactions for part measurements. Here, we demonstrate combinatorial testing of promoters and ribosomal binding sites (Fig. 4) and characterization of an inverter circuit using the transcriptional repressor TetR (Fig. 5). The activity of the regulatory parts tested is reported through the expression of a superfolder green fluorescent protein (sfGFP).

This section describes the assembly of TXTL reactions for measurements, which will yield a final reaction with the following composition: 40% v/v of *E. coli* autolysate, 10 mM ammonium glutamate; 7 mM Mg-glutamate; 70 mM K-glutamate; 10 mM oxalic acid; 1.5 mM ATP; 1.5 mM GTP; 0.9 mM UTP; 0.9 mM CTP; 0.068 mg/mL folinic acid; 1.5 mM amino acids; 30 mM 3-PGA; 0.33 mM NAD; 0.26 mM CoA; 1 mM putrescine; 1 mM spermidine; 0.75 mM cAMP; 1 mM DTT; 50 mM HEPES, 2% PEG 8000, 2 μM of Chi6 duplex DNA, and varying concentration of linear DNA. The cell-free components are dispensed with an Echo® 525 liquid handler, except for the lysate, which is manually dispensed at the end of the reaction. The use of the acoustic liquid handler is recommended for high-throughput experiments.

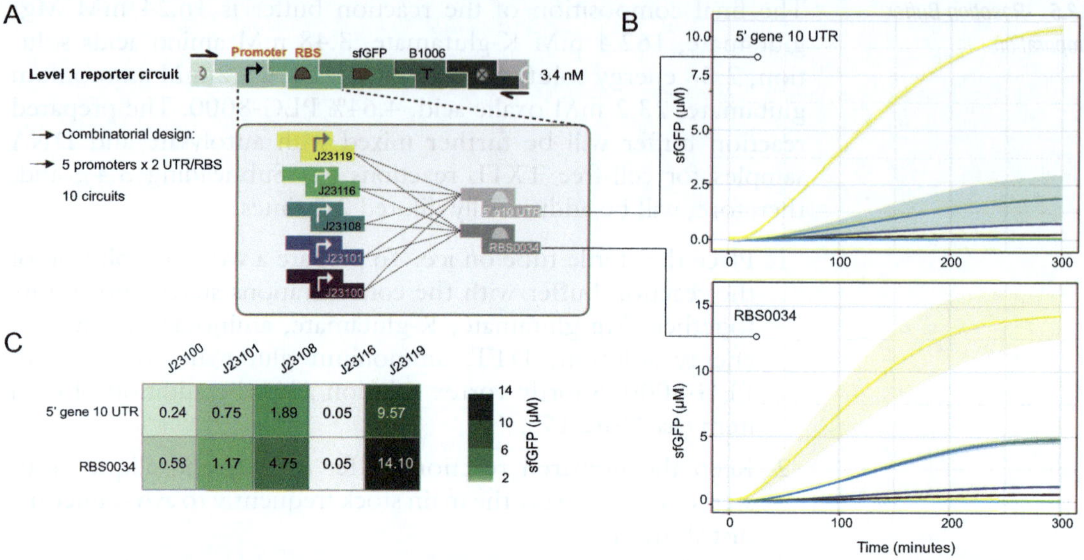

Fig. 4 Prototyping promoters and RBSs. (**a**) Schematic of parts generated and tested (combinations of multiple promoters and B0034 RBS/T7 phage 5′ gene 10 UTR). (**b**) Kinetics of sfGFP production in the cell-free reaction with constant reporter DNA (3.4 nM). Shaded regions show standard deviations of three replicates. (**c**) Heat map comparing endpoint sfGFP levels

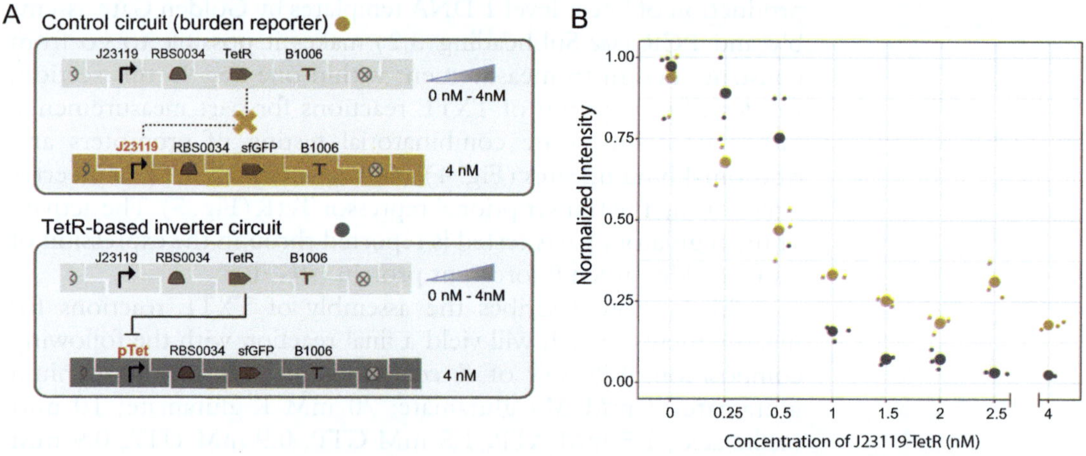

Fig. 5 Testing a promoter-transcription factor pair. (**a**) Schematic of the burden control circuit and the TetR-based inverter circuit. (**b**) Titration curves of varying the TetR production template DNA concentration in the presence of constant reporter DNA (4 nM). To allow comparison, endpoint fluorescence was normalized to the highest data point for each construct. Thick markers indicate the mean of three replicates

1. Each cell-free reaction is set up for a final volume of 5 μL and is composed of:
   (a) 2.15 μL reaction buffer.
   (b) 2 μL autolysate extract.
   (c) 0.10 μL Chi6 (at 100 μM).

(d) 0 μL to 0.75 μL of concentrated linear DNA.

(e) ddH_2O to 5 μL final reaction volume.

Calculate the number of autolysate extract and reaction buffer aliquots needed for your experiment. Take into account reaction replicates and add some spare volume (10%).

or

(*Optional*) Use the PyEcho script to generate the Echo® input protocol file according to your experimental design. Refer to the documentation from the GitHub repository for details (https://github.com/HN-lab/PyEcho). Once the experimental design Excel file is filled, the script will return the amounts of reagents needed. Do not include the lysate into the dispensing instructions. In our experience, autolysate is unsuitable for dispensing with an acoustic liquid handler. Autolysate should be added manually (*see* **step 9**).

2. Preheat the plate reader to 29 °C.
3. Thaw the cell-free reaction components on ice (DNA linear constructs, Chi6, and the appropriate number of reaction buffer and lysate aliquots).
4. Prepare 10× or 5× concentrated DNA solutions containing the linear DNA templates for the TXTL reaction. The final protein production of the TXTL reaction usually saturates at about 10 nM DNA template (final concentration). To determine the effects of promoters or ribosomal binding sites on the expression strength of a reporter gene, we recommend using a near-saturating DNA concentration (*see* Fig. 4). When characterizing the effect of a transcription factor on the expression of a downstream gene in a circuit composed of multiple expression constructs, care should be taken to account for resource competition effects [12]. To illustrate the effect of competition for resources resulting from the production of an additional protein, we included a burden reporter with the characterization of the TetR repressor—pTet promoter pair. The burden reporter consists of a strong nonrepressible, constitutive promoter driving the expression of the sfGFP reporter (Fig. 5a). Coexpression of the TetR construct at increasing concentrations decreases the output from the burden reporter, however, less than its target reporter with the pTet promoter (Fig. 5b).
5. (*Only for manual dispensing if Echo® dispenser is not used*) Prepare a master mix for (n + 1) 16.5 μL reactions (each 5 μL cell-free reaction is performed in triplicate and volume loss accounts for 10% of the final volume), consisting of the reaction buffer, the autolysate extract, and the Chi6 duplex. First, dispense the concentrated DNA solutions and water in separate reaction tubes. Second, dispense the master mix in each reaction tube. Third, for each reaction, mix well and pipette 3 replicates of 5 μL into the 384-well plate. Go to **step 12**.

6. Fill the Echo® source plate with the cell-free components according to your experimental design and the generated protocol. Proceed quickly with the next step.
7. Use an Echo acoustic liquid handler (we used the Echo® 525) with the generated input protocol to dispense the concentrated linear DNA constructs, the reaction buffer, and water into the 384-well target plate (*see* **Note 18**).
8. Seal the plate with a foil to prevent evaporation until the next step (*see* **Note 19**).
9. Remove the cover and manually dispense the autolysate extract into the 384-well plate (*see* **Note 20**).
10. Seal the plate with a transparent foil to prevent evaporation during the plate reader measurement.
11. Agitate the plate vigorously for a few seconds to ensure proper mixing of the reagents (a convenient method is to manually shake the plate between two tip boxes). Spin the plate to collect the liquid at the bottom of the wells.
12. Place the sealed plate into the plate reader and record fluorescence over time in a program that measures fluorescence at the appropriate wavelengths every 2 min for 5–8 h at 29 °C.
13. Using the data from the plate reader measurement, plot the kinetics of reporter production (Fig. 4b) and quantitate end-point levels (Figs. 4c and 5b) to prototype regulatory elements and transcription factors (*see* **Note 21**).

4 Notes

1. The start and stop codons should be omitted from the primer sequence when designing primers to amplify protein-coding sequences for level 0 construction. Start and stop codons are part of the overhang sequence (start codon) or appear during level 1 assembly (stop codon).
2. We advise collecting level 0 parts in plasmids and sequencing the level 0 plasmids prior to their use. Theoretically, it is also possible to use PCR products as level 0 parts if they contain the correct BsaI recognition and restriction sites. This works well in many cases, but we have seen that for certain parts, it leads to non-specific PCR products during the amplification of the level 1 constructs.
3. Leaving out the UDV backbone requires suitable amplification primers, e.g., primers binding on the 5′ and 3′ connectors. We observed a slight decrease in reaction efficiency when assembling without a backbone. We have tested our protocol with up to six level 0 fragments plus backbone (Fig. 2). However,

construct assemblies of up to 44 parts in a single reaction have been demonstrated [13]. We recommend prior testing when increasing the number of parts in the assembly.

4. No difference was observed in the efficiency of assembly, compared with the longer standard Golden Gate assembly protocol. The duration of. the improved protocol is about 80 min, whereas the longer standard protocol takes about 400 min (see Subheading 3.1.3 for the thermocycling program of the long protocol).
5. It does not matter if the Golden Gate reaction is diluted or not for the subsequent PCR reaction. Diluting the reaction gives more volume to be used in the future for PCR amplification. The diluted Golden Gate reaction can be stored at −20 °C.
6. Sequencing from the reverse direction through the terminator may sometimes lead to short sequencing reads. This can be solved by changing the placement of the sequencing primers.
7. Measurements of samples with $OD_{600} > 1$ are not accurate. Therefore, dilutions of these samples with 2xYTP medium must be prepared to ensure that measurements are performed within the spectrophotometer's linear response range.
8. This step is crucial for further autolysate extract preparation. Leftovers of supernatant will dilute the final autolysate, affecting transcription–translation activity. You can use aspiration with a vacuum to remove all leftovers of liquids if needed.
9. The freezing step is necessary even if you continue working on the same day. After freezing the cell suspension at −80 °C, the preparation procedure can be continued on the next day or even later.
10. You can use a warm water bath (not higher than 37 °C temperature) for 5–10 min.
11. After vortexing, partly-lysed cell biomass will be very viscous. Therefore, you can cut the end of a 1 mL pipette tip for easier pipetting to get a wider tip diameter.
12. Fill ¾ of the centrifugation tube volume to prevent the tube from breaking at high speed. For accurate balance, use a second centrifugal tube filled with water. Weigh the balance tube to use exactly the same weight as the corresponding biomass tube.
13. The high-speed centrifugation step could also be replaced by longer centrifugation at a lower speed (21,000–30,000× *g*) [10]. However, the higher centrifugation speeds compact the cell-pellet much better and significantly increase the amount of collected supernatant.
14. Briefly incubate at 37 °C for better solubility, if necessary. Cys may not fully dissolve.

15. The order of mixing the amino acids is very important. Vortex the Falcon tube after each addition of amino acid, and keep the solution on ice all the time. Cys can be added as a suspension. After addition, vortex until the solution is relatively clear, incubating at 37 °C if necessary. Cys may not fully dissolve.
16. Each component of the energy solution can be stored at −80 °C as stocks for later use.
17. Reaction buffers can be optimized for every batch of autolysate extract by adjusting the final concentrations of Mg-glutamate, K-glutamate, and PEG (*in that order*) to produce cell-free reactions with maximum expression levels. We noticed that DTT calibration does not significantly affect the activity of TXTL reactions. However, changes in PEG concentration can have a dramatic influence on activity, especially in autolysate-based TXTL reactions, therefore it should be optimized carefully.
18. For Echo® dispensing: Acoustic dispensing with the Labcyte Echo® does not work for solutions containing contamination with long genomic DNA. We suspect that acoustic dispensing of autolysate is impossible due to some remaining genomic DNA in the autolysate.
19. At this step, you can check the error report of the Echo®. If a reagent was not properly dispensed and the reason identified, you can dispense the missing reagents before proceeding to the next step.
20. A dispenser pipette or automated bulk dispenser can accelerate the manual dispensing of the autolysate. We advise reducing the duration of this step as much as possible to maximize the cell-free reaction's yield.
21. Alternatively, you can compute sfGFP production rates at the linear increase phase instead of endpoint measurements for benchmarking the performance of constructs. It can help with leaky expression, where proteins accumulate over time.

Acknowledgments

This work was supported by Deutsche Forschungsgemeinschaft (DFG, German Research Foundation) grant NI 2040/1-1. Figures created with BioRender.com.

References

1. Silverman AD, Karim AS, Jewett MC (2020) Cell-free gene expression: an expanded repertoire of applications. Nat Rev Genet 21:151–170. https://doi.org/10.1038/s41576-019-0186-3
2. Sun ZZ, Yeung E, Hayes CA, Noireaux V, Murray RM (2014) Linear DNA for rapid prototyping of synthetic biological circuits in an Escherichia coli based TX-TL cell-free system. ACS Synth Biol 3:387–397. https://doi.org/10.1021/sb400131a
3. Marshall R, Maxwell CS, Collins SP, Beisel CL, Noireaux V (2017) Short DNA containing χ sites enhances DNA stability and gene expression in E. coli cell-free transcription–translation systems. Biotechnol Bioeng 114:2137–2141. https://doi.org/10.1002/bit.26333
4. Geertz M, Rockel S, Maerkl SJ (2012) A high-throughput microfluidic method for generating and characterizing transcription factor mutant libraries. In: Weber W, Fussenegger M (eds) Synthetic gene networks: methods and protocols. Humana Press, Totowa, NJ, pp 107–123
5. Niederholtmeyer H, Stepanova V, Maerkl SJ (2013) Implementation of cell-free biological networks at steady state. Proc Natl Acad Sci 110:15985–15990. https://doi.org/10.1073/pnas.1311166110
6. Engler C, Kandzia R, Marillonnet S (2008) A one pot, one step, precision cloning method with high throughput capability. PLoS One 3: e3647. https://doi.org/10.1371/journal.pone.0003647
7. Weber E, Engler C, Gruetzner R, Werner S, Marillonnet S (2011) A modular cloning system for standardized assembly of multigene constructs. PLoS One 6:e16765. https://doi.org/10.1371/journal.pone.0016765
8. Stukenberg D, Hensel T, Hoff J, Daniel B, Inckemann R, Tedeschi JN, Nousch F, Fritz G (2021) The Marburg collection: a Golden Gate DNA assembly framework for synthetic biology applications in vibrio natriegens. ACS Synth Biol 10:1904–1919. https://doi.org/10.1021/acssynbio.1c00126
9. Moore SJ, Lai H-E, Kelwick RJR, Chee SM, Bell DJ, Polizzi KM, Freemont PS (2016) EcoFlex: a multifunctional MoClo kit for E. Coli synthetic biology. ACS Synth Biol 5:1059–1069. https://doi.org/10.1021/acssynbio.6b00031
10. Didovyk A, Tonooka T, Tsimring L, Hasty J (2017) Rapid and scalable preparation of bacterial lysates for cell-free gene expression. ACS Synth Biol 6:2198–2208. https://doi.org/10.1021/acssynbio.7b00253
11. Sun ZZ, Hayes CA, Shin J, Caschera F, Murray RM, Noireaux V (2013) Protocols for implementing an Escherichia coli based TX-TL cell-free expression system for synthetic biology. J Vis Exp e50762. https://doi.org/10.3791/50762
12. Marshall R, Noireaux V (2019) Quantitative modeling of transcription and translation of an all-E. coli cell-free system. Sci Rep 9: 11980. https://doi.org/10.1038/s41598-019-48468-8
13. Werner S, Engler C, Weber E, Gruetzner R, Marillonnet S (2012) Fast track assembly of multigene constructs using Golden Gate cloning and the MoClo system. Bioengineered 3: 38–43. https://doi.org/10.4161/bbug.3.1.18223

Chapter 12

Construction of Small Genes with Repetitive Elements Using Oligo Extension and Golden Gate Assembly

Michael T. A. Nguyen, Ebbe Sloth Andersen, and Georgios Pothoulakis

Abstract

Gene synthesis efficiency has greatly improved in recent years but is limited when it comes to repetitive sequences and results in synthesis failure or delays by DNA synthesis vendors. Here, we describe a method for the assembly of small synthetic genes with repetitive elements: First, a gene of interest is split in silico into small synthons of up to 80 base pairs flanked by Golden Gate-compatible overhangs. Then synthons are made by oligo extension and finally assembled into a synthetic gene by Golden Gate assembly.

Key words Gene synthesis, Molecular cloning, Golden Gate cloning, Synthetic biology, Repetitive DNA sequences

1 Introduction

Advances in DNA synthesis technologies have been pivotal for synthetic biology research. With the development of several DNA assembly methods, it is possible to construct almost any synthetic gene, biosynthetic pathway, or genome. Of the many methods, the one-pot restriction-ligation method, Golden Gate assembly (GGA), has become a standard method for synthetic biology workflows due to its utility for rapid and high-throughput cloning [1]. GGA utilizes type IIS restriction enzymes that cut distal to their recognition site, which allows overhangs to be designed with arbitrary sequence for scarless cloning. Although DNA synthesis technologies have advanced drastically, the synthesis of DNA sequences with high or low guanine-cytosine-content, stable secondary structures, and multiple repeats remain challenging [2], where especially repetitive DNA sequences result in synthesis failure or delays by vendors [3].

Here, we describe a two-step DNA assembly protocol for the construction of smaller synthetic genes with repetitive elements. An overview of the method can be seen in Fig. 1. The method consists

Daniel Schindler (ed.), *Golden Gate Cloning: Methods and Protocols*, Methods in Molecular Biology, vol. 2850,
https://doi.org/10.1007/978-1-0716-4220-7_12,

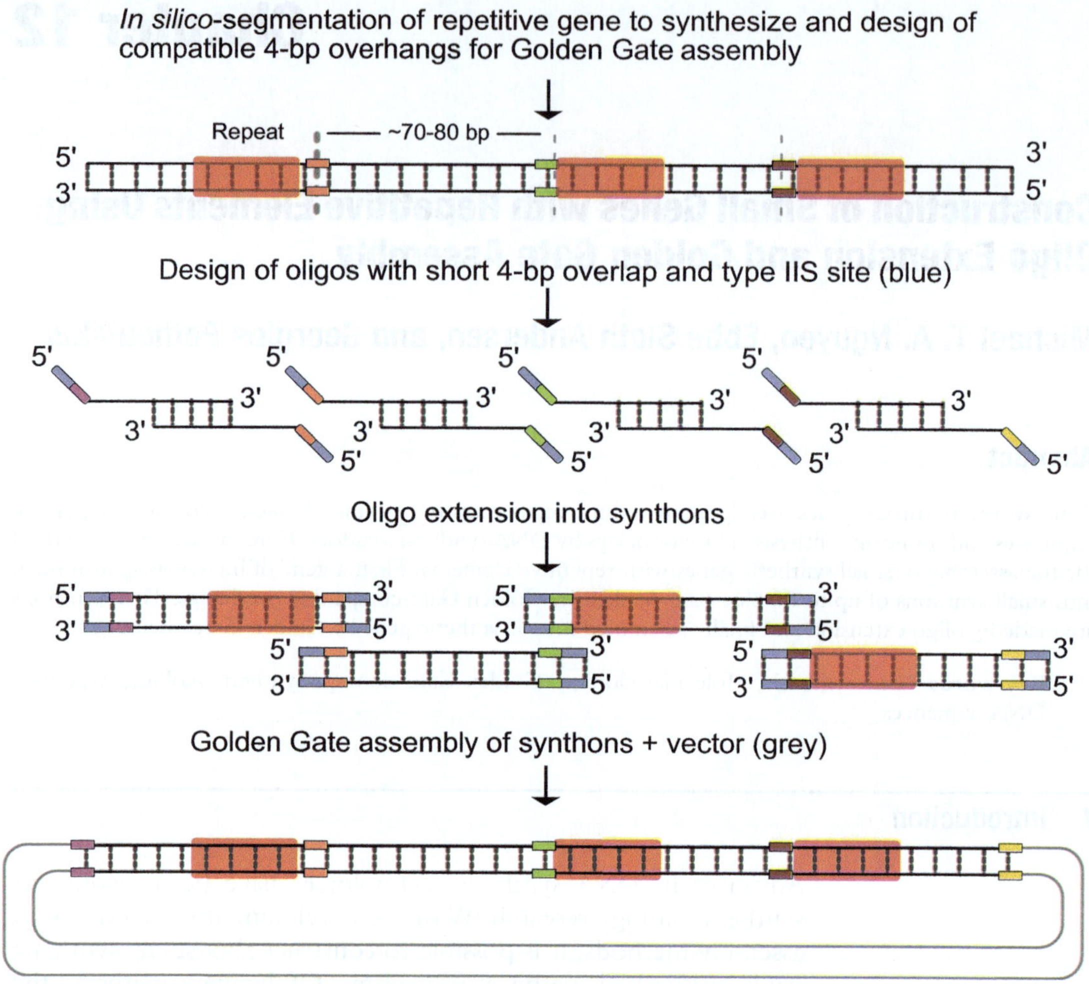

Fig. 1 Overview of the method for the construction of a small synthetic gene with repetitive elements. The designed gene with repeats (red) is split into 70–80 bp synthons in silico with 4-bp regions (orange, green, brown) that are used as overhangs for GGA. Oligos up to 60 bases are designed to cover the gene fragment and the oligos contain type IIS restriction enzyme sites in the 3′-ends (blue). Oligos covering the termini of the genes include overhangs (purple and yellow) compatible with an entry vector. Double-stranded DNA synthons are synthesized by oligo extension and inserted into a vector by GGA. Reproduced with permission from Nguyen et al. 2023 [10]

of splitting the gene of interest in silico into synthons (fragments) of up to 80 base pairs (bp) using molecular biology design software such as Benchling [4]. We used the GGA tool from Benchling to design compatible 4-bp overhangs. Other tools for overhang design exist, such as NEB Golden Gate tool [5] or the Golden Gate overhang designer from the Edinburgh biofoundry [6]. The genes are split into roughly equal-sized fragments depending on the sequence size starting from position 1 to position 70–80 and then from position 70–80 to position 140–150 and so on. The method is framed around GGA because of the use of small 4-bp overhangs that are designed to minimize undesired annealing, in contrast to homology-based approaches which require at least 15-bp homology [7], and the synthons can be directly assembled into a desired destination vector. Synthons are made by oligo extension with a DNA polymerase using oligos of up to 60 nucleotides (nt) that overlap with 18–25 bases in their 3′-ends for hybridization to reduce misannealing and allow for extension of short oligos to longer synthons. The synthetic gene is finally assembled and inserted into a desired entry vector by GGA.

2 Materials

Prepare all reactions using nuclease-free water. Keep enzymes cold.

2.1 Oligo Extension

1. Q5 DNA polymerase supplied with 10x Q5 PCR reaction buffer. Any high-fidelity DNA polymerase can be used.
2. dNTP mix (2.5 mM of each deoxyadenosine triphosphate, deoxyguanosine triphosphate, deoxythymidine triphosphate, deoxycytidine triphosphate).
3. DNA oligos resuspended in nuclease-free water or 1x IDTE buffer: 10 mM Tris-HCl, 0.1 mM ethylenediaminetetraacetic acid, pH 8.
4. Thermocycler.
5. PCR tubes (0.2 mL, sterile, DNAse-free).
6. A PCR-clean up purification kit.
7. Nanodrop spectrophotometer.

2.2 GGA Reagents

1. T4 DNA Ligase (400 U/μL) supplied with 10x T4 DNA ligase buffer with 10 mM ATP.
2. Type IIS restriction enzyme. We commonly use Esp3I or BsaI.
3. Thermocycler.
4. PCR tubes (0.2 mL, sterile, DNAse-free).
5. Nuclease-free water.
6. Purified synthons from the oligo extension (diluted to 100 nM in nuclease-free water)
7. GGA-compatible backbone.
8. Nanodrop spectrophotometer.

2.3 Chemical Transformation

1. Chemical-competent cloning strains such as Mach1 or DH5-alpha amongst others (*see* **Note 1**).
2. LB-agar plates supplemented with the appropriate antibiotic, e.g., 100 μg/mL carbenicillin or 34 μg/mL chloramphenicol.

3. PCR tubes (0.2 mL, sterile, DNase-free).
4. Thermocycler.
5. SOC outgrowth media: 20 g/L tryptone, 5 g/L yeast extract, 10 mM NaCl, 2.5 mM KCl, 10 mM MgCl2, 10 mM MgSO4, 20 mM glucose.

2.4 Plasmid Verification

1. Terrific broth: 24 g/L yeast extract, 20 g/L tryptone, 4 mL/L glycerol, 17 mM KH_2PO_4, 72 mM K_2HPO_4.
2. 24-deep well plate or 14-mL culture tubes.
3. Antibiotics.
4. Incubator with plate holder or thermoshaker with plate holder.
5. Miniprep kit.

3 Methods

Prepare all reactions using nuclease-free water. All work can be performed at room temperature. Add the enzymes last and keep them on ice, in a cold block, or in the freezer before use.

3.1 Design of Fragments and Oligos

1. With Benchling or another molecular biology software tool, divide genes of interest into 70–80 bp fragments in silico. Sequences need to be devoid of the specific type IIS restriction enzyme used for GGA, for example, Esp3I/BsmBI or BsaI sites. We annotated each fragment for easier selection with the assembly tool (see Fig. 2a).
2. Use the Assembly Wizard to design fragments with correct overhangs (see Fig. 2b). Compatible 4-bp overhangs and initial primers with restriction sites are designed with the Golden Gate assembly tool from Benchling (see Fig. 3).
3. Initial designed oligos are then manually extended in the Benchling sequence editor to overlap by 18–25 bp for optimal annealing (see Fig. 4). Oligos are kept at a maximum length of 60 bases to ensure the fastest turnaround and cheapest per base price from DNA synthesis vendors (*see* **Note 1**).

3.2 Oligo Extension for dsDNA Synthesis

Perform each oligo extension reaction in individual PCR tubes.

1. Construct synthons by oligo extension using, e.g., Q5 DNA polymerase (NEB) in 25 μL reactions consisting of 1x Q5 reaction buffer, 200 μM of each dNTP, 0.5 units Q5 DNA polymerase, 200 nM oligos (100 nM of each). Each reaction is made in individual PCR tubes (*see* **Note 2**).

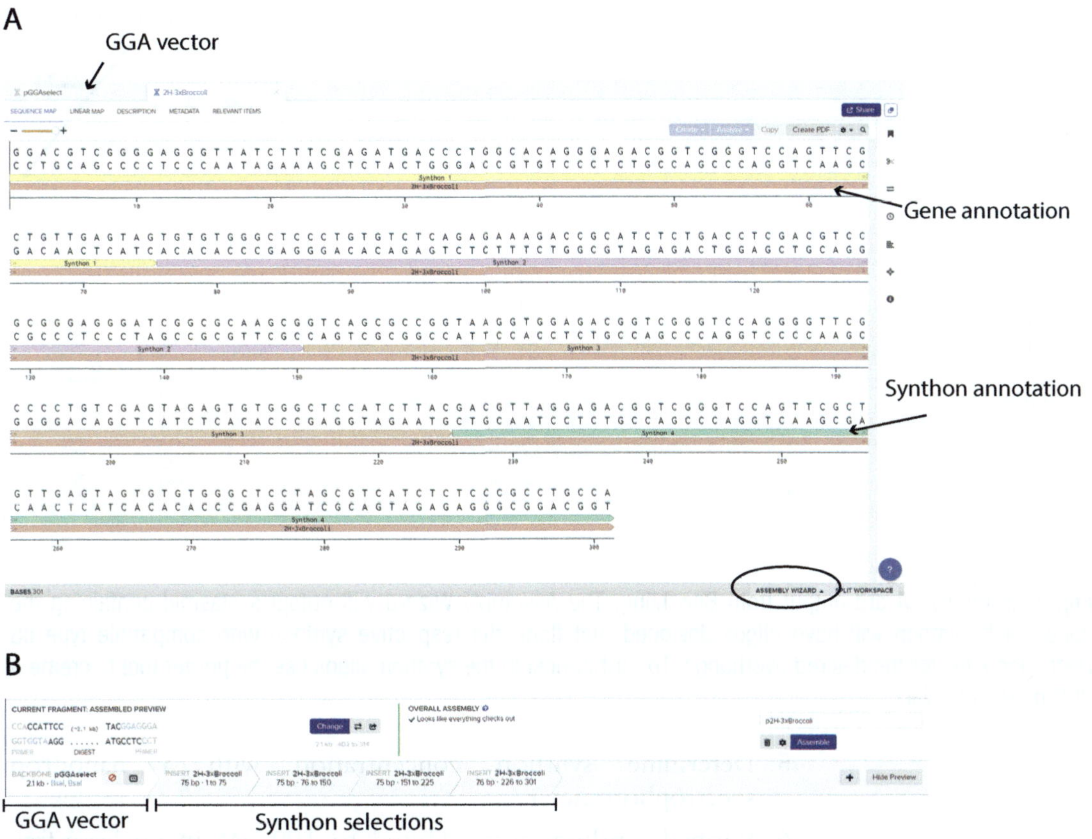

Fig. 2 Design of synthons and oligos with Benchling. (**a**) Overview of the gene of interest for construction. Using Benchling the gene of interest (301 bp) has been split into four synthons of lengths 75–76 bps as annotated. (**b**) The four synthons are set as input into the Assembly wizard with pGGAselect as the backbone. Use the Golden Gate tool in the Assembly wizard to simulate a GGA and design oligos for gene construction. The GGA vector can be in silico "digested" with a type IIS restriction enzyme while the regions of interest, that is, the synthons are "selected"

2. Thermocycling was performed as follows:
   (a) Initial denaturation at 98 °C for 30 s.
   (b) Then six cycles of:
      (i) Denaturation at 98 °C for 10 s
      (ii) Annealing at synthon-dependent temperatures for 20 s
      (iii) Elongation at 72 °C for 10 s
   (c) Final elongation at 72 °C for 2 min
3. Purify synthons with the PCR clean-up kit following the manufacturer's protocol (*see* **Note 3**).
4. Elution from the spin columns can be done with either nuclease-free water or elution buffer (10 mM tris-HCl, pH 8).

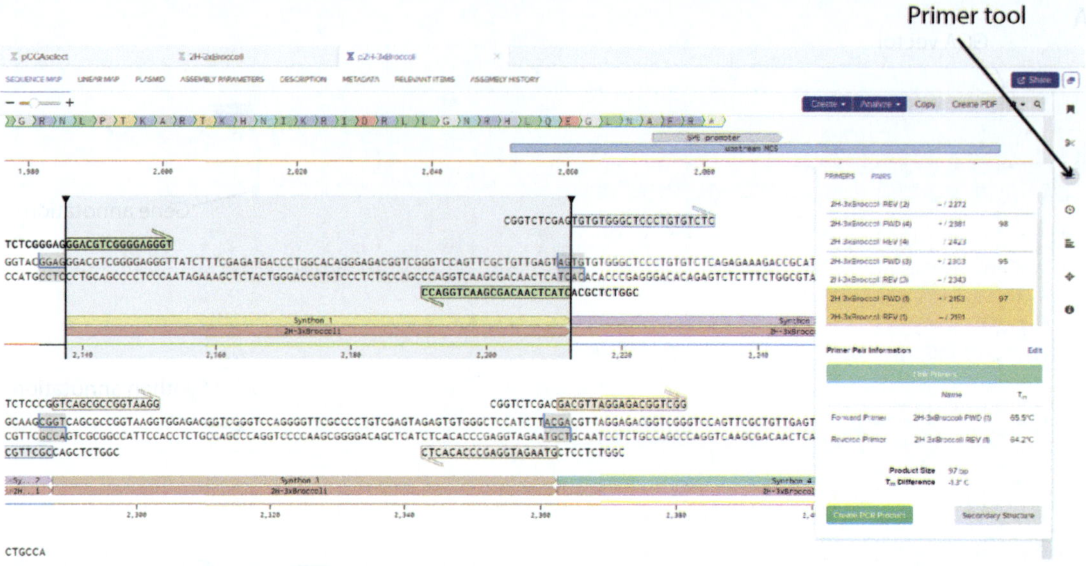

Fig. 3 Assembly Wizard output from Benchling. The Assembly Wizard will output a plasmid containing the gene. Each synthon will have oligos designed that flank the respective synthon with compatible type IIS restriction sites for the desired overhangs. To further design the synthon oligos use the primer tool to create a PCR product in silico

5. Determine synthon concentration with a nanodrop spectrophotometer.
6. Purified synthons were diluted to 100 nM in nuclease-free water (*see* **Note 4**).

3.3 Golden Gate Reaction

1. Golden Gate reactions (*see* **Note 5**) were performed with equimolar amounts of DNA using 25 or 50 fmol of synthon DNA, 25 fmol vector DNA, 0.5 μL T4 DNA ligase (NEB), 0.5 μL of either Esp3I (NEB) or BsaI (NEB) in 1x T4 DNA ligase buffer with 1 mM ATP (NEB) in 10 μL reactions.
2. Golden Gate reactions cycling:
   (a) 10 min at 37 °C
   (b) Followed by 15–25 cycles (*see* **Note 6**):
      (i) 5 min at 37 °C
      (ii) 5 min at 16 °C
   (c) 5 min incubation at 50 °C
   (d) 5 min at 80 °C

3.4 Chemical Transformation of *Escherichia Coli*

1. Transform the total volume of each GGA reaction in 30 μL Mach1 cells (*see* **Note 7**) by heat shock in a thermocycler:
   (a) 10 min at 4 °C
   (b) 45 s at 42 °C
   (c) Hold at 4 °C

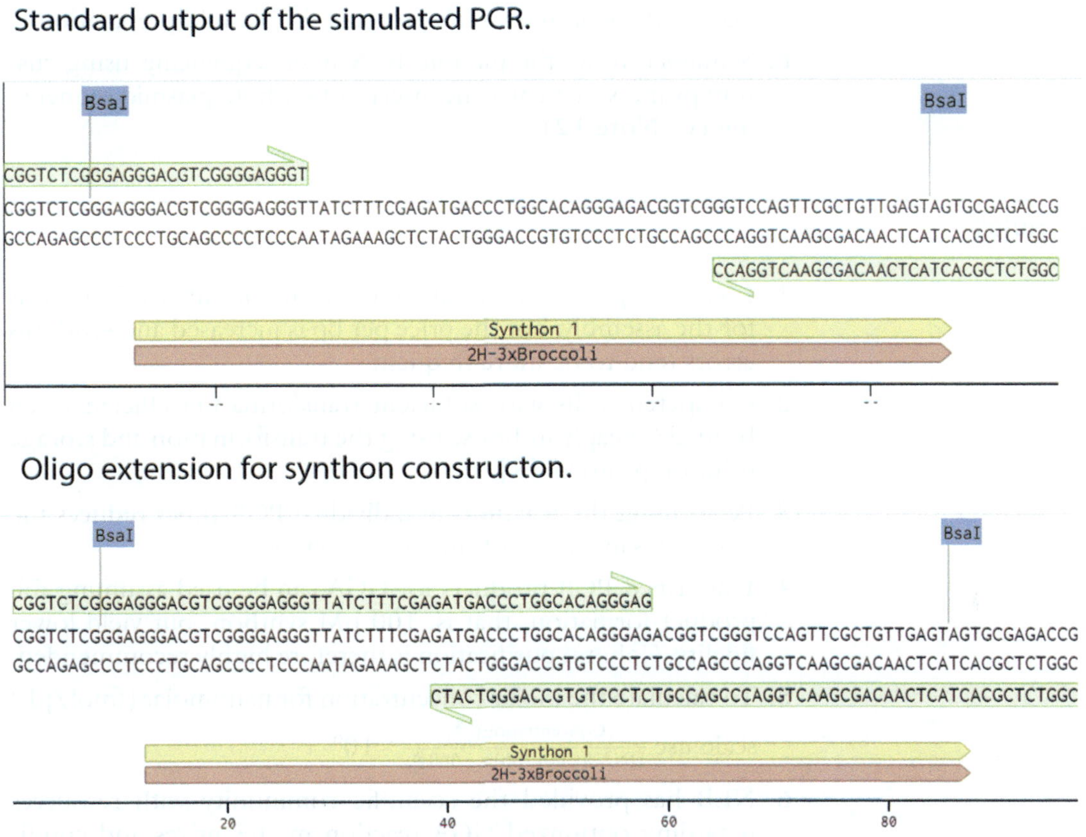

Fig. 4 Designing the synthon oligos with the simulated PCR product. The oligos from the standard output of the simulated PCR (top) are extended so they overlap and can be used for synthon construction (bottom). Each synthon oligo can be manually extended for the oligo extension. The forward oligo (left) is extended to 60 bases and the reverse oligo (right) is extended based on the melting temperature of the overlap (65 °C) resulting in 59 bases

2. Recover the cells in 160 μL SOC media and incubate for 1 h at 37 °C before plating (*see* **Note 8**).
3. Plate cells on selective LB-agar plates containing the appropriate amounts of antibiotics.

3.5 Plasmid Preparation

1. If using a fluorescent dropout vector for GGA (*see* **Note 9**) pick non-fluorescent colonies and inoculate them in 3 mL terrific broth (*see* **Note 10**) with antibiotic selection in 24 deep-well plate or 14 mL culture depending on the number of colonies picked (*see* **Note 11**).
2. Grow them at 37 °C with shaking at 220 rpm overnight.
3. Harvest 2 mL of the cultures by centrifugation at 4000 × *g* for at least 5 minutes. Save the remaining at 4 °C for making glycerol stocks after sequence verification.
4. Decant the media.

5. Follow the manufacturer's protocol for plasmid preparation.
6. Sequence-verify the plasmid by Sanger sequencing using custom primers that flank the insert or by whole-plasmid sequencing (*see* **Note 12**).

4 Notes

1. Longer oligos can be used to reduce the number of synthons for the assembly, but the price per bp is increased and synthesis errors tend to be more frequent.
2. Competent cells with sufficient transformation efficiency can be made cheaply in-house using the transformation and storage solution protocol [8].
3. Performing the reactions in individual PCR tubes reduces the risk of misannealing of the various oligos.
4. Unpurified PCR products for GGA can be used assuming full product formation, that is, 100 nM synthon, but yield lower fidelity. Column purification is therefore highly recommended.
5. To calculate the molar concentration for nanomolar (fmol/μL) scale use $\frac{\text{Concentration}\left(\frac{\text{ng}}{\mu\text{L}}\right)}{\text{DNA length (bp)} \times 660\frac{\text{g}}{\text{mol}}} \times 10^6$.
6. NEB has provided the scientific community with resources regarding optimized GGA reaction master mixes and conditions for improving the yield while reducing reaction time, see https://www.neb.com/en-us/protocols/2015/03/04/golden-gate-assembly-protocol-for-using-neb-golden-gate-assembly-mix-e1600.
7. These conditions worked in our hands and 15 cycles have been sufficient for the assembly 2–4 synthons. Higher cycle numbers can increase the fidelity when assembling more synthons.
8. The competent cells can be added directly to the reaction tube. Other standard transformation protocols can also be used. If electroporation is the preferred method, then desalt the reaction by drop dialysis or column purification.
9. For carbenicillin or ampicillin selection, recovery for an hour is not needed but plating without recovery will fewer colonies.
10. Fluorescent dropout vectors are commonly available in a range of MoClo toolkits from Addgene. We have made our own, but also commonly used a vector such as pYTK095 from the Yeast toolkit [9] (https://www.addgene.org/65202/).
11. Other common culture broths can also be used. We have used terrific broth to increase the cell density and plasmid yields.

12. We have picked up to 8 colonies that result in at least one positive clone.
13. We have observed around 87.5% fidelity for 2-synthon assemblies, 60% fidelity when assembling 4 synthons, and 12.5% fidelity when assembling 6 synthons [10].

References

1. Engler C, Kandzia R, Marillonnet S (2008) A one pot, one step, precision cloning method with high throughput capability. PLoS One 3(11):e3647–e3647
2. Czar MJ et al (2009) Gene synthesis demystified. Trends Biotechnol 27(2):63–72
3. Halper SM, Hossain A, Salis HM (2020) Synthesis success calculator: predicting the rapid synthesis of DNA fragments with machine learning. ACS Synth Biol 9(7):1563–1571
4. Benchling. Available from: https://www.benchling.com/
5. NEBridge Ligase Fidelity. Available from: https://ligasefidelity.neb.com/
6. Edinburgh Genome Foundry Overhang Designer. Available from: https://cuba.genomefoundry.org/design-overhangs/
7. Jeong JY et al (2012) One-step sequence- and ligation-independent cloning as a rapid and versatile cloning method for functional genomics studies. Appl Environ Microbiol 78(15):5440–5443
8. Chung CT, Niemela SL, Miller RH (1989) One-step preparation of competent Escherichia coli: transformation and storage of bacterial cells in the same solution. Proc Natl Acad Sci USA 86(7):2172–2175
9. Lee ME et al (2015) A highly characterized yeast toolkit for modular, multipart assembly. ACS Synth Biol 4(9):975–986
10. Nguyen MTA, et al (2022) Enzymatic assembly of small synthetic genes with repetitive elements. bioRxiv. 2022.11.23.517630.

Chapter 13

Simplifying Recombinant Protein Production: Combining Golden Gate Cloning with a Standardized Protein Purification Scheme

Sonja Zweng, Gabriel Mendoza-Rojas, Alexander Lepak, and Florian Altegoer

Abstract

Recombinant protein production is pivotal in molecular biology, enabling profound insights into cellular processes through biophysical, biochemical, and structural analyses of the purified samples. The demand for substantial biomolecule quantities often presents challenges, particularly for eukaryotic proteins. *Escherichia coli* expression systems have evolved to address these issues, offering advanced features such as solubility tags, posttranslational modification capabilities, and modular plasmid libraries. Nevertheless, existing tools are often complex, which limits their accessibility and necessitate streamlined systems for rapid screening under standardized conditions. Based on the Golden Gate cloning method, we have developed a simple "one-pot" approach for the generation of expression constructs using strategically chosen protein purification tags like hexahistidine, SUMO, MBP, GST, and GB1 to enhance solubility and expression. The system allows visual candidate screening through mScarlet fluorescence and solubility tags are removable via TEV protease cleavage. We provide a comprehensive protocol encompassing oligonucleotide design, cloning, expression, His-tag affinity chromatography, and size-exclusion chromatography. This method, therefore, streamlines prokaryotic and eukaryotic protein production, rendering it accessible to standard molecular biology laboratories with basic protein biochemical equipment.

Key words Golden Gate cloning, Modular cloning, Protein expression, His-tag affinity purification, Solubility tags, GB1, TEV-cleavage

1 Introduction

Recombinant protein production is one of the cornerstones of modern molecular biology. The characterization of proteins using biophysical, biochemical, and structural techniques provides a deep understanding of their functions and enables to unravel the intricate mechanisms of many cellular processes. Most of these techniques require large quantities of biomolecules, and their production using heterologous systems is often still a bottleneck, especially

Daniel Schindler (ed.), *Golden Gate Cloning: Methods and Protocols*, Methods in Molecular Biology, vol. 2850,
https://doi.org/10.1007/978-1-0716-4220-7_13,

considering eukaryotic proteins. *Escherichia coli* expression systems have been continuously advanced by, for example, introducing different solubility tags, enhancing the production of post-translationally modified proteins [1], or providing highly modular plasmid libraries for high-throughput screening [2]. Most of these tools are powerful for exhaustive expression screening of difficult-to-produce candidates and therefore likely used in laboratories with extensive biochemical expertise. However, many laboratories lacking this expertise rather require systems that provide easy access and rapid screening using standardized expression conditions.

Various molecular cloning methods are available to clone and assemble DNA fragments of interest. Golden Gate cloning offers a fast and reliable alternative to conventional restriction-insertion cloning relying on type IIS endonucleases that hydrolyze DNA outside their recognition sequence. This way, multiple fragments can be assembled hierarchically using pre-defined overhangs [3]. Furthermore, the Golden Gate cloning method is highly efficient and thus perfectly suited to assemble constructs in a single "one-pot" reaction.

Our approach, therefore, combines a simple collection of Golden Gate cloning-based expression constructs with a streamlined strategy of protein expression conditions (Fig. 1). We have included sequences for hexahistidine tags fused N- or C-terminally (6HN, 6HC) to the protein of interest (POI), classical solubility enhancing tags as the small ubiquitin-like modifier (SUMO), the maltose binding protein (MBP), and the glutathione-S-transferase (GST) that can also be used as an alternative affinity tag for, for example, interaction assays. In addition, we have identified the B1 domain of the Streptococcal Protein G (GB1) as a powerful tag to increase both expression and solubility of target proteins [4], even in the case of secreted plant and fungal proteins [5–7]. Tags can be removed from the POI by a tobacco etch virus (TEV) protease cleavage site separating the POI and the respective tag(s). In addition, the backbones contain an mScarlet under control of a stationary-phase promoter allowing visual inspection of positive *E. coli* candidates in which the mScarlet expression module has been replaced by the gene of interest (GOI). This protocol further includes a section on oligonucleotide design, cloning and a detailed description of expression conditions and his-tag-affinity chromatography using a simple peristaltic pump. We briefly discuss a size-exclusion chromatography (SEC) step to increase the purity and homogeneity of the sample. Our method therefore offers an easy and rapid production strategy for both prokaryotic and eukaryotic proteins. The whole described workflow can be performed in any standard molecular biology laboratory with basic protein biochemical equipment. As an example, we use the *Thecaphora* unique effector 16 (Tue16) of the smut fungus *Thecaphora thlaspeos*, a pathogenic fungus of Brassicaceae plants.

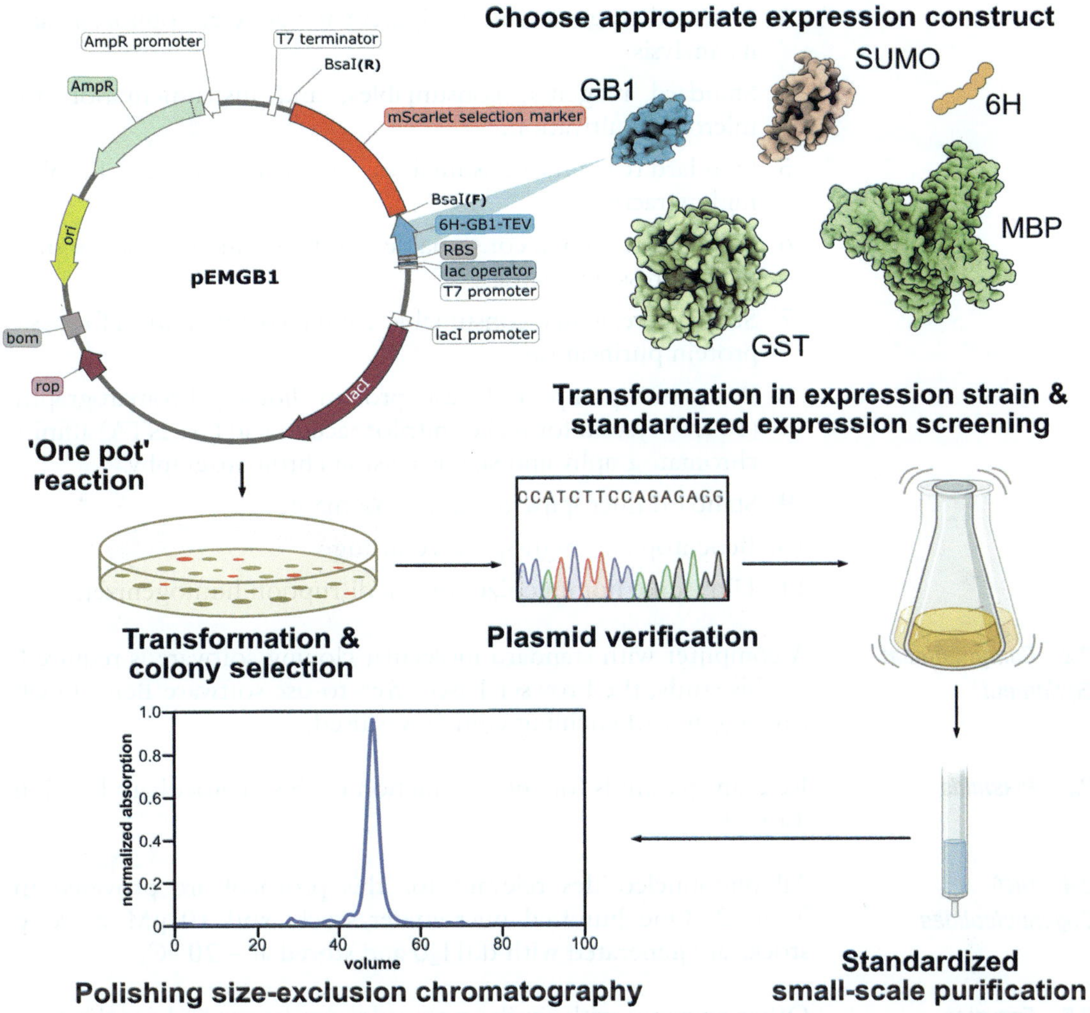

Fig. 1 Workflow for recombinant protein expression in *E. coli* using a simple Golden Gate cloning system. Shown is the workflow from choosing the appropriate assembly system for the GOI and the one-pot-reaction into the respective expression vector. Subsequent selection and verification, followed by test expressions of the protein of interest (POI) with different methods. The protocol ends with size-exclusion step to polish the purified POI for different downstream applications. The figure was generated using Snapgene, Biorender, and Chimera X [10]

2 Materials

The conduction of the described protocol requires the following standard laboratory equipment and consumables.

2.1 Standard Reagents and Laboratory Equipment

1. Standard reagents, consumables, and instrumentation for PCR reactions.
2. Standard reagents for molecular biology experiments (Golden Gate cloning).

3. Standard reagents and equipment for gel electrophoresis and its analysis.
4. Standard reagents, consumables, and instrumentation for microbial cultivation.
5. Standard reagents, consumables, and instrumentation for plasmid extraction.
6. Standard reagents, consumables, and instrumentation for protein expression analysis.
7. Standard reagents, consumables, and instrumentation for basic protein purification.
8. Peristaltic pump and fast protein liquid chromatography (FPLC) system for nickel-nitrilotriacetic acid (Ni-NTA) affinity chromatography and size-exclusion chromatography.
9. Standard micropipettes and consumables.
10. Benchtop and high-speed centrifuge.
11. Ultrasonic homogenizer or cell disruption homogenizer.

2.2 Computational Equipment

A computer with standard molecular cloning software is required. In this study, the browser-based, free-to-use software Benchling® (https://www.benchling.com/) was used.

2.3 Plasmids

Relevant plasmids for the conduction of this protocol are listed in Table 1.

2.4 DNA Oligonucleotides

All oligonucleotides relevant for this protocol are provided in Table 2. One hundred micrometer stocks and 10 μM working stocks are generated with ddH_20 and stored at −20 °C.

2.5 Enzymes

Other enzymes with similar properties can be applied for the conduction of this protocol (*see* **Note 1**).

1. BsaI (10,000 units / mL).
2. Phusion DNA polymerase (2000 units / mL).
3. T4 DNA ligase (400,000 units/mL).
4. TEV protease (10,000 units/mL) (if cleavage of respective tag is desired).

2.6 Chemicals, Buffers and Media

1. dYT liquid medium: 1.6% (w/v) Bacto-Tryptone, 1% (w/v) yeast extract, 1% (w/v) sodium chloride.
2. 10 × ligation buffer: 10 mM Tris(-HCl), pH 7.5–8.0, 50 mM NaCl and 1 mM ethylenediaminetetraacetic acid (EDTA).
3. 1 × tris-acetate-EDTA (TAE) buffer: 1 mM EDTA · Na_2 · 2 H_2O, 20 mM acetate, 40 mM Tris–HCl.
4. 6 × agarose gel loading dye.
5. Agarose (standard).

Table 1
Plasmids designed and used in this study

Name	Relevant feature	Parental plasmid	Reference
pEM-GST	Contains N-terminal hexahistidine tag followed by a GST and a TEV-cleavage site; Contains mScarlet to be released with BsaI; AmpR	–	F. Altegoer
pEM-6HN	Contains N-terminal hexahistidine tag; Contains mScarlet to be released with BsaI; AmpR	–	F. Altegoer
pEM-6HC	Contains C-terminal hexahistidine tag; Contains mScarlet to be released with BsaI; AmpR	–	F. Altegoer
pEM-GB1	Contains N-terminal hexahistidine tag followed by a GB1 and a TEV-cleavage site; Contains mScarlet to be released with BsaI; AmpR	–	F. Altegoer
pEM-SUMO	Contains N-terminal hexahistidine tag followed by a SUMO and a TEV-cleavage site; Contains mScarlet to be released with BsaI; AmpR	–	F. Altegoer
pEM-MBP	Contains N-terminal hexahistidine tag followed by a MBP and a TEV-cleavage site; Contains mScarlet to be released with BsaI; AmpR	–	F. Altegoer
pEM-GST_Tue16	pEMGST derivative containing *tue16* under T7 control, AmpR	pEM-GST	F. Altegoer
pEM-6HN_Tue16	pEM6HN derivative containing *tue16* under T7 control, AmpR	pEM-6HN	F. Altegoer
pEM-6HC_Tue16	pEM6HN derivative containing *tue16* under T7 control, AmpR	pEM-6HC	F. Altegoer
pEM-GB1_Tue16	pEMGB1 derivative containing *tue16* under T7 control, AmpR	pEM-GB1	F. Altegoer

6. DNA ladder (here used: λ/PstI).
7. YT solid medium: 0.8% (w/v) Bacto-Tryptone, 0.5% (w/v) yeast extract, 0.5% (w/v) sodium chloride, agar-agar 2% (w/v).
8. 1 M isopropyl-β-D-thiogalactopyranoside (IPTG).
9. Buffer A: 20 mM 4-(2-hydroxyethyl)-1-piperazineethanesulfonic acid (HEPES), pH 8.0, 20 mM KCl, 40 mM imidazole, 250 mM NaCl buffer B: 20 mM HEPES, pH 8.0, 20 mM KCl, 250 mM imidazole, 250 mM NaCl SEC-buffer: 20 mM HEPES pH 7.5, 20 mM KCl, 200 mM NaCl.
10. 3× Laemmli buffer: 150 mM Tris–HCl pH 6.8, 6% (v/v) SDS, 30% (v/v) glycerol, 15% (w/v) ß-mercaptoethanol, 0.003% (w/v) bromophenol blue.

Table 2
Listed are the oligonucleotides used in this study

Name	Sequence (5´-3´) *	Information
Tue16_BsaI_fwd	GGTCTCCCATGGGCACAAACCCCCCCTCCCCTCA	Amplification of *tue16* for cloning via BsaI
Tue16_BsaI_rev	GGTCTCCTCGAGGGGGCCAGGTCCCGC	Amplification of *tue16* for cloning via BsaI

[a]The color scheme is according to the scheme provided in Fig. 2

Table 3
Listed are the *E. coli* strains relevant for this study

Name	Relevant features	Reference
TOP10	*F-mcrA Δ(mrr-hsdRMS-mcrBC) Φ80lacZΔM15 ΔlacX74 recA1 araD139 Δ(araleu)7697 galU galK rps L (StrR) endA1 nupG*	ThermoScientific
BL21(DE3)	*Δlon protease* *ΔompT* *ΔhsdSB* contains the T7 RNAP gene under the *lacUV5* promoter	[3]

11. Buffers for SDS-PAGE preparation (as stated in the following protocol).
12. Antibiotics (ampicillin): Stock concentration of 100 mg / mL. The final concentration is 0.1 mg / mL.
13. SDS-PAGE resolving gel: 0.375 M Tris–HCl, pH 8.8, 10% (v/v) acrylamide, 0.1% (v/v) SDS, 1.25% (v/v) Glycerol, 0.05% (v/v) ammonium persulfate (APS), 0.1% (v/v) Tetramethylethylenediamine (TEMED).
14. SDS-PAGE stacking gel: 0.125 M Tris–HCl, pH 6.8, 5% (v/v) Acrylamide, 0.1% (v/v) SDS, 0.05% (v/v) Ammonium persulfate (APS), 0.1% (v/v) Tetramethylethylenediamine (TEMED).
15. Coomassie brilliant blue (CBB) staining solution: 0.05% (w/v) Coomassie brilliant blue R250 (CBB), 9% (v/v) glacial acetic acid, 45% (v/v) Methanol.
16. Destaining solution: 10% (v/v) glacial acetic acid, 10% (v/v) ethanol.

2.7 Strains

The relevant laboratory *E. coli* strains for the conduction of this protocol are provided in Table 3.

3 Methods

3.1 Construction of the Expression Plasmid by Golden Gate Cloning.

This part of the method section explains the primer design on the *tue16* gene of *T. thlaspeos*. However, this approach is suitable for any gene of any target organism.

3.1.1 Design of Oligonucleotides

1. Use an in silico cloning tool of choice (*see* **Note 2**).
2. Upload the coding sequence of interest. Eukaryotic genes can contain introns. Only use the exon sequences (cDNA sequence) for oligonucleotide design (*see* **Note 3**).
3. Design oligonucleotides with the overhangs shown in Fig. 2. Do not include the codon of the starting methionine as it is already present in the 5′-overhang (*see* **Note 4**).
4. Check for similar annealing temperatures and avoid primer dimer formation.

3.1.2 Amplification of the GOI (See **Note 5**)

1. If not ordered dissolved in water, apply the respective volume of ddH_2O to your oligonucleotides.
2. Mix the following components in a microcentrifuge tube for polymerase chain reaction (PCR):

Constituent	Volume
ddH_2O	33 μL
Oligonucleotides (5 μM each)	5 μL
5× Reaction buffer	10 μL
gDNA/ cDNA	1 μL
dNTP's	0.5 μL
DNA polymerase	0.5 μL

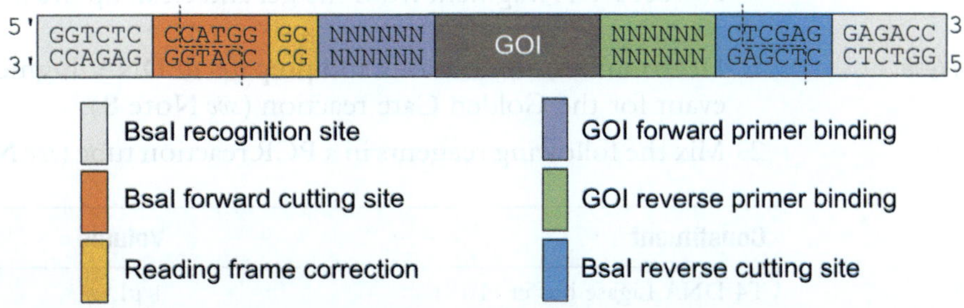

Fig. 2 Schematic depiction of the Golden Gate-based cloning approach. Shown are the precise DNA sequences required for amplification and subsequent Golden-Gate cloning of a gene of interest (GOI) maintaining the correct open-reading frame

3. Mix the reaction briefly by flipping the tube or pulse vortexing. Spin reactions quickly in a microcentrifuge ensuring the reaction mix is at the bottom of the tube. Place the reaction in a thermocycler running the following program:

Temperature	Time	Cycles
98 °C	2 min	1
95 °C	10–15 s	35
50–65 °C (depending on oligonucleotide) 72 °C	10–30 s 30 s for 1 kb	
72 °C	5 min	1
4 °C	Hold	1

4. Prepare an agarose gel for subsequent gel electrophoresis. Weigh in the respective amount of agarose (*see* **Note 6**), mix with the respective volume of 1× TAE buffer, and heat it in a microwave oven for 2–5 min until the solution appears translucent (*see* **Note 7**). Let it cool to approximately 60 °C and apply your DNA stain in the appropriate concentration. Pour the liquid into a leakproof gel tray and let it solidify for about 10–15 min. Place the gel tray in a chamber filled with 1× TAE buffer so that it is covering the gel completely.
5. Load a reference (size standard) and apply PCR samples mixed with 6× loading dye.
6. Apply appropriate voltage for DNA separation and take a photo of stained DNA with an imaging system.
7. Clean up the PCR reaction using an appropriate DNA purification kit and subsequently measure the DNA concentration using a UV-Vis spectrophotometer (e.g., NanoDrop, Thermo-Scientific). In the case of unspecific amplification, excise the correct DNA fragment from the gel and clean up the fragment.

3.1.3 Golden Gate Cloning

1. Determine concentrations and prepare all DNA fragments relevant for the Golden Gate reaction (*see* **Note 8**).
2. Mix the following reagents in a PCR reaction tube (*see* **Note 9**):

Constituent	Volume
T4 DNA Ligase buffer (10×)	1 μL
T4 DNA Ligase (400,000 units / mL)	1 μL
Type IIs restriction enzyme (here: BsaI)	1 μL
Destination plasmid	~20 fmol

(continued)

Constituent	Volume
GOI fragment	~20 fmol
In case of annealed oligonucleotides	20 fmol up to 2 pmol
ddH_2O	Ad 10 μL

3. Mix the reaction briefly by flipping the tube or pulse vortexing. Spin reactions quickly in a microcentrifuge ensuring the reaction mix is at the bottom of the tube. Place the reaction in a thermocycler running the following program:

Temperature	Time	Cycles
37 °C	2 min	10
16 °C	5 min	
37 °C (*see* **Note 10**)	5 min	1
80 °C	5 min	1
10 °C	Hold	1

4. Use the Golden Gate reaction immediately for *E. coli* transformation or store at −20 °C for later transformation (*see* **Note 11**).

3.1.4 Transformation of Competent E. coli Cells

For this protocol, in-house generated competent *E. coli* TOP10 cells were used. Chemically competent cells were prepared according to a RbCl-based method [8]. Nonetheless, other suitable methods for the preparation of competent *E. coli* cells can be used as well.

1. Thaw 20 μL of competent cells for 5 min on ice.
2. Add 5 μL of the Golden Gate reaction mixture and mix by gently snipping a finger against the microcentrifuge tube (*see* **Note 11**).
3. Incubate on ice for 5 min, while heating a water bath or heat block to 42 °C.
4. Perform a heat shock at 42 °C for 45 s and immediately put the microcentrifuge tube back on ice afterward.
5. After 5 min, add 300 μL of your preferred medium. (Ideally preheated to 37 °C.)
6. Incubate at 37 °C at a shaking frequency of 200 rpm for 30 min.
7. Plate 150–300 μL of the cell suspension on agar plates supplemented with the medium of preference and ampicillin using sterile glass beads/ pipettes or a cell spreader.
8. Incubate plates overnight at 37 °C.

3.1.5 Screening and Plasmid Verification

An mScarlet expression module inserted between the BsaI sites in the used vector set allows for visual inspection of transformants and identification of potentially positive candidates. A *glpT* promoter that is induced by low glucose concentrations (i.e., in stationary phase) is used for *mScarlet* gene expression based on the iGEM part BBa_J72163. Red colonies producing the mScarlet can be clearly differentiated from the potential positive "white" ones. In this protocol, we extract plasmid DNA and use a restriction digest followed by external sanger sequencing to confirm the correct DNA sequence.

1. Inoculate potentially correct "white" colonies (2–5 colonies; *see* **Note 12**) in 5 mL dYT media with ampicillin and incubate over night at 37 °C under continuous shaking at 200 rpm.
2. Verify the correct insert by colony PCR or directly purify the plasmid and choose appropriate restriction enzymes for a restriction digest (*see* **Note 13**).
3. Send potential candidates for Sanger sequencing using an appropriate sequencing primer.

3.2 Protein Expression of the Target Protein

In this protocol, we only describe protein production in *E. coli* BL21 DE3. Depending on the produced protein, other expression strains should be chosen (*see* **Note 14**).

3.2.1 Transformation of E. coli Expression Strain

1. To transform the respective *E. coli* strain with the verified plasmid, thaw 20 μL of the respective competent *E. coli* cells on ice for 5 min for each protein of interest.
2. Add 1 μL of the final plasmid (concentration 50–150 ng/μL) onto the thawed cells. Mix carefully by flipping your finger against the reaction tube. Proceed as described in Subheading 3.1.4.

3.2.2 Test Expression of the Target Protein

1. For test expression, three different conditions are tested. Of each culture, cell material of different colonies is gathered by wiping over the agar plate containing the freshly transformed expression strain with a sterile glass pipette (*see* **Note 15**).
2. For the autoinduction, inoculate your culture(s) of dTY-medium with appropriate antibiotic concentration and 1% (w/v) lactose (*Lac*) and incubate the culture at 28 °C for 20 h at 180–220 rpm.
3. For IPTG induction, inoculate pre-cultures (1/10 of your main culture volume) and after overnight incubation (12–16 h) upon shaking at 180–220 rpm, dilute to an optical density (OD_{600}) of 0.1–0.2. Incubate the cultures until they reach an optical density of 0.5–0.7. One of the cultures is induced with 0.5 mM IPTG and incubated at 37 °C for 3 h

at 180–220 rpm (*IPTG37*). The other culture needs to be cooled until it reaches 20 °C and is then induced with 0.5 mM IPTG to be incubated at 20 °C for 20 h, shaking at 180–220 rpm (*IPTG20*).

4. After the respective incubation time, harvest cells at 4000 × *g* for 15 min (*see* **Note 16**). Either freeze cells with liquid nitrogen and store at −20 °C or continue with the subsequent small-scale protein purification protocol.

3.3 Small-Scale Protein Purification

For a small-scale protein production (culture volumes <0.05 L), the following buffers should be prepared a priori. Each buffer should be adjusted to pH 8.0 and filtered through a 0.45 μm pore size filter (*see* **Note 17**).

During the protein purification procedure, three samples will be collected: Cell sample (*C*), cleared lysate sample (*L*), and elution sample (*E*). Exemplified purifications are shown in Fig. 3. All following steps are conducted on ice and the respective buffers have to be cooled to 4 °C prior to use.

1. Resuspend harvested cells in 10 mL ice cold buffer A to continue with the subsequent protein purification.
2. For cell lysis, prepare a 10 mg / mL lysozyme solution (dissolve in water or buffer A) and add 100 μL per 10 mL of culture

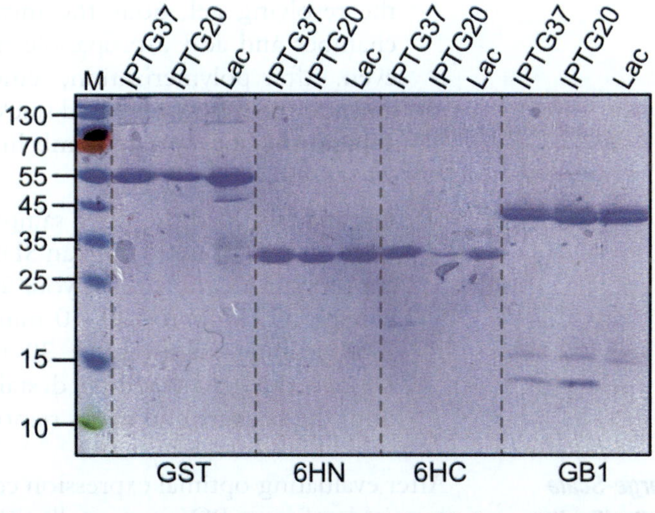

Fig. 3 Expression and purification of tagged Tue16 from different backbones using a predefined set of expression conditions. The polyacrylamide gel shows the elution fractions of Tue16 with four different tags expressed under the described three different standard conditions. In the case of Tue16, only slight differences in purified protein amounts are observed using the different expression conditions

(100 μg / mL final concentration). Sonicate each cell suspension for 5 × 30 s using an appropriate power (*see* **Note 18**) and take a sample of 20 μL into a new reaction tube (**C**). Add 10 μL 3 × Laemmli buffer to the sample.

3. Centrifuge the cells for 20 min at 16,000 × *g* at 4 °C to separate cell debris from the lysate, transfer the supernatant to a fresh 15 mL tube. Take 20 μL as a sample from the supernatant (*L*) and mix it with 10 μL 3 × Laemmli buffer.
4. Add 50 μL of Ni-NTA agarose (shake gently to resuspend the agarose resin before use) with a cut pipette tip and incubate at 4 °C for 15–20 min under continuous rotation (*see* **Note 19**).
5. Centrifuge the samples at 1500 × *g* and 4 °C for 5 min (*see* **Note 20**) and carefully discard the supernatant (do not touch the sedimented resin).
6. Add 1 mL of buffer A and transfer the resin to a fresh 2 mL reaction tube and wash the pellet thrice with 1 mL buffer A (centrifuge and remove the supernatant after each wash). After the last wash, remove the supernatant. For elution, add 50–100 μL buffer B and gently invert the tube to mix. After incubation for 5 min at 4 °C, centrifuge at 1500 × *g* for 5 min.
7. Take a 20 μL sample from the supernatant and mix with 10 μL 3 × Laemmli buffer (*E*).
8. For analysis, prepare an SDS polyacrylamide gel. First, prepare the resolving gel, pour the mixture into an appropriate gel chamber and add 2-propanole until the surface of the gel is even. After polymerization, remove the leftover 2-propanole and add the freshly prepared stacking gel mixture. Immediately append the appropriate comb into the stacking gel (*see* **Note 21**).
9. For analysis, load the three samples of each protein expression (**C**ell, **L**ysate, **E**luate) onto an SDS-PAGE with a protein ladder appropriate to the size of your POI. Let it run at a constant voltage of 250 V for 30–40 min and stain with a Coomassie Blue staining solution for 30 min (*see* **Notes 22** and **23**) Destain the gel by adding destaining solution multiple times until the background dye is removed.

3.4 Large-Scale Protein Purification

After evaluating optimal expression conditions providing the highest yields of your POI in a small-scale expression (*see* Subheading 3.3), a large-scale protein purification can be performed to obtain large yields of highly pure protein for biophysical, biochemical, or structural analysis. Follow the instructions of the respective expression method in Subheading 3.3 but increase the culture volume up to 6 L and elute the harvested cells in 30–35 mL buffer A per liter of expression culture. For the large-scale protein purification, you will typically need 1 L of buffer A, at least 50 mL of buffer B and 1 L of size exclusion chromatography (SEC) buffer. Each buffer should be previously filtered through a 0.45 μm filter. All steps are conducted on ice and all buffers are cooled prior to use.

3.4.1 Cell Lysis

1. After cultivation, cells have to be harvested in a centrifuge with the respective capacity for 15 min at 4000 × *g*. Eventually freeze the pellet at −20 °C until further use.
2. Take the harvested or thawed cells and resuspend them in 30–35 mL buffer A per liter of culture.
3. Meanwhile, cool the centrifuge to 4 °C with the respective rotor inside.
4. For cell lysis of larger culture volumes, a high-pressure cell disruptor is used (*see* **Note 24**). Flush the microfluidizer at least twice with cooled buffer A and pour in the resuspended sample through a sieve to avoid plugging the machine. Start disrupting the cells but discard the first bit of flow-through of buffer A as it does not contain your sample yet. Collect the disrupted cells and repeat once more but collect everything. Wash the system with up to 10–15 mL buffer A and collect the flow-through to avoid discarding some of the disrupted cells (*see* **Note 25**).

1. Transfer the lysed cells into 50 mL centrifugation tubes, take a sample (*C*—cell sample), and balance them out on a scale up to two decimal places by adding cooled buffer A to the lighter sample. Transfer into an appropriate rotor.
2. Centrifuge for 20 min at 4 °C at 40,000 × *g*.
3. In the meantime, calibrate your protein purification system for the later (SEC) by connecting the respective column and starting the system equilibration (*see* **Note 26**).
4. Also check your peristaltic pump and start equilibration of your his-tag affinity columns that you want to use. For that, connect the his-tag affinity columns to the peristaltic system and equilibrate with 10 column volumes (CV) pre-cooled buffer A. Depending on the flow rate, the equilibration approximately takes 10–20 min (*see* **Notes 27** and **28**).

3.4.2 His-tag Affinity Purification

During the his-tag affinity purification procedure, four different samples will be collected: cleared lysate sample (*L*), flow through sample (*FT*), wash sample (*W*), elution sample (*E*).

1. After the centrifugation, transfer the supernatant of each protein sample into a new 50 mL tube and take a sample (*L*) (*see* **Note 29**).
2. Stop the peristaltic pump, transfer the tubing into the tube containing the cleared lysate and start the pump. Put a new beaker underneath the opening of the columns to collect the

flow-through. Collect a sample of the flow-through (*FT*) by transferring 20 μL into a new reaction tube and adding 10 μL of 3 × Laemmli buffer.

3. Wash the columns for 5–10 CV with buffer A and collect the wash in a new beaker. Take a sample of 20 μL of the wash and add 10 μL of 3 × Laemmli buffer (*W*).
4. Disconnect the column(s) from the tubing and flush the tubing system with buffer B (*see* **Note 30**).
5. Reconnect the column(s) to the tubing, place a fresh 50 mL tube under the columns elute your protein of interest in 5 CV.
6. Take an elution sample (*E*) by transferring 20 μL of the eluate into a reaction tube and adding 10 μL of 3 × Laemmli buffer.

3.4.3 Reverse-his-Tag Affinity Purification (Optional)

For the cleavage of the respective tag from your protein of interest, His-tagged TEV protease is used. High imidazole concentrations can harm protein integrity, necessitating removal before TEV protease incubation. If no cleavage is planned, continue with Subheading 3.4.4. During the reverse affinity purification, again four different samples will be collected: TEV-cleaved sample (*TE*), TEV-cleaved flow through (*TF*), TEV-cleaved wash (*TW*), TEV-cleaved elution (*TE*).

1. Transfer the eluate into two appropriate Amicon concentrators (*see* **Note 31**). This depends on the size of your fusion protein. Balance out the Amicon concentrators using a precision scale (*see* **Note 32**).
2. Centrifuge at 5000–7500 × *g* with a fixed angle rotor or at 4000 × *g* with a swing-out rotor at 4–8 °C until the remaining volume reaches the 1–0.5 mL mark. This can take 30 min to several hours.
3. Fill the concentrator with precooled SEC buffer that does not contain imidazole and concentrate again to 1–0.5 mL.
4. Fill the concentrator up again to the 15 mL mark with SEC buffer and add an appropriate amount of TEV (use 1 A_{280} of TEV for 100 A_{280} of your protein; efficiency might be different for different proteins). Incubate for 2–3 h at room temperature or overnight at 4 °C while shaking on a rotary wheel or rocker.
5. Load your TEV-cleaved sample on a regenerated Ni-NTA column and take a sample by transferring 20 μL of the sample in a reaction tube and adding 10 μL of the 3 × Laemmli buffer (*TE*).
6. Start the peristaltic pump to load your protein onto the column. Collect the flow-through in a fresh 50 mL tube (*see* **Note 33**). Take a 20 μL sample and mix it with 10 μL 3 × Laemmli buffer (*TF*).

7. Wash the column with 5–10 CV of buffer A and collect it in a new 50 mL tube (*see* **Note 33**). Take a 20 μL sample and mix it with 10 μL 3 × Laemmli buffer (*TW*).
8. Elute the cleaved tag and TEV protease with 5 CVs and collect it as well in a fresh 50 mL tube. Take a 20 μL sample and mix it with 10 μL 3 × Laemmli buffer (*TE*).

3.4.4 Sample Analysis and Preparation for Size-Exclusion Chromatography

1. Transfer the eluate into an appropriate Amicon concentrator. Balance out the Amicon concentrators using a fine scale (*see* **Notes 31** and **32**).
2. Centrifuge at 5000–7500 × *g* with a fixed angle rotor or up to 4000 × *g* with a swing-out rotor at 4–8 °C until the remaining volume reaches approx. 1–0.5 mL. This can take 30 min to several hours depending on protein concentration (*see* **Note 34**).
3. Meanwhile, load 5 μL of the collected lysate, flow-through, wash, elution sample and potentially samples from the reverse-Ni-NTA onto an SDS-PAGE, adding 5 μL of an appropriate protein ladder into an additional lane and let it run at a constant voltage of 250 V for 30–40 min. Image your SDS-gel using a UV-imaging system and take a picture to check if your elution fraction contains your desired fusion protein (*see* **Note 35**).
4. Transfer the separating gel into an air-sealed box, cover the gel with Coomassie-Blue solution and incubate it on a rocker for at least 30 min. After incubation, remove the Coomassie-Blue solution and add destaining solution to the box, covering the gel completely. Change the destaining solution 3–5 times, each after at least 30 minutes of incubation. Once the gel is fully destained, use an imaging system to take a picture for documentation (*see* **Note 36**).
5. For regeneration of the Ni-NTA columns, wash them with ddH_2O, followed by 0.5 M EDTA solution, 0.5 M NaOH, and 0.2 M $NiSO_4$, with ddH_2O washing steps in between and at the end for 5 CVs each (*see* **Note 37**).
6. Once the elution fraction has been concentrated to 1–0.5 mL, resuspend the concentrated protein solution and transfer into a cooled 1.5 mL Reaction tube.
7. Clean the concentrator by filling it with SEC buffer and subsequent centrifugation until about 1 mL of the buffer is left on the membrane. Used, clean concentrators can be stored in ddH_2O (*see* **Note 38**).

3.4.5 Size-Exclusion Chromatography

This protocol aims at providing a simple and convenient methodology for protein purification. Size-exclusion chromatography is the method of choice for polishing purified proteins and retrieving homogenous samples. However, for the sake of simplicity, we will not introduce the pros and cons of choosing appropriate columns and buffer conditions here but rather give general advice on protein handling (*see* **Note 39**).

1. Spin down your concentrated protein at 8000 × *g* for 1 min to remove precipitants and measure the absorption at 280 nm to determine the protein concentration (*see* **Note 40**).
2. Wash the injection loop of the FPLC-system with 10 loop volumes of SEC buffer using a syringe (*see* **Note 41**).
3. Take up your concentrated protein solution with a syringe and inject into the injection loop.
4. Run an appropriate program depending on the choice of columns and the used buffer conditions. Collect the elution fractions using a fraction collector.

3.4.6 Sample Analysis

1. After the run has finished, open the respective evaluation program and choose up to 14 peak fractions for analysis (depending on the available SDS-PAGE setup), Of each fraction, take a 20 μL sample and mix it with 10 μL of 3 × Laemmli-Buffer. Run an SDS-PAGE at 250 V for 40 min with an appropriate protein size standard. Proceed with the SDS-PAGE as stated in Subheading 3.3. Once the SDS-PAGE is documented, protein purity can be evaluated, and appropriate fractions are chosen for concentration.
2. Pool the fractions containing your pure protein into one or several Amicon-concentrators and centrifuge at 5000–7500 × *g* with the fixed angle rotor or at 4.000 × *g* using a swing-out rotor at 4–8 °C until the volume reaches 0.5–1 mL. Measure the absorption at 280 nm to determine the protein concentration.
3. Prepare aliquots of your POI according to downstream analysis purposes and snap freeze in liquid nitrogen for long-term storage at −80 °C.

3.5 Summary and Perspectives

In conclusion, we here present an efficient and easily accessible approach to recombinant protein production that combines Golden Gate cloning with streamlined protein purification. It, therefore, might enable especially researchers from laboratories not specialized in biochemistry to explore protein functions and interactions. Researchers may also explore additional solubility tags and affinity purification strategies further than the ones presented here. In the future, we anticipate incorporating an additional set of plasmid backbones with different resistance markers and origins of replications to allow co-expression of two or more proteins to enhance the versatility of the system with respect to hetero-oligomeric complexes.

4 Notes

1. Enzymes should be stored at −20 °C and rigorously kept on ice during use.
2. In this study, the browser-based, free-to-use laboratory notebook Benchling® (https://www.benchling.com/) was used, which also features a comprehensive molecular cloning tool. However, any cloning program of choice can be used.
3. It is advisable to retrieve sequences from the respective genomic database of your organism of choice, if available. These databases can often be accessed via the UniProt database (https://www.uniprot.org).
4. Make sure to correctly design the overhangs to avoid a frameshift in the open reading frame. The overhangs include the codon ATG for starting methionine and a GC has to be added in the oligonucleotide sequence to correct the reading frame and bring the GOI *in frame* with the respective fusion tag.
5. In this protocol, the gene of interest was amplified from genomic DNA. However, it is possible to use other scaffolds for Golden Gate cloning such as (i) annealed oligonucleotides or (ii) synthesized DNA fragments with the respective overhangs.
6. We use a standard agarose concentration of 1% (w/v).
7. Handle carefully, risk of boiling retardation!
8. Using, for example, a nanodrop to determine the absorption at 260 nm is usually sufficient.
9. Enzymes should be kept at −20 °C and added last.
10. If the gene of interest has internal BsaI recognition sites, this step has to be omitted. The overall efficiency of the Golden Gate reaction will decrease but chances are your correctly assembled plasmid can be obtained.
11. The remaining reaction can be stored at −20 °C and used for another transformation with a larger volume in case of low transformation efficiency.
12. The maturation of the mScarlet can sometimes take up to 24 h. If distinguishing between white and red colonies is difficult due to pale red color, storage at 4 °C up to 24 h is recommended. The positive rate is usually high (<10% red colonies) but can vary in the case of long fragments or the assembly of an open reading frame from several fragments.
13. Ideally, choose a single enzyme that digests the backbone in 2–4 fragments that can be clearly distinguished by size and are not smaller than 500 bp for optimal resolution. The restriction pattern should be different from the hydrolysis of the ancestor

plasmid. Often, molecular cloning programs offer options for virtual digestion.

14. In our hands, BL21 DE3 retrieves/provides good results for the majority of proteins produced. In the case of proteins with disulfide bonds, we made good experience with using SHuffle T7 (NEB).
15. Always transform the expression strains freshly prior to performing expression tests. Transformation plates should not be older than 1 week.
16. Centrifugation time depends on the culture volume. A time of 15 minutes is usually sufficient to harvest culture volumes up to 1 L.
17. Buffers can be stored for several days to weeks at 4 °C. Filtration extends the shelf life of buffers and removes contaminations and non-dissolved precipitants. Carefully inspect your buffers prior to use.
18. The given values have to be adjusted to different ultrasonic homogenizer models. It is important to wear ultrasonic noise protective equipment in order to avoid long-term hearing damage.
19. An incubation time of 15–20 min is sufficient to ensure optimal protein binding to the Sepharose resin in our hands. However, the time can be extended or even shortened to optimize the results.
20. According to the manufacturer's instructions, the sedimentation of Ni Sepharose is usually performed at 500 × *g* for 5 min. However, we did not observe any issues in bead stability or protein binding by using higher *g* forces.
21. SDS polyacrylamide gels can be stored at 4 °C without the combs and wrapped in wetted paper towels for several weeks. In case of regular use, we recommend batch preparation of up to 12 gels to ensure homogenous separation properties and quality.
22. Running the polyacrylamide gels at 250 V is only possible when using a maximum of two gels as the increased current might lead to overheating. Always use fresh buffer to allow an equal electrolyte separation.
23. A stain-free approach visualizing separated proteins using UV light is possible when adding 2,2,2-trichloroethanol (0.5% final concentration) during polyacrylamide gel preparation [9].
24. An ultrasonic homogenizer can also be used for larger culture volumes by increasing homogenization cycles and time. However, if access to a cell disruptor is available, we always recommend this cell lysis method due to lower temperatures and a faster cell lysis preventing protein denaturation or aggregation.

25. Immediately continue with the next step to avoid degradation of your POI.
26. The time for column equilibration varies depending on the used column and the flow rates. We typically store our size-exclusion columns in filtered H_2O due to frequent (daily) use. Equilibration with 1 1/2 CV of SEC buffer therefore only takes approx. 2 h. In case of a less frequent use, long-time storage in 20% (v/v) ethanol increases the column lifetime.
27. We recommend using a 1 mL Ni-NTA column per liter of *E. coli* culture. However, depending on protein expression, usage of a 5 mL column might also be appropriate. Of note, Ni-NTA FF columns can also be stacked (e.g., 1 mL + 1 mL columns).
28. Ideally, the peristaltic pump should be located in a cooling cabinet or cold room (adjusted to 4 °C). Note the abbreviation on the columns: HP–HighPerformance (should not be stacked), FF–FastFlow (stackable). Avoid pumping air on the columns.
29. Only transfer cleared lysates without disturbing the cell debris pellet. In case of remaining debris, repeat the centrifugation step. Any remaining particles can potentially clog the Ni-NTA columns.
30. Put the tubing right to the bottom of your tube and stop the pump when only about 1 mL of lysate is left in the tube to avoid filling the tubing with air. We recommend measuring the exact flow rate and setting a timer for all upcoming steps as well on the peristaltic pump to minimize the risk of letting the columns run dry.
31. To calculate the molecular weight of your fusion protein we recommend using the ProtParam tool (https://web.expasy.org/protparam/). Keep in mind that the molecular weight of your TEV-cleaved product has changed and might require using a spin concentrator with a different MW cutoff.
32. Do not use a 50 mL tube to balance out the Amicon concentrator in the centrifuge since it has a different center of gravity. Use multiple concentrators for one protein construct to reduce centrifugation time. Whilst centrifugation, check the concentrators for leakage and balance them out once more to avoid imbalance of the centrifuge.
33. Your protein will most likely not bind to the column anymore and should be present in the flow-through. However, depending on the remaining concentration of imidazole in your buffer, it might also elute upon washing the column in the next step.

34. The appropriate end-volume of the concentration has also to be adjusted depending on protein stability and on the used injection loop (0.5–5 mL). The solubility limit of a given protein might be reached prior to this point, resulting in precipitation of the sample. In any case, we recommend using the smallest injection loop possible and stick to the manufacturer's instructions for the respective SEC column. Injection of large volumes on small columns will influence the separation on the SEC column.
35. Wear gloves while the handling of the SDS-PAGE, as well as the staining and destaining of the gel due to unpolymerized acryl-amid. Use a UV-detectable protein ladder for quick analysis under UV light. (e. g. not a prestained size standard). Note that the UV detection relies on a photoreaction of tryptophan with trichloroethanol and proteins that do not contain tryptophan residues will not be detectable.
36. Depending on the purpose, UV documentation might be sufficient. For quicker Coomassie-Blue staining and destaining, carefully heat the gel with the respective solution for 30 s in a microwave, and afterward, incubate it for 20 more minutes on a shaker. You can collect the used Coomassie-Blue solution for re-use.
37. Collect the waste fraction containing $NiSO_4$ separately and dispose appropriately.
38. Amicon spin concentrators can be re-used several times. Cleaning with 0.2 M NaOH allows to remove precipitated protein and cleans the membrane. For storage, we suggest filling the concentrators completely with sterile H_2O.
39. Depending on the protein purification purpose and the number of different constructs, for example, desalting in batch using gravity flow columns might be preferred over the described a FPLC-based SEC step.
40. To calculate the protein concentration on the basis of the measured absorption at 280 nm, use the following calculation: $A = \varepsilon \times c$ (A: measured absorption; ε: extinction coefficient of the protein; c: concentration in mmol/mL).
41. To prevent column clogging, we recommend to test protein stability in the used SEC buffer prior to application. Therefore, before loading a newly produced protein onto the SEC column, mix 50 μL of concentrated protein with 50 μL SEC buffer and check for precipitation. In case of visible precipitation, optimize the SEC buffer accordingly (e.g., by adding more salt or glycerol). As buffer optimization can be time-consuming and tedious, we will not cover this topic here.

Acknowledgments

We acknowledge Paul Weiland for providing the pEM-SUMO and pEM-MBP backbones. FA and SZ acknowledge funding through the German Research Foundation (Deutsche Forschungsgemeinschaft, DFG)—(Project ID 458090666 / CRC1535/B02). All materials are available from the corresponding author upon request.

Author Contributions FA designed the study with input from AL. FA and SZ wrote the paper with input from GM and AL. SZ and GM performed experiments and analyzed data. FA contributed funding and resources.

References

1. Rosano GL, Ceccarelli EA (2014) Recombinant protein expression in microbial systems. Front Microbiol 5:341. https://doi.org/10.3389/fmicb.2014.00341
2. Bentham AR, Youles M, Mendel MN, Varden FA, Concepcion JCDl, Banfield MJ (2021) pOPIN-GG: a resource for modular assembly in protein expression vectors. biorXiv. https://doi.org/10.1101/2021.08.10.455798
3. Engler C, Kandzia R, Marillonnet S (2008) A one pot, one step, precision cloning method with high throughput capability. PLoS One 3(11):e3647. https://doi.org/10.1371/journal.pone.0003647
4. Zhou P, Wagner G (2010) Overcoming the solubility limit with solubility-enhancement-tags: successful applications in biomolecular NMR studies. J Biomol NMR 46(1):23–31. https://doi.org/10.1007/s10858-009-9371-6
5. Altegoer F, Weiland P, Giammarinaro PI, Freibert SA, Binnebesel L, Han X, Lepak A, Kahmann R, Lechner M, Bange G (2020) The two paralogous kiwellin proteins KWL1 and KWL1-b from maize are structurally related and have overlapping functions in plant defense. J Biol Chem 295(23):7816–7825. https://doi.org/10.1074/jbc.RA119.012207
6. Han X, Altegoer F, Steinchen W, Binnebesel L, Schuhmacher J, Glatter T, Giammarinaro PI, Djamei A, Rensing SA, Reissmann S, Kahmann R, Bange G (2019) A kiwellin disarms the metabolic activity of a secreted fungal virulence factor. Nature 565(7741):650–653. https://doi.org/10.1038/s41586-018-0857-9
7. Weiland P, Dempwolff F, Steinchen W, Freibert SA, Tian H, Glatter T, Martin R, Thomma B, Bange G, Altegoer F (2023) Structural and functional analysis of the cerato-platanin-like protein Cpl1 suggests diverging functions in smut fungi. Mol Plant Pathol 24(7):768–787. https://doi.org/10.1111/mpp.13349
8. Hanahan D (1983) Studies on transformation of Escherichia coli with plasmids. J Mol Biol 166(4):557–580. https://doi.org/10.1016/s0022-2836(83)80284-8
9. Ladner-Keay CL, Turner RJ, Edwards RA (2018) Fluorescent protein visualization immediately after gel electrophoresis using an in-gel trichloroethanol photoreaction with tryptophan. Methods Mol Biol 1853:179–190. https://doi.org/10.1007/978-1-4939-8745-0_22
10. Pettersen EF, Goddard TD, Huang CC, Meng EC, Couch GS, Croll TI, Morris JH, Ferrin TE (2021) UCSF ChimeraX: structure visualization for researchers, educators, and developers. Protein Sci 30(1):70–82. https://doi.org/10.1002/pro.3943

Chapter 14

Generating Site Saturation Mutagenesis Libraries and Transferring Them to Broad Host-Range Plasmids Using Golden Gate Cloning

Niels N. Oehlmann and Johannes G. Rebelein

Abstract

Protein engineering is an established method for tailoring enzymatic reactivity. A commonly used method is directed evolution, where the mutagenesis and natural selection process is mimicked and accelerated in the laboratory. Here, we describe a reliable method for generating saturation mutagenesis libraries by Golden Gate cloning in a broad host range plasmid containing the pBBR1 replicon. The applicability is demonstrated by generating a mutant library of the iron nitrogenase gene cluster (*anfHDGK*) of *Rhodobacter capsulatus*, which is subsequently screened for the improved formation of molecular hydrogen.

Key words Directed evolution, Site saturation mutagenesis, Library generation, Broad host range plasmid, *Rhodobacter capsulatus*, Iron nitrogenase, H_2 evolution, Golden Gate cloning

1 Introduction

Site-directed mutagenesis has become one of the most powerful tools in biochemistry and enzymology [1, 2]. The possibility to create large mutant libraries of enzymes and recombinantly produce the enzymes paved the way for the field of enzyme engineering. Screening large libraries allows us to identify enzymes with increased stability, improved activities, or new reactivities [3]. For this, the gene of interest is usually introduced and mutated in a suitable plasmid, followed by the transformation of the protein-producing host strain [4]. One strategy for library design is site saturation mutagenesis, where the targeted residue is exchanged with every one of the 20 canonical amino acids [5]. The mutations can be introduced by mismatch primers with the degenerate NNK (N: any base, K: guanine or thymine) codon at the position of interest [6]. In cloning-free methods for library construction (e.g., QuickChange methods), [7] the NNK codon has to be located in the primer binding region, which can lead to codon

Daniel Schindler (ed.), *Golden Gate Cloning: Methods and Protocols*, Methods in Molecular Biology, vol. 2850, https://doi.org/10.1007/978-1-0716-4220-7_14,

bias during polymerase chain reaction (PCR) amplification. In contrast, the introduction of mutations with scarless cloning techniques such as Golden Gate cloning allows the introduction of mutations via the primer extensions with less influence on the primer binding properties [8, 9]. Golden Gate cloning relies on the reactivity of type IIS restriction enzymes that cleave outside their recognition site [10]. Type IIS restriction enzymes enable the introduction of any complementary 4 bp sticky ends at the DNA fragments for seamless ligation.

Plasmids are small circular DNA molecules with an origin of replication and a selection marker (e.g., an antibiotic resistance gene) [11]. They are used as cloning vehicles to transform organisms with DNA sequences of interest. The mechanisms of replicons are diverse and can have a limited range of compatible host organisms. Plasmids replicating in a wide range of host organisms are called broad host range plasmids. [12] A member of this group is the pBBR1 plasmid isolated from *Bordetella bronchiseptica* that replicates stably in gram-negative bacteria [13]. Here, we demonstrate the generation of a site saturation mutagenesis library in a broad host range plasmid based on the pBBR1 replicon (Fig. 1).

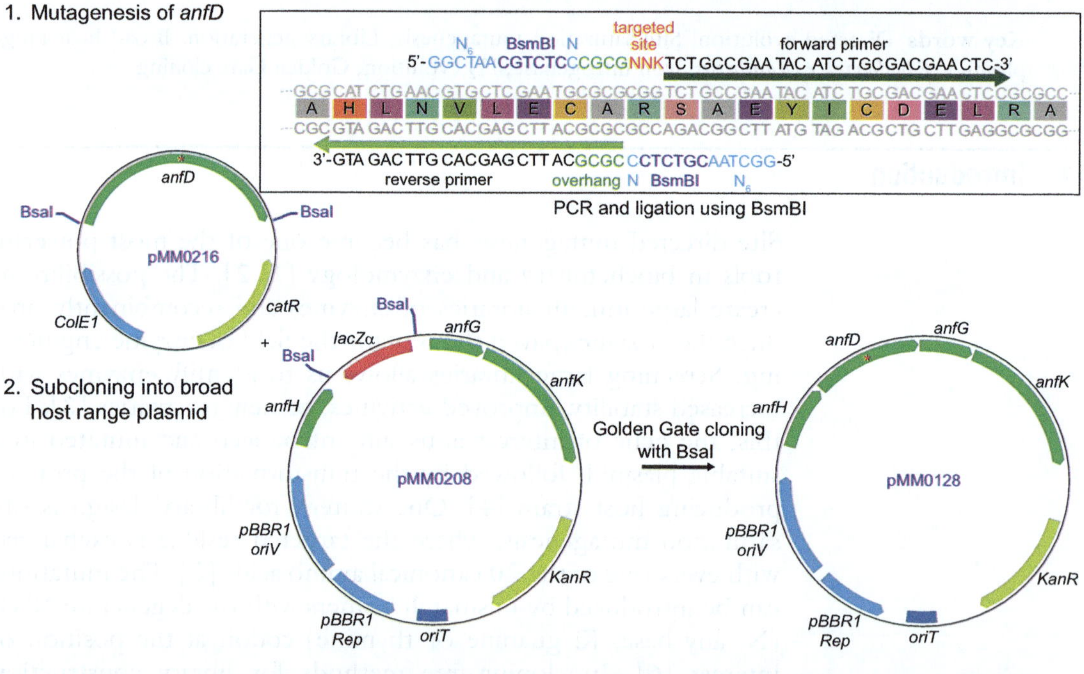

Fig. 1 Workflow for the saturation mutagenesis library generation of the Fe nitrogenase gene *anfD* in the broad host range plasmid pMM0128. The *E. coli-specific* high copy number plasmid pMM0216 containing *anfD* is mutagenized by PCR using a mismatched forward primer that introduces the NNK codon at the targeted site. The forward and reverse primer extensions introduce BsmBI cutting sites used to religate the linearized PCR amplicons. Subsequently, the *anfD* library on pMM0216 is cloned into a broad host range plasmid pMM0208 via Golden Gate cloning using BsaI cutting sites

The gene of interest, *anfD* (one of the structural genes of the Fe nitrogenase) from *Rhodobacter capsulatus*, is first mutagenized in the *Escherichia coli* cloning plasmid pMM0216 (based on pGGA-select) [14] by PCR with an NNK primer followed by seamless ligation of the amplicon using BsmBI restriction sites introduced by the primer extensions (*see* **Note 1**). In a second step, the generated saturation library of *anfD* is transferred into the broad host range plasmid pMM0208 (based on pOG024 created by Philip Poole; Addgene number #113991) [15] by using BsaI restriction sites to insert the mutated *anfD* gene into the *anfHDGK* gene cluster. Using this elaborate protocol, we obtained reproducibly excellent site-saturation libraries in broad host range plasmids with a medium copy number replicon. The library is then conjugated into *Rhodobacter capsulatus* via triparental mating, [16] followed by production and in vivo screening of the Fe nitrogenase mutants for hydrogen (H_2) formation.

2 Materials

2.1 Bacterial Strains, Media, and Antibiotics

1. *Escherichia coli*: DH5α is used for DNA cloning, plasmid amplification and transfer to *R. capsulatus* via conjugation. Competent cells of *E. coli* are prepared according to the Inoue method [17].
2. *R. capsulatus*: MM0167 (Δ*nifD*::*SpR*, Δ*modABC*, Δ*anfHDGK*::*GmR*, ΔdraTG mutant strain of B10S) [18] is used for expression of the *anfD* library in plasmid pMM0128. For general information on the handling and cultivation of *R. capsulatus*, see the more detailed description by Katzke et al. [16].
3. Luria–Bertani (LB) media: 10 g/L tryptone, 5 g/L yeast extract and 10 g/L NaCl dissolved in deionized water.
4. *R. capsulatus* minimal medium V (RCV): 30 mM DL-malic acid, 10 mM serine, 0.8 mM $MgSO_4$, 0.7 mM $CaCl_2$, 50 μM ethylenediaminetetraacetic acid disodium salt dihydrate (Na_2EDTA), 30 μM thiamine·HCl, 45 μM H_3BO_3, 9.5 μM $MnSO_4$, 0.85 μM $ZnSO_4$, and 0.15 μM $Cu(NO_3)_2$ with pH adjusted to 6.8. After autoclaving, 10 mM potassium phosphate buffer, 120 μM $FeSO_4$, and 10 mM serine are added from sterile filtered stock solutions. RCV is prepared as described by Katzke et al. [16].
5. PY medium: 10 g/L peptone, 0.5 g/L yeast extract, after autoclaving 2 mM $MgCl_2$, 2 mM $CaCl_2$, 80 μM $FeSO_4$ are added.
6. Super Optimal broth with Catabolite repression (SOC) medium: 2 g/L tryptone, 0.5 g/L yeast extract, 10 mM

NaCl, 2.5 mM KCl, 10 mM $MgCl_2$, 10 mM $MgSO_4$ and 20 mM glucose.

7. Agar plates: 1.5% (w/V) agar-agar is added to liquid media before autoclaving.
8. For blue-white screening, prepare a stock solution of 100 mM propan-2-yl 1-thio-β-D-galactopyranoside (IPTG) in ddH_2O and 20 mg/mL X-gal in dimethyl sulfoxide (DMSO) and store at −20 °C.
9. Autoclave 80% glycerol (V/V) for the generation of cryogenic stocks of libraries and picked clones.
10. Antibiotics: The culture media is supplemented to a final concentration of 50 μg/mL kanamycin (Km, dissolved in ddH_2O), 20 μg/mL streptomycin (Sm, dissolved in ddH_2O), or 34 μg/mL chloramphenicol (Cm, dissolved in ethanol), as indicated.

2.2 Plasmids

1. pMM0216 (Fig. 1): The cloning vector for *anfD* contains the high copy number ColE1 replicon for stable replication in *E. coli* and the *catR* gene conferring resistance to Cm. A silent point mutation in the codon for *anfD* residue S373 was introduced to remove a BsmBI restriction site. Two BsaI cutting sites flank the *anfD* gene to allow seamless subcloning into the *anfH-GK* gene cluster of pMM0208.
2. pMM0208 (Fig. 1): The *anfD* recipient plasmid pMM0208 contains the *anfH-GK* gene cluster of *R. capsulatus* with a *lacZα* reporter cassette flanked by two BsaI cutting sites as a placeholder for *anfD*, a broad host range replicon (pBBR1 rep) with medium copy number allowing replication in both *E. coli* and *R. capsulatus*, the *kanR* gene conferring resistance to Km and an *oriT* site for conjugation. A silent point mutation in the codon for *anfK* residue V280 was introduced to remove a BsaI restriction site.
3. pMM0128 (Fig. 1): A broad host range plasmid for the expression of the *anfHDGK* gene cluster.
4. pRK2013: Helper plasmid for conjugation, containing the RK2 transfer genes for conjugation, the ColE1 replicon allowing replication in *E. coli* and the *kanR* gene conferring Km resistance [19].

2.3 DNA Oligonucleotides

The used oligonucleotides are listed in Table 1. Stock solutions of 100 μM are prepared with ddH_2O and stored at −20 °C. Working solutions have a concentration of 10 μM oligonucleotide in ddH_2O.

Table 1
Oligonucleotide used in this method

name	Sequence (5′-3′)	Information
P1	GGCTAACGTCTCCCGCGNNKTCTGCCGAATACATC TGCGACGAACTC	Saturation mutagenesis of *anfD* residue R259
P2	GGCTAACGTCTCCCGCGCATTCGAGCACGTTCAGA TG	

2.4 Chemicals and Buffers for Agarose Gel Electrophoresis

1. Tris-acetate-EDTA (TAE) buffer: 1 mM Na_2EDTA, 20 mM acetate, 40 mM Tris–HCl.
2. Loading dye.
3. DNA ladder.
4. DNA dye: e.g., GelRed.
5. Agarose, 1% (w/V) in TAE buffer.

2.5 Enzymes

The enzymes used in this protocol were obtained from New England Biolabs (NEB) but may be replaced by enzymes with corresponding properties. They were applied with the provided buffers at the reaction conditions specified in the methods section.

1. Q5 polymerase (2000 u/mL), Q5 polymerase reaction buffer (×5), 10 mM deoxynucleotide triphosphates (dNTPs).
2. DpnI restriction enzyme (20,000 u/mL).
3. T4 ligase (400,000 u/mL), T4 ligase buffer (×10).
4. BsmBI restriction enzyme (10,000 u/mL).
5. BsaI restriction enzyme (20,000 u/mL).

2.6 Commercial Kits

1. Plasmid DNA extraction kit.
2. PCR purification kit.

2.7 Devices

1. Standard laboratory equipment for the cultivation of *E. coli* and DNA manipulation: Thermoshaker or waterbath for transformations via heat shock, 37 °C incubator for cultivation on agar plates, benchtop centrifuge, gel electrophoresis chambers, gel imaging system, etc.
2. Thermocycler to perform PCR reactions and incubation of enzymatic reactions.
3. For the anaerobic cultivation of *R. capsulatus* on PY agar plates, airtight anaerobic jars were used with gas packs to deplete atmospheric oxygen. Cells are illuminated by 2 × 3 60 W light bulbs at 30 °C for anaerobic phototrophic growth.

4. Custom-built LED panels (λ: 850 nm, 420 nm, and 470 nm, ~60 μmol photons m^{-2} s^{-1}) were used to cultivate liquid cultures of *R. capsulatus*.
5. Custom-built vacuum chambers for cultivating *R. capsulatus* in 96-well multitier plates.
6. A vacuum pump connected to a manifold allowing gas exchange of vacuum chambers and sealed gas chromatography (GC) vials.
7. Headspace sampling GC for gas analysis of culture headspace.

2.8 Consumables

1. Standard consumables of a molecular and microbiology laboratory: PCR reaction tubes, centrifugation tubes, Petri dishes, inoculation loops, etc.
2. 96-well microtiter plate and aluminum cover foil.
3. 20 mL GC vials, butyl rubber stoppers, aluminum caps to crimp the GC vials.
4. Disposable cannula and 0.2 μm syringe filter to connect GC vials to the vacuum manifold.
5. Gas cylinders of pure argon (Ar) and 8% CO_2 in Ar.

3 Methods

3.1 Cultivation of Strains

1. *E. coli* is cultivated aerobically on LB agar plates at 37 °C. Liquid cultures are grown in 10 mL LB media in 100 mL Erlenmeyer flasks, shaking at 180 rpm at 37 °C.
2. *R. capsulatus* is grown under phototrophic conditions: All cultivations are conducted under anaerobic conditions. To cultivate *R. capsulatus* on PY agar plates, the plates are placed in airtight anaerobic jars and commercially available gas packs are used to deplete the jar atmosphere from oxygen. The transparent jar is illuminated by light bulbs or LED panels at 30 °C. *R. capsulatus* liquid cultures are cultivated in RCV minimal media in sealed, airtight vials or chambers. Using a gas manifold, the gas phase of the vials or chambers is depleted from oxygen by three to five consecutive evacuation-refill cycles with pure Ar (*see* **Note 2**). The vials are connected to the manifold via cannulas attached to 0.2 μm syringe filters to remove contaminants from inflowing gas. The cultures are illuminated with LED panels at 30 °C.

3.2 Mutagenesis of *anfD*

1. Forward primer design (see Fig. 1): The forward primer binding region starts downstream of the targeted site and should have a melting temperature (T_m) of ~65 °C (calculated according to SantaLucia) [20]. The primer extension begins with six

Table 2
Thermocycler settings for mutagenesis by PCR using mismatched primers

Description	Cycles	Temperature	Time
Initial denaturation	1	98 °C	2:00 min
Denaturation	32	98 °C	30 s
Annealing		68.2 °C (−0.2 °C per cycle)	30 s
Elongation		72 °C	2:30 min
Final elongation	1	72 °C	4 min
Storage	1	12 °C	∞

nucleotides N_6 at the 5′-end, followed by the BsmBI recognition sequence CGTCTC. After a spacer nucleotide N, the 4 bp overhang is introduced, corresponding to the template sequence, followed by the degenerate codon NNK (*see* **Note 3**). In this example, site R259 was targeted (see Table 1 for example oligonucleotide).

2. Reverse primer design (*see* Fig. 1): The biding region of the reverse primer starts upstream of the targeted site and should have a T_m of ~65 °C. The primer extension contains six nucleotides N_6 at the 5′-end, followed by the BsmBI recognition sequence CGTCTC and a spacer nucleotide N.
3. For the PCR reaction (total volume 60 μL), add to ddH_2O (38.8 μL) the Q5 reaction buffer (12 μL), 10 mM dNTPs (1.2 μL), reverse and forward primer (3 μL each), template DNA (pMM0216, 4 ng), and Q5 DNA polymerase in a PCR reaction tube.
4. Incubate PCR in a thermocycler using the settings listed in Table 2 for a touchdown PCR.
5. Use 5 μL of the PCR reaction mix to conduct an agarose gel electrophoresis (*see* Subheading 3.3). Correct amplification of the vector DNA is verified by the DNA band of the appropriate size (corresponding to ~3.7 kbp; *see* **Note 4**).
6. Add DpnI (1 μL) to the PCR reaction mix. Incubate the reaction at 37 °C for 16–24 h.
7. Conduct a DNA purification of the reaction mix using a commercial DNA purification kit (*see* **Note 5**). Elute the DNA in ddH_2O (25 μL).
8. For the amplicon ligation via Golden Gate cloning, prepare a Golden Gate reaction mix (final volume 20 μL) containing T4 ligase buffer (2 μL), T4 ligase (0.5 μL), BsmBI (0.5 μL), and the PCR amplicon (150 ng) in ddH_2O on ice.

Table 3
Thermocycler settings for the Golden Gate reaction

Description	Cycles	Temperature	Time
Digestion	30	42 °C	5 min
Ligation		16 °C	5 min
Heat inactivation	1	60 °C	5 min
Storage	1	12 °C	∞

9. Incubate the reaction mix in a thermocycler using the settings listed in Table 3.
10. The Golden Gate reaction is transformed in *E. coli* DH5α cells via heat shock: An aliquot (100 μL) of chemocompetent DH5α cells is thawed on ice (15 min). The Golden Gate reaction (10 μL) is added and the cells are placed on ice (30 min). Next, the cells are heated to 42 °C (30 s) with a thermoshaker and then quickly placed in an ice bath (5 min). The cells are rescued by adding 1 mL SOC media prewarmed to 37 °C and incubation (37 °C, 400 rpm, 1 h). The suspension is centrifuged (6000 × *g*, 1 min), the supernatant is discarded, the pellet is suspended in the remaining liquid, and plated on an LB agar plate supplemented with Cm.
11. After overnight incubation of the LB agar plate, the single colonies are counted to ensure library oversampling (>96 for NNK library). Then, the colonies are spread on the LB plate with an inoculation loop to pool the library. After an additional ~6 h of incubation at 37 °C, the cells are harvested from the plate with an inoculation loop and resuspended in 1 mL LB media. The plasmids are isolated from the suspension using a commercial plasmid DNA preparation kit (0.25 mL of the cell mixture). The rest of the cell suspension is used to prepare a cryogenic stock (add 0.25 mL 80% glycerol, mix and freeze/store at −80 °C).
12. The library generation is verified using Sanger sequencing.

3.3 Agarose Gel Electrophoresis

1. Prepare a suspension of 1% (w/V) agarose in TAE buffer. Boil the suspension until the agarose has dissolved completely. Warning: The hot agarose solution can cause skin burns. Let the solution chill to ~60 °C, cast the gel, and add a DNA dye as indicated by the supplier.
2. Mix the sample (*see* Subheading 3.2, **step 5**) with loading dye.
3. Transfer the gel into the electrophoresis chamber with TAE buffer. Load the DNA ladder and samples into the gel chambers.

Table 4
Thermocycler settings for the subcloning of *anfD* into pMM0208

Description	Cycles	Temperature	Time
Digestion	30	37 °C	5 min
Ligation		16 °C	5 min
Heat inactivation	1	60 °C	5 min
Storage	1	12 °C	∞

4. The electrophoresis is conducted at 100 V for 20–40 min depending on equipment.
5. The stained DNA bands are visualized by a gel imaging system.

3.4 Subcloning of the *anfD* Library into pMM0208

1. Prepare a Golden Gate reaction mix (final volume 20 μL) containing T4 ligase buffer (2 μL), T4 ligase (0.5 μL), BsaI (0.5 μL), pMM0208 (100 ng), and the *anfD* library in pMM0216 (100 ng) in ddH_2O on ice.
2. Incubate the reaction mix in a thermocycler using the following settings (Table 4).
3. The reaction mix is used to transform DH5α cells via heat shock (*see* Subheading 3.2). The outgrowth is plated on an LB agar plate supplemented with Km, 20 μg/mL X-gal, and 100 μM IPTG for blue-white screening.
4. After overnight incubation of the plate at 37 °C, successful integration of the *anfD* library is assessed by the number of white and blue colonies (successful cloning results in hundreds of white and few blue colonies). The single colonies are spread on the plate using an inoculation loop to pool the library, the plate is incubated further for 6 h at 37 °C, and the cells are harvested and suspended in 1 mL LB media. The plasmid is isolated from the suspension (0.25 mL) using a commercial plasmid DNA preparation. The rest of the cell suspension is used to prepare a cryogenic stock (add 0.25 mL 80% glycerol, mix, and freeze/store at −80 °C).
5. Successful transfer of the *anfD* library is verified by pooled sequencing using Sanger sequencing.

3.5 Conjugation into *R. capsulatus*

1. Two days prior to the conjugation, cultivate *R. capsulatus* strain MM0167 (B10S *ΔanfHDGK::GmR*, *ΔmodABC* for plasmid-based Fe nitrogenase expression) on a PY agar plate supplemented with Sm. Incubate the plate under phototrophic conditions (*see* Subheading 3.1) at 30 °C.
2. The day before the conjugation, inoculate the DH5α strain carrying the conjugation helper plasmid pRK2013 and the

DH5α strain carrying the *anfD* library on pMM0128 on a separate LB agar plate supplemented with Km. Incubate the plates overnight at 37 °C.

3. For the conjugation, harvest the *R. capsulatus* cells with an inoculation loop and suspend them in 1 mL RCV media. In 0.5 mL PY media, suspend an equal amount (~1 inoculation loop) of the *E. coli* DH5α strains carrying pRK2013 or the pMM0128 *anfD* library and add the suspension to the *R. capsulatus* cells. Centrifuge the cell suspension at 13,000 × g for 10
4. Meanwhile, place an autoclaved cellulose acetate filter (*see* **Note 6**) on an LB plate (without antibiotics).
5. Decant the supernatant of the cell suspension and suspend the cell pellet in the remaining liquid (~200 μL) by gentle agitation using a pipette tip. Transfer the mixture to the cellulose acetate filter on the LB plate. Transfer the plate carefully to 30 °C incubator and incubate it for 24 hours.
6. Harvest the cells from the cellulose acetate filter. This can be done by transferring the filter into a 2 mL centrifuge tube filled with 1 mL RCV media and scraping the cell lawn into the media using a pipette tip. Remove the filter and suspend the cells using a pipette.
7. Plate out 100 μL of the cell suspension and 100 μL of a 1:10 suspension dilution in RCV media onto PY agar plates supplemented with Km and Sm.
8. Incubate the plates under phototrophic growth conditions until single colonies appear (~2 d).

3.6 In Vivo Screen of the *anfD* Library for H_2 Formation

As a proof of concept, a saturation mutagenesis library for *anfD* residue R259 was prepared according to the method described (*see* Subheadings 3.2 and 3.4). After conjugation of the library from *E. coli* strain DH5a into *R. capsulatus* strain MM0167 by triparental mating (described in Subheading 3.5), the library is investigated for the formation of H_2 in vivo according to the following procedure (Fig. 2).

1. Fill 250 μL of RCV media supplemented with Km into each well of a transparent 96-well microtiter plate.
2. Pick single colonies of *R. capsulatus* strain MM0167 carrying pMM0128 with the *anfD* library from the conjugation plate into the individual well of the 96-well microtiter plate (e.g., by using small pipette tips).
3. The 96-well plate is placed into a vacuum chamber. To minimize the edge effect (often-observed growth differences of cells cultured in the outer wells of a microtiter plate) [21] during the phototrophic growth of *R. capsulatus*, place a

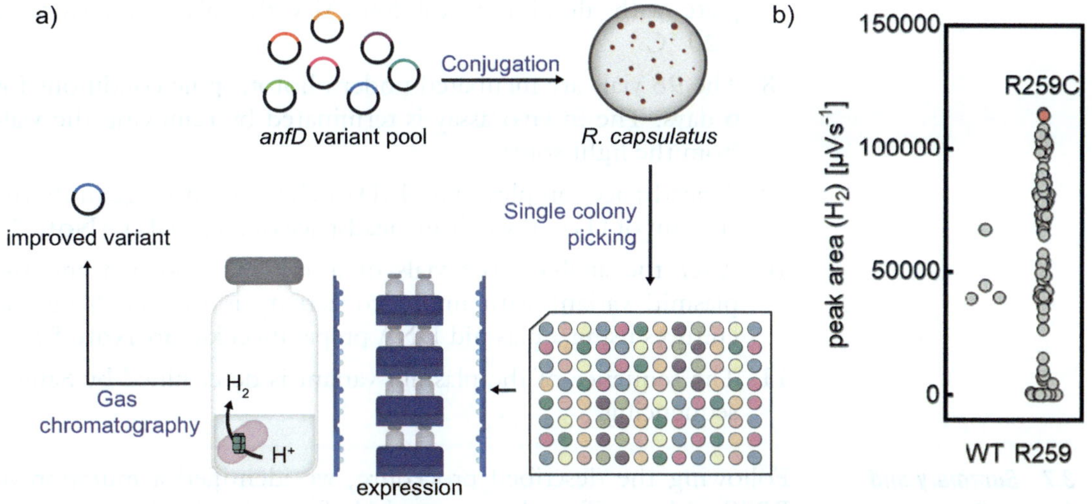

Fig. 2 (**a**) Workflow of the in vivo screening of *anfD* library in *R. capsulatus* strain MM0167 for the formation of H_2. (**b**) Result for the saturation mutagenesis library of *anfD* residue R259. (**a**) The variant pool of *anfD* in pMM0128 is conjugated by triparental mating into *R. capsulatus* strain MM0167. Single colonies of the conjugant *R. capsulatus* strain are picked into RCV media in a 96-well plate and cultivated phototrophically. Subsequently, the cultures in the microtiter plate are used to inoculate assay cultures in 20 mL GC vials. The assay cultures are cultivated phototrophically, which results in the recombinant expression of the *anfDGK* gene cluster on pMM0128 and the formation of H_2 by the Fe nitrogenase variants in vivo. Headspace gas analysis is used to determine the variant with the highest in vivo H_2 forming capabilities under the applied conditions. (**b**) The H_2 peak area of the headspace gas analysis for the R259 site saturation mutagenesis library and of strains carrying the wild type (WT) pMM0128 sequence are shown after 6 d of phototrophic growth

96-well microtiter plate filled with 250 μL ddH_2O in each well underneath the culture plate as a humidity reservoir.

4. The vacuum chamber is sealed airtight and the gas in the chamber is replaced with Ar by four evacuation-refill cycles using a gas manifold. In a final evacuation-refill cycle, the gas in the chamber is exchanged to 8% CO_2 in Ar to a final pressure of 1.2 atm.
5. The inoculated 96-well microtiter plate is incubated under phototrophic conditions at 30 °C for 5 days.
6. In 20 mL GC vials, fill 3.8 mL RCV media supplemented with Km. Open the anaerobic chamber and inoculate 200 μL of *R. capsulatus* cell cultures from each well of the 96-well plate into a GC vial. Close the vials with butyl rubber stoppers and crimp the vials airtight. Connect the GC vials with cannulas to a gas manifold and exchange the headspace to 8% CO_2 in Ar (*see* Subheading 3.1).
7. Add 75 μL 80% glycerol to the remaining liquid in each well of the 96-well plate to generate a cryogenic stock, and seal the

plate with aluminum seal foil. Mix the plate and store at −20 °C.

8. The 96 vials are incubated under phototrophic conditions for 6 days. The in vivo assay is terminated by removing the vials from the light source.
9. A headspace sampling GC-FID/TCD is used to quantify the amount of H_2 in the culture headspace of each vial (*see* **Note 7**).
10. After the analysis, the vials of interest are opened and the plasmid variant with improved activity is isolated from the cultures using a plasmid DNA preparation kit (*see* **Note 8**).
11. The sequence of the plasmid variant is determined by Sanger sequencing.

3.7 Summary and Perspective

Following the described procedure, we identified a mutation of R259 with significantly improved H_2 formation in vivo compared to the wild-type *anfD* sequence. The two culture vials with the highest amount of H_2 in the headspace expressed the R259C *anfD* mutant. The corresponding mutation R277C in the Mo nitrogenase of the model organism *Azotobacter vinelandii* has been described before and was found to still produce H_2 with a higher electron flux distribution towards H_2 under an N_2 atmosphere compared to the Mo nitrogenase wild-type sequence (WT: 70% NH_3, 30% H_2; R277C: 49% NH_3, 51% H_2) [22]. In the case of the Fe nitrogenase, a 2.4-fold increase of H_2 was detected in the vial headspace of the cultures expressing the R259C mutant compared to the WT *anfD* sequence.

4 Notes

1. In our experience, the direct mutagenesis of the broad host range plasmid pMM0128 (containing the *anfHDGK* gene cluster in plasmids based on the pBBR1 replicon) showed low efficiencies and lacked reproducibility. We solved this issue by introducing the mutations in pMM0216 (containing *anfD* in the well-established pGGAselect plasmid backbone), which is smaller and has a higher copy number than pMM0128.
2. *R. capsulatus* can grow aerobically using O_2 as a terminal electron acceptor during respiration. For the depletion of residual dissolved O_2 in culture media, it is usually sufficient to start the cultivation and let *R. capsulatus* consume the remaining O_2. For large culture volumes or, if necessary, the culture media can be degassed from O_2 by purging the media with Ar under vigorous stirring. The gas inlet needle should be connected to a 0.2 μm syringe filter to prevent contamination.

3. If codon bias is observed in the constructed libraries, different commercial suppliers for oligonucleotide synthesis should be tested. Depending on the commercial supplier, we experienced varying quality/bias of the NNK codons.
4. The appearance of strong unspecific amplification bands caused a drastic drop in efficiency for the following amplicon ligation. Optimizing the PCR conditions or purifying the correct DNA band from an agarose gel could help in such cases.
5. We observed an increase in amplicon ligation efficiency if the amplicon was purified using more washing buffer as described by the protocol of the commercial DNA purification kit supplier.
6. The cellulose acetate filters can be autoclaved but need a small weight on top (e.g., a small glass beaker) to keep their shape. The filters are placed on the LB agar plate with tweezers sterilized by ethanol.
7. This method description focuses on generating mutant libraries in a broad host range plasmid and their expression in *R. capsulatus*. A detailed description of the gas analysis can be found in Schmidt et al. [23].
8. Plasmids can be purified from *R. capsulatus* using a plasmid DNA preparation kit according to the protocol for *E. coli* cultures without adaptations.

Acknowledgments

J.G.R. thanks the Deutsche Foschungsgemeinschaft (DFG, German Research Foundation)—446841743 for funding. N.N.O. thanks the Fonds der Chemischen Industrie for a Kekulé fellowship. We thank B. Masepohl and T. Drepper for providing strains and plasmids.

References

1. Shortle D, DiMaio D, Nathans D (1981) Directed mutagenesis. Annu Rev Genet 15: 265–294. https://doi.org/10.1146/annurev.ge.15.120181.001405
2. Smith M (1985) In vitro mutagenesis. Annu Rev Genet 19:423–462. https://doi.org/10.1146/annurev.ge.19.120185.002231
3. Brannigan JA, Wilkinson AJ (2002) Protein engineering 20 years on. Nat Rev Mol Cell Biol 3(12):964–970. https://doi.org/10.1038/nrm975
4. Pourmir A, Johannes TW (2012) Directed evolution: selection of the host organism. Comput Struct Biotechnol J 2:e201209012. https://doi.org/10.5936/csbj.201209012
5. Wang Y, Xue P, Cao M, Yu T, Lane ST, Zhao H (2021) Directed evolution: methodologies and applications. Chem Rev 121(20):12384–12444. https://doi.org/10.1021/acs.chemrev.1c00260
6. Reetz MT, Carballeira JD (2007) Iterative saturation mutagenesis (ISM) for rapid directed evolution of functional enzymes. Nat Protoc 2(4):891–903. https://doi.org/10.1038/nprot.2007.72
7. Liu H, Naismith JH (2008) An efficient one-step site-directed deletion, insertion,

single and multiple-site plasmid mutagenesis protocol. BMC Biotechnol 8:91. https://doi.org/10.1186/1472-6750-8-91

8. Pullmann P, Ulpinnis C, Marillonnet S, Gruetzner R, Neumann S, Weissenborn MJ (2019) Golden Mutagenesis: an efficient multi-site-saturation mutagenesis approach by Golden Gate cloning with automated primer design. Sci Rep 9(1):10932. https://doi.org/10.1038/s41598-019-47376-1
9. Stein A, Chen D, Igareta NV, Cotelle Y, Rebelein JG, Ward TR (2021) A dual anchoring strategy for the directed evolution of improved artificial transfer hydrogenases based on carbonic anhydrase. ACS Cent Sci 7(11): 1874–1884. https://doi.org/10.1021/acscentsci.1c00825
10. Engler C, Kandzia R, Marillonnet S (2008) A one pot, one step, precision cloning method with high throughput capability. PLoS One 3(11):e3647. https://doi.org/10.1371/journal.pone.0003647
11. Couturier M, Bex F, Bergquist PL, Maas WK (1988) Identification and classification of bacterial plasmids. Microbiol Rev 52(3):375–395. https://doi.org/10.1128/mr.52.3.375-395.1988
12. Jain A, Srivastava P (2013) Broad host range plasmids. FEMS Microbiol Lett 348(2):87–96. https://doi.org/10.1111/1574-6968.12241
13. Antoine R, Locht C (1992) Isolation and molecular characterization of a novel broad-host-range plasmid from Bordetella bronchiseptica with sequence similarities to plasmids from gram-positive organisms. Mol Microbiol 6(13):1785–1799. https://doi.org/10.1111/j.1365-2958.1992.tb01351.x
14. Potapov V, Ong JL, Kucera RB, Langhorst BW, Bilotti K, Pryor JM, Cantor EJ, Canton B, Knight TF, Evans TC Jr, Lohman GJS (2018) Comprehensive profiling of four base overhang ligation fidelity by T4 DNA ligase and application to DNA assembly. ACS Synth Biol 7(11):2665–2674. https://doi.org/10.1021/acssynbio.8b00333
15. Geddes BA, Mendoza-Suarez MA, Poole PS (2018) A Bacterial Expression Vector Archive (BEVA) for flexible modular assembly of Golden Gate-compatible vectors. Front Microbiol 9:3345. https://doi.org/10.3389/fmicb.2018.03345
16. Katzke N, Bergmann R, Jaeger KE, Drepper T (2012) Heterologous high-level gene expression in the photosynthetic bacterium Rhodobacter capsulatus. Methods Mol Biol 824:251–269. https://doi.org/10.1007/978-1-61779-433-9_13
17. Sambrook J, Russell DW (2006) The inoue method for preparation and transformation of competent E. coli: "ultra-competent" cells. CSH Protoc 2006(1). doi:https://doi.org/10.1101/pdb.prot3944
18. Klipp W, Masepohl B, Puhler A (1988) Identification and mapping of nitrogen fixation genes of Rhodobacter capsulatus: duplication of a nifA-nifB region. J Bacteriol 170(2):693–699. https://doi.org/10.1128/jb.170.2.693-699.1988
19. Figurski DH, Helinski DR (1979) Replication of an origin-containing derivative of plasmid RK2 dependent on a plasmid function provided in trans. Proc Natl Acad Sci USA 76(4):1648–1652. https://doi.org/10.1073/pnas.76.4.1648
20. SantaLucia J Jr (1998) A unified view of polymer, dumbbell, and oligonucleotide DNA nearest-neighbor thermodynamics. Proc Natl Acad Sci USA 95(4):1460–1465. https://doi.org/10.1073/pnas.95.4.1460
21. Mansoury M, Hamed M, Karmustaji R, Al Hannan F, Safrany ST (2021) The edge effect: a global problem. The trouble with culturing cells in 96-well plates. Biochem Biophys Rep 26:100987. https://doi.org/10.1016/j.bbrep.2021.100987
22. Maskos Z, Fisher K, Sorlie M, Newton WE, Hales BJ (2005) Variant MoFe proteins of Azotobacter vinelandii: effects of carbon monoxide on electron paramagnetic resonance spectra generated during enzyme turnover. J Biol Inorg Chem 10(4):394–406. https://doi.org/10.1007/s00775-005-0648-2
23. Schmidt FV, Schulz L, Zarzycki J, Oehlmann NN, Prinz S, Erb TJ, Rebelein JG (2023) Structural insights into the iron nitrogenase complex. Nat Struct Mol Biol. https://doi.org/10.1038/s41594-023-01124-2

Chapter 15

Combining Oligo Pools and Golden Gate Cloning to Create Protein Variant Libraries or Guide RNA Libraries for CRISPR Applications

Alicia Maciá Valero, Rianne C. Prins, Thijs de Vroet, and Sonja Billerbeck

Abstract

Oligo pools are array-synthesized, user-defined mixtures of single-stranded oligonucleotides that can be used as a source of synthetic DNA for library cloning. While currently offering the most affordable source of synthetic DNA, oligo pools also come with limitations such as a maximum synthesis length (approximately 350 bases), a higher error rate compared to alternative synthesis methods, and the presence of truncated molecules in the pool due to incomplete synthesis. Here, we provide users with a comprehensive protocol that details how oligo pools can be used in combination with Golden Gate cloning to create user-defined protein mutant libraries, as well as single-guide RNA libraries for CRISPR applications. Our methods are optimized to work within the Yeast Toolkit Golden Gate scheme, but are in principle compatible with any other Golden Gate-based modular cloning toolkit and extendable to other restriction enzyme-based cloning methods beyond Golden Gate. Our methods yield high-quality, affordable, in-house variant libraries.

Key words Golden Gate cloning, Oligo pools, Mutational libraries, CRISPR libraries

1 Introduction

Constructing diverse libraries of a sequence of interest is a cornerstone of biological engineering, enabling the creation and examination of multiple variants of a biological function in one pot.

The original version of the chapter has been revised. A correction to this chapter can be found at https://doi.org/10.1007/978-1-0716-4220-7_28

Alicia Maciá Valero, Rianne C. Prins and Thijs de Vroet contributed equally. The authors have the right to change the order of their names appearing on the author list.

Supplementary Information The online version contains supplementary material available at https://doi.org/10.1007/978-1-0716-4220-7_15.

Daniel Schindler (ed.), *Golden Gate Cloning: Methods and Protocols*, Methods in Molecular Biology, vol. 2850,
https://doi.org/10.1007/978-1-0716-4220-7_15,

Protein variant libraries, for example, can be used in protein engineering to perform mutational scans, where each amino acid is systematically exchanged for several or all other 19 standard amino acids [1]. The resulting deep mutational scanning (DMS) libraries can subsequently be functionally characterized using various methods as outlined in available reviews [1, 2], allowing a protein scientist to create detailed sequence-function maps of a protein. Such maps have emerged as valuable tools to understand and engineer the substrate-specificity, solubility, and kinetics of enzymes [3–5], the allosteric behavior of proteins [6, 7], the clinical properties of therapeutic antibodies [8], or the attachment affinity of viruses to their host's receptor based on mutated capsid proteins [9, 10]. DMS data even allows us to predict protein structures by using experimentally determined epistatic interactions between pairs of residues and the fact that epistasis correlates with structural proximity [11].

Other types of mutational libraries can facilitate insertion or deletion scans, in which additional residues (e.g., encoding restriction sites or affinity tags) are systematically inserted after each amino acid position in a protein. These libraries can subsequently be harnessed to identify permissive sites for tags [12] or to create intermediate libraries for protein minimization strategies [13].

In addition, libraries for CRISPR applications often encompass a large number of different 20-base pair spacer regions that target specific DNA stretches in a host genome. These libraries can be combined with Cas9 and other Cas variants for CRISPRko (knockout), CRISPRi (inhibition), and CRISPRa (activation) screens. Such CRISPRx screens have been widely applied for functional genomics [14–19] and gene circuit design [20, 21] across bacteria, yeast, and mammalian cells. For instance, this has enabled the identification of essential gene functions [22], disease-related phenotypes, drug resistance profiles, and modes of action of drugs [23].

A key requirement for creating such libraries is access to synthetic DNA that encodes the numerous defined mutational variants. Although the cost of double-stranded DNA has significantly dropped by several orders of magnitude over the last decades, the substantial number of base pairs required for constructing comprehensive mutational libraries continues to pose a bottleneck in terms of cost.

Purchasing pools of single-stranded oligonucleotides is currently the most affordable option for obtaining synthetic DNA, with a cost per base approximately 100-fold lower than that of double-stranded DNA (<0.001 compared to 0.05–1 USD per base) [24].

However, oligo pools come with certain limitations, including a maximum length constraint (currently 350 bases), a higher error rate compared to alternative gene synthesis methods, and the presence of truncated variants of the oligos due to incomplete synthesis [24].

Here, we outline two methods that exemplify how oligo pools and Golden Gate cloning can be combined to create high-quality libraries for protein mutational scans or single guide RNA (sgRNA) libraries. We provide step-by-step methods for the design and creation of the libraries and provide recommendations for downstream screening and next-generation sequencing (NGS)-based analysis.

Our protocol overcomes the aforementioned limitations by working with a gene-tiling approach. This approach enables the use of relatively short oligos with an acceptably low error rate. Furthermore, we use a double-stranding protocol that minimizes complications caused by truncated oligos.

The first method (*Method 1*) describes a protocol for cloning mutational protein libraries, exemplified with a mutational scan, where each residue is replaced by a selected set of amino acids; and an insertional scan, where an epitope tag is introduced at each position within a given protein. Both examples cover the length of a full open reading frame (ORF), but the methods can also be used for a more focused mutagenesis of specific regions within an ORF.

The second method (*Method 2*) describes the cloning of a single guide RNA (sgRNA) library for CRISPRi applications.

The methods outlined herein make use of the genetic components from the Yeast Toolkit (YTK) [25], which has been developed for the yeast *Saccharomyces cerevisiae* and has been extended to other yeast such as Komagataella phaffii (*previously: Pichia pastoris*) [26] and Nakaseomyces glabratus (*previously: Candida glabrata*) [27]. However, the methods should be rapidly adjustable to other Golden Gate-based toolkits for bacteria [28, 29], plants [30], or mammalian cells [31].

We further note that, while we limit ourselves here to a few selected examples, the presented methods can be extended to any other Golden Gate scheme or even traditional restriction-enzyme-based cloning methods.

2 Materials

The execution of the described protocols requires the following standard laboratory equipment and consumables. While certain providers are mentioned in the manuscript, the materials described may be obtained from any other similar suppliers.

1. Standard reagents, consumables, and instrumentation for thermocycler reactions.
2. Standard reagents and equipment for agarose gel electrophoresis.
3. Standard reagents, consumables, and instrumentation for microbial culturing.
4. Standard reagents, consumables, and instrumentation for transformation of *Escherichia coli* and *S. cerevisiae.*

5. Standard reagents, consumables, and instrumentation for the purification of PCR products, extraction of DNA from agarose gels, and plasmid extraction.
6. Access to Sanger sequencing and NGS services.
7. Access to a transilluminator or transillumination goggles and a UV lamp.

2.1 Plasmids

All plasmids relevant to this protocol are listed in Table 1. These plasmids are either available via the YTK on Addgene (MoClo-YTK Plasmid Kit #1000000061) or upon reasonable request from the authors. The plasmids are organized by method and some mentions are redundant but necessary to follow the workflow. Several plasmids mentioned herein were only created in silico to exemplify the workflow and the links to the full sequences for those in silico plasmids can be found in Supplementary Table 1. Notably, the protocol described below can be adjusted to other (type IIS) restriction enzymes, overhang sequences, and antibiotic resistance markers.

2.2 Single-Stranded Oligonucleotides and Oligo Pools

Oligonucleotides used in this protocol are provided in Table 2. One-hundred micromolar stocks were generated by resuspending the lyophilized oligonucleotides in the appropriate amount of ddH_2O and subsequently stored at −20 °C, from which 10 μM working stocks were generated using ddH_2O. Examples of in silico-designed oligo pools are provided in Supplementary Tables 2, 3, and 4).' Ten micromolar stocks of oligo pools were generated in ddH_2O and stored at −20 °C.

2.3 Enzymes and Buffers

1. T7 Ligase.
2. BsaI-HFv2.
3. BsmBI-v2.
4. 2× Phire Hot Start II PCR Master Mix.
5. 2× Phusion high-fidelity PCR Master Mix.
6. 10× NEB buffer 3.
7. 10× T4 ligase buffer.

2.4 Media, Antibiotics, and Other Consumables

1. Ampicillin (100 mg/mL stock) and kanamycin (50 mg/mL stock) dissolved in sterile H_2O and stored in 1 mL aliquots at −20 °C. The stock concentration is 1000×.
2. *E. coli* growth medium: LB liquid and solid media: tryptone 10 g/L, yeast extract 5 g/L, NaCl, 5 g/L, add 20 g/L agar for solid media.
3. Yeast growth medium: Yeast extract peptone dextrose (YPD) liquid and solid media: yeast extract 10 g/L, peptone 20 g/L, dextrose 20 g/L, add 20 g/L agar for solid media; Synthetic complete (SC) dropout liquid and solid media: dextrose 20 g/

Table 1
List of plasmids

Name	Specification	Comment	Availability/Ref
Method 1			
pYTK002	ConLS, type 1	Used to assemble pRS413-pTEF2-Venus-tENO1	Addgene # 65109/
pYTK014	TEF2 promoter, type 2		Addgene # 65121/ [25]
pYTK033	Venus ORF, type 3		Addgene # 65140/ [25]
pYTK051	ENO1 terminator, type 4		Addgene # 65158/ [25]
pYTK072	ConRE, type 5		Addgene # 65179/ [25]
pYTK076	HIS3, type 6		Addgene # 65183/ [25]
pYTK081	CEN6/ARS4, type 7		Addgene # 65188/ [25]
pYTK083	AmpR-ColE1, type 8		Addgene # 65190/ [25]
pYTK047	GFP-dropout, type 234	Used as a template to clone the destination vectors destination vectors	Addgene # 65154/ [25]
pRS413-pTEF2-Venus-tENO1	pRS413-type vector containing the Venus gene under control of the TEF2 promoter and the ENO1 terminator (assembled via Golden Gate using the YTK)	Used to exemplify the segmentation for Method 1	Designed in silico. Links to sequences are given in Supplementary Table 1
Destination vector 1	pRS413-pTEF2-Venus-tENO1 with *Segment 1* replaced by a GFP-dropout		
Destination vector 2	pRS413-pTEF2-Venus-tENO1 with *Segment 2* replaced by a GFP-dropout		
Destination vector 3	pRS413-pTEF2-Venus-tENO1 with *Segment 3* replaced by a GFP-dropout		
Destination vector 4	pRS413-pTEF2-Venus-tENO1 with *Segment 4* replaced by a GFP-dropout		
Destination vector 5	pRS413-pTEF2-Venus-tENO1 with *Segment 5* replaced by a GFP-dropout		
Destination vector 6	pRS413-pTEF2-Venus-tENO1 with *Segment 6* replaced by a GFP-dropout		
Destination vector 7	pRS413-pTEF2-Venus-tENO1 with *Segment 7* replaced by a GFP-dropout		
Destination vector 8	pRS413-pTEF2-Venus-tENO1 with *Segment 8* replaced by a GFP-dropout		
Destination vector 9	pRS413-pTEF2-Venus-tENO1 with *Segment 9* replaced by a GFP-dropout		
Destination vector 10	pRS413-pTEF2-Venus-tENO1 with *Segment 10* replaced by a GFP-dropout		

(continued)

Table 1
(continued)

Name	Specification	Comment	Availability/Ref
Method 2			
pCgTK01	CgCEN/ARS, type 7	Used to assemble pRS414-type vector via Golden Gate	From authors; [27]
pCgTK02	TRP1, type 6		From authors; [27]
pYTK002	ConLS, type 1		Addgene # 65109/ [25]
pYTK047	GFP-dropout, type 234		Addgene # 65154/ [25]
pYTK072	ConRE, type 5		Addgene # 65179/ [25]
pYTK084	KanR-ColE1, type 8		Addgene # 65191/ [25]
pYTK50	sgRNA dropout, type 234	Used as a template to clone the gRNA expression cassette into the above pRS414-type vector, resulting in the gRNA Destination vector	Addgene # #65157/ [25]
gRNA Destination vector	pRS414 with gRNA expression cassette based on pYTK050 (20 bp spacer replaced by a GFP dropout)	Used for cloning the gRNA library	Unpublished. Link to sequence in Supplementary Table 1

Table 2
List of primers

#	Name	Sequence 5′-3′	Comment
Method 1			
1	segment 2–5′flank fw	GCAT **CGTCTCA** TATG tctaaaggtgaagaat	In silico-designed primers suggested to clone Destination Vector 2. Bold-capital letters: BsmBI restriction site. Underlined-capital letters: specific 4-bp overhangs. Lowercase letters: sequence for priming on the template
2	segment 2–5′flank rv	GCAT **CGTCTCA** ACCA ttaacatcaccatc	
3	segment 2–3′flank fw	GCAT **CGTCTCA** ACTG gtaaattgccag	
4	segment 2–3′flank rv	GCAT **CGTCTCA** GGAT ttatttgtacaattcatcc	
5	GFP dropout fw	GCAT **CGTCTCA** TGGT tgagaccgaaag	
6	GFP dropout rv	GCAT **CGTCTCA** CAGT tgagacctataaac	
7	oPool reverse	TGCCGTCTCAGGTCTCA	Reverse primer used for double-stranding the oligo pool in Method 1, binds to oligo pool-encoded landing pad (Supplementary Tables 3, 4, and 5 exemplify oligo pool sequences)
8	seq seg 2 fw (with partial Illumina adapter)	ACACTCTTTCCCTACACGACGCTCTTCCGATCT ctataattaactaaacagatct	Primers for amplification of Segment 2 for NGS. The amplicon size is 260 bp which is a suitable amplicon size for 250 bases paired-end reads. Capital letters: partial Illumina adapter. Lowercase letters: sequence for priming on the Destination vector 2. No barcodes for multiplexing have been added
9	seq seg 2 rv (with partial Illumina adapter)	GACTGGAGTTCAGACGTGTGCTCTTCCGATCT gtttcatatgatctgggta	
Method 2			
10	sgRNA oPool reverse	CTGCCGTCTCAAAAC	Reverse primer used for double-stranding the oligo pool in Method 2
11	YTK050_BsaI_fw	CTAG **GGTCTCG** TGCT **tatccactagacagaagtttgcgtt**	In silico-designed primers suggested to amplify sgRNA acceptor from YTK050. Bold-capital letters: BsmBI restriction site. Underlined-capital letters: specific 4-bp overhangs. Lowercase letters: sequence for priming on the template
12	YTK050_BsaI_rv	CTAG **GGTCTCG** CTGA **atgtgcttcagtattacatttttgcct**	
13	Illumina sequencing_adapter FWD2	AGACGTGTGCTCTTCCGATCTaaaacttcggtcaagtcatct	Primers for amplification of the 20 bp spacer region for NGS. Capital letters: Partial Illumina adapter. Lowercase letters: for priming on the gRNA Destination vector
14	Illumina sequencing_adapter REV2	CTACACGACGCTCTTCCGATCTgataacggactagccttattt	

Table 3
20× Amino acid mix

Compound	Amount (g) for making 500 mL 20× amino acid mix
Adenine Sulfate	0.2
Arginine HCl	0.2
Aspartic acid	1
Glutamic acid	1
Isoleucine	0.3
Lysine HCl	0.3
Methionine	0.2
Phenylalanine	0.5
Serine	4
Threonine	2
Tyrosine	0.3
Valine	1.5

Weigh in the components, add 500 mL ddH_2O, stir until dissolved, and sterile filter.

L, yeast nitrogen base 6.7 g/L, 50 mL/L amino acid dropout solution (20×, Table 3) supplemented with or without the auxotrophic selection components: 2 g/L uracil, 10 g/L tryptophan, 10 g/L leucine and 10 g/L histidine, add 20 g/L agar for solid media.

4. Yeast plate mixture (10 mL total): Mix 8.9 mL sterile-filtered 50% polyethylene glycol (PEG) 4000, 1 mL 1 M lithium acetate, 100 μL 1 M Tris-HCl (pH 7.5), 20 μL 0.5 M Ethylenediaminetetraacetic acid (EDTA)-NaOH (pH 8.0) and vortex.
5. PCR reaction tubes.
6. Eppendorf tubes.
7. Standard Petri dishes.

2.5 Strains

For cloning and plasmid maintenance, we use *E. coli* DH5α; for transformation of plasmids into yeast, we use *S. cerevisiae* BY4741 [32].

3 Methods

3.1 Method 1. Creating Protein Variant Libraries Using Oligo Pools and Golden Gate Cloning

We provide protocols for two types of protein variant libraries: (1) an insertional scan, where an epitope tag is inserted after every residue in a protein of interest; and (2) a mutational scan, where each amino acid is substituted with one of six other selected amino acid residues. The general workflow is illustrated in Fig. 1.

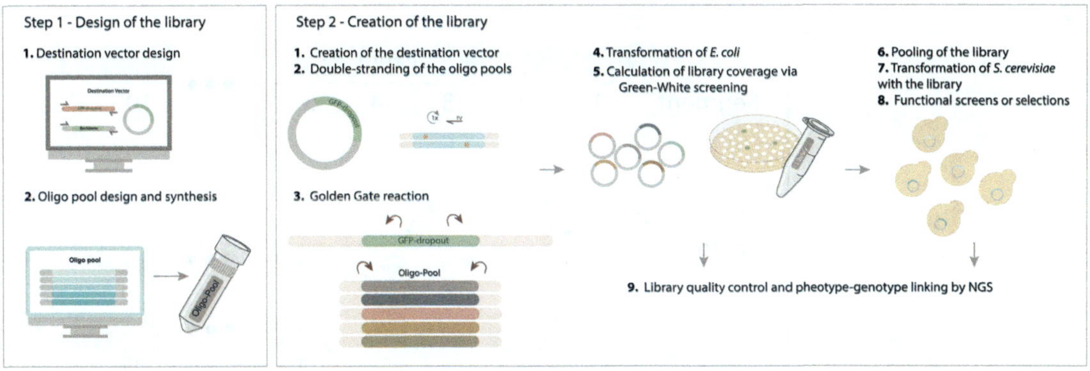

Fig. 1 Overview of the library creation workflow using oligo pools. We have divided our two example methods into two steps: Step 1 involves the in silico design of the library, and Step 2 describes the actual creation of the library. The numbered steps refer to the sub-steps outlined for Method 1, Subheading 3.1

Two important considerations should be kept in mind when using oligo pools for cloning such libraries: the length limitation of oligo synthesis and the single-stranded character of the oligonucleotides obtained.

First, the maximum oligo length possible for synthesis (350 bases) may often be shorter than the gene of interest. Consequently, our method involves tiling a gene into "segments" that are short enough to be covered by oligos (Fig. 2a). In fact, to reduce synthesis errors we keep them below 100 bases, as the chance for error increases with size. Subsequently, we prepare a destination vector for each segment, which can accept the corresponding oligos from the oligo pool and serve to clone each segment library (Fig. 2a, b). Eventually, all individual segment libraries can be pooled into a single final library. However, keeping individual segment libraries separate offers advantages in terms of quality control and troubleshooting per segment. For instance, if NGS-based sequencing reveals that a particular segment is incomplete or contains off-target mutations, the segment can be rapidly re-cloned. Moreover, the destination vectors are reusable for other purposes and this approach allows a swift creation of focused libraries in specific regions of a protein, e.g., as a follow-up after an initial round of full-gene mutagenesis.

The second consideration concerns the single-stranded nature of oligonucleotides, necessitating their conversion to a double-stranded form before use. Instead of using a conventional PCR approach for the double-stranding and amplification of the oligo pool, we have identified a crucial step: The double-stranding of the oligos through a one-step single primer extension reaction using a reverse primer (Fig. 3). This is crucial because not all oligos in a pool attain full-length synthesis due to technical limitations in the synthesis process (Fig. 3a). Consequently, an oligo pool contains truncated oligos, which may or may not encode an intended

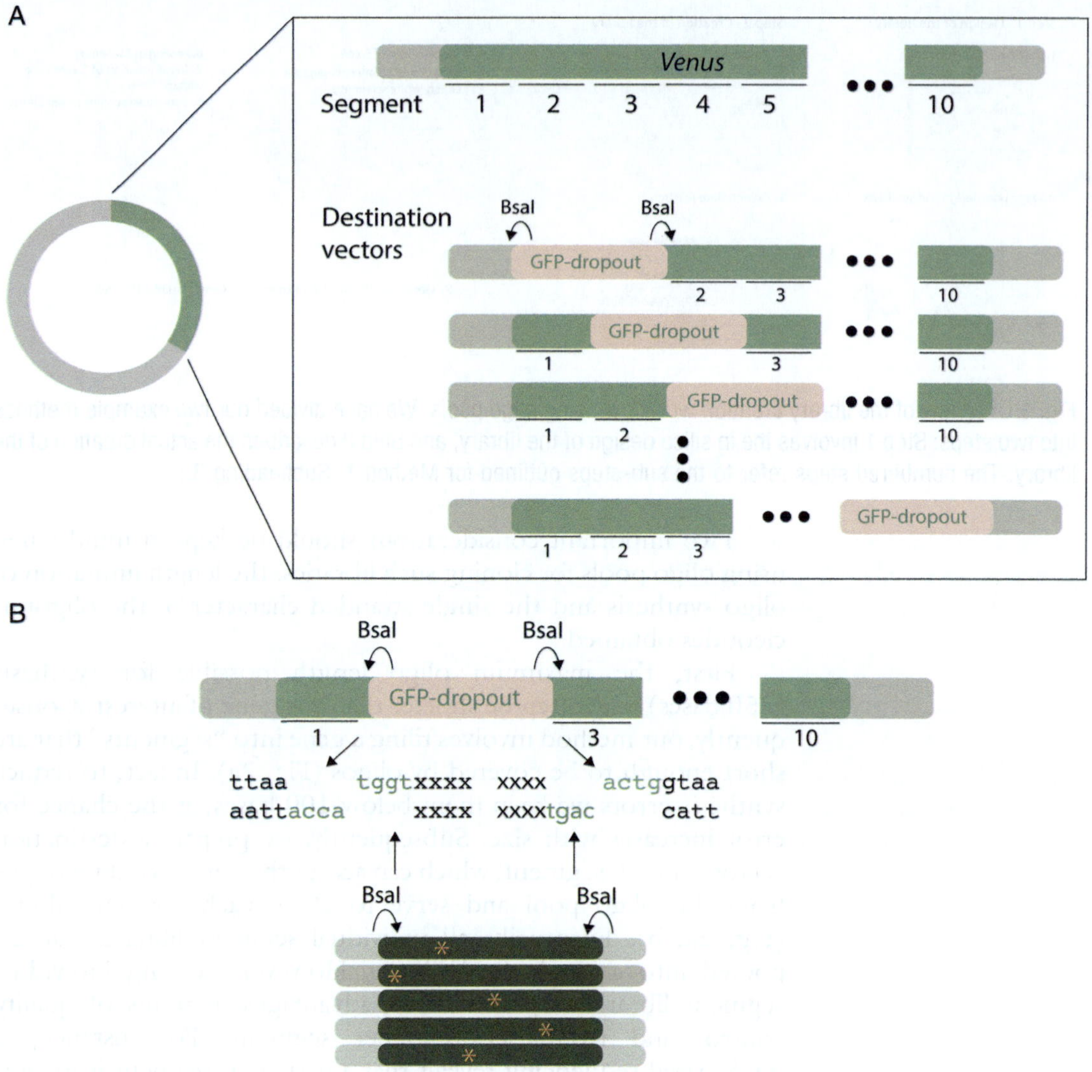

Fig. 2 Segmentation of a gene (as an example here Venus) and cloning of the destination vectors. (**a**) The gene is divided into segments. For each segment, a destination vector is cloned by replacing the segment with a GFP-dropout cassette. (**b**) The GFP-dropout cassette present in every destination vector already encodes outward-cutting BsaI sites that generate 4 bp overhangs. The overhangs of the destination vector are complementary to the overhangs similarly generated in the oligo pool of the respective segment library and are used to assemble the library constructs in a Golden Gate reaction

mutation. Given the high sequence similarity among oligos within an oligo pool designed for mutational scanning libraries, with variations mainly confined to the introduced tag or mutated position, these truncated versions can function as primers when using multiple PCR cycles, leading to crossovers between different oligos within the oligo pool. This, in our own experience, results in libraries that contain noise in the form of double-stranded oligos harboring non-designed mutations or even wild-type sequences

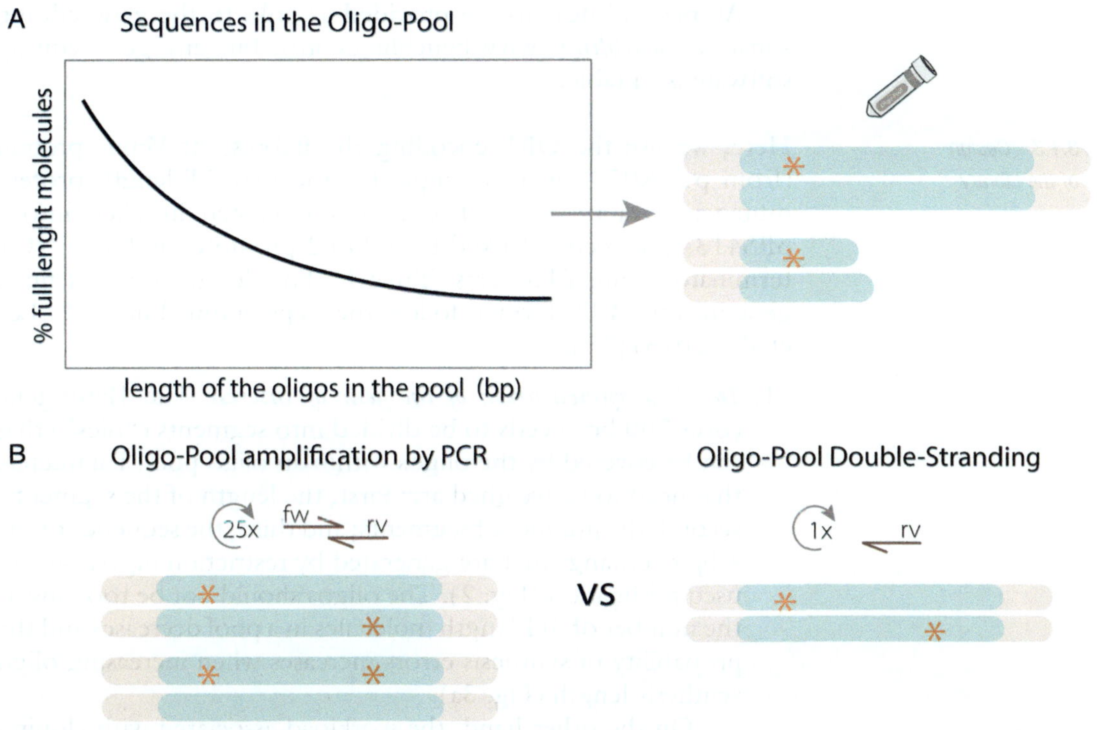

Fig. 3 Oligo pool amplification versus double-stranding. (**a**) Left panel: Sequences in an oligo pool are not all full-length due to current technical limitations in synthesis. As such, a mutagenesis oligo pool contains truncated oligos that may or may not encode an intended mutation (*). Right panel: Truncated oligos are present and those containing similar sequences may act as primers on complementary full-length oligos in a PCR reaction. (**b**) Comparison between oligo pool amplification by PCR and one-step extension with a single reverse primer. Amplification by PCR (left) can cause final products that may contain unintended combinations of designed mutations or even wild-type sequences. Oligo pool double-stranding (right) allows only the full-length synthesized oligos to get double-stranded and utilized in the subsequent Golden Gate reaction

(Fig. 3b). A single reverse primer extension ensures that only full-length oligos within the pool become double-stranded and are able to participate in the subsequent reactions.

We organized Method 1 into two steps (Fig. 1):

3.1.1 Design of the Library

This includes, (1) segmentation of the gene of interest; (2) design of the destination vectors based on the segmentation; and (3) design of the oligo pools.

3.1.2 Creation of the Library

This includes, (1) cloning of the destination vectors; (2) double-stranding the oligos; (3) performing the Golden Gate reaction; (4) transformation of *E. coli* with the mutational library; (5). evaluation of the library coverage and considerations for oversampling; (6) pooling the library; (7) transformation of the library into yeast (*S. cerevisiae*). Further, we provide suggestions for selection and/or screening assays (8); and for library quality control and genotype-phenotype linking via NGS (9).

All plasmid designs are provided as links to the gene editing software *Benchling* (www.benchling.com), but any gene editing software is suitable.

3.1.3 Design of the Library

Here, we use the ORF encoding the fluorescent Venus protein (from pYTK033) as an example for the two full-length protein mutagenesis approaches. The gene was cloned in silico into a pRS413-type vector flanked by a TEF2 promoter and an ENO1 terminator using YTK parts (Table 1). For cloning a user-defined gene into the YTK format, follow the steps outlined in M. E. Lee et al., (2015) [25].

1. *In silico segmentation of the gene of interest.* The *Venus* gene (over 700 bp) needs to be divided into segments ("tiles") that can be covered by the oligos within an oligo pool. Parameters that need to be weighed are: First, the length of the segments; second, the number of segments; and third, the sequence of the 4 bp overhangs that are generated by restriction digestion and used for ligation (Fig. 2). The oligos should not be too long as the number of full-length molecules in a pool decreases and the probability of synthesis errors increases when increasing oligo synthesis length (Fig. 3a).

   On the other hand, the workload associated with cloning individual destination vectors increases with an increasing number of segments. Taking these factors into account, we have had a positive experience working with segments of 15 amino acids (45 bp), 16 amino acids (48 bp), and 27 amino acids (81 bp) in length.

   Here, we divided the *Venus* gene into 10 segments, each spanning 23–27 amino acids. We adjusted the segment length based on the predicted ligation efficiency of the 4 bp overhangs. The 5′ and 3′ overhangs of a given segment are defined by the 4 bp flanking it at each side. Ligation efficiency for each overhang pair can be calculated using the NEB Fidelity Viewer® (https://ligasefidelity.neb.com/viewset/run.cgi). The resulting 4 bp overhangs for the Venus segmentation and their estimated ligation efficiencies are then as follows: Segment 1: ATCT and CACA (100%); Segment 2: TGGT and ACTG (100%); Segment 3: TACT and TACC (93%); Segment 4: TAGA and AAAG (100%); Segment 5: TTTC and GAAT (92%); Segment 6: AATC and AATG (99%); Segment 7: TCAC and GGTG (100%); Segment 8: AGAT and CATT (100%); Segment 9: CAAC and TTGT (100%); Segment 10: GGTC and ATCC (100%).

   Another secondary factor that should be considered is that, for the creation of the destination vectors, one will need to amplify the 3′ and 5′ regions flanking the segment (Supplementary Fig. 1a). As such, a functional primer binding site at these specific regions is also preferred (see Subheading 3.1.2, **step 1** Cloning of the destination vectors).

2. *In silico design of a destination vector for each segment.* Given the 10 segments, 10 destination vectors were designed. For each destination vector, the corresponding segment in the original gene was replaced with a green fluorescent protein (GFP)-dropout cassette. The GFP-dropout cassette can be expressed in *E. coli* and allows easy assessment of the Golden Gate efficiency via Green-White screening after transformation (*see* Subheading 3.1.2, **step 5**). We used the GFP-dropout cassette encoded on the YTK-vector pYTK047 (Table 1), which includes outward-cutting BsaI recognition sites. When replacing a gene segment, these sites generate specific overhangs in the original gene sequence, enabling subsequent seamless ligation of oligo pool sequences into the destination vector. The detailed sequences of the destination vectors can be found in Supplementary Table 1 and the concept is illustrated in Fig. 2. Note that, when using a cloning system other than YTK, precautions should be taken to ensure the absence of additional BsaI restriction sites in both the vector and the dropout cassette.
3. *Design of the oligo pools for each segment.* We will outline the design with two examples:

Example 1. Insertional scanning library: We exemplify the oligo pool design for an insertional scanning library with the insertion of an ALFA-tag after every amino acid position of segment 2 of the Venus protein. The ALFA-tag is a 15-residue epitope tag (SRLEEELRRRLTE) that is recognized by an engineered nanobody that can be used for imaging, immunoprecipitation, and protein purification [33]. Follow the steps of the oligo pool design (Fig. 4):

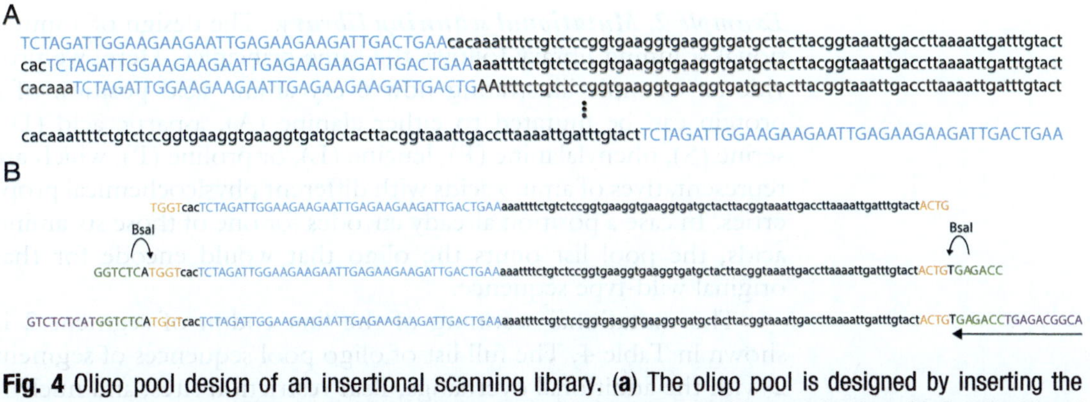

Fig. 4 Oligo pool design of an insertional scanning library. (**a**) The oligo pool is designed by inserting the desired tag (blue) after each codon in the segment. (**b**) Additional nucleotides need to be added at the 5′ and 3′ ends: First, the appropriate overhangs for the respective segment that allow specificity and directionality in ligation (orange); second, the BsaI-recognition sites in the correct orientation (green), plus one spacer nucleotide (black) as BsaI cuts one nucleotide from the recognition site in the 3′ direction. Last, additional bases at the oligos' 5′ and 3′ ends enhance cutting efficiency and generate a primer landing pad at the 3′ end of each oligo for double-stranding (purple). The black arrow at the right bottom marks the landing pad sequence that can be used as reverse primer to double-strand all the oligos in the oligo pool

1. Reverse translate the amino acid sequence of the ALFA-tag using codons optimized for the expression host—in our case *S. cerevisiae* (resulting in 5′-TCTAGATTGGAAGAAGAATT GAGAAGAAGATTGACTGAA-3′).
2. Insert this sequence in-frame after each codon of segment 2 in silico. Always verify that you did not create additional BsaI recognition sites at undesired positions.
3. Add the four segment-specific bases (four bases flanking the segment, yielding the overhangs after BsaI cleavage).
4. Add a BsaI-recognition site to the 5′ and 3′ ends of each oligo, making sure that the directionality of the cut is in the 3′ direction.
5. At the 5′ and 3′ end of each oligo, add at least three extra nucleotides to improve the digestion efficiency. Historically, we used the sequence 5′-GTCTCTCATC-3′. This also yields a primer landing pad that will be used for double-stranding (Fig. 4b and Table 2). When using the same landing pad for all oligos, only one primer will be required for double-stranding an entire oligo pool. However, in case a user wants to encode multiple libraries in one oligo pool and selectively double-strand a sub-pool of oligos, different orthogonal landing pads can be used.
6. Encode each variant on a separate oligo to obtain equimolar amounts of each variant. Stoichiometric ratios between variants can be shifted by duplicating (or triplicating, etc.) the number of certain oligos within the ordering list.

   The complete oligo pool list for segment 2 is available in Supplementary Table 2 (*see* **Note 1** for calculating the number of nucleotides and estimating the expected price of the pool).

Example 2. Mutational scanning library: The design of a mutational scanning library follows the same principles. Here, we provide an example illustrating how every amino acid position of a protein can be mutated to either alanine (A), aspartic acid (D), serine (S), phenylalanine (F), leucine (L), or proline (P), which are representatives of amino acids with different physicochemical properties. In case a position already encodes for one of those six amino acids, the pool list omits the oligo that would encode for that original wild-type sequence.

The mutational scanning of the first codon of segment 2 is shown in Table 4. The full list of oligo pool sequences of segment 2 with the additional overhangs, BsaI restriction sites, and nucleotides for better digestion efficiency (as outlined in Fig. 4b) are provided in Supplementary Table 3. This also generates a primer landing pad for double-stranding with the reverse primer (*see*

Table 4
Example of partial oligos designed for a mutational scan

Amino acid	Codon mutation	Sequence
H	-	cacaaatttctgtctccggtgaaggtgaaggtgatgctacttacggtaaattgaccttaaaattgatttgtact
H→A	gct	gctaaatttctgtctccggtgaaggtgaaggtgatgctacttacggtaaattgaccttaaaattgatttgtact
H→D	gat	gataaatttctgtctccggtgaaggtgaaggtgatgctacttacggtaaattgaccttaaaattgatttgtact
H→R	aga	agaaaatttctgtctccggtgaaggtgaaggtgatgctacttacggtaaattgaccttaaaattgatttgtact
H→S	tca	tcaaaatttctgtctccggtgaaggtgaaggtgatgctacttacggtaaattgaccttaaaattgatttgtact
H→F	ttt	tttaaatttctgtctccggtgaaggtgaaggtgatgctacttacggtaaattgaccttaaaattgatttgtact
H→L	ttg	ttgaaatttctgtctccggtgaaggtgaaggtgatgctacttacggtaaattgaccttaaaattgatttgtact
H→P	cca	ccaaaatttctgtctccggtgaaggtgaaggtgatgctacttacggtaaattgaccttaaaattgatttgtact

The first codon of segment 2 encodes for a histidine (H). This histidine was exchanged (in blue) for codons encoding either alanine (A), aspartic acid (D), arginine (R), serine (S), phenylalanine (F), leucine (L) or proline (P) using the codon with the highest usage frequency in *S. cerevisiae*.

Subheading 3.1.2, **step 2**). Note that in mutation S28L in Supplementary Table 3, the codon *TTA* was used instead of *TTG* in order to prevent the creation of additional BsaI recognition sites.

Depending on the gene length, a mutational library can quickly expand, resulting in increased costs. If necessary, the usage of degenerate bases can mitigate the cost of a library. Degenerate or mixed bases are equimolar mixtures of two, three or four bases at a certain position within a sequence. Supplementary Table 4 exemplifies the use of the mixed bases *M* (*C/A*), *K* (*G/T*) and *Y* (*C/T*). Consequently, rather than using the separate codons *gct* and *gat* for adenine and aspartic acid, respectively, the unified codon *gmt* can encode both. The use of degenerate bases does not affect the total library size (amount of possible different variants), but the varying probabilities associated with each base should be considered during subsequent experimental phases to ensure that all variants are present.

3.1.4 Creation of the Library

1. *Cloning of the destination vectors.*

   Here we provide recommendations and one example of how to clone the destination vectors. However, these can be cloned using any molecular cloning method of choice. The result should be 10 destination vectors in which each segment has been replaced by a GFP-dropout cassette. This cassette should contain 5′ directed BsaI restriction sites, such that digestion of the entry vector leaves compatible overhangs within the original sequence for ligation of the corresponding oligo pool and results in an in-frame replacement of the wild-type segment. The GFP-dropout cassette (from pYTK047) and the original gene regions flanking the segment of interest can be separately amplified by PCR using overlapping primers or primers encoding for restriction enzymes and assembled into a vector by Gibson assembly, yeast assembly, fusion PCR, or cloned via restriction digestion and ligation. In the case of scanning mutagenesis of an entire gene, one can consider keeping the start- and stop-codon in place.

An example for creating the Destination Vector 2 via restriction-ligation cloning using the type IIS restriction enzymes BsaI and BsmBI is given below and illustrated in Supplementary Fig. 1A:

1. Use primers 1 to 6 listed in Table 2, (*see* **Note 2**) to amplify the GFP-dropout cassette (template: pYTK047) and the 5′ and 3′ regions flanking Segment 2 (template: pYTK033).
2. Mix the YTK components required for assembly of the pRS413-type vector (pYTK002, pYTK072, pYTK076, pYTK081, pYTK083) and the TEF promoter (pYTK014) and ENO terminator (pYTK051) and digest with BsaI. Heat-inactivate BsaI after the digest.
3. Digest PCR-amplified segment 2-specific GFP dropout, and 5′ and 3′ flanking regions with BsmBI. Heat-inactivate BsmBI after the digest.
4. Perform a ligation reaction of all components, transform *E. coli* competent cells with the ligation mix and pick a few green colonies to verify the correct destination vector sequence. After sequence confirmation, this vector can be used for many applications of mutagenesis.

2. *Generation of double-stranded oligos.*

Oligonucleotides in the pool are single-stranded. In order to generate double-stranded DNA, we run one-step single primer extension using the reverse primer that binds to the landing pad (oPool reverse: 5′-TGCCGTCTCAGGTCTCA-3′, Table 2 and Fig. 4b). Follow the steps:

1. Mix the following components in a PCR tube:

Primer extension mix	
Reagent	**Volume**
Single-stranded oligos (1 μM)	1 μL
Primer oPool reverse (10 μM)	2.5 μL
2× Phire Hot Start II PCR Master Mix	25 μL
ddH_2O	Up to 50 μL

2. Run the following protocol in a thermocycler

Primer extension reaction		
Step	**Temperature**	**Time**
Denaturation	98 °C	1 min
Annealing[a]	59 °C	30 s
Extension	72 °C	30 s
Storage	4 °C	Hold

[a]To be adjusted depending on the primer used.

3. Purify the double-stranded oligonucleotides using a PCR clean-up kit and elute with 25 μL elution buffer or sterile ddH_2O.

3. *Golden Gate reaction.*

   A Golden Gate reaction needs to be prepared that uses the double-stranded oligonucleotides to insert them into the destination vector containing the GFP-dropout cassette in the corresponding segment of your gene. Follow the steps:

   1. Mix the following components in a PCR tube on ice:

Golden Gate reaction mix	
Reagent	**Volume**
10× T4 ligase buffer	2.5 μL
T7 ligase	1 μL
BsaI	1 μL
Destination vector (20 ng/μL)	1 μL
Double-stranded oligonucleotides (from previous reaction)	20 μL
ddH_2O	Up to 25 μL

   2. Run the Following Procedure in a Thermocycler

4. *Transformation of E. coli with the library.*

   We use an in-house protocol for making *E. coli* DH5α cells chemically competent [34]. For the transformation, follow the steps:

   1. Mix the entire 25 μL Golden Gate reaction with 150 μL of thawed competent cells in a 1.5 mL microcentrifuge tube.
   2. Incubate this mixture on ice for 30 min.
   3. Heat shock the cells for 45 s in a preheated water bath or thermomixer at 42 °C.
   4. Incubate the cells on ice for 2 min before adding 1 mL of LB medium.

5. Incubate the tube at 37 °C in a shaking or standing incubator for 1 h.
6. Centrifuge the tube at 7000 *g* for 3 min obtain a cell pellet.
7. Partially discard the supernatant to retain a volume of approximately 100 μL.
8. Resuspend the cell pellet by gently pipetting up and down. Plate the cell suspension on multiple LB agar plates with the corresponding antibiotic for plasmid maintenance (for our destination vectors 100 μg/mL ampicillin) using sterile glass beads or a cell spreader.
9. Incubate the plates overnight in a 37 °C stove.

5. *Evaluation of the library coverage and considerations for oversampling.*

The design of the destination vectors allows for a simple Green-White fluorescent screening to assess whether the Golden Gate reaction was successful. The green fluorescent colonies still harbor the original plasmid (with the GFP-dropout cassette), while the white colonies carry the plasmids in which the Golden Gate reaction was very likely successful.

1. The green color from colonies with an unsuccessful Golden Gate reaction is visible to the naked eye after incubation of transformants for several hours in the fridge but it can also be visualized using a transilluminator or transilluminating goggles and a flashlight.
2. The efficiency of the Golden Gate reaction is calculated via Eq. (1) and requires counting the total amount of colonies and the number of green colonies. We routinely reach 10^3 colonies per reaction and Golden Gate efficiencies of >97% (Supplementary Table 5 shows representative data from our library constructions). If the total colony number is lower than 100 or the Golden Gate efficiency is lower than 70%, *see* **Notes 3–5** for troubleshooting the Golden Gate reaction.

Golden Gate reaction			
Step	**Process**	**Temperature**	**Time**
1	Digestion	42 °C	2 min
2	Ligation	16 °C	5 min
	Steps 1–2 (25×)		
3	Final digestion	60 °C	10 min
4	Enzyme inactivation	80 °C	10 min
5	Storage	4 °C	Hold

$$\text{Golden Gate efficiency (\%)} = \left(1 - \frac{\text{green colonies}}{\text{total colonies (green + white)}}\right) \times 100 \tag{1}$$

3. In order to ensure the presence of at least one representative of each possible variant in the library, estimate the expected library completeness, or vice versa calculating the number of white transformants required to achieve a high likelihood that all variants are represented.

   Use the tool GLUE for calculating library completeness [35]. GLUE uses the Poisson distribution to estimate the fractional completeness of a library, where L is the number of white colonies, V is the number of equiprobable variants. The probability that a variant occurs at least once is calculated via Eq. (2):

$$1 - P(0) = 1 - e^{-L/V} \tag{2}$$

   Note that this calculation only serves as an estimate of library coverage.

   By simply using the number of white colonies for library size calculation, we likely overestimate the library coverage. This calculation relies on the assumption that every white colony contains a plasmid with a successful product from the Golden Gate reaction and that all possible variants are equally distributed. However, part of the white colonies may contain undesired constructs due to synthesis errors that were present in some oligos. In addition, the distribution of colonies is likely not equimolar due to stochastic differences in the number of full-length molecules of each oligo and potential differences in cloning efficiency for different oligos. Nevertheless, this method still constitutes a good estimation procedure.

Example 1 For an insertional ALFA-tag scan of segment 2, 25 insertions are created (an insertion after each residue). To reach a likelihood of >0.95 that each variant is represented, ≥73 white colonies (L) need to be picked; to reach a likelihood of >0.98, ≥96 colonies need to be pooled (Supplementary Fig. 2A).

Example 2 For the full *Venus* ORF 238 different variants are possible ($V = 238$).

To reach a likelihood of >0.95 that each variant is represented, ≥710 white colonies (L) need to be pooled; and to reach a likelihood of >0.98, ≥940 colonies need to be pooled (Supplementary Fig. 2B).

4. Typically, the protocol delivers in the order of 10^2–10^3 white colonies for a single round of Golden Gate assembly and transformation into *E. coli*. In case the required number of white colonies for high library coverage has not been reached after a single round of Golden Gate and transformation, one could perform more rounds in parallel.

6. *Pooling of the library.*

   The white *E. coli* colonies for each segment are counted and pooled segment-wise.

   1. For each segment, prepare a sterile microcentrifuge tube with 1 mL LB media.
   2. Use a sterile toothpick to scrape all white colonies, avoiding touching any green colonies. Dissolve the cells on the tip of the toothpick in the LB media.
   3. Use 200 μL to inoculate a 2 mL overnight culture, specifically 2 mL LB media supplemented with the corresponding antibiotic (for our destination vectors 100 μg/mL ampicillin), and incubate for 16–20 hours at 37 °C in a shaking incubator.
   4. From the remaining 800 μL, prepare a glycerol stock for future use by adding 200 μL 50% (v/v) glycerol, transferring it to a cryo-tube, and freezing it at −80 °C.
   5. The next day, extract the plasmids from the overnight culture using a standard plasmid Miniprep kit.

7. *Transformation of yeast with the library.*

   For yeast transformation, we use a modified version of the protocol Elble [36], but likely other yeast transformation protocols can be used. In terms of required yeast transformants (single colonies), oversample the number of variants in your library using the protocol provided in Subheading 3.1.2, **step 5**. We perform a separate transformation for each segment library, keeping a high level of control over the number of yeast colonies per segment, but mixing the segment libraries into one final library and performing a single transformation is also possible.

   1. Prepare an overnight culture (1 mL per transformation) by inoculating *S. cerevisiae* BY4741 in YPD media and incubate for 16–20 h at 30 °C with agitation.
   2. Transfer 1 mL of the overnight culture into a microcentrifuge, pellet the cells at 8000 *g* for 1 min, and decant the supernatant until ~50 μL of the liquid remains. Resuspend the cells gently.
   3. Add 2 μL of 10 mg/mL single-stranded carrier DNA (e.g., from salmon sperm).
   4. Add 1–2 μg of plasmid DNA (the library extracted in Subheading 3.1.2, **step 6**) and vortex briefly.

5. Add 500 μL of plate mixture and vortex until the cells and the viscous plate mixture have mixed.
6. Add 20 μL of 1 M DTT and vortex.
7. Incubate on your benchtop for 24 h. During this time, the cells will settle to the bottom of the tube.
8. Heat shock the cells for 10 min at 42 °C.
9. Pipette 75 μL from the settled cells and distribute them over four selective plates (in our case SC dropout media lacking histidine), spread with glass beads or a spatula, and incubate the plate at 30 °C for 2–3 days.

8. *Considerations for selection and screening assays.*

In order to evaluate the performance of each variant in the protein library, a relevant assay system is required that allows for sufficient throughput to assay a significant portion of the library. Having access to a suitable assay system that can be used to reach high library coverage is a crucial consideration before initiating the library creation.

1. The assay depends on the protein of interest and the research goal. Goals may range from identifying a limited number of variants showing improved protein activity to screening many variants to create a comprehensive sequence-fitness map: Appropriate assay systems are available for identifying new binders (e.g., yeast display), for enzyme function and enzyme kinetics and antimicrobial activity, among others [2].
2. One generally distinguishes between selections and screens (which range from low to high throughput). In a selection, one uses conditional differences in cell growth, allowing one to enrich fast-growing cells from a pool of non-growing or poorly growing cells over time. Thus, a large library can be processed in one pot without individual evaluation of the performance of each variant. In a screen, every variant is evaluated individually by a dedicated assay (e.g., measurement of enzymatic activity or antimicrobial activity) and the screenable library size is therefore often much smaller than in a one-pot growth selection assay.
3. Depending on the selection or screen that will be used to assay the variant library, the yeast colonies obtained in **step 2(7)** can either be pooled (for growth selections) and frozen in a similar way to the *E. coli* library (*see* Subheading 3.1.2, **step 6**) or arrayed into 96- or 384-well plates (for medium throughput screens). In the case of a screen, one needs to reach single colonies on a plate after transformation such that they can be individually picked and arrayed. In case colonies are too dense, they can be pooled, diluted 1:100 and 1:1000, and plated again.

4. Arraying is done by filling each well of a 96-well plate with 100 μL two-times concentrated growth media using a multichannel pipette, followed by adding 50 μL sterile H_2O. Then, inoculate single colonies into each well using sterile toothpicks, followed by incubation of the plate for 16–20 h at 30 °C. Pick enough colonies to reach a high likelihood that every variant will be assayed (see Subheading 3.1.2, **step 5**). The arrayed plate can be directly used for screening or stored at −80 °C by adding 50 μL sterile 50% (v/v) glycerol to each well.

9. *Considerations and resources for NGS and analysis.*

   1. After the creation of the library, we recommend using NGS in order to determine its quality and completeness. This step can be performed directly after the plasmid preparation from *E. coli* or after transformation of the yeast, by pooling all the yeast colonies that are going to be assayed.
   2. After screening or selection, the phenotype of the variants will need to be linked to their genotype. If only a few variants are selected, their genotype can be identified individually by Sanger sequencing. If the entire library has to be genotyped and linked to a phenotype, NGS approaches are more cost-efficient.
   3. Several points need to be considered when generating amplicons for NGS:

      (a) NGS can be outsourced and there are different providers that offer well-suited services for variant library sequencing that allow small PCR amplified regions to be paired-end sequenced (~250 bases per forward and reverse read, pay attention to the maximal read length offered by each provider). If variant libraries are longer than the offered maximal read length, a tiling approach can be used, where contiguous sequences are amplified and covered by more than one probe [37]. Amplify the entire region of interest (the segment where the oligo pool was introduced) such that the coverage and sequencing depth are high. This purified PCR product is the product sent for NGS.

      (b) Partial Illumina adapters as well as barcodes for indexing can be incorporated into the amplicons during the PCR. In Table 2, we provide example primers (#8 and #9) that could be used for amplification of Segment 2. These primers encode for partial Illumina adaptors and generate a 260 bp amplicon (excluding the adaptors) that can be covered by a 250 bp forward and reverse read. Each read generates 250 bp, meaning the amplicon can be covered with a large overlap (Supplementary Fig. 1B).

(c) Use a limited amount of cycles (~20) and high-fidelity polymerase to limit the introduction of nucleotide substitutions during PCR for NGS sample preparation.

(d) Use a fluorescence-based DNA quantification kit for accurate dsDNA quantification.

(e) If variants from different samples, cultures, or bins are multiplexed, it is common practice to determine the DNA concentration of the amplicons before mixing them together in equimolar concentrations, in order to keep the reads per variant as similar as possible.

(f) For analysis, nowadays no advanced bioinformatic experience is necessary; There are tools to analyze NGS sequencing data that enable inexperienced researchers to do such work [37].

4. In case only a low or medium throughput assay is available, several tools for "*sequencing before screening*" have been developed that reduce the screening effort by reducing the required oversampling size of the library [38, 39]. The tools use NGS before any screening effort has been made and rely on 96-well plate arrayed colonies and 96-well arrayed barcoded primers to amplify the region of interest in each variant. During analysis of the NGS reads, these double barcodes can link the variant sequences to the corresponding location in the plate. In this way, one can select variants of interest (e.g., three replicates) for screening and, as such, reduce the oversampling size required. These "*sequencing before screening*" approaches require more investment into setting up the technique, but have the advantage that no information is lost, as opposed to *sequencing after screening*, which often requires binning the colonies based on their phenotype, but likely requires a smaller number of primers.

3.2 Method 2. Cloning sgRNA Libraries Using Oligo Pools and Golden Gate Cloning

Here we outline a method for using oligo pools to create a single gRNA (sgRNA) library for CRISPRx applications. Of note, gRNA libraries are usually much larger than the mutational libraries outlined in Method 1. For example, a typical genome-wide gRNA interference library for *S. cerevisiae* that targets each promoter of each open reading frame (>5,000 ORFs) with at least 10 gRNAs comprises >50,000 independent variants [40, 41]. Usually several (more than 10) independent Golden Gate reactions and transformations need to be performed to reach the necessary library size.

To follow Subheading 3.2 (Method 2), we structured it in analogy to Subheading 3.1 (Method 1, Fig. 1). Several substeps of Subheading 3.2.2 are identical to Subheading 3.1.2, and we then refer to the respective substeps in Subheading 3.1.2.

3.2.1 Design of the library

This includes (1) The design of the destination vector and (2) the design of the oligo pool.

3.2.2 Creation of the Library

This includes (1) cloning of the destination vector; (2) double-stranding the oligos; (3) predigesting the destination vector; (4) performing the Golden Gate reaction; (5) transforming *E. coli* with the mutational library; (6) evaluating the library coverage and consideration for oversampling; (7) pooling the library; (8) transforming the library into yeast (*S. cerevisiae*); **steps 5–8** are identical to **steps 4–7** in Subheading 3.1.2.

Further, we provide suggestions for library quality control via NGS (9). This point strongly overlaps with **step 8** in Subheading 3.1.2.

3.2.3 Design of the sgRNA Library

1. *Design of the destination vector.*

   We use the gRNA expression cassette featured in the YTK part pYTK050 [25]. This expression cassette uses the well-established CRISPRm sgRNA architecture [42], where the gRNA expression is driven by a phenylalanine tRNA, followed by an HDV self-cleaving ribozyme that is linked to the sgRNA to stabilize it in the cell; the expression is terminated by an SNR52 terminator (Fig. 5a) [40, 41, 43].

2. *Design of the spacer-encoding oligo pool.*

   Here we outline the general design of the spacer-encoding oligo pool. Follow the steps (Fig. 5):

   1. Generate a list of 20 bp spacer sequences based on your intended CRISPR application and target organism; examples and resources have been described [40, 41, 43].
   2. Add the four-base overhangs GACT and GTTT to the 3′ and 5′ ends that allow for cloning into the destination vector.
   3. Add two thymines (Ts) between the overhangs and the 20 bp spacer. This is required as the 5′ sequence gRNA design has two thymine bases flanking the targeting region [25]. As this would lead to a 5′ CTTT overhang, which can potentially mis-ligate with the 3′GTTT overhang, the two thymines are incorporated in the oligo.
   4. Add 3′ cutting BsmBI-recognition sites to the 5′ and 3′ ends of each oligo.
   5. Add at least three extra nucleotides to improve the digestion efficiency. Here we add CAT at the 5′ and GCAG at the 3′ end, also yielding a primer landing pad that can be used for double-stranding the oligos from the library (Fig. 5b and Table 2). The resulting sequence of this double-stranding reverse primer is 5′—CTGCCGTCTCAAAAC—3′.

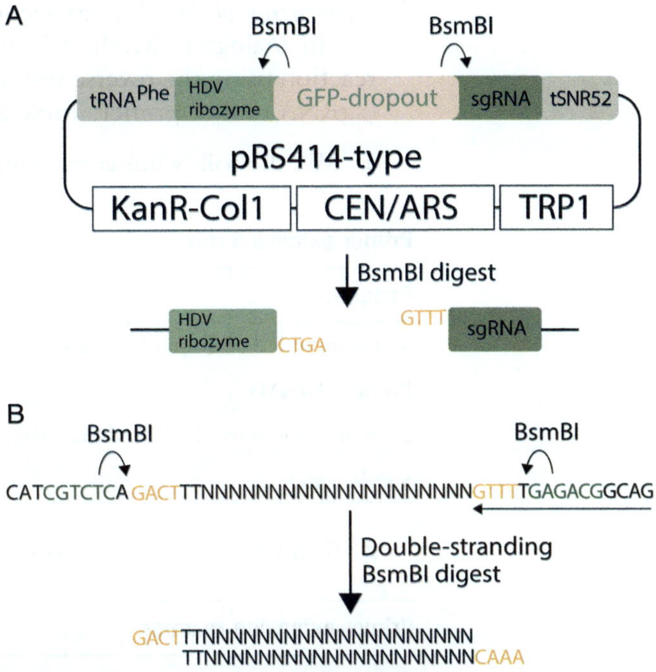

Fig. 5 Design of the gRNA destination vector and the spacer-encoding oligo pool. (**a**) The gRNA destination vector features an expression cassette based on a tRNA promoter, a self-cleaving HDV ribozyme followed by a GFP-dropout cassette that is replaced by the actual 20 bp targeting spacer during the Golden Gate reaction to yield a functional sgRNA and an SNR52 terminator. (**b**) Each oligo is designed by using a distinct 20 bp targeting sequencing (here depicted as N_x), by adding the appropriate 4 bp overhangs at the 5′ and 3′ ends (yellow), followed by adding a 3′ facing BsmBI recognition sites (green) and several nucleotides to enhance BsmBI-cutting. The bases also yield the reverse priming site (black arrow). After double stranding, the oligos are added to the Golden Gate reaction where the BsmBI digestions yield the correct overhangs for cloning into the destination vector

3.2.4 Creation of the sgRNA Library

1. *Cloning of the destination vector.*

   In analogy to Subheading 3.2.1, we provide one example of how to clone the destination vectors. However, these can be cloned using any molecular cloning method of choice.

   We assembled the destination vector as follows: The pRS414-type vector was assembled via Golden Gate using the parts pYTK002, pYTK047, pYTK072, pCgTK02, pCgTK01, and pYTK084 (Table 1). The gRNA expression cassette was amplified by PCR using pYTK050 as a template and primers #11 and #12 (Table 2) to add BsaI overhangs. The pRS414-type expression vector was digested with BsmBI, and the gRNA expression cassette-encoding PCR product was cloned in via a BsaI-based Golden Gate reaction.

2. *Generation of double-stranded oligos.*

   In analogy to Method 1, we run a single primer extension reaction using the reverse primer that binds to the landing pad (sgRNA reverse primer: Table 2). Follow the steps:

   1. Mix the following components in a PCR tube on ice:

Primer extension mix	
Reagent	**Volume**
Single-stranded oligo pool (30 μM)	1 μL
Primer (10 μM)	2.5 μL
2× Phire Hot Start II PCR Master Mix	12.5 μL
Sterile water	Up to 25 μL

   2. Run the following protocol in a thermocycler:

Primer extension reaction		
Step	**Temperature**	**Time**
Denaturation	98 °C	1 min
Annealing[a]	55 °C	30 s
Extension	72 °C	1 min (*see* **Note 7**)
Store	4 °C	Hold

[a]Adjust temperature based on the used primer.

   3. Purify the double-stranded oligonucleotides obtained using a PCR clean-up kit and elute with 25 μL sterile water.

3. *Pre-digestion of the destination vector.*

   We observed that digesting and gel-purifying the destination vector prior to the Golden Gate reaction greatly increased the Golden Gate efficiency for this protocol (*see* **Note 8**). Follow the steps:

   1. For digesting the vector mix the following components in a microcentrifuge tube on ice.

Vector digest mix	
Reagent	**Volume**
10× NEB Buffer 3.1	5 μL
Destination vector (100 ng/μL)	10 μL
BsmBI-v2	1 μL
Sterile water	Up to 50 μL

2. Incubate for 2 h at 55 °C.
3. Separate the digested vector by using standard agarose gel electrophoresis. Cut the band that corresponds to the destination vector with the GFP-dropout cassette being released and purify it using a standard Gel purification kit.
4. Measure the concentration of the Gel-purified digested destination vector using a spectrophotometer.

4. *Golden Gate reaction.*

The Golden Gate reaction needs to be prepared using the double-stranded oligonucleotides (insert) and the digested destination vector. Follow the steps:

1. Mix the following components in a PCR tube on ice:

Golden Gate reaction mix	
Reagent	**Volume**
10× T4 ligase buffer	2.5 μL
T7 ligase	1 μL
BsmBI-v2	1 μL
Digested destination vector (10 ng/μL)	1 μL
Double-stranded oligonucleotides (50 ng/μL)	1 μL
Sterile water	Up to 25 μL

2. Run the following procedure in a thermocycler. Note that this protocol uses 60 cycles of digestion and ligation instead of the 25 cycles outlined in Subheading 3.1.2, **step 3**. We found that increasing the cycle number increased the Golden Gate efficiency for this destination vector.

Golden Gate reaction			
Step	**Process**	**Temperature**	**Time**
1	Digestion	42 °C	2 min
2	Ligation	16 °C	5 min
	Steps 1–2 (60×)		
3	Final digestion	60 °C	10 min
4	Enzyme inactivation	80 °C	10 min
5	Storage	4 °C	Hold

5. *Transform E. coli with the library.*

Perform as outlined in Subheading 3.1.2, **step 4**.

6. *Evaluation of the library coverage and considerations for oversampling.*

   Perform as outlined in Subheading 3.1.2, **step 5**. Be aware that sgRNA libraries are usually much larger than mutational libraries.

7. *Pooling of the library.*

   Perform as outlined in Subheading 3.1.2, **step 6**.

8. *Transformation of S. cerevisiae or N. glabratus with the library.*

   For *S. cerevisiae*, perform as outlined in Subheading 3.1.2, **step** 7. For *C. glabrata*, use the transformation protocol described by Gietz et al. [44].

9. *Suggestions for library quality via NGS.*

   The recommendations are similar to those in Subheading 3.1.2, **step 8**.

A few colonies can be picked and Sanger sequenced. The completeness and quality of the library should then be checked by NGS. The 20 bp spacer region can be amplified by PCR using #13 and #14 (see Table 2), generating a ~250 bp amplicon. The primers already contain partial Illumina adapters that can be used as templates for i7 and i5 index primers (*see* **Note 6**). Single-end short-read sequencing (100 cycles) should be sufficient to capture the 20 bp spacer sequences within the amplicons.

4 Notes

1. The price of the oligo pool depends mainly on the total number of ordered nucleotides and vendors often specify the price per nucleotide. Two examples on how to calculate the number of nucleotides are given in Supplementary Notes 1 and 2.
2. Primers #1 and #2 amplify the 5′ region flanking segment 2, primers #3 and #4 amplify the 3′ region flanking segment 2, and primers #5 and #6 amplify the GFP dropout cassette (*see* Supplementary Fig. 1 and Table 2).
3. The transformation of a successful Golden Gate reaction should yield 10^2 to 10^3 colonies per reaction with an efficiency of at least 95%.
4. If few colonies appear on selective plates after transformation of the Golden Gate reaction, verify the following: (i) competence of the cells used. We recommend testing competence with a purified plasmid or designing an easy "control Golden Gate reaction" based on the well-established YTK. (ii) the functionality of the T7 ligase: zero or a low number of colonies can be caused by poor ligation efficiency regardless of the digestion. If the destination vector is digested but not ligated during the

Golden Gate reaction, the competent cells used will not be transformed, resulting in no or low number of colonies. Perform a "control Golden Gate reaction" as mentioned above or order fresh T7 ligase.

5. If a high number of green fluorescent colonies appear on selective plates after transformation of the Golden Gate reaction, do the following: (i) check functionality of the BsaI or BsmBI enzymes by performing a control digestion, or order fresh BsaI / BsmBI. Transformation of undigested destination vectors will lead to the growth of cells with intact plasmids containing the GFP-dropout cassette. (ii) Predigest the destination vector and gel-purify it before adding it to the Golden Gate reaction as described in Subheading 3.2.2, **step 3**. (iii) Increase the number of cycles from 25 to 60 during the Golden Gate reaction.

   The above tips were inspired by our own experience and resources on the BarrickLab website [45].

6. To minimize PCR bias during NGS sample preparation, it is recommended to run a qPCR on isolated library material (the prepped plasmid library) to determine a low PCR cycle number that is still within the dynamic range of detection. If too many PCR cycles are chosen, each guide will be amplified to saturation and no difference in abundance between the guides can be determined.

7. This double-stranding protocol was established with 1-min extension time. This differs from 30-second extension time of Method 1, but we have not specifically optimized the extension time in either protocol. Both extension times worked well for us.

8. This step is mainly recommended when one needs a high number of (white) colonies.

References

1. Fowler DM, Fields S (2014) Deep mutational scanning: a new style of protein science. Nat Methods 11:801–807. https://doi.org/10.1038/nmeth.3027
2. Wei H, Li X (2023) Deep mutational scanning: a versatile tool in systematically mapping genotypes to phenotypes. Front Genet 14: 1087267. https://doi.org/10.3389/fgene.2023.1087267
3. Markin CJ, Mokhtari DA, Sunden F et al (2021) Revealing enzyme functional architecture via high-throughput microfluidic enzyme kinetics. Science 373:eabf8761. https://doi.org/10.1126/science.abf8761
4. Wrenbeck EE, Azouz LR, Whitehead TA (2017) Single-mutation fitness landscapes for an enzyme on multiple substrates reveal specificity is globally encoded. Nat Commun 8: 15695. https://doi.org/10.1038/ncomms15695
5. Klesmith JR, Bacik J-P, Wrenbeck EE et al (2017) Trade-offs between enzyme fitness and solubility illuminated by deep mutational scanning. Proc Natl Acad Sci USA 114:2265–2270. https://doi.org/10.1073/pnas.1614437114
6. Leander M, Liu Z, Cui Q, Raman S (2022) Deep mutational scanning and machine learning reveal structural and molecular rules

governing allosteric hotspots in homologous proteins. elife 11:e79932. https://doi.org/10.7554/eLife.79932

7. Leander M, Yuan Y, Meger A et al (2020) Functional plasticity and evolutionary adaptation of allosteric regulation. Proc Natl Acad Sci USA 117:25445–25454. https://doi.org/10.1073/pnas.2002613117
8. Hanning KR, Minot M, Warrender AK et al (2022) Deep mutational scanning for therapeutic antibody engineering. Trends Pharmacol Sci 43:123–135. https://doi.org/10.1016/j.tips.2021.11.010
9. Starr TN, Greaney AJ, Hannon WW et al (2022) Shifting mutational constraints in the SARS-CoV-2 receptor-binding domain during viral evolution. Science 377:420–424. https://doi.org/10.1126/science.abo7896
10. Starr TN, Greaney AJ, Hilton SK et al (2020) Deep mutational scanning of SARS-CoV-2 receptor binding domain reveals constraints on folding and ACE2 binding. Cell 182:1295–1310.e20. https://doi.org/10.1016/j.cell.2020.08.012
11. Rollins NJ, Brock KP, Poelwijk FJ et al (2019) Inferring protein 3D structure from deep mutation scans. Nat Genet 51:1170–1176. https://doi.org/10.1038/s41588-019-0432-9
12. Billerbeck S (2018) Small functional peptides and their application in Superfunctionalizing proteins. In: Synthetic biology. John Wiley & Sons, Ltd, pp 217–235
13. Shams A, Higgins SA, Fellmann C et al (2021) Comprehensive deletion landscape of CRISPR-Cas9 identifies minimal RNA-guided DNA-binding modules. Nat Commun 12:5664. https://doi.org/10.1038/s41467-021-25992-8
14. Yan M-Y, Zheng D, Li S-S et al (2022) Application of combined CRISPR screening for genetic and chemical-genetic interaction profiling in Mycobacterium tuberculosis. Sci Adv 8:eadd5907. https://doi.org/10.1126/sciadv.add5907
15. Todor H, Silvis MR, Osadnik H, Gross CA (2021) Bacterial CRISPR screens for gene function. Curr Opin Microbiol 59:102–109. https://doi.org/10.1016/j.mib.2020.11.005
16. Sun L, Zheng P, Sun J et al (2023) Genome-scale CRISPRi screening: a powerful tool in engineering microbiology. Eng Microbiol 3:100089. https://doi.org/10.1016/j.engmic.2023.100089
17. Adames NR, Gallegos JE, Peccoud J (2019) Yeast genetic interaction screens in the age of CRISPR/Cas. Curr Genet 65:307–327. https://doi.org/10.1007/s00294-018-0887-8
18. Kampmann M (2018) CRISPRi and CRISPRa screens in mammalian cells for precision biology and medicine. ACS Chem Biol 13:406–416. https://doi.org/10.1021/acschembio.7b00657
19. Henriksson J, Chen X, Gomes T et al (2019) Genome-wide CRISPR screens in T helper cells reveal pervasive crosstalk between activation and differentiation. Cell 176:882–896.e18. https://doi.org/10.1016/j.cell.2018.11.044
20. Tickman BI, Burbano DA, Chavali VP et al (2022) Multi-layer CRISPRa/i circuits for dynamic genetic programs in cell-free and bacterial systems. Cell Syst 13:215–229.e8. https://doi.org/10.1016/j.cels.2021.10.008
21. Nielsen AAK, Voigt CA (2014) Multi-input CRISPR/Cas genetic circuits that interface host regulatory networks. Mol Syst Biol 10:763. https://doi.org/10.15252/msb.20145735
22. Liu X, Gallay C, Kjos M et al (2017) High-throughput CRISPRi phenotyping identifies new essential genes in Streptococcus pneumoniae. Mol Syst Biol 13:931. https://doi.org/10.15252/msb.20167449
23. Gonçalves E, Segura-Cabrera A, Pacini C et al (2020) Drug mechanism-of-action discovery through the integration of pharmacological and CRISPR screens. Mol Syst Biol 16:e9405. https://doi.org/10.15252/msb.20199405
24. Kuiper BP, Prins RC, Billerbeck S (2022) Oligo pools as an affordable source of synthetic DNA for Cost-effective library construction in protein- and metabolic pathway engineering. Chembiochem 23:e202100507. https://doi.org/10.1002/cbic.202100507
25. Lee ME, DeLoache WC, Cervantes B, Dueber JE (2015) A highly characterized yeast toolkit for modular, multipart assembly. ACS Synth Biol 4:975–986. https://doi.org/10.1021/sb500366v
26. Obst U, Lu TK, Sieber V (2017) A modular toolkit for generating Pichia pastoris secretion libraries. ACS Synth Biol 6:1016–1025. https://doi.org/10.1021/acssynbio.6b00337
27. Billerbeck S, Prins RC, Marquardt M (2023) A modular cloning toolkit including CRISPRi for the engineering of the human fungal pathogen and biotechnology *Host Candida glabrata*. ACS Synth Biol 12:1358–1363. https://doi.org/10.1021/acssynbio.2c00560
28. Moore SJ, Lai H-E, Kelwick RJR, Chee SM, Bell DJ, Polizzi KM, Freemont PS (2016) EcoFlex: a multifunctional MoClo kit for *E. coli*

synthetic biology. ACS Synth Biol 10:1059–1069. https://doi.org/10.1021/acssynbio.6b00031

29. Stukenberg D, Hensel T, Hoff J et al (2021) The Marburg collection: a Golden Gate DNA assembly framework for synthetic biology applications in *Vibrio natriegens*. ACS Synth Biol 10:1904–1919. https://doi.org/10.1021/acssynbio.1c00126
30. Engler C, Youles M, Gruetzner R et al (2014) A golden gate modular cloning toolbox for plants. ACS Synth Biol 3:839–843. https://doi.org/10.1021/sb4001504
31. Fonseca JP, Bonny AR, Kumar GR et al (2019) A toolkit for rapid modular construction of biological circuits in mammalian cells. ACS Synth Biol 8:2593–2606. https://doi.org/10.1021/acssynbio.9b00322
32. Brachmann CB, Davies A, Cost GJ et al (1998) Designer deletion strains derived from Saccharomyces cerevisiae S288C: a useful set of strains and plasmids for PCR-mediated gene disruption and other applications. Yeast 14:115–132. https://doi.org/10.1002/(SICI)1097-0061(19980130)14:2<115::AID-YEA204>3.0.CO;2-2
33. Götzke H, Kilisch M, Martínez-Carranza M et al (2019) The ALFA-tag is a highly versatile tool for nanobody-based bioscience applications. Nat Commun 10:4403. https://doi.org/10.1038/s41467-019-12301-7
34. https://www.qiagen.com/us/knowledge-and-support/knowledge-hub/bench-guide/plasmid/working-with-plasmids/transformation-of-dna
35. Patrick WM, Firth AE, Blackburn JM (2003) User-friendly algorithms for estimating completeness and diversity in randomized protein-encoding libraries. Protein Eng 16:451–457. https://doi.org/10.1093/protein/gzg057
36. Elble R (1992) A simple and efficient procedure for transformation of yeasts. BioTechniques 13:18–20
37. Wrenbeck EE, Faber MS, Whitehead TA (2017) Deep sequencing methods for protein engineering and design. Curr Opin Struct Biol 45:36–44. https://doi.org/10.1016/j.sbi.2016.11.001
38. Wittmann BJ, Johnston KE, Almhjell PJ, Arnold FH (2022) evSeq: Cost-effective amplicon sequencing of every variant in a protein library. ACS Synth Biol 11:1313–1324. https://doi.org/10.1021/acssynbio.1c00592
39. Appel MJ, Longwell SA, Morri M, Neff N, Herschlag D, Fordyce PM (2021) uPIC–M: efficient and scalable preparation of clonal single mutant libraries for high-throughput protein biochemistry. ACS Omega 45:30542–30554. https://doi.org/10.1021/acsomega.1c04180
40. McGlincy NJ, Meacham ZA, Reynaud KK et al (2021) A genome-scale CRISPR interference guide library enables comprehensive phenotypic profiling in yeast. BMC Genomics 22:205. https://doi.org/10.1186/s12864-021-07518-0
41. Momen-Roknabadi A, Oikonomou P, Zegans M et al (2020) An inducible CRISPR interference library for genetic interrogation of *Saccharomyces cerevisiae* biology. Commun Biol 3:723. https://doi.org/10.1038/s42003-020-01452-9
42. Ryan OW, Skerker JM, Maurer MJ et al (2014) Selection of chromosomal DNA libraries using a multiplex CRISPR system. eLife 3:e03703. https://doi.org/10.7554/eLife.03703
43. Hanna RE, Doench JG (2020) Design and analysis of CRISPR–Cas experiments. Nat Biotechnol 38:813–823. https://doi.org/10.1038/s41587-020-0490-7
44. Gietz RD, Schiestl RH (2007) Frozen competent yeast cells that can be transformed with high efficiency using the LiAc/SS carrier DNA/PEG method. Nat Protoc 2:1–4. https://doi.org/10.1038/nprot.2007.17
45. https://barricklab.org/twiki/bin/view/Lab/ProtocolsGoldenGateTroubleshooting

Chapter 16

Golden Gate Cloning of Synthetic CRISPR RNA Spacer Sequences

Selina Rust and Lennart Randau

Abstract

Prokaryotes use CRISPR-Cas systems to interfere with viruses and other mobile genetic elements. CRISPR arrays comprise repeated DNA elements and spacer sequences that can be engineered for custom target sites. These arrays are transcribed into precursor CRISPR RNAs (pre-crRNAs) that undergo maturation steps to form individual CRISPR RNAs (crRNAs). Each crRNA contains a single spacer that identifies the target cleavage site for a large variety of Cas protein effectors. Precise manipulation of spacer sequences within CRISPR arrays is crucial for advancing the functionality of CRISPR-based technologies. Here, we describe a protocol for the design and creation of a minimal, plasmid-based CRISPR array to enable the expression of specific, synthetic crRNAs. Plasmids contain entry spacer sequences with two type IIS restriction sites and Golden Gate cloning enables the efficient exchange of these spacer sequences. Factors that influence the compatibility of the CRISPR arrays with native or recombinant Cas proteins are discussed.

Key words CRISPR, Spacer, crRNA, Restriction enzymes, Genome editing, Golden Gate cloning

1 Introduction

The discovery of Clustered Regularly Interspaced Short Palindromic Repeats (CRISPR) and CRISPR-associated (Cas) proteins [1–3] has expanded the application range of genome manipulation tools, enabling programmable gene editing in many organisms [4, 5]. CRISPR-Cas systems provide adaptive immunity in many prokaryotes, as short DNA fragments of previously encountered foreign genomes can be incorporated as new spacer sequences in an extending CRISPR array [6, 7]. Transcription of the CRISPR array generates precursor CRISPR RNAs (pre-crRNAs) that are processed into noncoding, mature CRISPR RNAs (crRNAs). A single crRNA is bound by one or several Cas proteins to form a CRISPR ribonucleoprotein (crRNP) complex. Here, the target sequence (i.e., the protospacer) is complementary to the spacer sequence of the crRNA, that guides the crRNP and hybridizes to the target,

Daniel Schindler (ed.), *Golden Gate Cloning: Methods and Protocols*, Methods in Molecular Biology, vol. 2850,
https://doi.org/10.1007/978-1-0716-4220-7_16,

mediating CRISPR interference. Recognition of the target sequence relies not only on complementarity between the crRNA and protospacer, but also on the presence of a short sequence motif next to the protospacer. This protospacer adjacent motif (PAM) is specific for each CRISPR-Cas subtype and usually does not occur next to a spacer sequence found in the genomic CRISPR arrays of the host [8, 9].

To date, different CRISPR-Cas systems are used experimentally to efficiently edit genomes of many organisms [10] and they can be engineered to target specific sequences by using synthetic crRNAs. The expression of crRNAs depends on the presence of a CRISPR array characterized by a specific repeat-spacer-repeat motif. The length and sequence composition of these repeats and spacers are determined by the type of CRISPR-Cas system. Some repeat sequences show clear hairpin structure elements, while other repeats remain unstructured. For a single CRISPR array, targeting of specific genetic elements is achieved by highly variable spacer sequences, while its repeat sequences remain largely constant [11]. Only few CRISPR arrays contain degenerated repeats with mutations that often accumulate in promoter-distal regions. A minimal CRISPR array consists of a single repeat-spacer-repeat motif, which is rarely found in nature but allows for precise manipulation in genome engineering applications as the possibility of recombination of repeats is minimized.

To generate plasmid-based minimal CRISPR arrays, standard cloning procedures include restriction digests, annealing of oligonucleotides, and subsequent ligation [12]. In this protocol, we describe a Golden Gate assembly-based approach for precise and convenient sequence manipulation within CRISPR arrays using two type IIS restriction sites. A similar strategy was previously used to clone minimal CRISPR arrays of type I-C CRISPR-Cas systems [13, 14]. The restriction sites are strategically positioned at two repeat sequences of a specific CRISPR array and are part of an entry spacer sequence with no specific target. Transformed constructs are analyzed via polymerase chain reaction (PCR) and sequencing to verify efficient replacement of the entry spacer with the desired synthetic spacer. This protocol explains steps for efficient Golden Gate cloning of synthetic crRNA spacer sequences using the example of a type IV-A1 CRISPR-Cas system [15], while the method can be easily modified to be applied for any other CRISPR-Cas system.

2 Materials

2.1 *E. coli* Strains

For all cloning steps, standard *Escherichia coli* cloning strains (e.g., DH5α and TOP10) were used and *E. coli* BL21-AI was used for gene expression.

2.2 Cloning

1. DNA construct design tool (e.g., Benchling, https://www.benchling.com, free for academic use).
2. Oligonucleotide synthesis service (e.g., Eurofins; Microsynth).
3. 5 U/μL BseRI.
4. 10 U/μL T4 Polynucleotide Kinase (PNK), supplied with 10× T4 PNK reaction buffer: 700 mM Tris-HCl, 100 mM MgCl2, 50 mM DTT, pH 7.6 at 25 °C.
5. 400 U/μL T4 DNA Ligase, supplied with 10× T4 DNA ligation buffer: 500 mM Tris-HCl, 100 mM $MgCl_2$, 10 mM ATP, 100 mM DTT, pH 7.5 at 25 °C.
6. Spectrophotometer for measuring DNA concentration.
7. Thermocycler for PCR and further incubation steps.
8. Plasmid of interest for insertion of minimal CRISPR array (*see* **Note 1**).
9. Oligonucleotides carrying the appropriate spacer sequence with overhangs required for the assembly (see Table 1).
10. Luria-Bertani (LB) medium: 1% (w/v) Bacto tryptone, 0.5% (w/v) yeast extract, 1% (w/v) NaCl in deionized water, adjusted to pH 7.0 with 5 N NaOH. For plates, 1.5% (w/v) agar is added (*see* **Note 2**).
11. Antibiotic suitable for the plasmid of choice (*see* **Note 3**). For preparation of plates an appropriate amount of an antibiotic stock solution is added to autoclaved LB agar cooled down to ~55 °C (e.g., a final concentration of 50 μg/mL kanamycin was used in this study).

2.3 Screening of Colonies

1. 2× DreamTaq PCR Master Mix (e.g., Thermo Fisher Scientific).
2. Primers appropriate for insert verification and sequencing (see Table 1).
3. DNA ladder (e.g., 1 kb Plus DNA Ladder) is used as the marker for gel electrophoresis.
4. 50× TAE buffer: 242.0 g Tris base, 57.1 mL of acetic acid, and 100 mL of 0.5 M EDTA, pH 8.0, in 1 L of deionized water. Running buffer for agarose gels is 1× TAE.
5. 1–3% Agarose gels are prepared by melting the agarose in 1× TAE in a microwave. DNA dye is added according to manufacturer's instructions (e.g., 4 μL of MidoriGreen are added to 100 mL of melted agarose (cooled down to ~55 °C)).
6. Plasmid extraction kit (e.g., NucleoSpin® Plasmid) for preparation of plasmid DNA.

Table 1
Oligonucleotides used in this work

ID	Sequence 5′-3′	Comment
Primer1	GTGACTCTCCTCCGAAGAGG	Primer for Colony PCR, binding in the entry spacer, see Fig. 2
Primer2	CCCCAAGGGGTTATGCTAG	Primer for Colony PCR and sequencing, binding in the backbone of pRSFDuet-1, see Fig. 2
Oligo1	AATTCGTATTTCCCGCG TGCGCGGGGGTGAGCGGATG TGACTCTCCTC	Oligonucleotide for hybridization with Oligo2; Generation of minimal CRISPR array with entry spacer
Oligo2	TTCGGAGGAGAGTCACATCCGC TCACCCCCGCGCACGCGGGAAA TACG	Oligonucleotide for hybridization with Oligo1; Generation of minimal CRISPR array with entry spacer
Oligo3	CGAAGAGGAGGTAAGTACGTA TTTCCCGCGTGCGCGGGGG TGAGCGGGC	Oligonucleotide for hybridization with Oligo4; Generation of minimal CRISPR array with entry spacer
Oligo4	GGCCGCCCGC TCACCCCCGCGCACGCGGGAAA TACGTACTTACCTCCTC	Oligonucleotide for hybridization with Oligo3; Generation of minimal CRISPR array with entry spacer
Oligo5	ATCGCACTCCAGCCAGC TTTCCGGCACCGCTTGT	Oligonucleotide for hybridization with Oligo6; Generation of spacer fragment with GT overhang of type IV-A1 repeat
Oligo6	AAGCGGTGCCGGAAAGCTGGC TGGAGTGCGATCC	Oligonucleotide for hybridization with Oligo5; Generation of spacer fragment with CC overhang of type IV-A1 repeat

7. Gels are checked visually using an appropriate gel imaging system (e.g., Bio-Rad GelDoc EZ Gel Imaging System), while pictures are taken using the software included.

3 Methods

Enzymes and DNA samples should be stored at −20 °C and kept on ice during use. Reactions for PCR, digestion or assembly should be prepared on ice and then incubated as specified.

3.1 Compatibility of the CRISPR-Cas System

The crRNA originates from a repeat-spacer-repeat element, which is found in the CRISPR array in close proximity to the *cas* genes of a given CRISPR-Cas system. Repeat sequences vary in sequence, structure, and length for the different systems. Maturation of most crRNA relies on the specific recognition and cleavage activity of the associated Cas6 endonuclease. Repeat sequences and associated *cas* genes can be identified in host organisms using bioinformatic tools (e.g., CRISPRCasFinder [16]). These tools are capable of identifying CRISPR loci based on their distinctive repeat-spacer-repeat structure. It is often not trivial to determine the orientation of the CRISPR which defines the correct repeats sequence of pre-crRNA (*see* **Note 4**). The correct orientation of repeat and spacer sequences is important for the maturation of the crRNA and target interference. The orientation of a repeat is also important for finding leader regions, identifying target sites on foreign genetic elements, and characterizing PAM.

If in vivo experiments depend on the expression of mature crRNAs, correct repeat orientation ensures compatibility with the associated Cas6 endoribonuclease. Computational analyses of the repeat sequences can include screening of conservation of repeat orientation and PAM sequences in related organisms with published literature on crRNA maturation. In addition, Northern blots or nuclease activity assays with repeat-spacer-repeat transcripts and recombinant Cas6 can be used to verify crRNA maturation for previously uncharacterized systems.

3.2 Construction of an Editable CRISPR Plasmid with an Entry Spacer

An expression vector with a minimal CRISPR array needs to be designed and prepared to enable crRNA production and maturation. The vector should contain a minimal CRISPR array that carries an entry spacer sequence with two recognition sites for a type IIS restriction enzyme.

1. The vector is selected depending on the final expression host and desired mode of induction.
2. Choose two distinct restriction sites (*see* **Note 5**) and ensure transcription via an (inducible) promoter. A ribosome binding site is not necessary for expression of a noncoding crRNA. Digest the plasmid, heat-inactivate the restriction enzyme, and gel-purify the digested plasmid.
3. Design a repeat-spacer-repeat motif with recognition sites for BseRI in the spacer sequence (*see* **Note 6**), while cleavage sites of the enzyme occur in the repeat sequences at both ends of the spacer (Fig. 1). Plan the final structure of the vector accordingly using the preferred DNA analysis and editing software (e.g., Benchling; SnapGene).
4. Design and acquire complementary oligonucleotides that carry suitable overhangs of the linearized plasmid and the CRISPR motif (*see* **Note 7**).
5. Phosphorylate the oligos if necessary, while mixing 10 μL of the oligonucleotide (100 pmol/μL) with 1 μL T4 PNK, 2 μL 10× PNK reaction buffer, 2 μL 10 mM ATP and 5 μL sterile water. Incubate the samples for 1 h at 37 °C.
6. Anneal two complementary oligonucleotides mixing 1 μL of each phosphorylated oligo (50 pmol/μL) with 1 μL of 10× T4 DNA ligation buffer and 7 μL sterile water. Incubate the

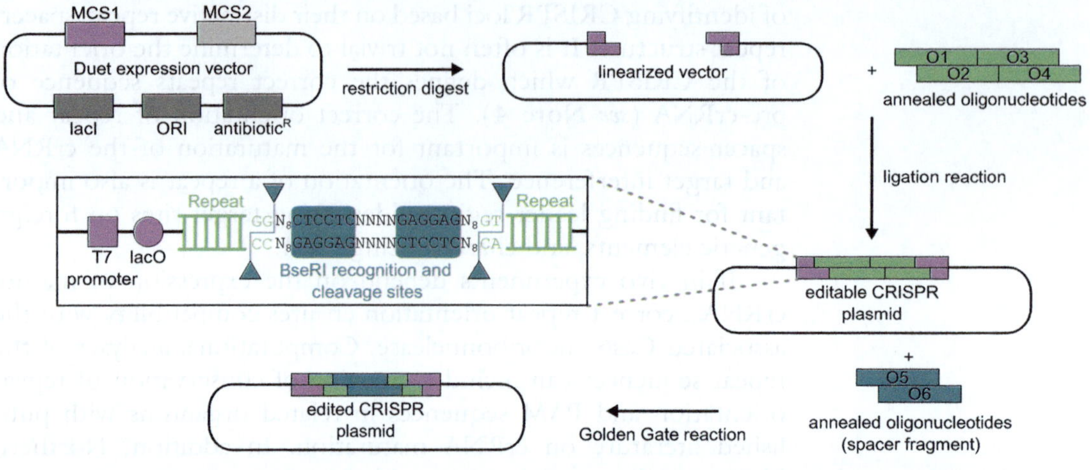

Fig. 1 Workflow for the generation of a plasmid carrying an editable minimal CRISPR array and the exchange of the spacer sequence. A spacer sequence with recognition sites for the type IIS restriction enzyme BseRI is created (Oligonucleotides 01-04, see Table 1 for example sequences) and ligated with a digested plasmid. Golden Gate reaction with the new plasmid results in cleavage of the repeats, generating specific overhangs. New spacer sequences (Oligonucleotides 05 and 06, see Table 1 for example sequences) can be used as inserts in the same reaction to create a new minimal CRISPR array, expressing a different synthetic crRNA

samples for 5 minutes at 95 °C and cool down the samples stepwise to 25 °C (*see* **Note 8**).

7. The created fragment(s) and the linearized vector are ligated using 1 μL of each fragment (10 pmol/μL), 25 ng of the linearized vector, 10 μL 10 mM ATP, 2 μL 10× T4 DNA ligation buffer, 1 μL T4 DNA ligase and filled up with sterile water to a total volume of 20 μL. The sample is incubated 15 min at room temperature or at 16 °C overnight.

8. Transform 5 μL of the ligation reaction into chemically competent *E. coli* cells.

9. Perform a colony PCR with appropriate primers (*see* **Note 9**, Fig. 2).

10. Screen colonies by performing an agarose gel electrophoresis and verify the insertion via sequencing. Here, preferred bioinformatics software can be used for in silico construction of assemblies, confirming that inserts match.

11. Use a plasmid mini preparation kit for DNA isolation, while growing the required *E. coli* strain overnight in LB medium supplied with the respective antibiotic. The spacer sequence of the editable CRISPR plasmid can now be exchanged using a Golden Gate reaction.

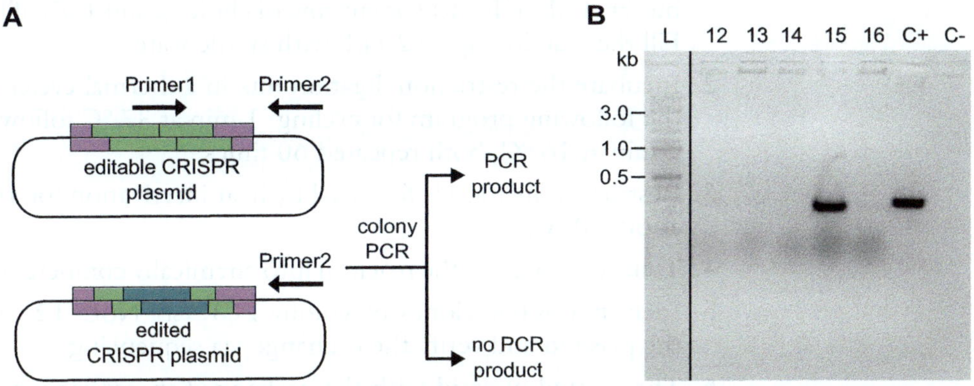

Fig. 2 Primer design and colony PCR after Golden Gate reaction. (**a**) Primers are designed to bind in the editable spacer sequence of the CRISPR plasmid (Primer 1) and the backbone of the plasmid (Primer 2). If the spacer sequence was successfully edited, no PCR product is expected, while the presence of a PCR products indicates an unsuccessful reaction. (**b**) Example gel image after colony PCR of different colonies (12–16). A total of 4 μL of 1 kb Plus DNA Ladder (NEB) was used (L). The positive control (C+) represents the editable CRISPR plasmid, while no DNA was used with regard to the negative control (C−). In total, 14 colonies showed no band, referring to an 87.5% efficiency regarding the exchange of the spacer sequence

3.3 Generation of Synthetic Spacer Sequences

Appropriate spacer sequences need to be designed to exchange the random spacer sequence of the plasmid with the entry spacer. The added spacer sequences will be defined by the desired target, while repeat sequences ensure compatibility with the investigated CRISPR-Cas system. The length of the spacer sequence is fixed for most single-subunit effector complexes, while multi-subunit effectors often display a higher degree of flexibility. It would be advisable to use synthetic spacers that mimic the length of natural spacers of the investigated CRISPR-Cas system. With respect to the type IV-A1 CRISPR-Cas system, a target sequence is sufficient if the protospacer is 32 bp in length and carries a 5′-AAN-3′ PAM.

1. Search for a distinct protospacer sequence (*see* **Note 10**) with an appropriate PAM and design the spacer motif accordingly. Pay attention to use the correct orientation of the spacer sequence and plan the final vector.
2. Design and acquire appropriate oligonucleotides that contain overhangs, matching the overhangs generated by the type IIS restriction enzyme (Fig. 2, *see* **Note 11**).
3. Phosphorylate the oligonucleotides if necessary and anneal them (Subheading 3.2, **steps 5** and **6**). Annealed oligonucleotides now serve as (spacer) fragment for Golden Gate reaction and are stable for up to 1 year if stored at −20 °C.

3.4 Exchanging Spacer Sequences Via Golden Gate Reaction

1. Set up a one-pot restriction–ligation reaction by mixing 40 fmol of the plasmid carrying the editable minimal CRIRSPR array, with 40 fmol of the spacer fragment, 2 μL 10× ligation

buffer, 1 μL of BseRI (or enzyme of choice), and 1 μL of ligase. Fill the reaction up to 20 μL with sterile water.

2. Incubate the restriction–ligation mix in a thermal cycler using the following program for cycling: 1 min at 37 °C followed by 1 min at 16 °C, both repeated 60 times.
3. Restriction-ligation is followed by heat inactivation for 10 min at 60–80 °C.
4. Transform 5 μL of the reaction into chemically competent cells.
5. Identify positive clones by colony PCR (*see* **Note 12**), purify the plasmid and verify the exchange via sequencing.
6. The created plasmid with the exchanged spacer sequence can now be used for further assays; however, it cannot undergo another Golden Gate reaction, as recognition sites of the type IIS restriction enzyme are lost when the entry spacer sequence is exchanged.

4 Notes

1. The plasmid needs to be analyzed regarding further cleavage sites of the type IIS restriction enzyme. If type IIS restriction sites are present in the selected plasmid, they can be eliminated by using a mutagenesis strategy of choice. Duet expression vectors are carrying two distinct multiple cloning sites, while missing recognition sites of the type IIS restriction enzyme BseRI. Expression is based on the presence of a T7 polymerase and the addition of arabinose. An expression strain must be selected accordingly. Based on our experience, the use of *E. coli Bl21-AI* is recommended.
2. LB medium can be prepared using commercially available mixes. Here, we prepare liquid LB using 25 g of LB medium mix (e.g., Carl Roth, Karlsruhe, Germany) in 1 L of water. For LB solid medium, we add 15 g of LB Agar High salt (e.g., Carl Roth) to 1 L of water. Media is then autoclaved.
3. This study focused on the use of a derivative of the pRSFDuet-1 vector (Novagen), carrying type IV-A1 *cas* genes.
4. Studies show that repeats often start with the motif 5′-GT-3′ [17]. However, further computational or experimental analyses are necessary if the direction of a repeat is uncertain.
5. If the final expression vector is needed to express crRNA only, we recommend using restriction sites in both multiple cloning sites. Thus, the presence of an unnecessary promoter is prevented. However, Duet vectors can be used to express Cas proteins as well as the crRNA using only one vector. Here, *cas* genes need to be checked with respect to the compatibility with

the type IIS restriction enzyme, while eventually only one MCS is used for the minimal CRISPR array.

6. The minimum length of the spacer sequence using BseRI is 28 bp due to the recognition and cleavage sites of the restriction enzyme. However, using a total length that corresponds to the native spacer length of the specific CRISPR-Cas system is recommended as the plasmid eventually serves as a nontargeting control. In this study, the minimal CRISPR array has a total length of 90 bp, with repeats of 29 bp and a spacer length of 32 bp.
7. Due to the presence of two (usually) identical repeats, we recommend designing two (or more) fragments of the CRISPR array to avoid problems during annealing of the oligonucleotides. Oligo 1–4 (see Table 1) were used to create 2 inserts carrying a type IV-A1 minimal CRISPR array with overhangs to EcoRI and NotI of the first MCS of the pRSFDuet-1 vector.
8. Using a PCR cycler for a controlled cool down is recommended. However, samples can also be placed in a heater that is turned off after 5 minutes, while samples cool down ~2 h to room temperature.
9. We recommend using a primer that specifically binds in the editable, inserted spacer sequence in combination with a primer binding in the plasmid backbone (e.g., Primer1 & 2, see Table 1, Fig. 1). Thus, primer can be used for further analyses once the spacer was exchanged.
10. The protospacer should not exceed a GC content of 65% and occur with a single copy, while not having similarities to an undesirable target.
11. It is critical to pay attention on the overhangs generated by distinct restriction enzymes. For BseRI, both oligonucleotides require a 2 bp overhang at the 3′-end (e.g., Oligo 5 & 6, see Table 1).
12. The same primers as in **Note 9** (No. 5–6, see Table 1) can be used, while a positive clone results in the absence of a PCR product (Fig. 2b). However, it is recommended to also amplify the entire region to avoid false-positive clones, as homologous recombination can result in losing one repeat and the spacer motif. If primers result in very short PCR products, agarose gels with high percentages (2–3%) are recommended.

Acknowledgments

This work was supported by DFG grant SPP 2141 to L.R.

References

1. Barrangou R, Fremaux C, Deveau H et al (2007) CRISPR provides acquired resistance against viruses in prokaryotes. Science 315: 1709–1712. https://doi.org/10.1126/science.1138140
2. Garneau JE, Dupuis M-È, Villion M et al (2010) The CRISPR/Cas bacterial immune system cleaves bacteriophage and plasmid DNA. Nature 468:67–71. https://doi.org/10.1038/nature09523
3. Jansen R, van Embden JDA, Gaastra W et al (2002) Identification of genes that are associated with DNA repeats in prokaryotes. Mol Microbiol 43:1565–1575. https://doi.org/10.1046/j.1365-2958.2002.02839.x
4. Barrangou R, Doudna JA (2016) Applications of CRISPR technologies in research and beyond. Nat Biotechnol 34:933–941. https://doi.org/10.1038/nbt.3659
5. Hidalgo-Cantabrana C, Barrangou R (2020) Characterization and applications of Type I CRISPR-Cas systems. Biochem Soc Trans 48: 15–23. https://doi.org/10.1042/BST20190119
6. Mojica FJ, Díez-Villaseñor C, García-Martínez J et al (2005) Intervening sequences of regularly spaced prokaryotic repeats derive from foreign genetic elements. J Mol Evol 60:174–182. https://doi.org/10.1007/s00239-004-0046-3
7. Makarova KS, Grishin NV, Shabalina SA et al (2006) A putative RNA-interference-based immune system in prokaryotes: computational analysis of the predicted enzymatic machinery, functional analogies with eukaryotic RNAi, and hypothetical mechanisms of action. Biol Direct 1:7. https://doi.org/10.1186/1745-6150-1-7
8. Bhaya D, Davison M, Barrangou R (2011) CRISPR-Cas systems in bacteria and archaea: versatile small RNAs for adaptive defense and regulation. Annu Rev Genet 45:273–297. https://doi.org/10.1146/annurev-genet-110410-132430
9. Mojica FJM, Díez-Villaseñor C, García-Martínez J et al (2009) Short motif sequences determine the targets of the prokaryotic CRISPR defence system. Microbiology (Reading) 155: 733–740. https://doi.org/10.1099/mic.0.023960-0
10. Jinek M, Chylinski K, Fonfara I et al (2012) A programmable dual-RNA-guided DNA endonuclease in adaptive bacterial immunity. Science 337:816–821. https://doi.org/10.1126/science.1225829
11. Makarova KS, Wolf YI, Iranzo J et al (2020) Evolutionary classification of CRISPR-Cas systems: a burst of class 2 and derived variants. Nat Rev Microbiol 18:67–83. https://doi.org/10.1038/s41579-019-0299-x
12. Pausch P, Müller-Esparza H, Gleditzsch D et al (2017) Structural variation of type I-F CRISPR RNA guided DNA surveillance. Mol Cell 67: 622–632.e4. https://doi.org/10.1016/j.molcel.2017.06.036
13. Csörgő B, León LM, Chau-Ly IJ et al (2020) A compact Cascade–Cas3 system for targeted genome engineering. Nat Methods 17:1183–1190. https://doi.org/10.1038/s41592-020-00980-w
14. Lammens E-M, Volke DC, Schroven K et al (2023) A SEVA-based, CRISPR-Cas3-assisted genome engineering approach for Pseudomonas with efficient vector curing. bioRxiv:2023.06.29.547033. https://doi.org/10.1101/2023.06.29.547033
15. Guo X, Sanchez-Londono M, Gomes-Filho JV et al (2022) Characterization of the self-targeting Type IV CRISPR interference system in Pseudomonas oleovorans. Nat Microbiol 7: 1870–1878. https://doi.org/10.1038/s41564-022-01229-2
16. Couvin D, Bernheim A, Toffano-Nioche C et al (2018) CRISPRCasFinder, an update of CRISRFinder, includes a portable version, enhanced performance and integrates search for Cas proteins. Nucleic Acids Res 46:W246–W251. https://doi.org/10.1093/nar/gky425
17. Santiago-Frangos A, Henriques WS, Wiegand T et al (2023) Structure reveals why genome folding is necessary for site-specific integration of foreign DNA into CRISPR arrays. Nat Struct Mol Biol. https://doi.org/10.1038/s41594-023-01097-2

Chapter 17

Golden Gate Cloning of Expression Plasmids for Synthetic Small RNAs in Bacteria

Sophie Dittmar and Bork A. Berghoff

Abstract

Bacterial small RNAs (sRNAs) are well known for their ability to modulate gene expression at the post-transcriptional level. Their rather simple and modular organization provides the user with defined building blocks for synthetic biology approaches. In this chapter, we introduce a plasmid series for *Escherichia coli* and describe protocols for fast and efficient construction of synthetic sRNA expression plasmids based on Golden Gate assembly. In addition, we present the G-GArden tool, which assists with the design of oligodeoxynucleotides and overhangs for scarless assembly strategies. We propose that the presented procedures are suitable for many applications in different bacteria, which are related to *E. coli* and beyond.

Key words Synthetic biology, Golden Gate assembly, Small RNAs, Gene expression, Post-transcriptional regulation

1 Introduction

In basic and applied biosciences, control over gene expression is imperative to study gene functions and modulate phenotypes. While many strategies aim at controlling transcription initiation, gene expression may also be modulated at the post-transcriptional level by the use of regulatory RNAs. A prominent class of regulatory RNAs in prokaryotes are small RNAs (sRNAs), which regulate their target mRNAs via base-pairing, thereby affecting mRNA stability and/or translation [1, 2]. Prototype sRNAs often have a length of less than 100 nucleotides (nts) and can be divided into at least two defined building blocks: the seed region and the scaffold [3, 4]. This modular organization serves as blueprint for the design of customized sRNAs in the field of synthetic biology. The seed region, which is often located at the very 5′-end of the sRNA, represents a stretch of nucleotides that is used for binding of mRNA targets (Fig. 1a). The scaffold usually contains a transcriptional terminator (hairpin structure followed by a poly-uridine

Daniel Schindler (ed.), *Golden Gate Cloning: Methods and Protocols*, Methods in Molecular Biology, vol. 2850,
https://doi.org/10.1007/978-1-0716-4220-7_17,

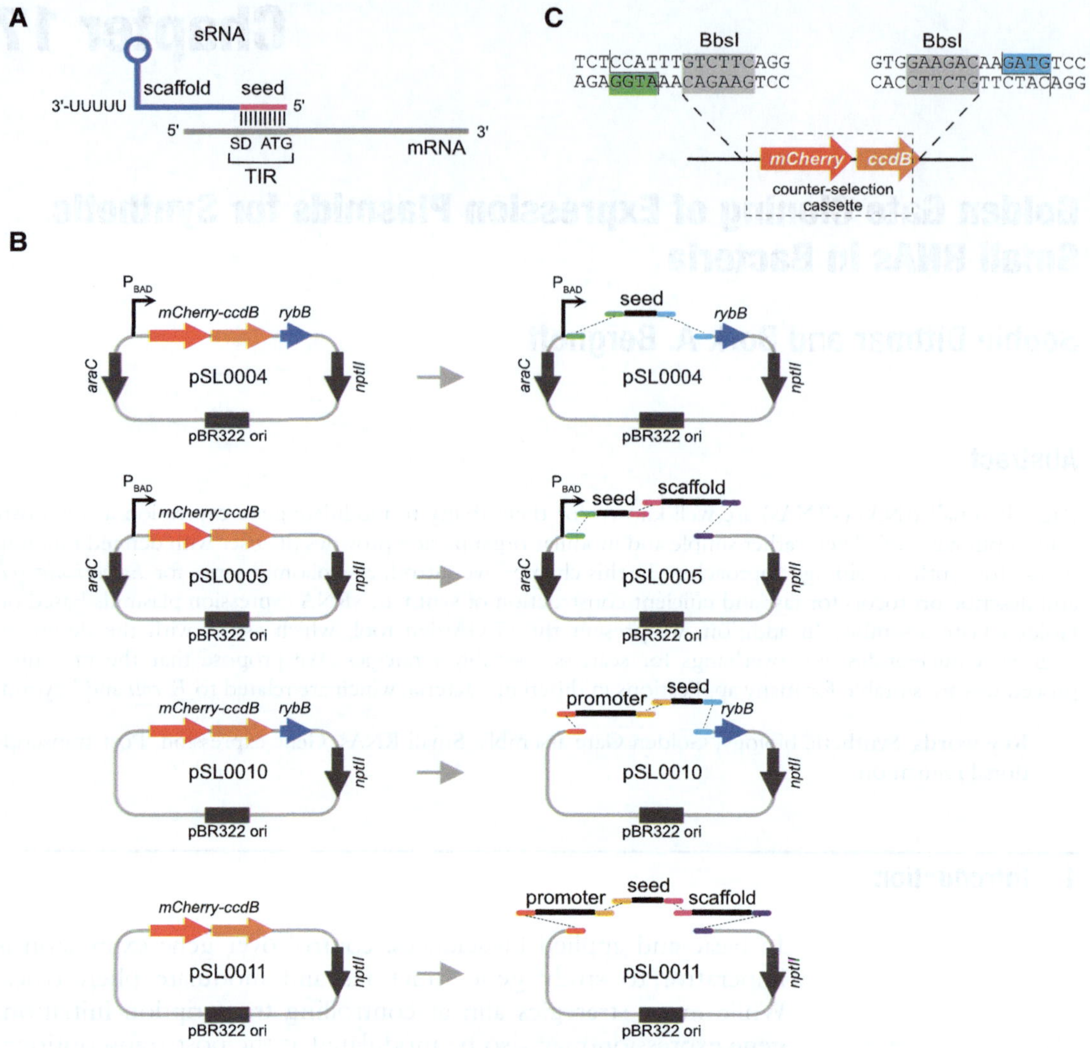

Fig. 1 Plasmids for Golden Gate cloning of synthetic sRNAs. (**a**) Synthetic sRNAs consist of a seed region (pink) and a scaffold region (blue). The scaffold has a terminator hairpin (lollipop structure) that is followed by a poly-uridine (poly-U) stretch. The seed region base pairs with the translation initiation region (TIR) of an mRNA of choice to suppress translation. SD: Shine-Dalgarno sequence, ATG: start codon. (**b**) Plasmid series for Golden Gate assembly of sRNA transcription units (TUs). Plasmids pSL0004 and pSL0005 are used for inducible expression of synthetic sRNAs from the P_{BAD} promoter, and plasmids pSL0010 and pSL0011 are used for sRNA expression from a promoter of choice. Plasmids pSL0004 and pSL0010 contain a DNA sequence for the RybB scaffold. The *mCherry-ccdB* counter-selection cassette (left-hand side) is replaced by DNA parts (promoter, seed, and/or scaffold) to construct sRNA TUs (right-hand side). Ovals of the same color are connected by dotted lines to indicate matching 4-nt overhangs for assembly. pBR322 ori: medium-copy origin of replication, *araC*: gene encoding the AraC repressor, *nptII*: gene encoding neomycin phosphotransferase II for selection with kanamycin. (**c**) Recognition sites for the type IIS restriction enzyme BbsI are present upstream and downstream of the *mCherry-ccdB* counter-selection cassette. Upon BbsI cleavage, the counter-selection cassette and the BbsI recognition sites are removed, and the resulting 4-nt overhangs (green and blue) enable insertion of DNA parts with matching overhangs. pSL0004 is shown as an example

stretch) at the 3′-end (Fig. 1a) and additional interaction sites for RNA chaperones, such as Hfq [5, 6]. Naturally occurring sRNAs support adaptation to adverse conditions by modulating gene expression of multiple targets through imperfect base-pairing [2]. In contrast, synthetic sRNAs are designed to regulate one specific mRNA of interest through perfect base-pairing, hence providing a regulatory tool for phenotypic modulation and metabolic engineering [7–9].

In order to construct synthetic sRNA expression plasmids, the user needs to specify three parts: (i) the promoter, (ii) the seed region, and (iii) the sRNA scaffold, which together form the sRNA transcription unit (TU). The seed region is designed to specifically bind to the mRNA of choice. The seed region length may range from 15 to 30 nts for efficient regulation [10, 11], but shorter/longer seed regions can be appropriate as well [7, 11]. Seed regions are usually complementary to the translation initiation region (TIR) of the target mRNA to repress translation and deplete protein levels (Fig. 1a). Since the design of seed regions is not in the scope of this chapter, we would like to suggest further reading on this topic [7, 11, 12]. Here, we describe protocols for fast and efficient construction of synthetic sRNA expression plasmids based on Golden Gate cloning. The benefit of Golden Gate cloning is the simplicity of combining multiple DNA parts (promoter, seed, and scaffold) without generating scars at the promoter-seed or seed-scaffold fusion sites (i.e., scarless assembly strategy). Hence, the introduction of additional nucleotides (scars), which may interfere with the structure and/or function of the resulting sRNAs, is avoided. We have recently designed a series of acceptor plasmids that are compatible with Golden Gate cloning [7]. The plasmid series derives from the pBAD system that is commonly used in *Escherichia coli* for inducible expression from the P_{BAD} promoter using L-arabinose. All plasmids contain the pBR322 origin of replication, yielding approximately 15–20 plasmid copies per cell, and the *nptII* gene (encoding neomycin phosphotransferase II) for selection with kanamycin (Fig. 1b). However, the plasmid series is also available with the *bla* gene (encoding β-lactamase) for selection with ampicillin [7]. If the P_{BAD} promoter is used for inducible expression, a gene for the AraC repressor is present as well (Fig. 1b). All plasmids contain a *mCherry-ccdB* counter-selection cassette that is removed upon cleavage with the type IIS restriction enzyme BbsI (Fig. 1c). Due to the *ccdB* gene (encoding a gyrase-inhibiting toxin), cloning plasmids are only propagated in suitable strain backgrounds, such as *E. coli* DB3.1, but are toxic to recipient strains, such as *E. coli* TOP10 or MG1655. The *mCherry* gene (encoding a red fluorescent protein) serves as marker to detect false positives during cloning procedures. We would like to refer the reader to a recent publication for a detailed description of the counter-selection cassette [7]. During Golden Gate assembly, the counter-selection cassette is replaced by DNA parts with appropriate 4-nt overhangs to generate an sRNA TU (Fig. 1b). The DNA parts are either generated via annealing of oligodeoxynucleotides or via PCR, depending on fragment length. If needed, the design step is assisted by the G-Garden tool [12]. The plasmid series is built for different levels of cloning complexity. The pSL0004 plasmid (low complexity) contains the P_{BAD} promoter and the DNA sequence for the well-established RybB scaffold from *E. coli*. Hence, insertion of a seed sequence is sufficient to construct the TU for an inducible, synthetic RybB sRNA. At medium complexity, the user specifies the seed region and scaffold sequence (e.g., the MicA scaffold from *E. coli*) for inducible expression from the P_{BAD} promoter (pSL0005). Alternatively, the user selects a customized promoter (e.g., the constitutive $P_L$lacO-1 promoter) and the seed region for expression of a synthetic RybB sRNA (pSL0010). Finally, plasmid pSL0011 (high complexity) gives the highest degree of freedom, since all three DNA parts can be chosen by the user. In the following protocols, we showcase the design/cloning procedures by constructing plasmids for inducible and constitutive expression of synthetic RybB and MicA sRNAs targeting the *acrA* gene in *E. coli*. We also describe a Northern blot protocol for sRNA expression analysis.

2 Materials

2.1 Golden Gate Acceptor Plasmids

Golden Gate acceptor plasmids need to match the specific requirements of the user, such as origin of replication, resistance marker, and restriction sites for the desired type IIS restriction enzyme. In this protocol, the following Golden Gate acceptor plasmids are used for cloning with the type IIS restriction enzyme BbsI and expression of synthetic sRNAs in *E. coli*:

Name	Size [bp]	Features	Purpose
pSL0004	5736	*araC*, P_{BAD}, *rybB* scaffold, *mCherry-ccdB*, pBR322 ori, *nptII*	Seed region acceptor plasmid for inducible expression of RybB sRNAs from the P_{BAD} promoter
pSL0005	5664	*araC*, P_{BAD}, *mCherry-ccdB*, pBR322 ori, *nptII*	Seed region and scaffold acceptor plasmid for inducible expression of customized sRNAs from the P_{BAD} promoter
pSL0010	4545	*rybB* scaffold, *mCherry-ccdB*, pBR322 ori, *nptII*	Promoter and seed region acceptor plasmid for expression of RybB sRNAs from a promoter of choice

(continued)

Name	Size [bp]	Features	Purpose
pSL0011	4473	*mCherry-ccdB*, pBR322 ori, *nptII*	Promoter, seed region, and scaffold acceptor plasmid for expression of customized sRNAs from a promoter of choice
pSL0009	2671	pBR322 ori, *nptII*	Empty control plasmid

2.2 Bacterial Strains and Growth Media

1. In this protocol, the following *E. coli* strains are used:

Strain	Genotype	Purpose
MG1655	K-12 F^- λ^- *ilvG*⁻ *rfb-50 rph-1*	Recipient strain for synthetic sRNA expression plasmids
DB3.1	F^- *gyrA462 endA1 glnV44* Δ(*sr1-recA*) *mcrB mrr hsdS20*(r_B^-, m_B^-) *ara14 galK2 lacY1 proA2 rpsL20*(Str^R) *xyl5* Δ*leu mtl1*	Propagation of Golden Gate acceptor plasmids containing the *mCherry-ccdB* counter-selection cassette

2. All strains are cultivated in lysogeny broth (LB) medium with 50 μg/mL kanamycin.
3. LB medium: Mix 10 g tryptone, 5 g yeast extract, 10 g NaCl, and 1 L distilled water and autoclave. For preparation of LB agar plates, add 15 g agar to 1 L LB medium before autoclaving.

2.3 Operating System and Availability of G-GArden

1. The G-GArden tool is written in Python and bundled separately for both Linux and Windows operating systems. It is implemented as a graphical user interface (GUI) and can be easily launched by an executable file that is available for both operating systems.
2. G-GArden is freely available at https://github.com/DIGGER-Bac under the CC BY-NC-SA 4.0 license.

2.4 Oligodeoxynucleotides for Golden Gate Cloning

The following oligodeoxynucleotides are used for Golden Gate cloning of sRNA expression plasmids (bold letters indicate overhangs for Golden Gate cloning and underlined letters indicate BbsI recognition sites):

Name	Sequence (5′ to 3′)	Purpose
pSL10-11 PLlac-fwd	**ACCA**TTGTGAGCGGATAACAATTGACATTGTGAGCGGATAACAAGATACTGA	Forward primer: $P_L$lacO-1 promoter for pSL0010 and pSL0011

(continued)

Name	Sequence (5′ to 3′)	Purpose
pSL10-11 PLlac-rev	**GTGC**TCAGTATCTTG TTATCCGCTCACAA TGTCAATTGTTA TCCGCTCACAA	Reverse primer: $P_L$lacO-1 promoter for pSL0010 and pSL0011
TKo0107	**CCAT**CATATGTAAACC TCGA	Forward primer: s8 seed region for pSL0004
TKo0155	**CATC**TCGAGGTTTACA TATG	Reverse primer: s8 seed region for pSL0004
pSL05 s8-fwd	**CCAT**CATATGTAAACC TCGA	Forward primer: s8 seed region for pSL0005
pSL05-11 s8-rev	**GATG**TCGAGGTTTACA TATG	Reverse primer: s8 seed region for pSL0005 and pSL0011
pSL10-11 s8-fwd	**GCAC**CATATGTAAACC TCGA	Forward primer: s8 seed region for pSL0010 and pSL0011
pSL10 s8-rev	**CATC**TCGAGGTTTACA TATG	Reverse primer: s8 seed region for pSL0010
pSL05-11 MicA-fwd	CGAAGACCC**CATC**CC TGAATTCAGAGAT	Forward primer: amplification of MicA scaffold for pSL0005 and pSL0011
pSL05-11 MicA-rev	CGAAGACCC**A** **TCC**GAAGAGGA TTTTGTAGCCG	Reverse primer: amplification of MicA scaffold for pSL0005 and pSL0011

2.5 Phosphorylation of Oligodeoxynucleotides

1. Forward oligodeoxynucleotide (100 μM stock).
2. Reverse oligodeoxynucleotide (100 μM stock).
3. 10× T4 DNA Ligase Reaction Buffer (500 mM Tris-HCl, pH 7.5, 100 mM $MgCl_2$, 10 mM ATP, 100 mM DTT).
4. T4 Polynucleotide Kinase (10,000 units/mL).
5. Sterile water.

2.6 Annealing of Oligodeoxynucleotides

1. Phosphorylated oligodeoxynucleotides.
2. 10× annealing buffer (100 mM Tris-HCl, pH 7.5–8.5, 500 mM NaCl, 10 mM ethylenediaminetetraacetic acid (EDTA)).
3. Sterile water.

2.7 Golden Gate Assembly

1. Acceptor plasmid.
2. All DNA parts (fragments) in equimolar concentrations to the acceptor plasmid.
3. 10× T4 DNA Ligase Reaction Buffer (500 mM Tris-HCl, pH 7.5, 100 mM $MgCl_2$, 10 mM ATP, 100 mM DTT).
4. Type IIS restriction enzyme BbsI (20,000 units/mL).
5. T4 DNA Ligase (400,000 units/mL).
6. Sterile water.

2.8 Primers for Sanger Sequencing

The following primers are used for Sanger sequencing of sRNA expression plasmids:

Name	Sequence (5′ to 3′)	Target plasmid
Mult-targ-scr-rev	CAGGCAAATTCTGTTTTATCAGACC	pSL0004 and pSL0005
Mult-targ-scr-fw	CTGTCAAATGGACGAAGCAG	pSL0010 and pSL0011

2.9 Total RNA Isolation

1. Stop solution (5% (v/v) phenol and 95% (v/v) ethanol).
2. Lysis buffer (100 mM Tris-HCl, pH 7.5, 40 mM EDTA, 200 mM NaCl, 0.5% SDS).
3. Water-saturated phenol (pH 4–5).
4. Chloroform-isoamyl alcohol (24:1).
5. Isopropyl alcohol.
6. Sodium acetate (3 M, pH 4.5).
7. Ethanol (75%).
8. RNase-free water.

2.10 Reagents for Northern Blot Analysis

1. Polyacrylamide-urea gel: urea, acrylamide/bis-acrylamide (40%, 37.5:1), 10× Tris-borate-EDTA (TBE) buffer, distilled water, 10% ammonium persulfate (APS), and tetramethylethylenediamine (TEMED).
2. Formamide-urea (FU) mix: 80% deionized formamide, 6 M urea, 1× TBE, 0.1% bromophenol blue, and 0.1% xylene cyanol in distilled water.
3. Semi-dry electroblotting: blotting paper, nylon membrane (positively charged), 1× TBE.
4. Sodium-phosphate (Na-P) buffer (2×): 500 mM Na_2HPO_4 in distilled water, adjusted to pH 7.2 with H_3PO_4 (85%).
5. Church hybridization buffer (low stringency): 1% (w/v) BSA, 7% SDS, 1× Na-P buffer, and 1 mM EDTA in distilled water.

6. Saline-sodium citrate (SSC) stock solution (20×): 3 M NaCl and 300 mM sodium citrate in distilled water, adjusted to pH 7.0 with HCl.
7. Wash buffer: 5× SSC and 0.01% SDS in distilled water.
8. Stripping buffer: 0.1% SDS in distilled water.

2.11 Oligodeoxynucleotides for Northern Blot Analysis

The following oligodeoxynucleotides are used for detection of synthetic sRNAs:

Name	Sequence (5′ to 3′)	Purpose
RybB-probe-2	GAAATGGCGGGGTTGATGGGCTCCACAAAATGGGGACATC	Detection of synthetic RybB sRNAs
MicA-probe-2	CTCGTGAGTGGCCAAAATTTCATCTCTGAATTCAGGGATG	Detection of synthetic MicA sRNAs
5S probe-2	CCTGGCAGTTCCCTACTCTCGCATGAGGAG	Detection of 5S rRNA

3 Methods

3.1 G-GArden: Oligodeoxynucleotide Design for Golden Gate Assembly

3.1.1 Launching of G-GArden

1. Since G-GArden was created with PyInstaller, users do not need to preinstall anything.
2. Use the following link for Windows operating systems: https://github.com/DIGGER-Bac/G-GArden-Windows. For Linux, use this link: https://github.com/DIGGER-Bac/G-GArden-Linux.
3. Use the "Code" dropdown menu on GitHub to download the ZIP folder and unzip the folder to any directory of choice.
4. Navigate through folder *dist* into *ggarden* and start the *ggarden* executable file (*ggarden.exe*) by double-clicking. The GUI should open as a new window.

3.1.2 Preparing the Database File

1. The database file (*saved_data.csv*) is maintained by the user. All DNA sequences that should be used for Golden Gate assembly must be uploaded to this file.
2. The user needs to specify the following information for each DNA sequence: the fragment name, fragment type, sequence of the fragment, and a path to a GenBank (*.gbk*) file (the GenBank file is optional; *see* **Note 1**).
3. There are three options for adding new DNA sequences to the database file.

4. *Option A*: Open the *saved_data.csv* file via Excel or another editor of your choice and edit the file. The *saved_data.csv* file can be found in folder *data* (*dist* > *ggarden* > *data*). Type in the fragment name (column "Name"), fragment type (column "Fragment"; choose "plasmid," "promoter," "seed," or "scaffold"), sequence of the fragment (column "Sequence"), and a path to a GenBank file (column "File"; optional; *see* **Note 1**). If an editor different than Excel is applied, please use a semicolon as separator. Excel will take care of the separators on its own. Save the *saved_data.csv* file and close it before you open the G-GArden GUI.
5. *Option B*: Within the GUI you can upload new DNA sequences via the "Insert fragment sequence" widget at the bottom left (Fig. 2a). Note that the widget will only add one fragment at a time. Simply type the fragment name, fragment type, sequence of the fragment, and a path to a GenBank file (optional; *see* **Note 1**), all separated by comma, and click on "Submit."
6. *Option C*: Multiple fragments can be added at once by using the "Load File" button in the lower left (Fig. 2a). In the popup window, you will see your file system. Navigate to a FASTA file containing the fragments that should be added. The header of each fragment sequence must contain the fragment name, fragment type and GenBank path (optional; *see* **Note 1**), all separated by comma. Once a FASTA file is selected, click on the "Load" button to add the fragments to the database. This option may also be used to upload output files from the SEEDling tool (*see* **Note 2**).

3.1.3 G-GArden Parameter Settings

1. The parameter settings are found in the lower middle of the G-GArden GUI (Fig. 2a). Default settings are displayed and can be adjusted according to the user's specifications.
2. Widget "Restriction Enzyme": Click on the widget to open the dropdown menu and select the type IIS restriction enzyme that should be used for the cloning procedure.
3. "Overhang Length": The overhang length will be displayed automatically depending on the selected type IIS restriction enzyme.
4. Widget "Maximum Annealing Length": The user can specify the maximum fragment length (including overhangs) that is allowed to be created by annealing of oligodeoxynucleotides. If the fragment length exceeds the specified length, G-GArden will design primers to create the fragment via PCR.
5. Widget "PCR Primer Length": The user can specify the length of the PCR primers for amplification of larger fragments. If Primer3 is applied for primer search (*see* **Note 3**), this value is allowed to deviate by 5 bases from the specified length value.

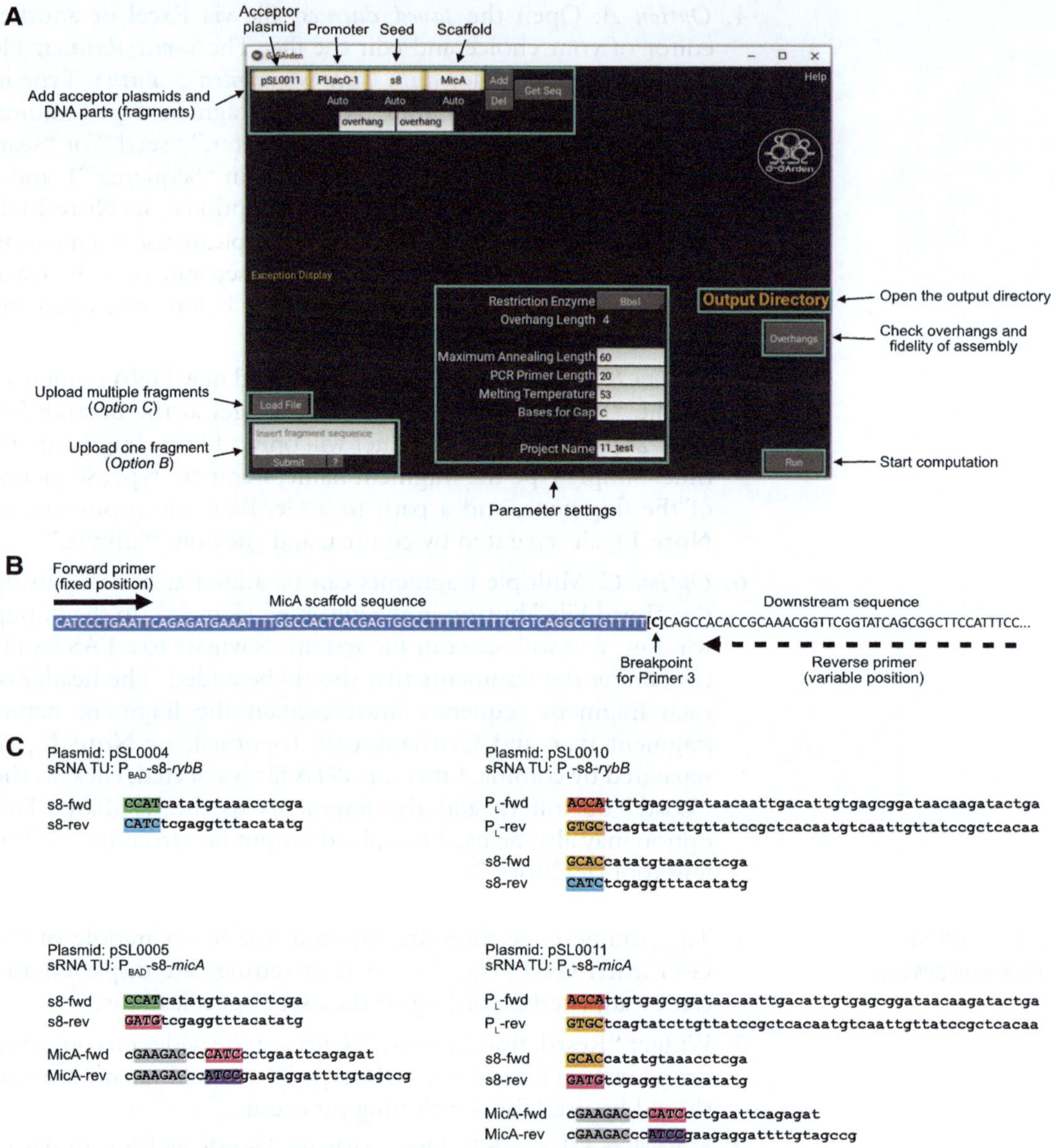

Fig. 2 Design of oligodeoxynucleotides for Golden Gate assembly using G-Garden. (**a**) Shown is a screenshot of the G-GArden graphical user interface (GUI). The most important buttons/widgets are highlighted (see text for detailed explanations). (**b**) Scaffold sequences in the database file (*saved_data.csv*) must contain an additional downstream sequence, if Primer3 should be applied for the design of PCR primers. The MicA scaffold (highlighted in blue) is shown as an example. The "C" in square brackets indicates the breakpoint. Primer3 uses the downstream sequence to search for a reverse primer (variable position) that matches the forward primer (fixed position). (**c**) Exemplary G-GArden output of oligodeoxynucleotides/primers used for construction of the indicated sRNA transcription units (TUs). The 4-nt overhangs used for Golden Gate cloning are highlighted. Matching overhangs are indicated by the same color (compare to Fig. 1b). BbsI recognition sites (highlighted in grey) are present at the 5′-end of PCR primers for generation of MicA fragments with suitable overhangs. P_L: $P_L$lacO-1 promoter, s8: seed region targeted against the *acrA* gene in *E. coli*, fwd: forward, rev: reverse

6. Widget "Melting Temperature": If Primer3 is applied for primer search (*see* **Note 3**), the user can specify the melting temperature of the PCR primers. Primer3 is allowed to deviate by 5 °C from the specified value.
7. Widget "Bases for Gap": The user can specify the bases (A, C, G, T) that are added at the 5′-end of the PCR primers in front of the recognition site for the type IIS restriction enzyme. A subset of these bases will be used to fill the gap between the recognition site and the cut site.
8. Widget "Project Name": Choose a unique name that is used for all output files and the directory of the project.

3.1.4 Running a Cloning Project

1. The "Add" button in the upper left is used to add widgets for selection of acceptor plasmids and DNA parts (fragments). The corresponding fragments are retrieved from the database file (*saved_data.csv*). Use the "Del" button to remove previously added widgets.
2. Click on the first widget to select an acceptor plasmid from the dropdown menu (Fig. 2a).
3. Click on the subsequently added widgets to select promoter, seed, and scaffold fragments (Fig. 2a; *see* **Note 4**).
4. Optional: Click on the "Get Seq" button to access the sequences of all selected fragments. A new window will open, in which you can copy individual sequences.
5. Optional: For promoter, seed, and scaffold fragments, a second widget is available for specification of the fragment creation mode ("PCR" or "Anneal"). The default option "Auto" will determine the fragment creation mode based on the "Maximum Annealing Length" parameter. If "PCR" is selected, Primer3 may be applied for primer search (Fig. 2b; *see* **Note 3**).
6. Optional: As soon as fragments are selected, a third widget will appear ("overhang"; text input field) that allows the user to customize overhangs for fusion of two fragments (*see* **Note 5**).
7. Click on the "Overhangs" button in the middle right to check whether any problems are reported regarding overhangs (Fig. 2a). A new window will open, in which the overall fidelity of the assembly reaction is displayed (*see* **Note 6**). The quality of individual overhangs is indicated by a color code. Green indicates that the overhang can be used without further refinement (plasmid overhangs are green by default). Yellow indicates low-fidelity overhangs, which can be optimized if desired. Red indicates duplicate overhangs, an issue that must be fixed (*see* **Note 6**).
8. Click on the "Run" button in the bottom right to start computation of oligodeoxynucleotides/primers. If the cloning strategy is feasible, the "Output Directory" button will appear

in the middle right, which can be used to open the output directory and navigate to the folder of the cloning project (as specified in the "Project Name" widget; Fig. 2a).

3.1.5 G-GArden Output Files

1. In the folder of the cloning project, several output files are available: *x_gb*, *x_seq*, *x_vis*, *x_coll*, and *x_fin_coll* (*x* indicates the project name).
2. Plasmid maps for the final constructs (*x_gb*; *.gbk* files) can be used for visualization and documentation of the cloning strategy.
3. Text files contain the oligodeoxynucleotide/primer pairs needed for each fragment of a construct (*x_seq*; *.txt* files).
4. Overhangs for Golden Gate assembly are visualized in text files (*x_vis*; *.txt* files). These files should be used to verify the overhangs and fusion sites of the cloning strategy.
5. One table summarizes all oligodeoxynucleotides/primers needed for cloning (*x_coll*; *.csv* file).
6. One table summarizes all non-redundant oligodeoxynucleotides/primers needed for cloning (*x_fin_coll*; *.csv* file). This file should be used for ordering of oligodeoxynucleotides/primers.

3.1.6 Survey of G-GArden

1. The downloaded *saved_data.csv* file already contains several sequences that are used in this chapter: three acceptor plasmids (pSL0004, pSL0010, and pSL0011), one promoter ($P_L$lacO-1), two seed regions (s8 and s28; targeted against the *acrA* mRNA in *E. coli*), and two scaffolds (RybB and MicA from *E. coli*). These sequences can be used to explore the performance of G-GArden before individual cloning projects are initiated.
2. Follow the steps described above to generate constructs P_{BAD}-s8-*rybB*, P_L-s8-*rybB*, and P_L-s8-*micA* using pSL0004, pSL0010, and pSL0011, respectively.
3. Compare the output to the pre-computed oligodeoxynucleotides/primers used for this chapter (Fig. 2c).

3.2 Golden Gate Assembly and Validation of Expression Constructs

3.2.1 Phosphorylation of Oligodeoxynucleotides

1. Phosphorylation may be required for oligodeoxynucleotides before the annealing step is performed (*see* **Note 7**). 5′-ends of oligodeoxynucleotides are phosphorylated using T4 Polynucleotide Kinase (PNK). The T4 DNA Ligase Reaction Buffer contains ATP, which is needed for phosphorylation (*see* **Note 8**).
2. Prepare the following 20-μL reaction for each oligodeoxynucleotide pair:

Reagent	Volume
Sterile water	15 μL
Forward oligodeoxynucleotide (100 μM)	1 μL
Reverse oligodeoxynucleotide (100 μM)	1 μL
10× T4 DNA Ligase Reaction Buffer	2 μL
T4 Polynucleotide Kinase (10,000 units/mL)	1 μL

3. Incubation should be performed in a PCR thermocycler with the following settings: 37 °C for 1 h, 65 °C for 20 min (inactivation of T4 Polynucleotide Kinase), storage at 8 °C.
4. The phosphorylated oligodeoxynucleotides can be stored at −20 °C until use.

3.2.2 Annealing of Oligodeoxynucleotides

1. Add 2.5 μL of 10× annealing buffer to the 20-μL phosphorylation reactions from Subheading 3.2.1.
2. Add 2.5 μL sterile water to reach a final annealing buffer concentration of 1× (*see* **Note 9**).
3. Incubation should be performed in a PCR thermocycler with the following settings: 95 °C for 2 min, decrement to 25 °C over a period of 45 min (i.e., a drop of 0.026 °C per second; *see* **Note 10**), storage at 8 °C.
4. The annealed oligodeoxynucleotides can be stored at −20 °C until use.

3.2.3 Amplification of Larger Fragments via PCR

1. If fragments (e.g., scaffolds) exceed a length of ~60 bp, generation via PCR is recommended (*see* **Note 3**).
2. Use a proof-reading polymerase and appropriate DNA (chromosomal or plasmid) as template for amplification of the respective fragments according to standard PCR protocols.
3. Purify the amplified fragments before proceeding to Golden Gate assembly.

3.2.4 Golden Gate Assembly

1. For Golden Gate assembly reactions, 40 fmol of the acceptor plasmid is used (*see* **Note 11**).
2. All other DNA parts (fragments) must be equimolar to the acceptor plasmid (*see* **Note 12**).
3. After phosphorylation and annealing, the concentration of DNA parts (fragments) is ~250 ng/μL. Since only small amounts are needed for the assembly reaction (0.3–2.5 ng), fragments are typically diluted to a concentration of ~1 ng/μL (*see* **Note 13**).

4. Prepare a 15-μL Golden Gate assembly reaction according to the following scheme (example for pSL0011):

Reagent	Volume
Sterile water	Ad 15 μL
Acceptor plasmid pSL0011	~40 fmol (*see* **Note 12** and **13**)
Promoter fragment	~40 fmol (*see* **Note 12** and **13**)
Seed fragment	~40 fmol (*see* **Note 12** and **13**)
Scaffold fragment	~40 fmol (*see* **Note 12** and **13**)
10× T4 DNA Ligase Reaction Buffer	1.5 μL
BbsI (20,000 units/mL)	1 μL
T4 DNA Ligase (400,000 units/mL)	1 μL

5. Incubation should be performed in a PCR thermocycler with the following settings: 16 °C for 4 min and 37 °C for 3 min (repeat these two steps 50 times), 50 °C for 10 min (inactivation of T4 DNA Ligase), 80° for 10 min (inactivation of BbsI), storage at 8 °C.
6. Use directly for transformation.

3.2.5 Transformation into Recipient Strain

1. Prepare electrocompetent cells of the recipient strain (here, *E. coli* MG1655) according to standard protocols (*see* **Note 14**).
2. Add 5 μL of the Golden Gate assembly reaction from Subheading 3.2.4 to 50 μL ice-cold electrocompetent cells and incubate for 20 min on ice.
3. Use an electroporator to transform the assembled expression plasmids. We recommend using 0.2-cm electroporation cuvettes and applying an electrical pulse of 2.5 kV for 5 ms.
4. Add 1 mL LB medium and incubate for 1 h at 37 °C.
5. Plate 100 μL of the cell suspension on an LB agar plate with the appropriate antibiotic (here, 50 μg/mL kanamycin).
6. Spin down the remaining cell suspension (10,000× *g*, 1 min), discard most of the supernatant, resuspend the pellet in the remaining volume (~100 μL), and plate on a LB agar plate with the appropriate antibiotic (here, 50 μg/mL kanamycin).
7. Incubate the agar plates overnight at 37 °C to receive colonies.

3.2.6 Validation of Expression Constructs

1. Inspect the agar plates from Subheading 3.2.5. White colonies potentially contain the correct construct and should be considered for validation. Red colonies should be disregarded because they express *mCherry* and are likely false positives.

2. Pick and transfer individual white colonies to 5 mL LB medium in 50-mL Erlenmeyer flasks with the appropriate antibiotic (here, 50 μg/mL kanamycin). In parallel, the same colonies should be re-streaked on a fresh LB agar plate with the appropriate antibiotic (here, 50 μg/mL kanamycin).
3. Liquid cultures are incubated overnight at 37 °C and shaken at 180 rpm. Use overnight cultures to isolate the plasmid DNA according to a protocol of your choice.
4. Measure the plasmid concentration and prepare samples for Sanger sequencing according to the recommendations of your sequencing service. For pSL0004 and pSL0005, primer Mult-targ-scr-rev, and for pSL0010 and pSL0011, primer Mult-targ-scr-fw is used for Sanger sequencing.
5. Use re-streaks for cryopreservation of clones that contain the correct constructs.

3.3 Expression Analysis of Synthetic sRNAs

3.3.1 Cultivation of Strains and Collection of Samples for RNA Isolation

1. The required strains (*see* **Note 15**) are streaked on LB agar plates with the appropriate antibiotic (here, 50 μg/mL kanamycin) and incubated overnight at 37 °C.
2. Prepare 5 mL LB medium in 50-mL Erlenmeyer flasks with the appropriate antibiotic (here, 50 μg/mL kanamycin) and inoculate single colonies from the step before.
3. Incubate liquid cultures overnight at 37 °C and shaking at 180 rpm.
4. Over-night cultures are diluted 1:100 in 10 mL LB medium in 100-mL Erlenmeyer flasks without antibiotic.
5. Incubate cultures at 37 °C and shake at 180 rpm until an OD_{600} of ~0.4 is reached (exponential phase).
6. Harvest 3 mL cell suspension from each culture and mix immediately with 600 μL stop solution on ice. Spin down the samples in a cold centrifuge at maximum speed for 2 min, remove supernatants, and store cell pellets at −20 °C for RNA isolation.
7. The remaining cultures are treated with 0.2% L-arabinose to induce sRNA expression from the P_{BAD} promoter.
8. After 30 min, harvest 3 mL cell suspension from each culture and collect cell pellets as before.

3.3.2 Isolation of Total RNA

1. Thaw cell pellets on ice.
2. Preheat water-saturated phenol to 65 °C (*see* **Note 16**).
3. Resuspend cell pellets in 200 μL lysis buffer and keep at room temperature.
4. Incubate samples at 65 °C for 5 min in a heating block or water bath.

5. Add 200 μL preheated, water-saturated phenol and mix thoroughly by vortexing (*see* **Note 16**).
6. Incubate samples at 65 °C for 3 min in a heating block or water bath.
7. Snap-freeze samples in liquid nitrogen.
8. Spin down samples in a microcentrifuge at 13,000 rpm for 10 min.
9. Move 160 μL of the aqueous phase (upper phase) to a new reaction tube.
10. Add 160 μL chloroform-isoamyl alcohol (24:1) and mix thoroughly by vortexing (*see* **Note 16**).
11. Spin down samples in a microcentrifuge at 13,000 rpm for 3 min.
12. Move 120 μL of the aqueous phase (upper phase) to a new reaction tube.
13. Add 120 μL isopropyl alcohol and 12 μL sodium acetate (3 M, pH 4.5) and mix thoroughly by vortexing.
14. Keep samples on ice for 15 min to allow precipitation of RNA.
15. Spin down samples in a microcentrifuge at maximum speed for at least 20 min and remove supernatants.
16. Add 200 μL ethanol (75%).
17. Spin down samples in a microcentrifuge at maximum speed for 10 min and remove supernatants.
18. Dry RNA pellets in a vacuum concentrator.
19. Resuspend in 30 μL RNase-free water.
20. Measure RNA concentrations (e.g., by using a spectrophotometer at 260 nm).
21. For Northern blot analysis, up to 5 μg of total RNA should be used.

3.3.3 Polyacrylamide Gel Electrophoresis

1. Add 4 mL 10× TBE (final concentration 1×) and 10 mL acrylamide/bis-acrylamide (final concentration 10%) to 16.8 g urea (final concentration 7 M). Use distilled water to increase the volume to 40 mL.
2. Stir at room temperature until the urea has completely dissolved (~15 min).
3. Assemble glass plates (20 × 20 cm) and spacers (thickness 1 mm). Silanization may be applied for one glass plate.
4. Add 120 μL APS (10%) and 20 μL TEMED to the gel mixture and pour the mixture between the glass plates. Apply a comb and allow the gel to polymerize for 1 h.

5. Mix the required volume of a total RNA isolate (containing up to 5 μg of total RNA) with the 0.7-fold volume of FU mix on ice (e.g., add 7 μL FU mix to 10 μL RNA isolate).
6. After incubation at 65 °C for 10 min, immediately place the samples on ice.
7. After polymerization of the gel, carefully remove the comb and place the assembly in a gel apparatus for vertical electrophoresis. Fill the electrophoresis chamber with 1× TBE.
8. Thoroughly rinse the wells of the gel with a syringe to remove any remaining urea.
9. Load the samples and start the gel run (100 V for 10 min followed by 300 V for ~2 h and 30 min, until the bromophenol blue has run out).

3.3.4 Semi-dry Electroblotting

1. Use a blotting apparatus for semi-dry electroblotting and 1× TBE as blotting buffer.
2. Prepare six layers of blotting paper according to the size of the gel area that should be blotted. Prepare a nylon membrane with the same size.
3. If the bottom plate of the blotting apparatus represents the anode, prepare a sandwich in the following order: three blotting papers—nylon membrane—gel—three blotting papers. The blotting papers and nylon membrane are soaked in 1× TBE.
4. Assemble the blotting apparatus and start the blotting (250 mA for 2–3 h).
5. Recommended: Apply UV light to cross-link the RNA to the nylon membrane.
6. The membrane can be wrapped in a dry blotting paper and stored at room temperature.

3.3.5 Hybridization to Radioactively Labeled Probes

1. Transfer the nylon membrane to a glass tube and add 20 mL Church hybridization buffer.
2. Place the glass tube in a rotating oven and pre-hybridize for 1 h at 42 °C.
3. Add a radioactively labeled probe for overnight hybridization at 42 °C (*see* **Note 17**).
4. Wash the membrane with wash buffer until the unspecific background signal is removed. It is usually sufficient to wash twice for 30 s at room temperature.
5. Seal the nylon membrane in a polypropylene bag.

3.3.6 Phosphorimaging

1. Expose the sealed nylon membrane to a phosphorimaging screen. For sRNAs, an exposure time of one or 2 days is recommended. For 5S rRNA, an exposure time of ~4 h is usually sufficient.
2. Use an appropriate imaging system and software for data acquisition.

3.3.7 Reprobing of Membranes

1. After data acquisition, transfer the membrane to a glass bowl and remove the radioactive probe by incubation in stripping buffer for 20 min at 96 °C and gentle shaking.
2. The membrane can be used for another hybridization procedure as described above (*see* **Note 18**).

3.4 Benchmarking of the Described Procedures

Here, we used G-GArden and Golden Gate assembly for construction of the following plasmids: p-P_{BAD}-s8-*rybB* (inducible expression of s8-RybB), p-P_{BAD}-s8-*micA* (inducible expression of s8-MicA), p-P_L-s8-*rybB* (constitutive expression of s8-RybB), and p-P_L-s8-*micA* (constitutive expression of s8-MicA). Northern blot analysis showed that inducible expression from the P_{BAD} promoter results in lower steady-state sRNA levels as compared to expression from the strong, constitutive $P_L$lacO-1 promoter (Fig. 3a). Synthetic RybB and MicA sRNAs containing the s8 seed region (Fig. 3b) can be used to repress translation of the *acrA* gene in *E. coli* to increase oxacillin sensitivity [7].

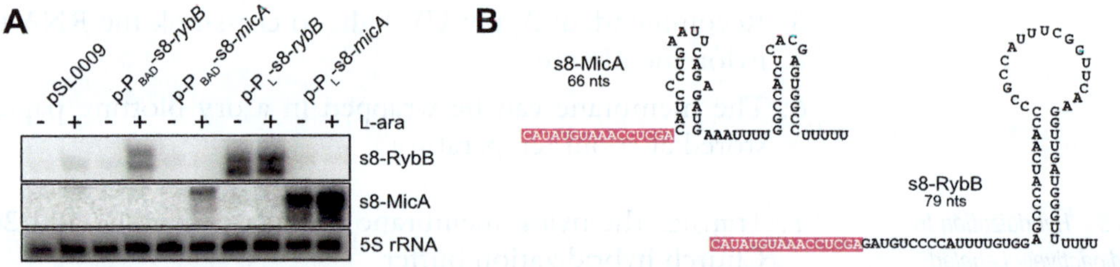

Fig. 3 Expression analysis of synthetic RybB and MicA sRNAs. (**a**) Northern blot results for inducible (P_{BAD} promoter) and constitutive expression (P_L: $P_L$lacO-1 promoter) of synthetic RybB and MicA sRNAs containing the s8 seed region. Exponential-phase cultures containing the indicated plasmids were either treated with L-arabinose (L-ara) for 30 min (+) or left untreated (−). pSL0009 is an empty control plasmid. Total RNA (5 μg) was isolated and separated on 10% polyacrylamide gels containing 7 M urea. After transfer to a nylon membrane by semi-dry electroblotting, RybB and MicA sRNAs were hybridized to specific, radioactively labeled probes. 5S rRNA was probed as loading control. (**b**) Secondary structures of synthetic RybB and MicA sRNAs containing the s8 seed region. The s8 seed region is highlighted in pink

4 Notes

1. It is recommended that GenBank files (*.gbk*) for all acceptor plasmids be added to the database (*saved_data.csv*). If these files are provided, G-GArden will create GenBank files for the resulting cloning constructs as part of the output, which can then be used for documentation/visualization purposes. Examples of GenBank files for acceptor plasmids can be found in folder *data* (*dist* > *ggarden* > *data*) alongside with the *saved_data.csv* file. Examples of output GenBank files can be found in folder *example* (*dist* > *ggarden* > *data* > *output* > *example*).
2. SEEDling is a tool that predicts optimal seed regions for the construction of synthetic sRNAs and regulation of specified target mRNAs. SEEDling can be found alongside the G-GArden tool within the DIGGER-Bac project (https://github.com/DIGGER-Bac/SEEDling). We would like to refer the reader to a detailed documentation on GitHub and a recent publication for further information [12].
3. Larger fragments (> 60 bp) should not be created by annealing of oligodeoxynucleotides and may be ordered as synthesized DNA. Alternatively, isolated DNA (plasmid or chromosomal) can be used as template to amplify the desired fragment via PCR. Since seed region fragments and many promoters (e.g., $P_L$lacO-1) are usually smaller than 60 bp, PCR is mainly used for scaffold fragments. If Primer3 should be applied for primer design, an additional downstream sequence must be added to the corresponding scaffold sequence in the database file (*saved_data.csv*). Primer3 will search for optimal pairs of PCR primers using this additional sequence space. In order to activate Primer3, a square bracket must be present to mark the breakpoint of the fragment sequence and the additional downstream sequence (an example for the MicA scaffold is illustrated in Fig. 2b).
4. For pSL0004, only a seed sequence needs to be selected. For pSL0005, seed and scaffold sequences are selected. For pSL0010, promoter and seed sequences are selected. For pSL0011, promoter, seed, and scaffold sequences are selected. In each widget, multi-selection is also possible. If multiple sequences are selected in more than one widget, G-GArden will compute all possible combinations. The resulting output files are indexed by consecutive numbers.
5. The "overhang" widget only appears between fragments that are allowed to be fused by customized overhangs. Note that overhangs of acceptor plasmids cannot be customized. Please be aware that the G-GArden tool avoids the creation of scars in

the final constructs, but specification of custom overhangs within the "overhang" widget will create scars. Hence, customized overhangs should only be used if problems with a scarless strategy are reported.

6. The overall fidelity depends on parameters that can be selected from a dropdown menu at the top (T4 ligase reaction for 1 or 18 h at 25 or 37 °C). The fidelity computation is based on experimental data [13, 14]. Three modes are available for optimization of overhangs, which can be selected by using the slider at the top. In the left slider position (grey), all overhangs are optimized. In the middle slider position (red), only duplicate overhangs are replaced. In the right slider position (red/yellow), both duplicate overhangs and overhangs with low fidelity are optimized. After mode selection, click on the text line to store the problematic overhangs in the clipboard (a successful copy is indicated by random color change). Use the "Optimize" button to access the NEBridge GetSet™ Tool (https://ggtools.neb.com/getset/run.cgi). After selecting the "Overhang length," the "Ligation conditions," and the total "Number of overhangs," paste the copied overhangs from G-GArden into the "Required overhangs" field and press "Submit." The NEBridge GetSet™ Tool will suggest optimized overhangs, which can then be copy-pasted in the corresponding "overhang" widgets in the main window of the G-GArden GUI (please be aware that these overhangs will be present as scars in the final constructs). Finally, click on the "Overhangs" button to check whether the problem has been solved.

7. Oligodeoxynucleotides are usually ordered without 5′-monophosphates. Hence, the resulting fragments do not contain 5′-monophosphates after the annealing step. If two or more annealed fragments are used for Golden Gate assembly, the phosphorylation step is mandatory to enable ligation of these fragments. However, acceptor plasmids (and PCR fragments) have 5′-monophosphates after BbsI treatment. Hence, the phosphorylation step can be skipped if only one annealed fragment is used for Golden Gate assembly.

8. If the 10× T4 DNA Ligase Reaction Buffer is fresh, it has a sulfurous smell. If you cannot sense this sulfurous smell, you may prefer to not use the buffer. We recommend to aliquot the 10× T4 DNA Ligase Reaction Buffer in multiple tubes to reduce freeze-thaw cycles.

9. If phosphorylation of oligodeoxynucleotides is not performed, the following annealing reaction is suggested: 4.5 μL forward oligodeoxynucleotide (100 μM), 4.5 μL reverse oligodeoxynucleotide (100 μM), and 1 μL annealing buffer (10×).

10. Depending on the precision of the PCR thermocycler, a decrement of 0.026 °C per second might not be feasible. Alternatively, a decrement of 0.1 °C per second may be applied (~12 min are needed in this case), which performed well in our hands.

11. Acceptor plasmids containing the *mCherry-ccdB* counter-selection cassette can only be propagated in *ccdB*-resistant strains, such as *E. coli* DB3.1 harboring the *gyrA462* allele. Grow the respective strains overnight to reach high cell densities and isolate the plasmid DNA according to a protocol of your choice.

12. We recommend mole-to-weight calculator that is freely available online (e.g., https://www.bioline.com/media/calculator/01_07.html). As an example for calculation: the acceptor plasmid pSL0004 has a size of 5736 kb. For 40 fmol, ~149 ng is needed. For 40 fmol of a 20-bp seed fragment, only ~0.5 ng is needed. However, to improve the assembly efficiency, the amount of small DNA parts (especially seed fragments) should be increased to 80 fmol.

13. It is recommended to prepare the dilution of DNA parts (fragments) immediately before the Golden Gate assembly reaction is performed. To reach a concentration of ~1 ng/μL, a serial dilution might be preferable. Note that for larger DNA parts (fragments) dilution may not be necessary at all.

14. Chemically competent cells may be used as well. However, we had a better experience with electrocompetent cells.

15. Here, we used *E. coli* K-12 wild-type MG1655 containing synthetic RybB and MicA expression plasmids and isolated the total RNA during exponential phase. The native *rybB* and *micA* genes, which are still present in the MG1655 chromosome, are only weakly expressed during exponential phase. Hence, negligible interference occurs when synthetic RybB and MicA sRNAs are analyzed on Northern blots. However, it might be necessary to transform expression plasmids to strains, in which the native sRNA genes have been deleted.

16. Phenol and chloroform are toxic compounds. Please adhere to common safety precautions, such as the use of a fume hood, safety glasses, and gloves. The corresponding waste should be disposed according to local regulations.

17. To generate radioactively labeled probes, oligodeoxynucleotides, which are complementary to the RNA species of interest, are phosphorylated at their 5′-ends using T4 Polynucleotide Kinase (10,000 units/mL) and [γ-^{32}P]-ATP according to standard protocols. RybB-probe-2 and MicA-probe-2 are used for detection of synthetic RybB and MicA sRNAs, respectively. Both probes are complementary to the respective sRNA

scaffold sequence; hence, synthetic sRNAs with variable seed region can be detected with the very same probe. 5S-probe-2 is used for detection of 5S rRNA, which serves as loading control.

18. Membranes can be hybridized at least three times. We recommend starting with the probe that gives the weakest signals. In our case, the RybB-probe-2 is used first, followed by the MicA-probe-2. The 5S-probe-2 is used at last.

Acknowledgments

We would like to acknowledge Tania S. Köbel and Daniel Schindler for construction of the pSL plasmid series. We are grateful to Niklas Philipp for developing the G-GArden tool. This work was supported by the Boehringer Ingelheim Foundation (Exploration Grant to B.A.B.).

References

1. Waters LS, Storz G (2009) Regulatory RNAs in bacteria. Cell 136:615–628
2. Wagner EGH, Romby P (2015) Small RNAs in bacteria and archaea: who they are, what they do, and how they do it. Adv Genet 90:133–208
3. Storz G, Vogel J, Wassarman KM (2011) Regulation by small RNAs in bacteria: expanding frontiers. Mol Cell 43:880–891
4. Papenfort K, Bouvier M, Mika F, Sharma CM, Vogel J (2010) Evidence for an autonomous 5′ target recognition domain in an Hfq-associated small RNA. Proc Natl Acad Sci 107:20435–20440
5. Schu DJ, Zhang A, Gottesman S, Storz G (2015) Alternative Hfq-sRNA interaction modes dictate alternative mRNA recognition. EMBO J 34:2557–2573
6. Vogel J, Luisi BF (2011) Hfq and its constellation of RNA. Nat Rev Microbiol 9:578–589
7. Köbel TS, Melo Palhares R, Fromm C, Szymanski W, Angelidou G, Glatter T, Georg J, Berghoff BA, Schindler D (2022) An easy-to-use plasmid toolset for efficient generation and benchmarking of synthetic small RNAs in bacteria. ACS Synth Biol 11:2989–3003
8. Sharma V, Yamamura A, Yokobayashi Y (2012) Engineering artificial small RNAs for conditional gene silencing in *Escherichia coli*. ACS Synth Biol 1:6–13
9. Na D, Yoo SM, Chung H, Park H, Park JH, Lee SY (2013) Metabolic engineering of *Escherichia coli* using synthetic small regulatory RNAs. Nat Biotechnol 31:170–174
10. Yoo SM, Na D, Lee SY (2013) Design and use of synthetic regulatory small RNAs to control gene expression in *Escherichia coli*. Nat Protoc 8:1694–1707
11. Hoynes-O'Connor A, Moon TS (2016) Development of design rules for reliable antisense RNA behavior in *E. coli*. ACS Synth Biol 5: 1441–1454
12. Philipp N, Brinkmann CK, Georg J, Schindler D, Berghoff BA (2023) DIGGER-Bac: prediction of seed regions for high-fidelity construction of synthetic small RNAs in bacteria. Bioinformatics 39:btad285
13. Potapov V, Ong JL, Langhorst BW, Bilotti K, Cahoon D, Canton B, Knight TF, Evans TC, Lohman GJS (2018) A single-molecule sequencing assay for the comprehensive profiling of T4 DNA ligase fidelity and bias during DNA end-joining. Nucleic Acids Res 46:e79
14. Potapov V, Ong JL, Kucera RB, Langhorst BW, Bilotti K, Pryor JM, Cantor EJ, Canton B, Knight TF, Evans TC, Lohman GJS (2018) Comprehensive profiling of four base overhang ligation Fidelity by T4 DNA ligase and application to DNA assembly. ACS Synth Biol 7:2665–2674

Chapter 18

Utilizing Golden Gate Assembly to Streamline CRISPR-Cas/NgTET-Based Phage Mutagenesis

Nadiia Pozhydaieva and Katharina Höfer

Abstract

Phage engineering is an emerging technology due to the promising potential application of phages in medical and biotechnological settings. Targeted phage mutagenesis tools are required to customize the phages for a specific application and generate, in addition to that, so-called designer phages. CRISPR-Cas technique is used in various organisms to perform targeted mutagenesis. Yet, its efficacy is notably limited for phage mutagenesis due to the highly abundant phage DNA modifications. Addressing this challenge, we have developed a novel approach that involves the temporal removal of phage DNA cytosine modifications, allowing for effective CRISPR-Cas targeting and subsequent introduction of mutations into the phage genome. The removal of cytosine modification relies on the catalytic activity of a eukaryotic ten-eleven translocation methylcytosine (TET) dioxygenase. TET enzymes iteratively de-modify methylated or hydroxymethylated cytosines on phage DNA. The temporal removal of cytosine modification ultimately enables efficient DNA cleavage by Cas enzymes and facilitates mutagenesis. To streamline the application of the coupled TET-CRISPR-Cas system, we use Golden Gate cloning for fast and efficient assembly of a vector that comprises a TET oxidase and a donor DNA required for scarless site-specific phage mutagenesis. Our approach significantly advances the engineering of modified phage genomes, enabling the efficient generation of customized phages for specific applications.

Key words Phage engineering, Designer phage, DNA modifications, CRISPR-Cas-mediated mutagenesis, Cytosine modifications, Glycosylation, TET dioxygenase, Golden Gate assembly

1 Introduction

Phage engineering has emerged as a technology of increasing interest due to its constantly growing relevance in medical and biotechnological applications [1–3]. Prior to targeted phage application, a comprehensive understanding of the phage infectious cycle is required to tailor their properties to specific needs. Targeted and efficient phage mutagenesis is key in this context, allowing not only the investigation of the biological role of specific phage proteins [4] but also the precise adjustment of phage properties for desired applications [5, 6].

Daniel Schindler (ed.), *Golden Gate Cloning: Methods and Protocols*, Methods in Molecular Biology, vol. 2850,
https://doi.org/10.1007/978-1-0716-4220-7_18,

The CRISPR-Cas (clustered regularly interspaced short palindromic repeats and CRISPR-associated protein) system-based techniques have revolutionized the genome editing field and have been successfully applied for engineering various organisms [7]. Recently, the application was extended to phage engineering [8, 9]. CRISPR-Cas-based phage engineering relies on targeting a specific position within the phage genome by a Cas nuclease during the infection of the host. The targeting results in a DNA double-strand break, preventing phage replication. Subsequently, the double-strand break can be repaired through homologous recombination with donor DNA, which is harbored within the mutagenesis strain as a plasmid DNA. Nevertheless, the first step of the mutagenesis—the CRISPR-Cas-induced DNA double-strand break—is frequently hindered by phage DNA modifications, which are highly abundant among phage genomes and serve as protection toward host nucleases, including Cas nucleases (Fig. 1) [10, 11]. The negative impact of DNA modifications on CRISPR-Cas-mediated mutagenesis was demonstrated in several studies with model phage T4 [8–10], resulting in a low mutagenesis efficiency.

Phage T4 encodes three proteins being responsible for extensive DNA modification: deoxycytidylate 5-hydroxymethyltransferase (protein 42 responsible for hydroxymethylation on cytosine and formation of 5-hydroxymethyl-2′-deoxycytidine, 5hmdC), and α- or β-glycosyltransferases (α−/β-gt, responsible for formation of 5-α−/β-glycosyl hydroxymethyl-2′-deoxycytidine, 5ghmdC) [11, 12]. Modification-deficient phage T4 strains, where genes encoding DNA-modifying proteins are deleted, have been previously utilized to enhance mutagenesis efficiency.

In the genome of $\Delta\alpha-/\beta$-*gt* phage T4 strain, all cytosines are hydroxymethylated, lacking glycosylation, which makes phage DNA more susceptible to CRISPR-Cas targeting [8, 9]. However, the permanent lack of phage DNA modifications presents advantages and disadvantages. While the modification-deficient strains are beneficial for targeted phage mutagenesis, they also adversely affect the phage phenotype, as DNA modifications are crucial for phage DNA stability during the infection [12, 13]. In addition to studying the phenotype linked to the desired mutation, the effect of missing DNA modifications might also be studied.

Recently, we introduced the application of eukaryotic TET dioxygenases to modulate phage DNA modification levels temporarily [14]. This technology increases the susceptibility of phage DNA to Cas nuclease cleavage. Our CRISPR-Cas-mediated TET-based approach eliminates the need to apply modification-deficient strains for efficient phage mutagenesis (Fig. 2). To increase the efficiency of mutagenesis on phage DNA, we temporarily reduce the abundance of cytosine modifications in phage DNA by expressing the eukaryotic TET dioxygenase in the

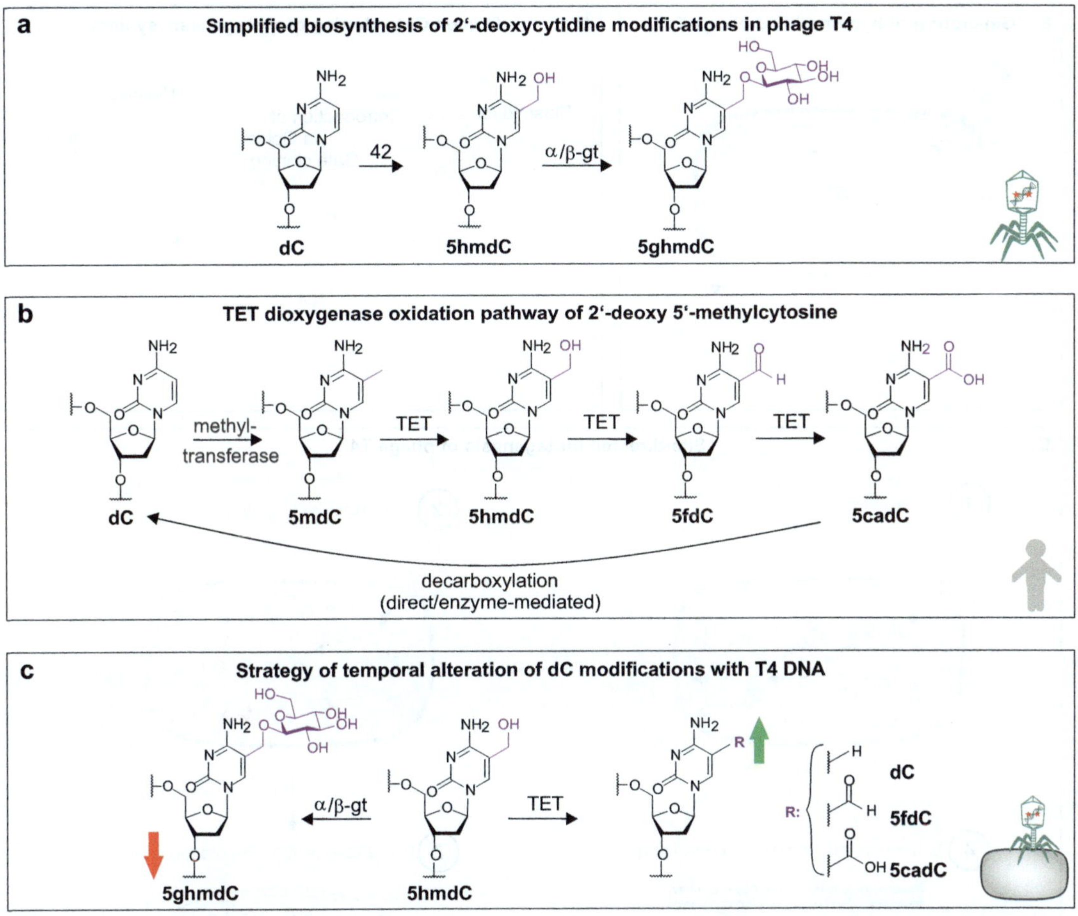

Fig. 1 In phage T4 DNA, cytidines (dC) undergo extensive modifications: (**a**) Phage T4 protein 42 modifies dC to 5-hydroxymethyl-2′-deoxycytidine (5hmdC), while α–/β-gt enzymes further convert it to 5-α–/β-glycosyl hydroxymethyl-2′-deoxycytidine (5ghmdC). (**b**) Eukaryotic TET dioxygenases oxidize 5-methyl-2′--deoxycytidine (5mdC) through successive stages to 5hmdC, 5-formyl-2′-deoxycytidine (5fdC), and eventually to 5-carboxyl-2′-deoxycytidine (5cadC). 5cadC reverts spontaneously or enzymatically to dC. (**c**) When eukaryotic TET dioxygenases are expressed in bacteria during phage infection, they may oxidize phage-derived 5hmdC. This oxidation leads to the demodification of cytidine, preventing the addition of glycosyl residues

bacterial host during phage infection and the mutagenesis process. TET dioxygenases are eukaryotic epigenetic regulators and iteratively oxidize 5-methyl-2′-deoxycytidine (5mdC) to 5-hydroxymethyl-2′-deoxycytidine (5hmdC), 5-formyl-2′-deoxycytidine (5fdC), and 5-carboxy-2′-deoxycytidine (5cadC). 5cadC can spontaneously or enzymatically decarboxylate to unmodified 2′--deoxycytidine (dC) [15, 16]. We express TET dioxygenase from *Naegleria gruberi* (NgTET) in *Escherichia coli* during phage T4 infection to oxidize 5hmdC, a product of the phage T4 protein 42's activity [14]. The oxidation to 5fdC and 5cadC and subsequent spontaneous decarboxylation to dC occur on T4 DNA and

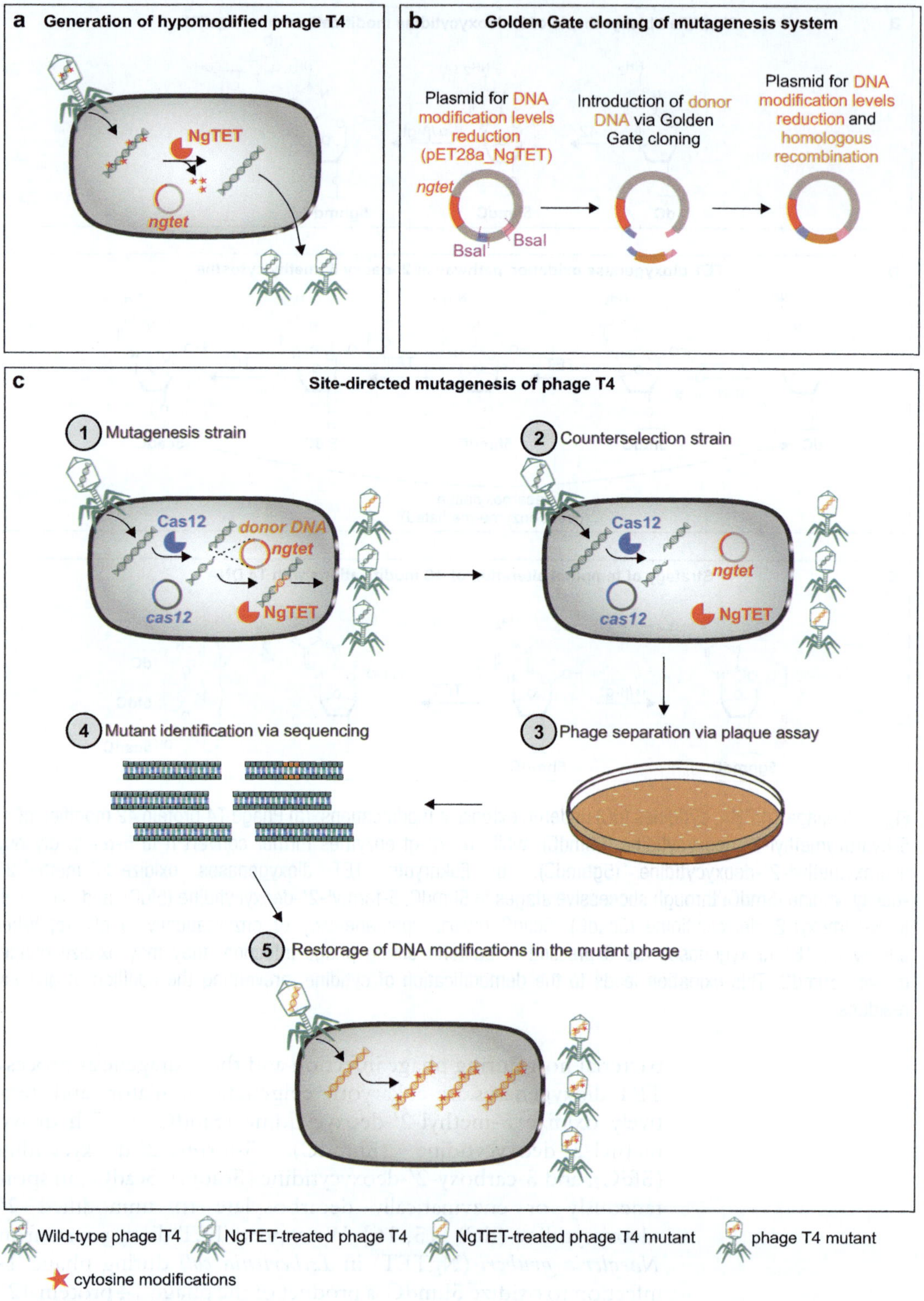

Fig. 2 Visualization of the single steps of phage T4 mutagenesis workflow, including generation of hypomodified phage T4 (**a**) mutagenesis plasmid generation via Golden Gate cloning (**b**) CRISPR-Cas-based site-directed mutagenesis of phage T4 (**c**)

effectively prevent cytosine glycosylation by phage glycosyltransferases. As a result, the relative abundance of 5ghmdC in phage DNA is significantly reduced (by ~45%), enhancing phage DNA's susceptibility to CRISPR-Cas targeting and consequently improving mutagenesis efficiency. Crucially, the phage DNA modifications can be restored entirely in the subsequent phage generation produced in the absence of NgTET [14].

We use Golden Gate cloning to exchange donor DNA sequences for homologous recombination to streamline the mutagenesis approach and rapidly adapt it for different targets. The donor DNA is co-located in the same plasmid with the gene encoding for NgTET dioxygenase. Our approach facilitates efficient and scarless CRISPR-Cas-based mutagenesis of phages with extensive cytosine modifications, as demonstrated here with CRISPR-Cas12 nuclease. This method circumvents the drawbacks linked to modification-deficient phage strains [13] and their influence on subsequent phage generations, as illustrated in Fig. 2.

2 Materials

Prepare all solutions using ultrapure water (double-distilled water (ddH_2O)) and analytical-grade reagents for use in molecular biology. Sterile-filter all prepared solutions by applying a 0.45 μm filter, and store them at room temperature (RT) if not stated otherwise. Store commercial reagents as indicated by the manufacturer. Some solutions or reagents overlap between subsections and are listed only once.

2.1 Cloning of pET28a_NgTET_donor-DNA and pCpf1-sp_spacer Plasmids

- Primers of interest.
- 1 M ATP, store in aliquots at −20 °C. Avoid freeze-thaw cycles.
- pCpf1-sp plasmid (CRISPR-Cas12 system), addgene accessory number 122186.
- pET28a_NgTET (plasmid is available upon request, the plasmid map can be found at the following https://zenodo.org/records/10615143) (Fig. 3).
- T4 DNA Ligase and T4 DNA Ligase buffer.
- BsaI.
- High-fidelity (HF) DNA polymerase and high GC buffer.
- dNTPs mix.
- PCR purification kit.
- Dimethyl sulfoxide (DMSO).
- 10× Tris-borate-EDTA (TBE) buffer: 1 M Tris base, 1 M boric acid, 25 mM Ethylene diamine tetraacetic acid (EDTA).

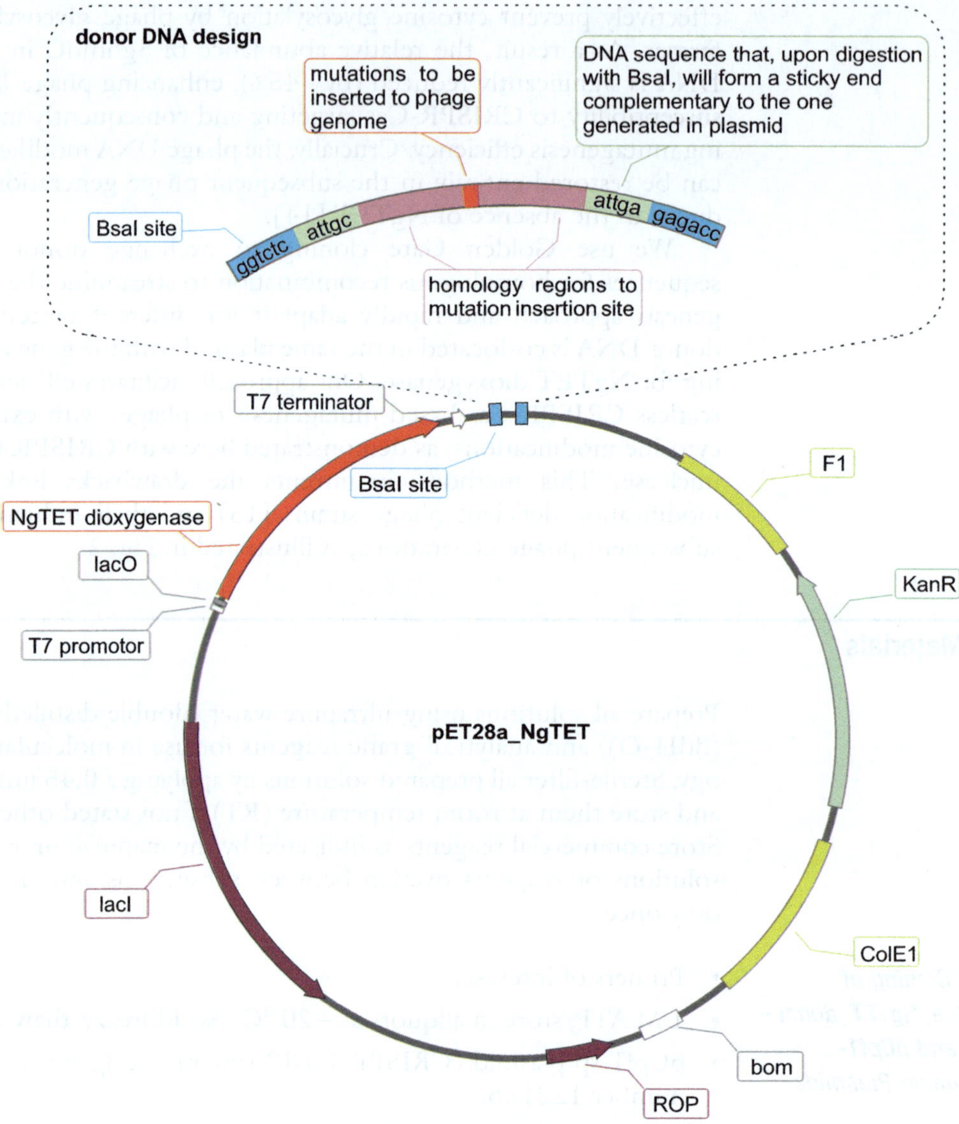

Fig. 3 Design of pET28a_NgTET_donor-DNA plasmid. Key elements for the design of the donor DNA insert are highlighted in the figure

- DNA stain for gel electrophoresis.
- DNA loading dye.
- DNA ladder.
- 1% (w/v) agarose in 1× TBE buffer, supplemented with DNA dye.
- Chemically competent *E. coli* cells, *E. coli* BL21 (DE3) strain must be used for Subheadings 3.3, 3.5, and 3.6.
- Kanamycin 30 mg/mL.

- Streptomycin 50 mg/mL.
- *Luria-Bertani* (LB)-medium: [10 g/L tryptone, 5 g/L yeast extract, 10 g/L sodium chloride, pH 7] For 1× LB-agar the media is supplemented with 15 g/L agar.
- Plasmid Miniprep kit.

2.2 Phage Propagation

- Phage of interest.

2.3 Phage Mutagenesis

- $MgCl_2$ solution: 1 M $MgCl_2$.
- $CaCl_2$ solution: 1 M $CaCl_2$.
- IPTG solution: 1 M isopropyl-β-thiogalactoside.

2.4 Plaque Assay and Phage Separation

- Bottom agar: 1× LB-agar.
- Top agar: 0.5× LB-agar.

2.5 Phage Mutants Screening

- Pi-Mg buffer: 26 mM Na_2HPO_4, 68 mM NaCl, 22 mM KH_2PO_4, 1 mM $MgSO_4$, pH 7.5, 2% (v/v) $CHCl_3$ (*see* **Note 1**).

2.6 General Consumables

- Conical tubes, 15 mL and 50 mL.
- 1.5 mL reaction tubes.
- Petri plates.
- PCR reaction vessels.
- 0.45 μm filters.

3 Methods

If necessary, carry out all procedures under sterile conditions using a fume hood and at RT unless otherwise specified. Perform the work with the phages carefully to avoid any contaminations (*see* **Note 2**).

3.1 Golden Gate Cloning of pET28a_NgTET_donor DNA Plasmids

To create donor DNA for insertion into pET28a_NgTET plasmid, you can use the following strategies: (1) Order the desired sequence as a gene block; (2) amplify the sequence of interest from a genome, e.g., phage genome, and introduce any necessary mutations (*see* **Note 3**). Figure 3 illustrates all the critical components that must be present in the donor DNA.

1. The donor DNA sequence is inserted downstream of the NgTET gene and terminator within the pET28a_NgTET plasmid via Golden Gate cloning. For the assembly, prepare the reaction by pipetting the following components:

Reagent	Stock concentration	Volume
pET28a_NgTET	–	X μL for 70 fmol
Donor DNA	–	X μL for 140 fmol
BsaI	20 U/μL	1 μL
T4 DNA Ligase	5 U/μL	5 U
T4 DNA Ligase buffer	10×	2 μL
ddH_2O	–	Ad 20 μL

2. Transfer the reaction to the PCR cycler and apply the following settings:

Step	Settings	Cycles
Restriction	37 °C for 1.5 min	25
Ligation	16 °C for 3 min	
Final restriction	37 °C for 5 min	–
Denaturation	60 °C for 10 min	–

3. Thaw 100 μL of chemically competent *E. coli* cells on ice for 10 min. Add 6 μL of the reaction product from **step 2** to the cells. Incubate on ice for 10 min. Then, perform a 45-s heat shock at 42 °C. Immediately cool the cells on ice for 1 min, followed by adding 750 μL of lysogeny broth (LB) medium. Incubate the cells at 37 °C, shaking at 500 rpm, for 1 h (*see* **Note 4**). Finally, plate the cells on agar containing kanamycin (30 μg/mL) as a selection marker. Incubate the plates overnight (o/n) at 37 °C.
4. Pick a colony from the plate and transfer it to 20 mL of LB-medium supplemented with 30 μg/mL kanamycin. Incubate o/n at 37 °C and 160 rpm.
5. Take 5 mL of the o/n culture, spin the cells down, and perform plasmid isolation from the cells using the plasmid extraction method of choice. Elute plasmid in 40 μL ddH_2O water.
6. Verify the cloning by sequencing the insertion site of the plasmid using Sanger sequencing.

3.2 Introduction of Spacer to CRISPR-Cas Plasmid (Exemplified on pCpf1-sp Plasmid [17])

The plasmid described in this chapter is essential for the initial step of mutagenesis: introducing a site-specific double-strand break to the phage DNA. Here, we demonstrate the mutagenesis procedure using the CRISPR-Cas12 system. However, alternative CRISPR-Cas systems, e.g., CRISPR-Cas9, can be applied.

1. Design a spacer (20 nt) targeting the mutagenesis site (PAM: TTTV) (*see* **Note 5**).
2. Order two 5′-phosphorylated DNA oligos, each containing 10 nt of spacer sequence (totaling a 20 nt spacer sequence split between the two primers) and 10 nt complementary to the pCpf1-sp plasmid at the site of insertion.
3. Integrate the CRISPR-Cas spacer into the plasmid by amplifying it through PCR, using the primers designed in **step 2**. Prepare the following reaction:

Reagent	Stock concentration	Volume
GC buffer	10×	5 μL
dNTPs	10 mM	1 μL
Forward primer	10 μM	1 μL
Reverse primer	10 μM	1 μL
DMSO	100%	2.5 μL
pCpf1-sp	–	× μL for 2 fmol
Phusion Polymerase	2 U/μL	1 μL
ddH_2O	–	ad 50 μL

4. Load the tubes to the PCR cycler and apply the following settings:

Step	Settings	Cycles
Initial denaturation	98 °C for 30 s	–
PCR	98 °C for 10 s Primer annealing (X °C) for 30 s 72 °C for 6 min	25
Final extension	72 °C for 10 min	–
Hold	4 °C	–

5. Mix 5 μL of the PCR reaction with 10× DNA loading buffer and load on 1% agarose.
6. Run the agarose gel at 130 V for 20 min and visualize it on a UV-transmitter.
7. If the amplification is successful, clean up the PCR product using a commercial kit. Elute the purified product in 20 μL of ddH_2O.
8. Perform the circularization of the plasmid via ligation by mixing 100 ng of amplified product with 1× T4 DNA Ligase buffer

and 60 U T4 DNA Ligase in 15 μL total volume. Incubate for 1 h at RT.

9. Transform chemically competent *E. coli* (100 μL) with the product of circularization reaction as described in Subheading 3.1, **step 3**. Finally, plate the cells on LB agar plates containing 50 μg/mL streptomycin. Incubate the plates o/n at 37 °C.
10. Pick a clone from the plate and transfer it to 20 mL of LB-medium supplied with 50 μg/mL streptomycin. Incubate o/n at 37 °C and 160 rpm.
11. Take 5 mL of the o/n culture, spin the cells down, and perform plasmid isolation from the cells using a plasmid extraction method of choice. Elute/resuspend plasmid in 40 μL ddH_2O water.
12. Verify the cloning by sequencing the plasmid at the insertion site using Sanger sequencing.

3.3 Phage DNA Pretreatment with NgTET

Reducing the abundance of phage DNA modifications through NgTET pre-treatment in this step is essential for the effective targeting of Cas nucleases during subsequent mutagenesis.

1. Inoculate *E. coli* BL21 (DE3) pET28a_NgTET strain into 10 mL of LB-media supplemented with 30 μg/mL kanamycin and incubate o/n at 37 °C and 160 rpm.
2. The next day, use the o/n culture to inoculate a main culture to an initial OD_{600} 0.1 in 50 mL of LB-media containing 30 μg/mL kanamycin.
3. Grow the culture until OD_{600} 0.8.
4. Add phage T4 to the multiplicity of infection (MOI) of 0.1. Perform infection for 4 h at RT and 120 rpm.
5. Spin the cells down by centrifugation at 4000× *g* for 20 min.
6. Filter the supernatant via a 0.45 μm filter. The generated NgTET-treated phages are the phages with hypomodified DNA. Store the filtrate at 4 °C until further use.

3.4 Plaque Assay

Plaque assay is carried out to determine phage titer or to separate the phage population.

1. Set up an *E. coli* culture in 10 mL LB media. Incubate the culture o/n at 37 °C and 160 rpm.
2. The following day, aliquot the o/n culture into sterile 1.5 mL reaction tubes, distributing 300 μL into each tube.
3. Prepare a dilution series of the phages generated the previous day. Transfer 100 μL of phage solution into 900 μL of LB media. Repeat this step-by-step to create serial dilutions, reaching up to a 10^8-fold dilution.

4. Infect each aliquot of *E. coli* cells with 50 μL from each phage dilution. Incubate the mixture for 6 min at 37 °C.
5. Transfer the mixture to 4 mL of melted top agar, already aliquoted into 15 mL conical tubes and pre-warmed to 50–55 °C (*see* **Note 6**). Gently mix by slowly inverting the tubes.
6. Pour the top agar mixture over plates precovered with bottom agar. Allow the top agar to cool and solidify.
7. Incubate the plates o/n at 37 °C in an inverted position (top-down). The following day, calculate the number of plaques on the plates with the best dilutions (plates with three- and two-digit numbers are best for the calculation) to determine the phage concentration (Subheading 3.3).

3.5 Phage Mutagenesis

In this step, target mutations are introduced into the genome of NgTET-treated T4 upon its infection of *E. coli*. During the mutagenesis, NgTET dioxygenase is expressed to ensure consistently low levels of DNA modifications in the T4 genome.

1. Transform *E. coli* BL21(DE3) with two plasmids—pET28a_NgTET_donor-DNA and the pCpf1-sp_spacer. Therefore, mix 1 μL of each plasmid with 100 μL of chemically competent *E. coli* BL21 (DE3) cells. Therefore, follow the protocol in Subheading 3.1, **step 3**. Finally, plate the cells on agar containing 50 μg/mL streptomycin and 30 μg/μL kanamycin. Incubate the plates o/n at 37 °C.
2. Inoculate *E. coli* BL21 (DE3) pET28a_NgTET_donor-DNA/pCPf1-sp_spacer strain into 10 mL of LB-media supplemented with 30 μg/mL kanamycin and 50 μg/mL streptomycin and incubate o/n at 37 °C and 160 rpm.
3. The next day, use the o/n culture to inoculate a main culture to an initial OD_{600} 0.1 in 50 mL of LB-media containing 30 μg/mL kanamycin and 50 μg/mL streptomycin.
4. Incubate the main culture at 37 °C with agitation at 160 rpm until the OD_{600} 0.8. Reduce the temperature to RT and lower the agitation speed to 130 rpm. Add $MgCl_2$ and $CaCl_2$ to achieve a final concentration of 1 mM each. Infect the culture with NgTET-pretreated phage T4 at an MOI of 0.1. Incubate the culture further at RT for an additional 4 h.
5. Transfer the culture to a 50 mL conical tube. Centrifuge the culture at 4000× *g* and 4 °C for 20 min to pellet the cells. To remove residual bacterial cells, filter the supernatant through a 0.45 μm filter. Use the filtrate for the plaque assay (Subheading 3.4) to determine the concentration or to perform a counterselection.

3.6 Counterselection for the Wild-Type Phages

Counterselection aims to reduce the ratio of wild-type phages in the phage suspension after mutagenesis. Therefore, phages resulting from the mutagenesis (Subheading 3.5, **step 5**) are used to infect *E. coli* expressing the same CRISPR-Cas system with the identical spacer as used in the mutagenesis. Thus, only the wild-type phage DNA will be targeted by Cas nucleases and not that carrying the mutations. The effective targeting prevents the propagation of wild-type phages and consequently enriches the total population of mutated phages.

1. For the counterselection, set up an o/n culture of *E. coli* BL21 (DE3) pCPf1-sp_spacer/pET28a_NgTET (*see* **Note** 7) strain in 10 mL LB-media supplied with 50 μg/mL streptomycin and 30 μg/mL kanamycin. Grow o/n at 37 °C, 160 rpm.
2. The following day, use the o/n culture to set up a main culture to start OD_{600} of 0.1 in 50 mL LB-media containing 50 μg/mL streptomycin and 30 μg/mL kanamycin. Incubate at 37 °C and 160 rpm until OD_{600} 0.8 is reached. Transfer the culture to RT and 130 rpm. Add $MgCl_2$ and $CaCl_2$ to 1 mM final concentration each. Infect with wild-type phage T4 to the MOI of 0.5. Incubate further at RT for 4 h.
3. Transfer the culture to a 50 mL conical tube and spin down the cells by centrifugation at 4000× *g*, 4 °C for 20 min. To remove residual bacterial cells, filter the supernatant through a 0.45 μm filter. Use the filtrate to determine the phage concentration and separate the phages via plaque assay, described in Subheading 3.4.

3.7 Phage Mutants Screening

The step aims the identification and isolation of the mutant phages. To isolate mutant phages, the phages must be separated on an agar plate via plaque assay (Subheading 3.4) that allows for the screening of single phages.

1. Use a plate from the plaque assay with a suitable dilution (max. 100 plaques).
2. Prepare the 1.5 mL reaction tubes containing 198 μL Pi-Mg buffer and 2 μL $CHCl_3$.
3. Pick a plaque from the plate with a sterile 200 μL pipette tip and wash the tip by pipetting up and down in the aliquoted Pi-Mg- $CHCl_3$ solution.
4. Vortex the reaction tube and incubate at RT and shaking (300 rpm) for 1 h. Afterwards, store at 4 °C until further use.
5. For the screening, transfer 1 μL from the solution from Subheading 3.7, **step 4**. to a PCR tube (*see* **Note 8**). Spin the solution down and heat it to 95 °C for 10 min to denature the phages.

6. Take 1 μL of phage suspension to perform amplification of the putative mutated region via PCR. The amplification product can be purified via a PCR purification method of choice or submitted to a multiplexed Nanopore sequencing procedure (as for example described by Ramírez-Rojas et al. in this book).

3.8 Propagation of the Phage Mutant

Once the phage mutant of interest is identified, it is propagated in wild-type *E. coli* to recover DNA modification levels akin to those of the wild-type phage T4. Once DNA modifications are restored, the phage mutant can be used for the respective biological experiments without being compromised except for the introduced mutation.

1. Set up an *E. coli* B strain culture in 10 mL LB media. Incubate the culture o/n at 37 °C and 160 rpm.
2. The next day, use the o/n culture to inoculate a main culture to an initial OD_{600} 0.1 in 50 mL of LB-media.
3. Grow to OD_{600} 0.8 at 37 °C and 160 rpm.
4. Infect with 50 μL of the mutant phage from respective Pi-Mg-$CHCl_3$ stock (Subheading 3.7, **step 4**). Carefully avoid pipetting of $CHCl_3$.
5. Perform infection o/n at RT and 120 rpm.
6. Remove the cells via centrifugation of the culture at 4000× *g*. Filter the supernatant through a 0.45 μm filter.
7. Store the mutant phage at 4 °C in a glass bottle until further use (*see* **Note 9**).

4 Notes

1. $CHCl_3$, being volatile, should be added immediately before using the buffer. Work should be performed under a fume hood according to local safety regulations.
2. To minimize contamination risks in phage experiments, we recommend dedicating separate working areas for bacteria- and phage-related experiments and using dedicated equipment sets for each (e.g., pipettes). After each experiment, disinfect the equipment and working area with 70% ethanol. Additionally, decontamination of the pipettes by exposing them to UV light is advised. Should contamination occur, treat affected areas with a 2% solution of sodium hypochlorite for 5 min. Autoclave reusable equipment, such as flasks and glass bottles, separate from other glassware to prevent cross-contamination.
3. When generating donor DNA for insertion into phage DNA, ensure that homologous regions flanking the insertion site are present on both sides of the donor DNA (Fig. 3). For point

mutations, 200 bp on each side is sufficient. For larger insertions, the size of the homologous regions might be increased. Additionally, the final linear insert should include BsaI restriction sites. Upon cleavage, these sites should generate sticky ends complementary to those in the target plasmid. To verify the cloning strategy, we recommend to perform in silico cloning to confirm the design.

4. The transformation protocol may vary depending on the chemically competent cells used.
5. To simplify the identification of a suitable protospacer and the design of the corresponding spacer, tools such as Geneious can be utilized.
6. The top agar for the plaque assay must be maintained at 50–55 °C. Temperatures above this range can kill *E. coli* cells, while lower temperatures may cause the agar to solidify, hindering homogeneous plating. Utilize a water bath to ensure stable temperature control of agar.
7. Some Cas nucleases, including Cas12 used in this protocol, may accept a small number of mismatches between the spacer sequence and its corresponding protospacer [18]. This results in low counterselection efficiency. To increase the efficiency of counterselection processes, it can be considered to incorporate silent mutations in the protospacer region within the target phage DNA.
8. Before pipetting the phage for sequencing, allow the organic phage ($CHCl_3$) to separate completely. Then, cautiously withdraw 1 μL of the phage suspension from the surface, taking care to avoid transferring any $CHCl_3$. This is crucial as $CHCl_3$ can inhibit the PCR reaction needed to amplify the mutated region within the phage genome.
9. We advise sequencing the phage mutants after each propagation to confirm the presence of the introduced mutation and to ensure the mutation was not reverted to the wild-type phage T4 sequence.

Acknowledgements

K.H. is supported by funding from the Max Planck Society and the German Research Council (DFG SPP 2330, Project number 464500427). Katharina Höfer and Nadiia Pozhydaieva filed a PCT application for "Engineering of Phages", European Patent Application No. 23 175 257.7.

References

1. Vandamme EJ, Mortelmans K (2019) A century of bacteriophage research and applications: impacts on biotechnology, health, ecology and the economy! J Chem Technol Biotechnol 94(2):323–342. https://doi.org/10.1002/jctb.5810
2. Moye ZD, Woolston J, Sulakvelidze A (2018) Bacteriophage applications for food production and processing. Viruses 10(4). https://doi.org/10.3390/v10040205
3. Elois MA, Silva RD, Pilati GVT, Rodriguez-Lazaro D, Fongaro G (2023) Bacteriophages as biotechnological tools. Viruses 15(2). https://doi.org/10.3390/v15020349
4. Wolfram-Schauerte M, Pozhydaieva N, Viering M, Glatter T, Höfer K (2022) Integrated omics reveal time-resolved insights into T4 phage infection of E. Coli on proteome and transcriptome levels. Viruses 14(11). https://doi.org/10.3390/v14112502
5. Meile S, Du J, Staubli S, Grossmann S, Koliwer-Brandl H, Piffaretti P, Leitner L, Matter CI, Baggenstos J, Hunold L, Milek S, Guebeli C, Kozomara-Hocke M, Neumeier V, Botteon A, Klumpp J, Marschall J, McCallin S, Zbinden R, Kessler TM, Loessner MJ, Dunne M, Kilcher S (2023) Engineered reporter phages for detection of Escherichia coli, enterococcus, and Klebsiella in urine. Nat Commun 14(1):4336. https://doi.org/10.1038/s41467-023-39863-x
6. Du J, Meile S, Baggenstos J, Jaggi T, Piffaretti P, Hunold L, Matter CI, Leitner L, Kessler TM, Loessner MJ, Kilcher S, Dunne M (2023) Enhancing bacteriophage therapeutics through in situ production and release of heterologous antimicrobial effectors. Nat Commun 14(1):4337. https://doi.org/10.1038/s41467-023-39612-0
7. Mohanraju P, Makarova KS, Zetsche B, Zhang F, Koonin EV, van der Oost J (2016) Diverse evolutionary roots and mechanistic variations of the CRISPR-Cas systems. Science 353(6299):aad5147. https://doi.org/10.1126/science.aad5147
8. Tao P, Wu X, Tang WC, Zhu J, Rao V (2017) Engineering of bacteriophage T4 genome using CRISPR-Cas9. ACS Synth Biol 6(10):1952–1961. https://doi.org/10.1021/acssynbio.7b00179
9. Liu Y, Dai L, Dong J, Chen C, Zhu J, Rao VB, Tao P (2020) Covalent modifications of the bacteriophage genome confer a degree of resistance to bacterial CRISPR systems. J Virol 94(23). https://doi.org/10.1128/JVI.01630-20
10. Bryson AL, Hwang Y, Sherrill-Mix S, Wu GD, Lewis JD, Black L, Clark TA, Bushman FD (2015) Covalent modification of bacteriophage T4 DNA inhibits CRISPR-Cas9. MBio 6(3):e00648. https://doi.org/10.1128/mBio.00648-15
11. Weigele P, Raleigh EA (2016) Biosynthesis and function of modified bases in bacteria and their viruses. Chem Rev 116(20):12655–12687. https://doi.org/10.1021/acs.chemrev.6b00114
12. Miller ES, Kutter E, Mosig G, Arisaka F, Kunisawa T, Ruger W (2003) Bacteriophage T4 genome. Microbiol Mol Biol Rev 67(1):86–156. https://doi.org/10.1128/MMBR.67.1.86-156.2003
13. Kutter EM, Wiberg JS (1969) Biological effects of substituting cytosine for 5-hydroxymethylcytosine in the deoxyribonucleic acid of bacteriophage T4. J Virol 4(4):439–453. https://doi.org/10.1128/JVI.4.4.439-453.1969
14. Pozhydaieva N, Billau FA, Wolfram-Schauerte-M, Rojas AAR, Paczia N, Schindler D, Höfer K (2024) Temporal epigenome modulation enables efficient bacteriophage engineering and functional analysis of phage DNA modifications. bioRxiv:2024.2001.2028.577628. https://doi.org/10.1101/2024.01.28.577628
15. Ito S, D'Alessio AC, Taranova OV, Hong K, Sowers LC, Zhang Y (2010) Role of Tet proteins in 5mC to 5hmC conversion, ES-cell self-renewal and inner cell mass specification. Nature 466(7310):1129–1133. https://doi.org/10.1038/nature09303
16. Tahiliani M, Koh KP, Shen Y, Pastor WA, Bandukwala H, Brudno Y, Agarwal S, Iyer LM, Liu DR, Aravind L, Rao A (2009) Conversion of 5-methylcytosine to 5-hydroxymethylcytosine in mammalian DNA by MLL partner TET1. Science 324(5929):930–935. https://doi.org/10.1126/science.1170116
17. Niu TC, Lin GM, Xie LR, Wang ZQ, Xing WY, Zhang JY, Zhang CC (2019) Expanding the potential of CRISPR-Cpf1-based genome editing Technology in the Cyanobacterium Anabaena PCC 7120. ACS Synth Biol 8(1):170–180. https://doi.org/10.1021/acssynbio.8b00437
18. Murugan K, Seetharam AS, Severin AJ, Sashital DG (2020) CRISPR-Cas12a has widespread off-target and dsDNA-nicking effects. J Biol Chem 295(17):5538–5553. https://doi.org/10.1074/jbc.RA120.012933

Chapter 19

Golden Gate-Assisted Gene Doctoring for Streamlined and Efficient Recombineering in Bacteria

Nicholas M. Thomson

Abstract

Gene Doctoring is a genetic modification technique for *E. coli* and related bacteria, in which the Red-recombinase from bacteriophage λ mediates chromosomal integration of a fragment of DNA by homologous recombination (known as recombineering). In contrast to the traditional recombineering method, the integrated fragment for Gene Doctoring is supplied on a donor plasmid rather than as a linear DNA. This protects the DNA from degradation, facilitates transformation, and ensures multiple copies are present per cell, increasing the efficiency and making the technique particularly suitable for strains that are difficult to modify. Production of the donor plasmid has, until recently, relied on traditional cloning techniques that are inflexible, tedious, and inefficient. This protocol describes a procedure for Gene Doctoring combined with Golden Gate assembly of a donor plasmid, using a custom-designed plasmid backbone, for rapid and simple production of complex, multi-part assemblies. Insertion of a gene for superfolder green fluorescent protein, with selection by tetracycline resistance, into *E. coli* strain MG1655 is used as an example but in principle the method can be tailored for virtually any modification in a wide range of bacteria.

Key words Recombineering, λ-Red recombinase, Golden Gate assembly, Gene knock-out, Gene knock-in, Synthetic biology, Metabolic engineering, *Escherichia coli*

1 Introduction

Recombineering is a widely used and adaptable technique for genome engineering in *Escherichia coli* and other bacteria, in which an exogenous DNA molecule is integrated into the chromosome by homologous recombination [1]. In the most commonly used implementation, the Red recombinase from bacteriophage λ (λ-Red; composed of three proteins, known as Exo, Beta, and Gamma) is expressed from a helper plasmid and mediates recombination at 30–50 bp homologous regions at either end of a linear, double-stranded DNA introduced by electroporation (Fig. 1) [2]. The homologous regions determine the exact genomic location to be targeted, and the sequence between them can be

Daniel Schindler (ed.), *Golden Gate Cloning: Methods and Protocols*, Methods in Molecular Biology, vol. 2850,
https://doi.org/10.1007/978-1-0716-4220-7_19,

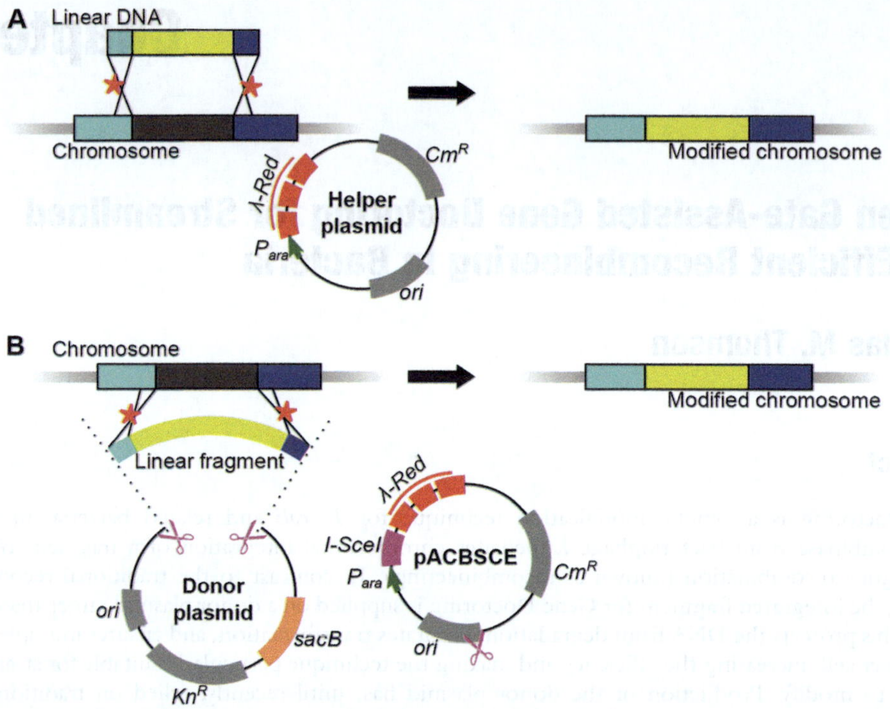

Fig. 1 Comparison of traditional recombineering with Gene Doctoring. (**a**) In traditional recombineering, a linear DNA carrying the modification (yellow) is generated and introduced directly into the cells to be modified. A helper plasmid expresses λ-Red (red stars), which mediates recombination between the linear DNA and the chromosome, guided by homologous regions at either end (light and dark blue). (**b**) In Gene Doctoring, the modification is carried on a donor plasmid. The helper plasmid pACBSCE expresses both λ-Red and I-SceI when induced with arabinose. I-SceI recognizes specific sequences (purple scissors) to release a linear fragment from the donor plasmid, that undergoes homologous recombination, and cuts pACBSCE. Thus, both plasmids are removed from the cells. To assist screening for recombinants, *sacB* on the donor plasmid renders cells that still contain the plasmid sensitive to sucrose

customized to delete, substitute, or insert virtually any desired sequence. Recombineering is rapid and efficient in many instances, but linear DNAs are susceptible to degradation by nucleases and often do not efficiently enter electroporated cells.

A variation in recombineering called Gene Doctoring overcomes these challenges and is particularly suitable for pathogenic isolates, which are often more difficult to engineer [3]. The key feature is that in Gene Doctoring, the mutagenesis cassette is supplied on a plasmid, flanked by recognition sites for the I-SceI meganuclease (Fig. 1). The plasmid protects the DNA from degradation, facilitates transformation, and ensures that multiple copies of the template are present in each cell. The helper plasmid (pACBSCE), in addition to λ-Red, expresses I-SceI, which can therefore excise the mutagenic fragment and make it available for homologous recombination at the same time as λ-Red is produced. I-SceI also cuts a target site on pACBSCE, so this plasmid is

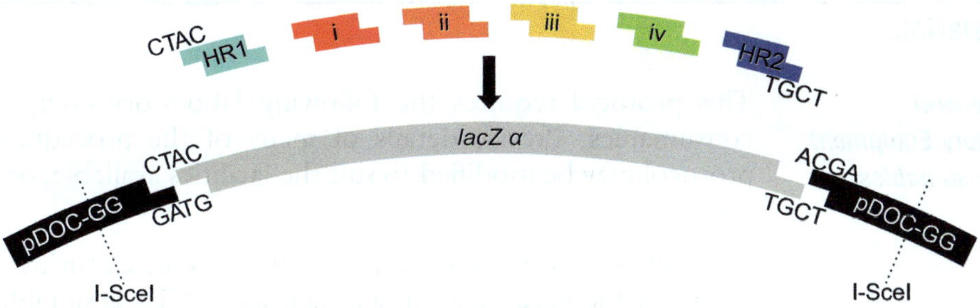

Fig. 2 Summary of the Golden Gate assembly strategy to produce donor plasmids for Gene Doctoring. Only the cloning region of pDOC-GG is shown. During a Golden Gate assembly reaction, BsaI releases a *lacZα* dropout cassette leaving CTAC and ACGA overhangs. Compatible genetic parts (colored pieces), including the two homologous regions, are inserted in place of the dropout cassette, guided by the compatibility of their own BsaI-generated overhangs. In this way, complex, multi-part donor plasmids can be assembled in a single reaction

self-destructive. This means that λ-Red is only produced in a short burst, which is sufficient to mediate desired recombination but reduces secondary, off-target recombination events that can arise from prolonged expression.

Until recently, Gene Doctoring donor plasmids were time-consuming and cumbersome to produce because they relied on traditional restriction-ligation cloning. With colleagues, I therefore designed a new plasmid (pDOC-GG) to facilitate Golden Gate assembly of mutagenesis cassettes flanked by homologous regions, which is available from the Addgene repository [4]. This greatly simplifies rapid assembly of complex, multi-part cassettes in a single reaction and makes it easier to automate the process for large-scale mutagenesis projects such as metabolic engineering (Fig. 2).

This protocol provides a simple example of the construction and use of a donor plasmid to insert the *sfgfp* gene (for superfolder green fluorescent protein expression) into the phenotypically-neutral insertion site of transposon Tn7, downstream of the *glmS* gene [5] in *E. coli* strain MG1655. Selection for recombinants is achieved by including a gene for tetracycline resistance. By substituting the necessary homologous regions and other genetic parts, pDOC-GG can function as the backbone for virtually any genomic modification of any *E. coli* or related species. With appropriate choices of replication origin, selection cassettes, and promoters, the tools described here could also be adapted for use in a much wider range of species.

The power of synthetic biology lies in the development and sharing of genetic tools and I therefore strongly encourage the reader to build their own compatible Gene Doctoring plasmids and genetic parts, and to share them with the scientific community as a resource for adaptable, streamlined, and efficient genetic engineering.

2 Materials

2.1 General Laboratory Equipment and Consumables

This protocol requires the following laboratory equipment and consumables. Precise details of many of the procedures in the protocol may be modified to suit the facilities available, or personal preference:

1. Facilities, equipment, and consumables for culturing of bacteria at the necessary containment level (CL1 is suitable for this protocol).
2. Standard micropipettes and tips.
3. Standard (micro)centrifuge and tubes.
4. Equipment and consumables for PCR.
5. Equipment for agarose gel electrophoresis.
6. Water bath for heat shock transformation.

2.2 Specific Laboratory Equipment

1. Qubit fluorimeter (or similar) and consumables for fluorescence-based quantification of DNA.
2. Nanodrop or similar micro-volume spectrophotometer.
3. Shaking heat block.
4. Blue light transilluminator.
5. Clean, sterile razor blades or scalpels for cutting gel bands.
6. Genomic DNA isolation kit.
7. Plasmid mini-prep kit.
8. PCR purification and gel extraction kit.
9. Reagents and equipment for Sanger sequencing and whole-genome shotgun sequencing, or access to suitable services.

2.3 Chemicals and Growth Media

Growth media are sterilized by autoclaving; other chemicals used in bacterial growth are sterilized by filtration through a 0.22 μm cellulose acetate syringe filter.

1. Lysogeny broth (LB) medium: 10 g/L tryptone, 10 g/L NaCl, 5 g/L yeast extract. Add 15 g/L of agar for lysogeny agar (LA).
2. Lysogeny agar, no salt, 10% sucrose (LA-SS) medium: as above but omit NaCl; after autoclaving combine 750 mL base medium with 250 mL of sterile 40% (w/v) sucrose.
3. Super optimal broth with catabolite repression (SOC) medium: 20 g/L tryptone, 5 g/L yeast extract, 10 mM NaCl, 10 mM $MgCl_2$, 10 mM $MgSO_4$, 2.5 mM KCl, 20 mM glucose. Combine and autoclave all ingredients except glucose, which is added after the medium has cooled.

4. Carbenicillin (or ampicillin) stock in water (100 mg/mL).
5. Kanamycin stock in water (50 mg/mL).
6. Chloramphenicol stock in 100% ethanol (30 mg/mL).
7. Tetracycline stock in 75% ethanol (5 mg/mL).
8. Isopropyl β-d-1-thiogalactopyranoside (IPTG) stock in water (100 mM).
9. 5-bromo-4-chloro-3-indolyl-β-D-galactopyranoside (X-gal) stock in dimethyl sulfoxide (20 mg/mL).
10. L-arabinose stock in water (20% (w/v)).
11. Agarose for DNA electrophoresis.
12. Fluorescent DNA dye (10,000× solution).
13. 100 bp and 1 kb DNA ladders.
14. Glycerol (50% (v/v) in ultrapure water), autoclaved to sterilize.

2.4 Enzymes and Buffers

1. High-fidelity DNA polymerase; 2× master mix.
2. Taq DNA polymerase; 2× master mix.
3. BsaI (20,000 units/mL).
4. T4 DNA ligase (400,000 units/mL).
5. T4 DNA ligase buffer.
6. Bovine serum albumin (2 mg/mL).
7. DNA loading buffer (6× stock).
8. Enzymatic PCR clean-up kit containing exonuclease I and phosphatase.

2.5 Bacteria Strains

1. *E. coli* strain MG1655—strain to be modified.
2. *E. coli* strain DH5α (or similar)—chemically competent cells for use in cloning.

2.6 Plasmids

Table 1 details the plasmids that are either required for this protocol or will be generated during the experiment. Plasmids are propagated in *E. coli* strain DH5α (stocks maintained at −80 °C in 15% glycerol). Purified plasmids should be stored at 4 °C short-term and − 20 °C long-term, and protected from multiple freeze-thaw cycles:

2.7 PCR Primers

The PCR primers required to conduct this protocol are given in Table 2.

Table 1
Plasmids required for this protocol or produced during the experiment

Name	Purpose	Relevant features	Source
pDOC-GG	Destination plasmid for production of Gene Doctoring donor plasmids by Golden Gate assembly	Kn^{R}; *lacZα* dropout cassette; BsaI sites; I-SceI sites; *sacB*	Addgene No. 149377
pACBSCE	Helper plasmid for Gene Doctoring	Cm^{R}; *λ-Red; I-SceI*; I-SceI site	Ref. [3]
mUAVbla: tetR 5′	Contains *tetA* cassette flanked by BsaI sites for assembly into pDOC-GG and FRT sites for excision using Flp recombinase	Tc^{R}; Ap^{R}; BsaI sites; FRT sites	Addgene no. 149379
pMA_Level2A	Template to amplify *sfgfp*	Cm^{R}; *sfgfp*	Addgene No. 102705
pCP20	Optional removal of *tetA* cassette by Flp recombinase	Ap^{R}; Cm^{R}; *flp*; temperature-sensitive *ori*	Ref. [2]
pINS:sfgfp	Donor plasmid for insertion of *sfgfp* and *tetA* into *E. coli* MG1655 by Gene Doctoring	Kn^{R}; Tc^{R}; Homologous regions for insertion in Tn7 target site; *sfgfp*, *tetA*, I-SceI sites	This protocol

Table 2
PCR primers required for this protocol

Name	Sequence
HR1_F	CACCAC*GGTCTC*ACTAC**AAGTTCGCGCGCGTGGC**
HR1_R	CACCAC*GGTCTC*ACTCC**CCCCGCTTACGCAGGGC**
HR2_F	CACCAC*GGTCTC*CAGGT**CATTTTTCTTCCTGTTATGTTTTTAATC**
HR2_R	CACCAC*GGTCTC*CTCGT**GGCTACGAGAAGCAAAATAG**
J23106_F	CACCACA*GGTCTC*GACCT**TTTACGGCTAGCTCAGTCC**
sfGFPterm_R	CACCACA*GGTCTC*GCGCT**GTGACTCTAGTAGAGAGCG**
pDOC-K_F	CATGATTACGCCAAGCTCTAG
pDOC-K_R	GGGTTTTCCCAGTCACGACGT
glmS-screen_F	ACAAACGCATTGAAGCGCTG
glmS-screen_R	GAACCGATACCCTGGTAGTTA

For cloning primers, BsaI sites are indicated in italics, 4 bp BsaI overhangs are underlined and binding regions are indicated in bold

3 Methods

3.1 Preparation of Parts for Plasmid Assembly

1. Examine an annotated genome sequence for *E. coli* strain MG1655 (NC_000913.3) and identify the region to be modified (Fig. 3). Design PCR primers to amplify 250 bp homologous regions immediately upstream (HR1) and downstream (HR2) of the insertion site, between a G and a C at bp 3,911,813 (*see* **Note 1**).
2. Add 5′ extensions to the PCR primers to include a buffer region for efficient restriction digestion, a BsaI recognition sequence, and appropriate nucleotides to create the necessary 4 bp overhangs to direct assembly of the mutagenesis plasmid (*see* **Note 2**). The homologous regions will define where homologous recombination will take place. The sequences of HR1 and HR2 will be unchanged in the final mutant strain, but any sequence between HR1 and HR2 will be replaced by the sequence contained in the donor plasmid.
3. Order the PCR primers designed above.
4. Purify genomic DNA from a culture of *E. coli* strain MG1655 using a commercial genomic DNA preparation kit, according to the manufacturer's instructions.
5. Conduct PCR reactions using a high-fidelity, proofreading DNA polymerase to amplify HR1 (primers HR1_F and HR1_R) and HR2 (primers HR2_F and HR2_R). A typical reaction mix would contain:

2× DNA polymerase pre-mix:	25 μL
Forward primer (10 μM):	2.5 μL
Reverse primer (10 μM):	2.5 μL
MG1655 gDNA template:	10 ng
Sterile ultrapure water:	To final 50 μL reaction volume

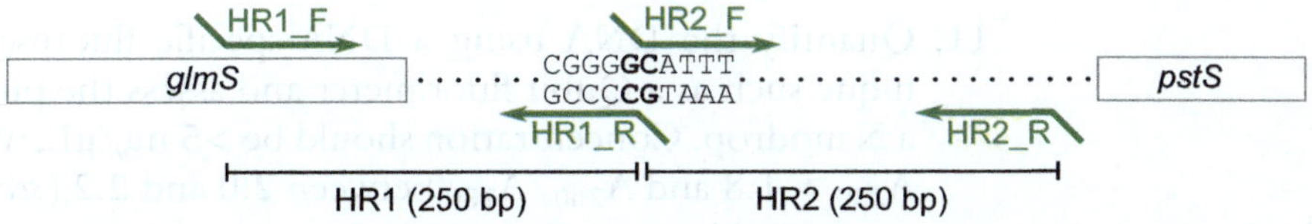

Fig. 3 Schematic overview of the targeted insertion site (between the two residues in bold) at bp 3,911,813 of the *E. coli* strain MG1655 chromosome. Green arrows represent primers to amplify 250 bp homologous regions upstream and downstream of the insertion site, and add appropriate extensions to introduce BsaI sites with compatible overhangs. The positions of the *glmS* and *pstS* genes are shown in boxes

For this example, run a PCR program as follows:

Initial denaturation:	98 °C for 10 s
35 cycles of:	
Denaturation:	98 °C for 10 s
Annealing:	60 °C for 20 s
Extension:	72 °C for 10 s
Final extension:	1 min at 72 °C
Storage:	Hold at 10 °C

Optimal conditions will depend on the polymerase used, the sequence to be amplified, and the sequence of the primers.

6. While the PCR is running, prepare a 1.5% (w/v) agarose gel containing a fluorescent DNA dye (*see* **Note 3**).
7. Add 10 μL of a 6× DNA sample loading buffer to the final PCR reactions and load the entire volume of each reaction to the agarose gel (*see* **Note 4**). Also load 0.5 μg of a 100 bp DNA ladder.
8. Run the gel at 100 V for 30 min.
9. View the gel using a blue light transilluminator (*see* **Note 5**). Each reaction should have produced a single fragment, which should show as a bright band at 250 bp. Poor amplification, contaminating bands, or primer dimers (fuzzy bands at 50–100 bp) indicate that optimization of the PCR reaction is necessary, although might be acceptable if sufficient desired product can be recovered without co-purifying contaminating products (*see* **Note 6**).
10. Cut out the desired bands with a clean, sterile razor blade or scalpel and recover the DNA using a commercial DNA gel extraction kit. Elute the DNA twice, using 30 μL of elution buffer each time, then combining both, for maximum product recovery (*see* **Note 7**).
11. Quantify the DNA using a DNA-specific fluorescent technique such as a Qubit fluorimeter and assess the purity using a Nanodrop. Concentration should be >5 ng/μL, with $A_{260}/A_{280} > 1.8$ and A_{260}/A_{230} between 2.0 and 2.2 (*see* **Note 8**).
12. Follow a similar procedure to **steps 1–11** to prepare any other DNA parts that will be produced by PCR. For this example, use pMA_Level2A as the template to design primers and amplify *sfgfp*. Figure 4 shows the order of parts to be assembled and the required overhangs to direct their correct assembly. The part for *tetA* is provided on plasmid mUAVbla:tetR 5′ so does not require PCR (*see* **Note 9**).

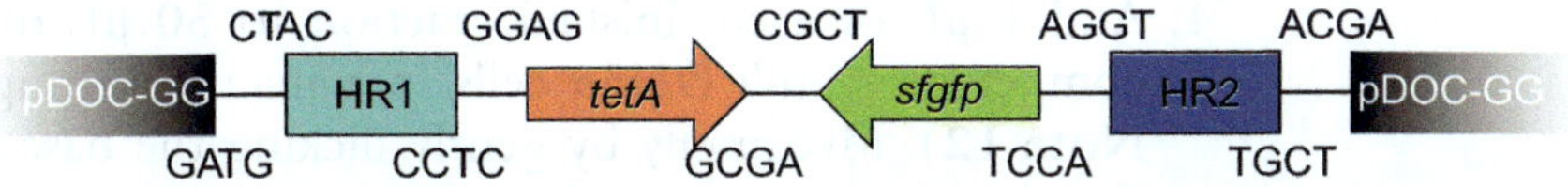

Fig. 4 Summary of the construction of the mutagenesis cassette for the example in this protocol. The fragments for HR1, HR2 and *sfgfp* are produced by PCR, while the *tetA* fragment is supplied on a plasmid. Fragments are shown in the order in which they will be assembled, between the arms of pDOC-GG following digestion with BsaI. The complementary 4 bp overhangs generated by BsaI are shown for each fragment. Arrows indicate gene expression cassettes (promoter, coding sequence and terminator, with FRT sites in the case of *tetA*) and the direction of transcription

13. Prepare stocks of pDOC-GG, mUAVbla:tetR 5′ (and any other DNA parts provided on plasmids), and pACBSCE by performing plasmid minipreps using a commercial kit.

3.2 Assembling the Donor Plasmid

1. In a PCR tube, prepare a Golden Gate assembly reaction to construct the donor plasmid. This consists of (in this order):

Sterile ultrapure water:	to final 10 μL reaction volume
T4 DNA ligase buffer:	1 μL
Bovine serum albumin:	1 μL
mUAVbla:tetR 5′:	10 fmol
pDOC-GG:	10 fmol
HR1 PCR product:	20 fmol
HR2 PCR product:	20 fmol
sfgfp PCR product:	20 fmol
T4 DNA ligase:	0.5 μL (200 units)
BsaI:	0.5 μL (10 units)

2. Also prepare a negative control reaction, which is identical except does not contain BsaI (*see* **Note 10**).
3. Run both reactions on a PCR machine with the following program (*see* **Note 11**):

   30 cycles of:

   37 °C for 1 min.

   16 °C for 1 min.

Final incubation:	60 °C for 10 min
Denature enzymes:	80 °C for 20 min
Storage:	Hold at 10 °C

4. Add 1 μL of each finished reaction to 50 μL of chemically competent *E. coli* DH5α cells in a microcentrifuge tube (*see* **Note 12**). Mix briefly by gently flicking the base of the tube then incubate on ice for 30 min (*see* **Note 13**).
5. Meanwhile, pre-warm a water bath to 42 °C.
6. Quickly transfer the tubes in a floating rack to the water bath for 45 s before placing back on ice for 2 min. Add 950 μL of pre-warmed (37 °C) SOC medium and incubate the tubes on a shaking heat block at 37 °C and 1000 rpm for 45 min.
7. Spread 100 μL from each tube on lysogeny agar (LA) plates containing 50 μg/mL kanamycin, 15 μg/mL tetracycline, 100 μM IPTG, and 20 μg/mL X-gal (*see* **Note 14**).
8. Incubate the plates overnight at 37 °C. Following incubation, the donor plasmid plate should contain 50–500 colonies. Most should be white but there may be some blue colonies arising from mis-assemblies that contain the *tetA* gene but have retained the *lacZα* fragment. There should be no colonies on the control plate.
9. Using sterile toothpicks or pipette tips, pick 6 well-isolated, white colonies, transferring the cells into 50 μL of sterile, ultrapure water by gentle agitation (*see* **Note 15**). Enough cells should remain on the toothpick or pipette tip so that each colony can then be streaked onto one of eight segments drawn on an LA plate containing 50 μg/mL of kanamycin, 15 μg/mL of tetracycline, 100 μM of IPTG, and 20 μg/mL of X-gal (*see* **Note 16**). Prepare a positive control using a colony of cells containing pDOC-GG and a negative control using only water.
10. Incubate the plate overnight at 37 °C to obtain single colonies of each candidate (*see* **Note 17**). Both controls should not grow. Keep this plate, sealed with parafilm, at 4 °C, for use in **step 17**.
11. Meanwhile, use the cells suspended in water as templates for colony PCR reactions to identify correctly assembled donor plasmids by amplifying a region spanning the assembly with primers pDOC-K_F and pDOC-K_R (*see* **Note 18**). Each reaction mixture should contain:

Sterile ultrapure water:	4.2 μL
2× pre-mix of Taq polymerase:	10 μL
pDOC-K_F (10 μM):	0.4 μL
pDOC-K_R (10 μM):	0.4 μL
Template suspension:	5 μL

Prepare a master mix with sufficient volume of all components, except template, for 9 reactions (8 to be used + an extra dead volume). Pipette 15 μL of the master mix into each PCR tube before adding the template suspensions individually. Run a PCR program consisting of:

Initial cell lysis step:	95 °C for 3 min
25 cycles of:	
Denaturation:	95 °C for 15 s
Annealing:	53 °C for 20 s
Extension:	72 °C for 3 min
Final extension:	72 °C for 5 min
Storage:	Hold at 10 °C

12. While the PCR is running, prepare a 0.8% (w/v) agarose gel containing a fluorescent DNA dye.
13. Add 5 μL of DNA loading dye to each reaction, mix, then load 3 μL of each to the 0.8% agarose gel. Also load 0.5 μg of a 1 kb DNA ladder.
14. Run the gel at 100 V for 30 min.
15. View the gel using a blue light or UV transilluminator. Successful assemblies should have a single band of the expected size (for this example, 2924 bp). Additional or differently-sized bands indicate errors in the assembly or problems with the PCR, but typically 4 or 5 of the screened colonies will be correct. The pDOC-GG positive control should have two bands, at 218 bp and 256 bp, as the pDOC-K_R primer binds in two locations. The negative control should have no bands.
16. For all correct colonies, treat the remaining PCR reaction with an enzymatic PCR clean-up kit containing exonuclease I and phosphatase (*see* **Note 19**). Combine 5 μL of PCR reaction with 2 μL of enzyme mix, incubate at 37 °C for 15 min then 80 °C for 15 min, cool to room temperature, then add 13 μL of sterile ultrapure water. Submit this mixture as the template for Sanger sequencing using primers pDOC-K_F and pDOC-K_R to verify the sequence of the insert (*see* **Note 20**).
17. Select one correct assembly and pick a single colony from the plate prepared in **step 9**. This is the donor plasmid, pINS:*sfgfp*, for use in Gene Doctoring. Inoculate a 5 mL overnight culture in LB containing 50 μg/mL kanamycin and 15 μg/mL tetracycline. Incubate at 37 °C with shaking at 200 rpm. Purify pINS:*sfgfp* from the overnight culture using a commercially available plasmid miniprep kit.

3.3 Gene Doctoring

1. Transform the strain to be modified with both the donor plasmid and pACBSCE (*see* **Note 21**). Select for cells containing both plasmids by growing on an LA plate containing 30 μg/mL chloramphenicol and 15 μg/mL tetracycline. Also transform the same strain with only the donor plasmid for use as a control. Recover these cells on LA plates containing 30 μg/mL kanamycin and 15 μg/mL tetracycline.
2. Following transformation, restreak a few colonies onto a fresh LA plate containing the same antibiotics as before, to ensure pure cultures of plasmid-carrying cells are obtained. Grow at 37 °C overnight.
3. The next day, transfer a single, large colony of the strain containing the donor plasmid and pACBSCE into 500 μL of LB containing 30 μg/mL chloramphenicol and 15 μg/mL tetracycline in a 2 mL microcentrifuge tube (*see* **Note 22**). Resuspend thoroughly by pipetting.
4. Incubate on a shaking heat block at 37 °C and 1000 rpm for 3 h or until the culture is turbid (*see* **Note 23**). Do the same, in parallel, for the strain containing only the donor plasmid (using 30 μg/mL kanamycin and 15 μg/mL tetracycline).
5. Centrifuge the cultures at 13,000× *g* for 1 min. Remove the supernatants without disturbing the pellets, then resuspend the pellets in 1 mL of 0.1× LB.
6. Repeat the centrifugation and resuspension twice more. The washing removes any trace of antibiotics, while the 0.1× LB supports homologous recombination but discourages growth.
7. Centrifuge again, then resuspend the pellets a final time in 500 μL of 0.1× LB containing 0.3% L-arabinose to induce expression of λ-Red and I-SceI.
8. Incubate the cultures at 37 °C without shaking for 3 h to allow homologous recombination to proceed.
9. For each culture, spread 100 μL on a plate of LA-SS containing 15 μg/mL tetracycline, and 100 μL on a plate containing LA with no salt and 15 μg/mL tetracycline (*see* **Note 24**). Incubate the plates at 30 °C overnight. Comparison between the number of colonies on plates with and without sucrose will provide a measure of stringency for the sucrose selection. Comparison between test and control plates will provide an indication of the number of recombinants versus nonrecombinants that still contain the plasmid.
10. Pick 6 candidate colonies from the test plate and streak them on plates containing LA-SS and 15 μg/mL tetracycline to ensure pure cultures of candidate recombinants are obtained. Grow overnight at 30 °C.

11. Follow Subheading 3.2, **steps 9–15** to perform colony PCR analysis of the candidate recombinants, but use primers glmS-screen_F and glmS-screen_R. Successful recombinants should produce a band at 3291 bp. Include a wild-type control colony, which should produce a band at 977 bp, and a no-template control.
12. As an additional screen, recombinants should now constitutively express sfGFP. As only a single copy of the gene is present, the fluorescence may not be visible by the naked eye but cells should appear green when viewed under a fluorescence microscope set up for GFP fluorescence.
13. Use the remaining template suspension from the colony PCR to check for the presence of donor plasmid and/or pACBSCE by "picking and patching." Dip a sterile toothpick into one of the template mixtures then draw a line, 1–2 cm long, on each of three LA plates: first with 30 μg/mL of kanamycin, next with 30 μg/mL of chloramphenicol, and last with 15 μg/mL of tetracycline.
14. Repeat for each template suspension, ensuring the plates are well labelled so that each line corresponds to the correct colony number.
15. Incubate the plates at 37 °C overnight. Recombinant colonies that have lost the plasmids will not grow on the first two plates but will grow on the tetracycline plate.
16. Select three recombination-positive, plasmid-free colonies and inoculate into 5 mL of LB containing 15 μg/mL of tetracycline. Grow overnight at 37 °C with shaking at 200 rpm.
17. The next day, purify the genomic DNA from these cultures using a commercially available DNA extraction kit and submit the DNA for whole genome sequencing to verify the sequence of the modified region and to check for off-target recombination events or other mutations that may have been introduced during the process (*see* **Note 25**).
18. Select one correct colony for use in future experiments. Dispose of any plates used during the previous procedures. The final colony may be stored long-term by growing an overnight culture in LB containing 15 μg/mL of tetracycline and mixing 700 μL of the culture with 300 μL of sterile 50% (v/v) glycerol in a screw-capped tube. Store the stock tube at −80 °C.

3.4 (Optional) Removal of tetA by Flp Recombinase Expression

After using the *tetA* gene for selection of recombinants, it can be maintained in the chromosome for future selections or to help prevent contamination in subsequent experiments. However, it might also be desirable to remove *tetA*, for example, to allow transformation with a Tc^R plasmid or before conducting further rounds of Gene Doctoring. The *tetA* cassette in mUAVbla:tetR 5′

is flanked by Flp Recombination Target (FRT) sites. These short direct repeats are recognized by Flp recombinase, expressed from pCP20, which recombines the two sites, deleting any sequence between them and leaving an 18 bp genomic scar [2, 6].

1. Transform the (sequence-verified) modified strain with pCP20. Recover cells on LA containing 100 μg/mL of ampicillin or carbenicillin at 30 °C (*see* **Note 26**).
2. Pick 10 single colonies and grow each colony in 5 mL of LB without antibiotics, overnight at 42 °C with shaking at 200 rpm (*see* **Note 27**).
3. Check for loss of pCP20 and *tetA* by picking and patching from the overnight cultures, as in Subheading 3.3, **step 13**. Patch colonies to LA plates containing first 100 μg/mL of carbenicillin or ampicillin, then 15 μg/mL of tetracycline, and finally with no antibiotics. Colonies that have lost pCP20 and *tetA* will only grow on the plate with no antibiotics.
4. Perform colony PCR to confirm the loss of *tetA* by following Subheading 3.2, **steps 9–15**. Using primers glmS-screen_F and glmS-screen_R, colonies that still contain *tetA* will produce a band at 3291 bp, while those that have lost *tetA* will produce a band at 1892 bp. Use the untreated *tetA*$^+$ strain from Subheading 3.3, **step 17** as a control.
5. Select one correct colony and inoculate into 5 mL of LB with no antibiotics. Grow overnight at 37 °C with shaking at 200 rpm. The next day, purify the genomic DNA from this culture using a commercially available DNA extraction kit and submit the DNA for whole genome sequencing to verify the removal of *tetA* and the absence of unintended mutations.

4 Notes

1. As the insertion site in this example is given with reference to *glmS*, the sequences are designed relative to the *glmS*-coding strand, as seen in Fig. 3. This is the reverse strand in NC_000913.3.
2. PCR is simple, quick, and cheap but the homologous regions may also be purchased, either as double-stranded gene fragments or as fully sequence-verified parts ligated into a suitable plasmid vector. Homologous regions can be as short as 30 bp [1], but I find that longer regions provide greater efficiency of homologous recombination and are easier to purify when generated by PCR.
3. If possible, use a comb with wide, thick teeth when casting the gel to form a well with sufficient volume (60 μL) to hold the

entire reaction. If such a comb is not available, you can stick small pieces of masking tape across multiple teeth to form a larger well.

4. Although I prefer to gel-purify PCR products, it is not strictly necessary, particularly if the PCR reaction was efficient and specific. Run 3 μL of each PCR reaction on a gel to check that a single product of the correct size was formed. If so, you may purify the remaining reaction volume using a commercial PCR purification spin column kit, and proceed to **step 6**.
5. It is essential to use blue light illumination for gel purification. Never expose the gel to UV light as it will damage the DNA and drastically reduce assembly efficiency [7].
6. Excessive contaminating bands and/or primer dimers can usually be avoided by increasing the annealing temperature. Online tools to predict suitable annealing temperatures are available, such as that provided by New England Biolabs (https://tmcalculator.neb.com/#!/main). If no, or very little, PCR product is formed, check that:
   (i) All components of the reaction were added and no reagents remain inside pipette tips.
   (ii) The correct PCR program was run.
   (iii) The primer sequences are correct and the annealing temperature was not too high.
   (iv) The genomic DNA template has not degraded.
   (v) The DNA polymerase is active by running a known control PCR reaction.
7. DNA recovery yields from gel extraction can be very low [8]. To maximize recovery: begin with at least 50 μL of an efficient PCR reaction; minimize both the percentage of agarose used to cast the gel and the quantity of gel surrounding the cut band; ensure that the correct ratio of gel band to recovery buffer is used; ensure that the gel is completely melted and well mixed before loading to the column; incubate the column for 5–10 min at 50 °C after adding elution buffer to help the DNA to dissolve. I have found the NucleoSpin Gel and PCR Clean-up kit produced by Macherey-Nagel to be particularly reliable.
8. High-quality DNA, with no contaminants, is important for efficient assembly of DNA parts. Deviation from the quality ratios indicated here can usually be dealt with by an extra purification step. The quickest and easiest method is to use a commercial PCR purification spin column kit, eluting a single time with 40 μL of elution buffer.

9. Modern proofreading DNA polymerases are generally accurate enough to amplify short DNA parts, such as the homologous regions, without too much concern for introducing errors. Therefore, I do not usually sequence-verify the homologous regions at this stage. The sequences will be checked as part of the assembled donor plasmid in Subheading 3.2. Longer parts, sequences that are difficult to amplify, or parts that will be used ltiple times may be sequenced at this stage, and could also be cloned into a plasmid to facilitate storage and further production.
10. For inexperienced users, or to assist with troubleshooting, it might be beneficial to also run a positive control assembly that is known to work efficiently. Parts for such a reaction might be available from previous work in your laboratory, or you could use the sequences provided in this protocol. Alternatively, commercially available Golden Gate assembly mixtures are often provided with positive control reagents and have the convenience of being supplied in a pre-mix format.
11. A generic assembly program is provided here, which should work well for the majority of assemblies to produce Gene Doctoring donor plasmids. Difficult assemblies, such as those with many parts, may require optimization, and often benefit from increasing both the duration of the incubation steps and the number of cycles [9]. As it is often convenient to run assemblies overnight, I, therefore, frequently extend the program to 99 cycles of 37 °C for 3 min and 16 °C for 5 min to provide the greatest opportunity for success at the first attempt.
12. High-efficiency chemically competent cells can be purchased or produced in-house. A variety of protocols to produce chemically competent cells can easily be found online, and specific modifications might be necessary for your chosen strain or species. For standard *E. coli* strains, I find the following procedure works well: inoculate 500 μL of an overnight culture into 50 mL of 2× YT medium (16 g/L tryptone, 10 g/L yeast extract, 5 g/L NaCl; pH 7.0) in a 250 mL conical flask. Incubate at 37 °C with shaking at 200 rpm for approximately an hour, or until the $OD_{600} = 0.3–0.4$. Transfer to a 50 mL centrifuge tube, cool on ice then recover cells by centrifugation at 4000× *g* for 10 min at 4 °C. Resuspend in 30 mL of ice-cold buffer containing 80 mM $MgCl_2$ and 20 mM $CaCl_2$. Centrifuge again and re-suspend in 10 mL of ice-cold 100 mM $CaCl_2$. Incubate on ice for at least 1 h. Repeat the centrifugation and re-suspension in 100 mM $CaCl_2$, then perform a final centrifugation and resuspend the cells in 100 μL of 100 mM $CaCl_2$. Volumes can be scaled proportionally to produce greater quantities if required. If preferred, electrocompetent

cells may be used instead, with transformation achieved by electroporation [10].

13. I recommend preparing a positive control for transformation efficiency using 1 μL of a known concentration of a standard plasmid such as pUC19 (in this case, use LA plates containing 100 μg/mL ampicillin or carbenicillin). This helps both to rule out problems with competent cells in case no colonies are recovered and to compare efficiencies between batches.
14. You can centrifuge the remaining cell suspension (1 min at 13,000 × *g*), remove some supernatant to leave approximately 100 μL, and then re-suspend the cell pellet in the remaining supernatant. This suspension can be spread onto a second plate (with the same additions) as insurance against low efficiency of assembly or transformation.
15. Too many cells resuspended here will inhibit the PCR reaction. Lightly touch the center of a single colony to pick up some cells—it is not necessary to pick the entire colony. Sufficient cells will be transferred to the water by gentle agitation against the side of the tube. If the solution appears cloudy or clumps of cells are visible, too many cells have been transferred and the solution should be further diluted with ultrapure water before use as a PCR template.
16. Colony PCR is a fast and inexpensive way to screen colonies and select a few promising candidates for verification by sequencing. However, you may prefer to screen directly by sequencing. In this case, restreak colonies as described here but omit the suspension in water. Grow overnight cultures from single colonies on the restreaked plate, purify the donor plasmids using a commercial plasmid miniprep kit, and submit aliquots for Sanger sequencing using primers pDOC-K_F and pDOC-K_R. Keep the remaining plasmid solutions until one correct sample has been identified. You may then proceed from Subheading 3.3 using the selected donor plasmid.
17. The purpose of re-streaking the colonies is to ensure that pure cultures are screened, in which all cells contain identical versions of the plasmid. Colonies tested directly from the transformation plate may arise from more than one cell lying close to each other, with plasmids that may contain different sequences.
18. Longer assemblies might require a different strategy if the PCR is unable to span the required length. In this case, use pDOC-K_F and design a complementary reverse primer approximately 3 kb downstream, within the inserted region. A product of the correct length will suggest that the construct is correct, but sequence verification will be more important as the PCR will not indicate the size of the entire assembly. Note, too, that the

pDOC-GG control reaction in this case will not produce a PCR product.

19. Enzymatic PCR clean-up is fast and convenient compared to alternative methods such as spin columns or magnetic beads, and has 100% product recovery. The exonuclease I degrades single-stranded DNA, including primers, while the phosphatase removes unincorporated deoxynucleotide triphosphates (dNTPs). This leaves a solution of double-stranded DNA that is free from contaminants and ready for sequencing. If a different purification method is preferred, simply substitute your method of choice at this stage.
20. This protocol assumes the use of Sanger sequencing for the analysis of plasmid sequences as it is generally the fastest and most convenient method at the time of writing. Other sequencing technologies may be preferred and can be substituted where required, here and later in the protocol.
21. For standard strains of *E. coli*, transformation efficiencies should be high enough to allow co-transformation of both plasmids simultaneously. Competent cells can be produced as in **Note 12** and heat shock performed as in Subheading 3.2, **steps 4–6**. Wild-type strains or other species may be harder to transform, and it might be necessary to transform with each plasmid sequentially.
22. Transfer a large colony or a few small colonies, to allow growth within a reasonable time. It is also possible to prepare an overnight culture and use 5–10 μL as inoculum.
23. A shaking heat block is convenient and helps with rapid growth but cultures can also be grown in a standard shaking incubator. The aim is to reach the late-exponential growth stage. Growth without shaking is possible but not recommended as it will take significantly longer to reach the late-exponential stage.
24. Counterselection based on *sacB* works because the gene product (levansucrase) converts sucrose to levansucrose, which cannot be metabolized and is toxic [11]. Lower concentrations of sucrose may still be effective, but it is recommended to use at least 10% to ensure efficacy. The use of salt-free LA potentiates the selective effect [12]. Although *sacB* counterselection is widely applicable in Gram-negative bacteria, some strains are more sensitive than others. If *sacB* selection is not effective, consider modifying pDOC-GG to use an alternative negative selection system [13].
25. As a minimum, the edited region of the genome should be verified by Sanger sequencing; however, whole genome sequencing is strongly advised to ensure that no off-target recombination events have occurred. Check with your whole genome sequencing service provider for sample preparation

requirements. If you wish to remove the antibiotic resistance marker (see Subheading 3.4), whole genome sequencing may be delayed until the completion of that procedure. However, I still recommend some form of sequence verification at this stage.

26. Plasmid pCP20 has a temperature-sensitive origin of replication. Therefore, growth at 30 °C is required to maintain and propagate the plasmid in the cells. Do not grow at higher temperatures until Flp recombinase expression is required.
27. Growth at 37 °C might expose the cells to less stress, but 42 °C is recommended for more stringent inhibition of plasmid replication. In either case, the shift to a higher temperature both prevents plasmid replication and induces expression of Flp recombinase. Overnight incubation allows sufficient time for many cell generations, so that the plasmid is diluted and lost from the majority of daughter cells.

References

1. Muyrers JPP, Zhang Y, Testa G, Stewart AF (1999) Rapid modification of bacterial artificial chromosomes by ET-recombination. Nucleic Acids Res 27:1555–1557. https://doi.org/10.1093/nar/27.6.1555
2. Datsenko KA, Wanner BL (2000) One-step inactivation of chromosomal genes in *Escherichia coli* K-12 using PCR products. Proc Natl Acad Sci USA 97:6640–6645. https://doi.org/10.1073/pnas.120163297
3. Lee DJ, Bingle LE, Heurlier K et al (2009) Gene doctoring: a method for recombineering in laboratory and pathogenic *Escherichia coli* strains. BMC Microbiol 9:252. https://doi.org/10.1186/1471-2180-9-252
4. Thomson NM, Zhang C, Trampari E, Pallen MJ (2020) Creation of Golden Gate constructs for gene doctoring. BMC Biotechnol 20:54. https://doi.org/10.1186/s12896-020-00648-5
5. McKenzie GJ, Craig NL (2006) Fast, easy and efficient: site-specific insertion of transgenes into enterobacterial chromosomes using Tn7 without need for selection of the insertion event. BMC Microbiol 6:39. https://doi.org/10.1186/1471-2180-6-39
6. Cherepanov PP, Wackernagel W (1995) Gene disruption in *Escherichia coli*: TcR and KmR cassettes with the option of Flp-catalyzed excision of the antibiotic-resistance determinant. Gene 158:9–14. https://doi.org/10.1016/0378-1119(95)00193-A
7. Gründemann D, Schömig E (1996) Protection of DNA during preparative agarose gel electrophoresis against damage induced by ultraviolet light. BioTechniques 21:898–903. https://doi.org/10.2144/96215rr02
8. Matsumura I (2015) Why Johnny can't clone: common pitfalls and not so common solutions. BioTechniques 59:4–13. https://doi.org/10.2144/000114324
9. Pryor JM, Potapov V, Kucera RB et al (2020) Enabling one-pot Golden Gate assemblies of unprecedented complexity using data-optimized assembly design. PLoS One 15: e0238592. https://doi.org/10.1371/journal.pone.0238592
10. Sambrook J, Russell DW (2001) Molecular cloning: a laboratory manual, 3rd edn. Cold Spring Harbour Press, Cold Spring Harbour
11. Ried JL, Collmer A (1987) An *nptI-sacB-sacR* cartridge for constructing directed, unmarked mutations in Gram-negative bacteria by marker exchange-eviction mutagenesis. Gene 57:239–246. https://doi.org/10.1016/0378-1119(87)90127-2
12. Blomfield IC, Vaughn V, Rest RF, Eisenstein BI (1991) Allelic exchange in *Escherichia coli* using the *Bacillus subtilis sacB* gene and a temperature-sensitive pSC101 replicon. Mol Microbiol 5:1447–1457. https://doi.org/10.1111/j.1365-2958.1991.tb00791.x
13. Khetrapal V, Mehershahi K, Rafee S et al (2015) A set of powerful negative selection systems for unmodified *Enterobacteriaceae*. Nucleic Acids Res 43:e83. https://doi.org/10.1093/nar/gkv248

Chapter 20

Protocol for NT-CRISPR: A Method for Efficient Genome Engineering in *Vibrio natriegens*

Daniel Stukenberg, Josef Hoff, Anna Faber, and Anke Becker

Abstract

Vibrio natriegens is a gram-negative bacterium, which has received increasing attention due to its very fast growth with a doubling time of under 10 min under optimal conditions. To enable a wide range of projects spanning from basic research to biotechnological applications, we developed NT-CRISPR as a new method for genome engineering. This book chapter provides a step-by-step protocol for the use of this previously published tool. NT-CRISPR combines natural transformation with counterselection through CRISPR-Cas9. Thereby, genomic regions can be deleted, foreign sequences can be integrated, and point mutations can be introduced. Furthermore, up to three simultaneous modifications are possible.

Key words *Vibrio natriegens*, Golden Gate cloning, CRISPR-Cas9, Natural transformation, Genome engineering, Synthetic biology, Biotechnology, Chassis

1 Introduction

Vibrio natriegens was first isolated in 1958 from salt marshes at the coast of Georgia, USA [1]. Shortly after, its super-fast growth with a doubling time of less than 10 min was described [2]. This was also confirmed by more recent measurements [3]. In recent years, *V. natriegens* was proposed as a new chassis organism for synthetic biology and biotechnology [4, 5]. A first generation of genome engineering techniques has been developed [6–8]. The MuGENT method [7] served as the starting point for the development of NT-CRISPR. MuGENT takes advantage of the ability of *V. natriegens* to take up free DNA from the environment and integrate it into its genome in a process known as natural transformation [7]. Natural transformation is under control of several intertwined regulatory pathways such as quorum sensing and carbon catabolite repression [9,10]. In addition, the master regulator TfoX is necessary for the expression of many genes required for natural transformation. While it is known that *tfoX* is induced in the

Daniel Schindler (ed.), *Golden Gate Cloning: Methods and Protocols*, Methods in Molecular Biology, vol. 2850,
https://doi.org/10.1007/978-1-0716-4220-7_20,

presence of chitin in *Vibrio cholerae* [11], the natural inducer of *tfoX* in *V. natriegens* is not known. However, heterologous expression of *tfoX* from an inducible promoter allows induction of natural transformation in *V. natriegens* [7]. NT-CRISPR relies on the counterselection against nonmodified cells after a first natural transformation step. This is achieved by targeting the nonmodified DNA sequence with the clustered regularly interspaced short palindromic repeats (CRISPR)-CRISPR associated protein 9 (Cas9) system CRISPR-Cas9. Originally discovered as an adaptive immune system of prokaryotes against bacteriophages [12, 13], CRISPR-Cas9 has since been developed into an extremely powerful and widely applicable genome engineering method [14, 15]. This system is based on two components: The Cas9 protein and a guide RNA (gRNA) [14]. The gRNA directs the Cas9 protein to a specific DNA sequence, where the Cas9 introduces a DNA double-stranded break [14]. This DNA cleavage leads to cell death [16].

Here we provide a step-by-step protocol to use the previously published NT-CRISPR method [17]. NT-CRISPR achieves a high editing efficiency through two temporally distinct steps. First, the genomic modification is introduced through natural transformation. This is successful in only a subset of cells, while the majority of cells remain unaffected. Therefore, CRISPR-Cas9 is induced in a second step to select against nonmodified cells. This leads to an enrichment of correctly edited cells. With this method, we achieved high editing efficiencies for deletions, integrations, and point mutations. Furthermore, we could demonstrate simultaneous deletion of up to three genomic regions [17]. The NT-CRISPR method relies on a single plasmid, which carries all necessary components (Fig. 1). First, *tfoX, the gene encoding the natural competence master*

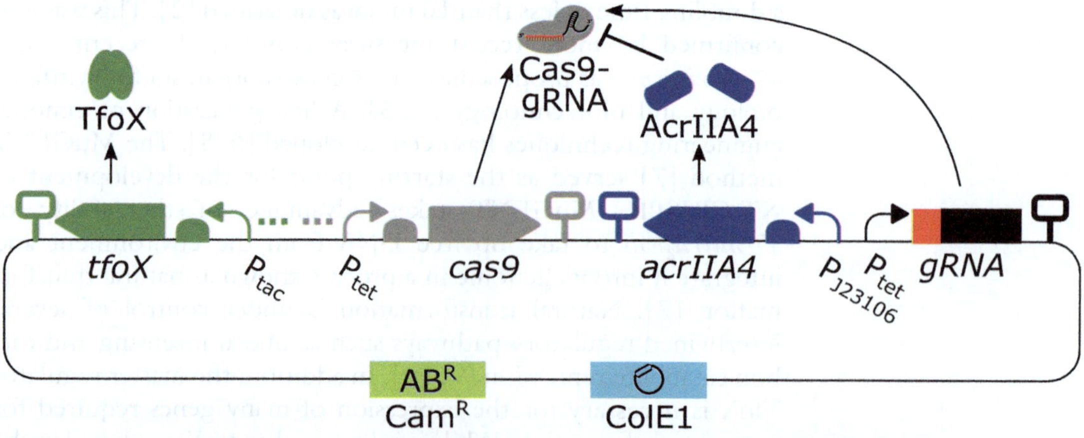

Fig. 1 NT-CRISPR plasmid. NT-CRISPR plasmid carries *tfoX* (P_{tac}, green), *cas9* (P_{tet}, gray), *acrIIA4* (J_{23106}, blue), and *gRNA*, consisting of spacer (red) and scaffold (black), controlled by P_{tet}. The backbone carries a chloramphenicol resistance marker (Cam^R) and a ColE1 origin of replication. Synthetic biology open language (SBOL) symbols for omitted detail (three points) represent the transcriptional unit for the regulatory proteins LacI and TetR for P_{tac} and P_{tet}, respectively

regulator TfoX, is under control of the inducible promoter P_{tac}, which can be induced with isopropyl-β-D-thiogalactopyranoside (IPTG). Expression of *tfoX* induces natural transformation. Second, the *cas9* and *gRNA* sequences are under control of the anhydrotetracycline (ATc) inducible P_{tet} promoter. Addition of ATc allows counterselection against nonmodified cells. Last, the sequence coding for the anti CRISPR protein AcrIIA4. This prevents untimely cell death through leaky expression of *cas9* and *gRNA*. This allows *V. natriegens* cells to stably maintain the NT-CRISPR plasmid.

2 Materials

2.1 Standard Equipment

- Standard equipment for microbial cultivation.
- Standard equipment for PCR and DNA analysis.

2.2 Strains

- *V. natriegens* strain (e.g., ATCC14048), a *Δdns* strain is recommended for efficient plasmid transformation.
- A suitable *Escherichia coli* cloning strain.

2.3 Media

- Lysogeny broth (LB) medium: 10 g/L NaCl, 10 g/L tryptone, 5 g/L yeast extract (with media supplements when indicated).
- LB agar plates: 10 g/L NaCl, 10 g/L tryptone, 5 g/L yeast extract, 15 g/L agar (with media supplements when indicated).
- LBv2 medium: 10 g/L NaCl, 10 g/L tryptone, 5 g/L yeast extract, 100 mL/L 10× V2 solution (with media supplements when indicated).
- LBv2 agar plates: 10 g/L NaCl, 10 g/L tryptone, 5 g/L yeast extract, 15 g/L agar, 100 mL/L 10× V2 solution (with media supplements when indicated).
- Sea salt medium: 28 g/L in H_2O.

2.4 Media Supplements

- 10× V2 solution: 2.04 M NaCl, 42 mM KCl, and 231.4 mM MgCl2.
- Chloramphenicol: 2 mg/mL in ethanol (working concentration for *V. natriegens*: 2 μg/mL).
- Chloramphenicol: 25 mg/mL in ethanol (working concentration for *E. coli*: 25 μg/mL).
- Kanamycin: 50 mg/mL in H_2O (working concentration for *E. coli*: 50 μg/mL).
- Isopropyl-β-D-thiogalactopyranoside (IPTG): 100 mM in H_2O (working concentration: 100 μM).

- Anhydrotetracycline (ATc): 1 mg/mL in ethanol (working concentration: 200 ng/mL).
- 50% glycerol (w/w).

2.5 Enzymes and Buffers

- BsaI.
- Esp3I (only for multiple simultaneous modifications).
- T4-DNA-Ligase.
- T4-DNA-Ligase Buffer (10×).
- PCR reaction mix (high fidelity, for construction of tDNAs).
- Gibson Assembly reaction mix.

2.6 DNA

Plasmids can be obtained from Addgene: Kit #1000000189:

- pST_116.
- pST_140 (only for alternative PAM sequences, *see* **Note 1**).
- pST_119.
- gRNA position vectors (Fig. 2).

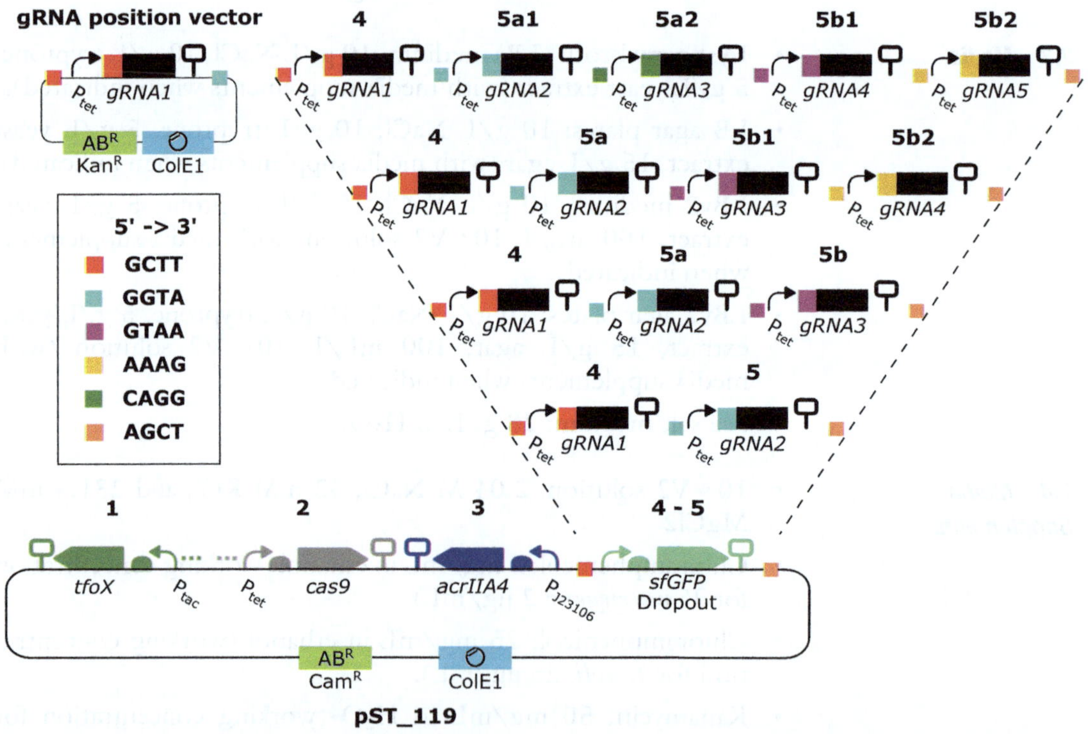

Fig. 2 Assembly of NT-CRISPR plasmid encoding multiple gRNAs. gRNA spacer sequences are first introduced into different gRNA position vectors and are then combined with the pST_119 plasmid to generate NT-CRISPR plasmids with multiple gRNAs. pST_119 carries a *sfGFP* dropout part, which is replaced with the respective *gRNA* expression cassettes. Thereby, correct assembly is indicated by loss of the sfGFP signal yielding white instead of green colonies after transformation. Different gRNA position vectors are used, depending on the number of desired gRNAs in the final construct

- Forward and reverse oligonucleotides for each gRNA (100 μM).
- Primers for amplification of fragments for the assembly of the tDNA.
- Template DNA for PCR reaction (e.g., genomic DNA of *V. natriegens* for homologous region, plasmid with suitable backbone, plasmid with insert for integration).

3 Methods

3.1 Design of gRNAs

The gRNA defines the sequence which is targeted by CRISPR-Cas9. In the NT-CRISPR method, a gRNA is designed to target the unmodified sequence. This leads to cell death of unmodified cells.

1. Use a suitable tool (e.g., Geneious Prime, Benchling, Chop-Chop) to identify potential gRNAs with an NGG protospacer adjacent motif (PAM) sequence. In some cases, a designed gRNA is not functional. Therefore, designing multiple gRNAs for each target can be wise.

   Refer to **Note 1** in case no NGG PAM sequence is available and alternative PAM sequences are needed.
2. Design and order forward and reverse oligonucleotides for the gRNA with the correct overhangs added for the subsequent cloning step. The PAM is **NOT** part of the gRNA sequence.

   Example for a gRNA targeting *wbfF*:

   DNA target sequence: TTAGCCAAGATCAGTCACGT**AGG** (**PAM in bold**)

   Forward oligonucleotide: GTCC TTAGCCAAGATCAGT CACGT (overhangs underlined)

   Reverse oligonucleotide: AAAC ACGTGACTGATCTTG GC TAA (overhangs underlined)

3.2 Construction of NT-CRISPR Plasmids

The final NT-CRISPR plasmids carry all the necessary components to perform NT-CRISPR. Each application only requires the exchange of the gRNA sequence to specify the target site. All other components are already present in the preassembled NT-CRISPR plasmids pST_116, pST_119, and pST_140. Construction of NT-CRISPR plasmid is different when a single or multiple modifications are performed.

3.2.1 Construction of NT-CRISPR Plasmid (Single gRNA)

1. Use plasmid pST_116 or pST_140 (*see* **Note 1**).
2. Anneal oligonucleotides for gRNA. Mix 1.5 μL of each oligonucleotide (100 μM) with 5 μL 10× T4-DNA Ligase buffer and 42 μL H_2O in a 1.5 mL reaction tube. Incubate in a heat block at 95 °C for 10–15 min. Switch off the heat block to let the reaction cool down slowly to room temperature (~ 1 h).

3. Prepare the Golden Gate Assembly reaction for the integration of the gRNA sequence into pST_116: ~200 ng of pST_116, 3 μL of the annealed oligonucleotide reaction, 0.5 μL BsaI, 0.5 μL T4-DNA Ligase, 1 μL 10× T4-DNA Ligase buffer, add H_2O to a final volume of 10 μL.
4. Run Golden Gate assembly reaction: 30 cycles of 37 °C (2 min) and 16 °C (5 min), followed by a final digestion step at 37 °C for 30 min and an enzyme denaturation step at 80 °C for 10 min.
5. Introduce cloning reaction directly into *V. natriegens* or a suitable *E. coli* cloning strain. Plasmid transformation with *V. natriegens* can be done through heat shock transformation [18] or electroporation [6]. Use LB agar plates with chloramphenicol (2 μg/mL for *V. natriegens* and 25 μg/mL for *E. coli*). Colonies carrying a correctly assembled plasmid will appear white. Green colonies are the parental pST_116. In case an *E. coli* strain was used, isolate the plasmid and introduce it into *V. natriegens.*

3.2.2 Construction of NT-CRISPR Plasmid (Multiple gRNAs)

1. Use plasmid pST_119 and the correct gRNA position vectors, depending on the number of gRNA sequences in the final design (Fig. 2).
2. Anneal oligonucleotides (*see* Subheading 3.2.1, **Step 2**).
3. Prepare and run the Golden Gate Assembly reactions (*see* Subheading 3.2.1, **steps 3** and **4**). Use plasmids of the gRNA position vectors instead of pST_116.
4. Introduce cloning reactions into a suitable *E. coli* cloning strain. Use LB agar plates with kanamycin (50 μg/mL). Colonies carrying a correctly assembled plasmid will appear white.
5. Isolate plasmids.
6. Prepare Golden Gate Assembly reaction to integrate gRNA expression cassettes into pST_119: ~200 ng of pST_119, ~70 ng of each gRNA position vector, 0.5 μL Esp3I (instead of BsaI), 0.5 μL T4-DNA Ligase, 1 μL 10× T4-DNA Ligase buffer, add H_2O to a final volume of 10 μL.
7. Introduce cloning reaction directly into *V. natriegens* or a suitable *E. coli* cloning strain. Use LB agar plates with chloramphenicol (2 μg/mL for *V. natriegens* and 25 μg/mL for *E. coli*). Colonies carrying a correctly assembled plasmid will appear white. Green colonies are the parental pST_119. In case an *E. coli* strain was used, isolate the plasmid and introduce it into *V. natriegens.*

3.3 Generation of tDNAs

tDNAs are used as linear double-stranded DNA fragments with homology to the genomic region flanking the desired change. Any method capable of generating such a DNA fragment can be used. Here, only one approach using Gibson Assembly is described.

1. Design tDNA: A homology of 3 kb is recommended for high efficiency. For deletions, the tDNA is composed of the upstream and downstream regions flanking the intended deletion. For a point mutation, the desired modification can be introduced at the fusion site between the upstream and downstream fragments. For integrations, the tDNA consists of the regions flanking the integration site and the fragment with the sequence to be integrated.
2. Amplify all necessary fragments by PCR, for example, upstream, downstream, insert. Additionally amplify one backbone fragment, carrying an origin of replication and an antibiotic resistance marker. Assemble plasmids via Gibson Assembly.
3. Introduce cloning reaction into a suitable *E. coli* cloning strain.
4. Isolate and verify plasmid DNA (e.g., test digest, PCR, Sanger sequencing). This is the tDNA template plasmid.
5. Amplify tDNA fragment via PCR from the tDNA template plasmid.

3.4 NT-CRISPR: Natural Transformation + Counterselection

In this step, the genomic modification is performed. In a first step, the natural transformation leads to a genomic modification in a subset of cells. In a second step, CRISPR-Cas9 is induced and leads to the death of unmodified cells, thereby increasing the fraction of modified cells. We suggest including controls in every NT-CRISPR procedure (*see* **Note 2**).

1. Inoculate an overnight culture of *V. natriegens* carrying the previously constructed NT-CRISPR plasmid (*see* Subheading 3.2): 5 mL in a glass tube, LBv2 with 4 µg/mL chloramphenicol and 100 µM IPTG, 30 °C, 200 rpm.
2. Prepare natural transformation reaction: 350 µL sea salt medium, 3.5 µL *V. natriegens* overnight culture with NT-CRISPR plasmid from overnight culture, 100 ng tDNA (multiple tDNAs when multiple simultaneous modifications are performed).
3. Mix briefly.
4. Incubate at 30 °C for 5 h (*see* **Note 3**).
5. Add 1 mL LBv2 with 200 ng/mL ATc.
6. Mix briefly.
7. Incubate for 1 h at 30 °C and 300 rpm.
8. Mix briefly.

9. Prepare suitable dilutions (*see* **Note 4**).
10. Plate 100 μL on LBv2 plates with 2 μg/mL chloramphenicol and 200 ng/mL ATc (*see* **Note 5**).
11. Incubate overnight.

3.5 Screening of Correctly Modified Cells by Colony PCR (cPCR)

1. Estimate editing efficiency based on the number of colonies in comparison to control without tDNA to approximate the number of colonies that should be tested (*see* **Note 6**).
2. Resuspend cells from colonies in 20 μL H_2O in a 1.5 mL reaction tube.
3. Incubate cell suspension at 95 °C for 15 min.
4. Pellet cell debris by centrifugation at max. Speed for 3 min.
5. Use 1 μL of supernatant in a 12.5 μL cPCR reaction.
6. Identify colonies with a correct band pattern.

3.6 Curing of Modified Cells from NT-CRISPR Plasmid

1. Inoculate 5 mL antibiotic-free LBv2 with cells from the colonies tested by cPCR (this can be done in parallel to running the cPCR to safe time).
2. Incubate for 5–7 h at 37 °C and 200 rpm.
3. Prepare a suitable dilution (e.g., 10^{-6} or 10^{-7}) in LBv2 (without antibiotic).
4. Plate 100 μL of dilution on LBv2 agar plates (without antibiotic).
5. Incubate overnight.

3.7 Checking for Plasmid Loss

1. Streak colonies from curing (*see* Subheading 3.8) on LBv2 plates with and without 2 μg/mL chloramphenicol (two to three colonies per created strain should be sufficient).
2. Incubate for ~6 h at 37 °C.
3. Identify colonies growing on plates without but not with chloramphenicol. Those colonies have lost the NT-CRISPR plasmid.

3.8 Preparing Generated Strain for Long-Term Storage

1. Inoculate 5 mL LBv2 (without antibiotic) with cells from the colonies tested for plasmid loss (this can be done in parallel to checking for plasmid loss (*see* Subheading 3.7)).
2. Incubate for 5–7 h at 37 °C and 200 rpm. Alternatively, a culture can be grown overnight at 30 °C.
3. Prepare a glycerol stock by mixing the grown culture in equal amounts with 50% glycerol.
4. Store at −80 °C.

4 Notes

1. In some cases, it is not possible to design a gRNA with an NGG PAM sequence. This is especially the case for integrations and point mutations as these modifications require binding of the gRNA directly on the site of integration or point mutation. For these cases, we created plasmid pST_140. This plasmid carries the spG Cas9, which only requires NGN as the PAM [19]. This plasmid can be used identically to pST_116. We found alternative gRNAs with alternative PAMs to be less reliable than NGG PAMs.
2. We suggest two controls, which should be included in every NT-CRISPR procedure. This helps to identify the most probable causes for failure of the NT-CRISPR method.
   (a) A reaction without tDNA: This control checks for efficient cell killing. Plate a 10^{-1} dilution of this reaction on LBv2 plates with 2 μg/mL chloramphenicol and 200 ng/mL ATc. A 10^{-1} dilution of the NT-CRISPR reaction should result in less than 300 colonies. If significantly more colonies are formed, it is likely that the gRNA is not functional or the ATc is degraded.
   (b) A reaction without ATc: This control checks for the number of viable cells with the NT-CRISPR plasmid in the NT-CRISPR reaction. Run the protocol as described but use LBv2 without ATc (*see* Subheading 3.4, **steps 5** and **10**). Plate a 10^{-4} dilution of this control reaction on LBv2 plates with 2 μg/mL chloramphenicol but without ATc. We usually obtain approximately 100–400 colonies. If significantly less colonies are formed, it is possible that the preculture was not grown long enough (incubate longer) or that cells have lost the plasmid (check chloramphenicol stock).
3. The dilution that should be plated depends on the anticipated number of colonies and therefore on the editing efficiency. In our hands, for most modifications plating 100 μL of a 10^{-2} or 10^{-3} dilution resulted in a good number of colonies (50–500). If a low efficiency is anticipated or if multiple modifications are performed simultaneously, plating undiluted samples or a 10^{-1} dilution should be considered.
4. Incubation times in the natural transformation step shorter than 5 h are possible (e.g., 2 or 3 h). However, this will result in reduced efficiency, resulting in a lower number of correctly edited cells. Incubation times longer than 5 h (including overnight incubation) did not result in a higher efficiency in our hands.

5. ATc is light and temperature sensitive. LBv2 agar plates with ATc can be used for at least 1 week when stored cold and dark, e.g., wrapped in aluminum foil or in nontransparent boxes. When ATc is added to LBv2 agar plates, it should be added immediately before pouring of the plates and it should be avoided to pour the plates too hot.
6. The editing efficiency of an NT-CRISPR run can be estimated when compared to the control without the addition of tDNA (*see* **Note 2.a**). Example: 500 colonies at a 10^{-1} dilution of the samples with tDNA and 50 colonies at the same dilution without tDNA. The 50 colonies from the sample without tDNA provide an estimate for the rate of escapee formation (different for each gRNA!). The difference between these colony counts (500–50 = 450) suggests that out of the 500 colonies obtained with tDNA, approximately 450 colonies are correct, resulting in an editing efficiency of 90%. This estimation can be used to decide on the number of colonies that should be tested through cPCR.

Acknowledgments

This work was funded by the State of Hesse (Germany) through the LOEWE research cluster MOSLA and the European Union through the BioRoboost project (H2020-NMBP-TR-IND-2018-2020/BIOTEC-01-2018 (CSA), Project ID 210491758).

References

1. Payne WJ (1958) Studies on bacterial utilization of uronic acids. III. Induction of oxidative enzymes in a marine isolate. J Bacteriol 76: 301–307. https://doi.org/10.1128/JB.76.3.301-307.1958
2. Eagon RG (1962) Pseudomonas natriegens, a marine bacterium with a generation time of less than 10 minutes. J Bacteriol 83(4):736–737. https://doi.org/10.1128/jb.83.4.736-737.1962
3. Hoffart E, Grenz S, Lange J et al (2017) High substrate uptake rates empower vibrio natriegens as production host for industrial biotechnology. Appl Environ Microbiol 83:1–10. https://doi.org/10.1128/AEM.01614-17
4. Hoff J, Daniel B, Stukenberg D et al (2020) Vibrio natriegens: an ultrafast-growing marine bacterium as emerging synthetic biology chassis. Environ Microbiol 22:4394–4408. https://doi.org/10.1111/1462-2920.15128
5. Thoma F, Blombach B (2021) Metabolic engineering of *Vibrio natriegens*. Essays Biochem 65:381–392. https://doi.org/10.1042/EBC20200135
6. Weinstock MT, Hesek ED, Wilson CM, Gibson DG (2016) Vibrio natriegens as a fast-growing host for molecular biology. Nat Methods 13:849–851. https://doi.org/10.1038/nmeth.3970
7. Dalia TN, Hayes CA, Stolyar S et al (2017) Multiplex genome editing by natural transformation (MuGENT) for synthetic biology in Vibrio natriegens. ACS Synth Biol 6:1650–1655. https://doi.org/10.1021/acssynbio.7b00116
8. Lee HH, Ostrov N, Wong BG et al (2019) Functional genomics of the rapidly replicating bacterium Vibrio natriegens by CRISPRi. Nat Microbiol 4:1105–1113. https://doi.org/10.1038/s41564-019-0423-8
9. Jaskólska M, Stutzmann S, Stoudmann C, Blokesch M (2018) QstR-dependent regulation of natural competence and type VI secretion in

vibrio cholerae. Nucleic Acids Res. https://doi.org/10.1093/nar/gky717

10. Shin J, Rychel K, Palsson BO (2023) Systems biology of competency in vibrio natriegens is revealed by applying novel data analytics to the transcriptome. Cell Rep 42:112619. https://doi.org/10.1016/j.celrep.2023.112619
11. Meibom KL, Blokesch M, Dolganov NA et al (2005) Chitin induces natural competence in Vibrio cholerae. Science 310:1824–1827. https://doi.org/10.1126/science.1120096
12. Bolotin A, Quinquis B, Sorokin A, Ehrlich SD (2005) Clustered regularly interspaced short palindrome repeats (CRISPRs) have spacers of extrachromosomal origin. Microbiology 151: 2551–2561. https://doi.org/10.1099/mic.0.28048-0
13. Mojica FJM, Díez-Villasenor C, García-Martínez J, Soria E (2005) Intervening sequences of regularly spaced prokaryotic repeats derive from foreign genetic elements. J Mol Evol 60:174–182. https://doi.org/10.1007/s00239-004-0046-3
14. Jinek M, Chylinski K, Fonfara I et al (2012) A programmable dual-RNA–guided DNA endonuclease in adaptive bacterial immunity. Science 337:816–821. https://doi.org/10.1126/science.1225829
15. Jiang W, Bikard D, Cox D et al (2013) RNA-guided editing of bacterial genomes using CRISPR-Cas systems. Nat Biotechnol 31:233–239. https://doi.org/10.1038/nbt.2508
16. Cui L, Bikard D (2016) Consequences of Cas9 cleavage in the chromosome of Escherichia coli. Nucleic Acids Res 44:4243–4251. https://doi.org/10.1093/nar/gkw223
17. Stukenberg D, Hoff J, Faber A, Becker A (2022) NT-CRISPR, combining natural transformation and CRISPR-Cas9 counterselection for markerless and scarless genome editing in vibrio natriegens. Commun Biol 5:265. https://doi.org/10.1038/s42003-022-03150-0
18. Stukenberg D, Hensel T, Hoff J et al (2021) The Marburg collection: a Golden Gate DNA assembly framework for synthetic biology applications in *Vibrio natriegens*. ACS Synth Biol 10:1904–1919. https://doi.org/10.1021/acssynbio.1c00126
19. Walton RT, Christie KA, Whittaker MN, Kleinstiver BP (2020) Unconstrained genome targeting with near-PAMless engineered CRISPR-Cas9 variants. Science 368:290 LP – 296. https://doi.org/10.1126/science.aba8853

Chapter 21

Golden Gate Cloning in Actinobacteria: Opportunities and Challenges

Anna Christina R. Ngo, Melody Haarmann, Nils Weindorf, David Angelo V. Guanzon, Vivian Linke, Joe Smitka, and Dirk Tischler

Abstract

As we exploit biological machineries and circuits to redesign nature, it is just important to use efficient cloning strategies and methods to heterologously express the resulting DNA constructs. Golden Gate cloning allows the assembly of multiple fragments in a single reaction, making the process efficient and seamless. Although Golden Gate strategies have already been employed for different organisms, it is still not well-established for Actinobacteria. Here, we describe methods for Golden Gate cloning and how it can be utilized for Actinobacteria.

Key words Golden Gate assembly, Actinobacteria, *Rhodococcus*, Electroporation

1 Introduction

Synthetic biology is a growing field in biology that combines molecular biology and engineering to manipulate cell systems. Through the concepts employed in this field it becomes possible to engineer and improve microorganisms deliberately to produce valuable compounds [1, 2], to circumvent infections [3, 4], and to be used as a biosensor [5, 6], among others. Nowadays, there is even a rising demand for microbial cell factories especially with the shift toward green biotechnology than traditional chemical synthesis [7, 8].

Traditional expression systems rely on organisms such as *Pichia pastoris* for proteins of fungal origin [9] and *Escherichia coli* for heterologous expression [10, 11]. However, these systems can be difficult to exploit once the target or conditions become more intricate. *E. coli* systems have been the gold standard for the past decades, but certain proteins can be toxic to produce or the available tRNA pool is not optimal [12]. This is why the search for alternative expression systems also increases. Actinobacteria such as

Daniel Schindler (ed.), *Golden Gate Cloning: Methods and Protocols*, Methods in Molecular Biology, vol. 2850, https://doi.org/10.1007/978-1-0716-4220-7_21,

Table 1
List of Actinobacteria species where Golden Gate assembly was applied to

Organisms	Source
Corynebacterium glutamicum	[17]
Rhodococcus sp.	[18–20]
Streptomyces sp.	[21–23]

Streptomyces spp. and *Rhodococcus* spp. are seen as potential alternatives as they can tolerate harsher conditions such as higher solvent conditions and are known to produce powerful enzymes or systems that can be engineered for valuable products [13–16]. This is especially true for a variety of natural products synthesized by the complex machinery of enzymes functioning in a concerted mode. Here, the controlled gene expression and thus assembly of functional systems is of major importance.

However, molecular tools and plasmids for Actinobacteria are still not yet fully established, and available tools are limited. The utilization of Golden Gate assembly, for example, can be tricky as most type IIS restriction sites are present in plasmids often used in *Rhodococcus* spp. and related Actinobacteria. Moreover, there are only a few publications about Golden Gate assembly on Actinobacteria in general (Table 1). In this chapter, we will describe how to assemble fragments via Golden Gate assembly and describe the procedures on how to insert these plasmids into Actinobacteria such as *Rhodococcus* sp.

2 Materials

2.1 Standard Enzymes Are Needed to Perform the Following Procedures (See Note 1)

1. DpnI.
2. PaqCI + PaqCI enhancer (here, we use the one from New England Biolabs that includes the PaqCI enhancer).
3. T4 DNA ligase.
4. Taq polymerase.

2.2 Plasmids

All plasmids used in this study are listed in Table 2.

2.3 DNA Oligonucleotides

All oligonucleotides used in this study are listed in Table 3. The stock solutions for the oligonucleotides are prepared using a 100 μM concentration and are stored in −20 °C. For PCR reactions, working solutions of 10 μM concentration are prepared and stored in −20 °C.

Table 2
List of plasmids used in the study

Plasmids	Features	Source
pDD57[a]	Constitutive expression of GFP+ with kanamycin resistance from pAL5000/pMB1 backbone	[14]
pNit_QC	Constitutive expression using nit promotor with chloramphenicol resistance	[24]

[a]The plasmid pDD57 was a gift from Tae Seok Moon (Addgene plasmid # 119882; http://n2t.net/addgene:119882 ; RRID:Addgene_119,882)

Table 3
List of oligonucleotides used to amplify fragments for the Golden Gate assembly

Oligos	Sequence	Features
Vector + nit overhang	TATCACCTGCACTATGTACACCCGAGAAGCTCC	For amplification of vector
Vector + GFP overhang	TATCACCTGCACTAGACTAGTCGACCCACCG	For amplification of vector
Nit Fragment Fw	TATCACCTGCACTATACATATCGAGGCGGGC	For amplification of promoter
Nit Fragment Rv	TATCACCTGCACTACATA TATCTCCTTCTTAAAGTTAAACAAAATTATTT	For amplification of promoter
His Sumo Fw	TATCACCTGCACTATATGCATCATCATCATCA TCACAGC	For amplification of tags
His Sumo Rv	TATCACCTGCACTATCATATGTCCCTGGAAG TACAGG	For amplification of tags
GFP Fragment Fw	TATCACCTGCACTAATGAGCAAAGGAGAAGAAC TTTTCAC	For amplification of gene of interest
GFP Fragment Rv	TATCACCTGCACTAAGTC TAGTATTTGTAGAGCTCATCCATGC	For amplification of gene of interest

2.4 Chloramphenicol (30 mg/mL Stock Concentration)

This antibiotic is resuspended in absolute ethanol and is distributed in 1 mL aliquot. The tubes are stored in −20 °C. The prepared stock solution should be diluted 1000×.

2.5 Chemicals, Buffers, and Media Components

1. 1× Tris-acetate-EDTA (TAE) buffer (40 mM Tris base, 20 mM acetic acid, 1 mM EDTA [ethylenediaminetetraacetic acid], pH 8.5).
2. DNA stain (SERVA DNA Stain G).
3. DNA ladder.
4. Agarose gel.
5. Lysogeny broth (LB) medium (10% tryptone, 5% NaCl, 5% yeast extract).

6. Super optimal broth with catabolite repression (SOC) medium (2% tryptone, 0.05% NaCl, 2.5 mM KCl, 10 mM $MgCl_2$, 10 mM $MgSO_4$, 20 mM glucose).
7. Solid medium for bacterial cultivation is prepared using 1.5% (w/v) agar.
8. LB broth with 1.5% (w/v) glycine.
9. 10% (v/v) glycerol and 80% (v/v) glycerol.

2.6 Consumables

1. PCR reaction tubes.
2. Standard petri dishes.
3. 0.2 mm electroporation cuvettes.
4. Cryogenic tubes.

2.7 Devices

1. Thermocycler.
2. Electroporator.
3. Thermal shaker, incubator.

3 Methods

The first part of the methods section (Subheadings 3.1, 3.2 and 3.3) explains the needed steps and preparations for constructs specifically used for Actinobacteria using Golden Gate assembly. Here, a pDD57 plasmid containing green fluorescent protein (GFP) will be used to amplify the target gene fragment where the amplicon will be assembled altogether with other fragments consisting of the Nit promotor, His-Sumo-Tev, and transfer it to a new vector, pNit_QC (Fig. 1a). Meanwhile, the second part of the methods section (Subheadings 3.4 and 3.5) talks about the preparations needed for making electrocompetent cells of *Rhodococcus* sp. and the processes needed when transforming them.

Overall, using GFP is just a proof-of-concept, but this can be replaced by any target gene that one may want to express and produce in nontraditional expression systems such as *Rhodococcus* sp., which this protocol can serve as a guide for.

3.1 Primer Design for Vector and Fragments

1. Search for the proper type IIS restriction sites to be used. Here, we made use of PaqCI restriction enzyme to allow the insertion of overhangs for each fragment. This enzyme uses the following recognition site [5′- . . . .CACCTGC(N)$_4$. . . .-3′] to cut the designed sequence to fuse the DNA fragments (*see* **Note 2**).
2. Design primers accordingly. Here, we used the following order:

   W3-**PaqCI recognition site** - extra 4 nucleotides - **4 nucleotides complementary overhang** - annealing parts of the target fragment (*see* **Note 3**).

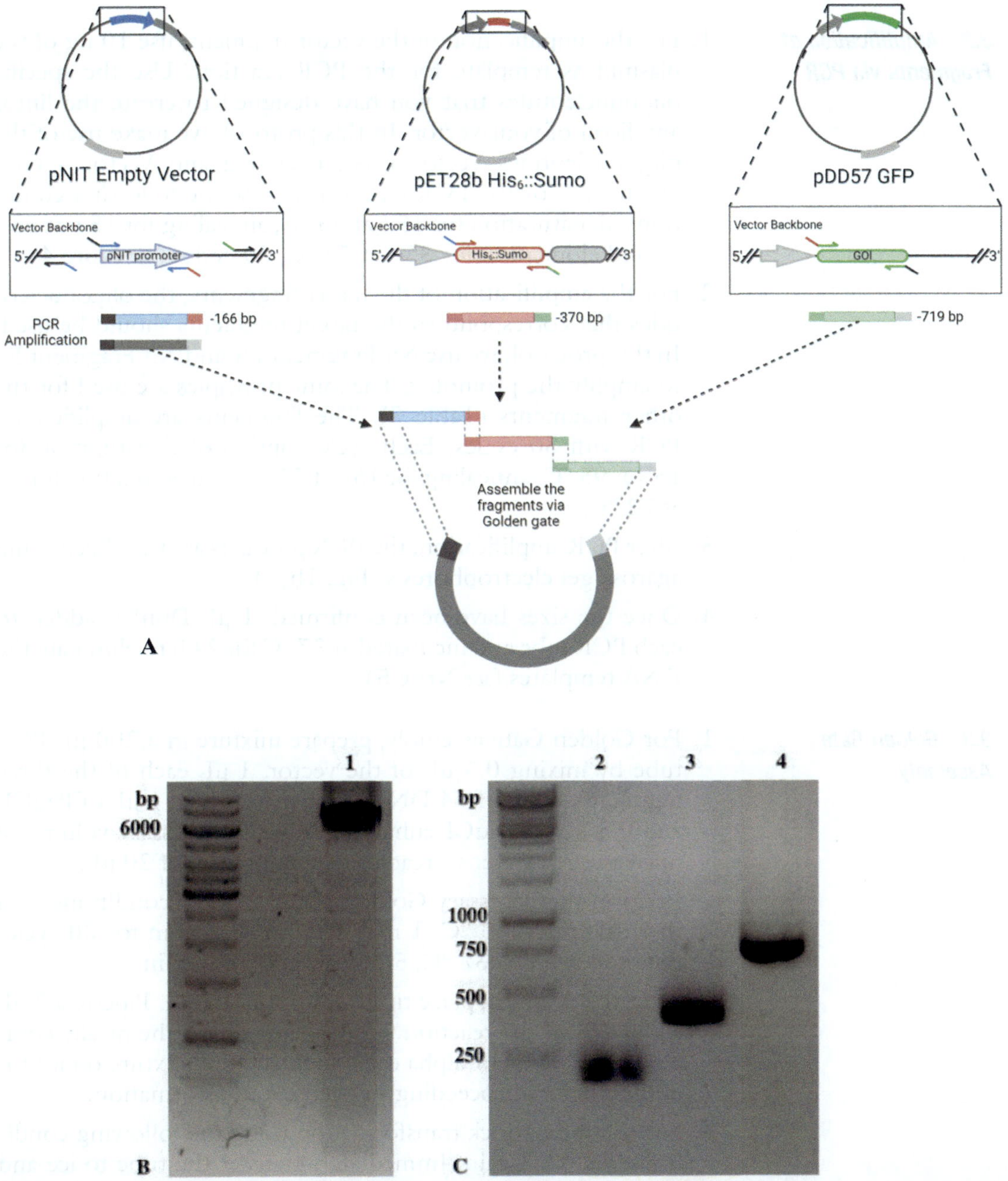

Fig. 1 (**a**) The workflow for this protocol consists of amplifying four fragments that are assembled via Golden Gate assembly. The colors represent the fragment (gray = vector, blue = promoter, red = His-Sumo tag, and green = GFP) and each fragment amplified contains the specific overhang (also indicated by the colors) complementary to the overhang of the gene it should anneal with. The vector pDD57 contains a kanamycin resistance cassette. When GFP gets amplified using this plasmid and transferred to a pNit, a vector that has an ampicillin/chloramphenicol resistance, the possibility of having false-positive clones is slim. (**b**) The resulting amplicon of pNit empty vector (1). (**c**) The resulting amplicons for the inserts: nit promoter (2), His-Sumo tag (3), and GFP (4)

3.2 Amplification of Fragments via PCR

1. For the amplification of the vector fragment, use 10 ng of the plasmid as template for the PCR reaction. Use the specific oligonucleotides that you have designed to create the linear amplicon of your vector. In this protocol, we make use of the oligonucleotides Vector + Nit overhang and Vector + GFP overhang. For the PCR reaction, we use the following conditions: denaturation for 15 s at 95 °C, annealing for 15 s at 55 °C, and elongation for 35 s at 72 °C, 30 cycles (*see* **Note 4**).
2. For the amplification of the insert fragments, the oligonucleotides that correspond to the target fragments should be used. In this protocol, we use Nit Fragment Fw and Nit Fragment Rv to amplify the promotor. The same principles are used for the other fragments (Table 3). The fragments are amplified via PCR with 30 cycles. Each cycle consists of denaturation for 15 s at 95 °C, annealing for 15 s at 57 °C, and elongation for 5 s at 72 °C.
3. After PCR amplification, the PCR products are visualized using agarose gel electrophoresis (Fig. 1b, c).
4. Once the sizes have been confirmed, 1 μL DpnI is added to each PCR tube and incubated at 37 °C for 24 h to eliminate the DNA templates (*see* **Note 5**).

3.3 Golden Gate Assembly

1. For Golden Gate assembly, prepare mixture in a 200 μL PCR tube by mixing 0.5 μL of the vector, 1 μL each of the three fragments, 5 μL of T4 DNA ligase master mix, 1 μL of PaqCI, and 0.5 μL of PaqCI enhancer. Add the necessary volume of nuclease-free water to reach the final volume of 20 μL.
2. Perform the necessary Golden Gate assembly conditions on a thermal cycler: 37 °C, 1 min and 16 °C, 1 min for 30 cycles then followed by 37 °C, 5 min and 60 °C, 5 min.
3. After the reaction, place tube on ice for 10 min. Pipette a 2 μL aliquot from the reaction to a vial containing chemically competent *E. coli* DH5-alpha cells. Incubate the mixture on ice for 30 min before proceeding to bacterial transformation.
4. Perform heat shock transformation using the following conditions: 42 °C, 1 min. Immediately transfer the tube to ice and rest for 10 min. Add 900 μL of SOC medium or LB medium (without antibiotics) for recovery and incubate the tubes at 37 °C with shaking for 1 h.
5. Centrifuge the tubes at 5000× *g*, 7 min. Discard the supernatant and resuspend the cell pellet with 100 μL SOC or LB medium and then plate the whole suspension on a LB agar plate with the appropriate antibiotics.

6. Check for GFP colonies under UV light. Since the pNit promoters are constitutively expressed, the *E. coli* colonies will glow once illuminated with UV light. Validate using sequencing primers. Sequencing primers are recommended to be designed upstream of the open reading frame.

3.4 Preparation of Electrocompetent Rhodococcus sp. Cells

1. Prepare preculture of *Rhodococcus* sp. and incubate at 30 °C, overnight.
2. Prepare the main culture with 100 mL LB with 1.5% w/v glycine on a 500 mL flask (*see* **Note 6**).
3. Incubate at 30 °C with shaking until it reaches the $OD_{600} = 0.5$ to 0.6.
4. Harvest cells via centrifugation at 5000× *g*, 30 min, 4 °C.
5. Discard the supernatant and put 35 mL of ice-cold water aseptically (*see* **Note 7**).
6. Vortex the tube and repeat Step #4.
7. Discard the supernatant and put another 35 mL of ice-cold water aseptically and.
8. Vortex the tube and repeat Step #4.
9. After centrifugation, discard the supernatant and wash with 35 mL of ice-cold 10% v/v glycerol.
10. Vortex the tube and repeat Step #4.
11. Discard the supernatant.
12. Resuspend the cells with 1 mL 10% v/v glycerol and measure the OD_{600}.
13. Adjust the cells to $OD_{600} = 5$ by pipetting the appropriate amount of 10% v/v glycerol to reach the desired density.
14. Aliquot the suspension in 100 μL fractions.
15. Freeze the tubes in liquid nitrogen and store immediately at −80 °C.

3.5 Electroporation, Validation, and Cryopreservation of Rhodococcus sp.

1. Slowly thaw electrocompetent *Rhodococcus* sp. for 20 min on ice.
2. Put electroporation cuvettes (2 mm gap) on ice.
3. Add 10 ng plasmid to cells and gently pipette up and down.
4. Transfer the cells to the electroporation cuvette and make sure to gently tap the cuvette so the cells settle down at the bottom.
5. Keep cuvettes on ice for 30 min.
6. Perform the electroporation using the following parameters: 1.25 kV, 10 μF, 600 Ω. (*see* **Note 8**).
7. Immediately add 900 μL of SOC medium to the cuvette and mix gently.

8. Transfer the suspension to a 1.5 mL Eppendorf tube.
9. Incubate the tube at 30 °C for 4 h, 250 rpm.
10. Plate the solution at selection plate and leave at 30 °C for 48–72 h.
11. Observe for colonies and validate using colony PCR or restriction digestion.
12. Send the positive clones for sequencing.
13. Once validated, grow positive clones on 5 mL LB with the appropriate antibiotics at 30 °C, overnight with shaking.
14. Harvest the cells aseptically in a sterile 1.5 mL Eppendorf tube at 5000× *g*, 10 min. Discard the supernatant, collect the remaining cultures, and centrifuge once more. Discard the supernatant again.
15. Resuspend the cells with 500 μL LB with the appropriate antibiotics. Mix the bacterial solution with 500 μL sterile 80% v/v glycerol solution. Pipette the mixture gently until the glycerol solution mixes up properly.
16. Transfer the solution to a cryogenic tube and store at −80 °C.

4 Notes

1. All enzymes should be stored at −20 °C and during usage, it should be always kept on ice.
2. Search for existing type IIS restriction sites on the plasmid of interest. This is necessary to check because existing type IIS restriction sites will also cause the plasmid of interest to be cut. If possible, the plasmid can also be mutated to remove the respective restriction sites with the method of your choice (e.g., site-directed mutagenesis).
3. It is recommended to design primers with an annealing temperature of 60 °C and the 3′ end should have GC clamps.
4. In this protocol, the DNA polymerase used has an extension time of 5 s/kb. DNA polymerases can vary depending on the source and the supplier. It is important to always check the recommended conditions from the supplier, especially the extension part.
5. It is not necessary to digest the PCR products with DpnI, but it is a recommended step to remove any DNA templates and to avoid false-positive candidates.
6. The use of glycine aids in weakening the cell wall of *Rhodococcus* sp. This can also be used for other actinobacteria. However, it is best to check other additives such as mannitol and sorbitol. The

additives should be optimized for the specific actinobacteria that you want to investigate.

7. When making electrocompetent cells, make sure to wash properly, leftover salts are conductive and can lead to short circuits. Thus, the transformation will fail.
8. After incubation of the electroporation cuvette on ice, make sure to wipe it dry first with tissue before proceeding for electroporation.

Acknowledgments

The project was supported in the frame of the MERCUR cooperation grant Ko-2022-0008. A part of this research was funded by the DFG Research Training Group GRK 2341 "Microbial Substrate Conversion (MiCon)." D.A.V. Guanzon was supported by a pre-doctoral scholarship from Katholischer Akademischer Ausländer Dienst (KAAD). We thank Prof. Tae Seok Moon for providing us with the vector pDD57 and other constructs and Mr. Selvapravin Kumaran for providing the plasmid pET28b His-Sumo-Tev.

References

1. Breitling R, Takano E (2015) Synthetic biology advances for pharmaceutical production. Curr Opin Biotechnol 35:46–51
2. Jullesson D, David F, Pfleger B, Nielsen J (2015) Impact of synthetic biology and metabolic engineering on industrial production of fine chemicals. Biotechnol Adv 33:1395–1402
3. Khan A, Ostaku J, Aras E, Safak Seker UO (2022) Combating infectious diseases with synthetic biology. ACS Synth Biol 11:528–537
4. Koh E, Hwang IY, Lee HL, De Sotto R, Lee JWJ, Lee YS, March JC, Chang MW (2022) Engineering probiotics to inhibit *Clostridioides difficile* infection by dynamic regulation of intestinal metabolism. Nat Commun 13:1–13
5. Bereza-Malcolm LT, Mann G, Franks AE (2015) Environmental sensing of heavy metals through whole cell microbial biosensors: a synthetic biology approach. ACS Synth Biol 4: 535–546
6. Kim HJ, Jeong H, Lee SJ (2018) Synthetic biology for microbial heavy metal biosensors. Anal Bioanal Chem 410:1191–1203
7. Haas R, Nikel PI (2023) Challenges and opportunities in bringing nonbiological atoms to life with synthetic metabolism. Trends Biotechnol 41:27–45
8. Kordesedehi R, Asadollahi MA, Shahpiri A, Biria D, Nikel PI (2023) Optimized enantioselective (S)-2-hydroxypropiophenone synthesis by free- and encapsulated-resting cells of *Pseudomonas putida*. Microb Cell Factories 22:1–11
9. Karbalaei M, Rezaee SA, Farsiani H (2020) Pichia pastoris: a highly successful expression system for optimal synthesis of heterologous proteins. J Cell Physiol 235:5867–5881
10. Chen R (2012) Bacterial expression systems for recombinant protein production: *E. coli* and beyond. Biotechnol Adv 30:1102–1107
11. Peleg Y, Prabahar V, Bednarczyk D, Unger T (2017) Heterologous gene expression in *E.coli*. https://doi.org/10.1007/978-1-4939-6887-9
12. Kaur J, Kumar A, Kaur J (2018) Strategies for optimization of heterologous protein expression in *E. coli*: roadblocks and reinforcements. Int J Biol Macromol 106:803–822
13. Guan C, Cui W, He X, Hu X, Xu J, Du G, Chen J, Zhou Z (2015) Construction and development of a novel expression system of *Streptomyces*. Protein Expr Purif 113:17–22
14. Delorenzo DM, Moon TS (2019) Construction of genetic logic gates based on the T7 RNA polymerase expression system in

Rhodococcus opacus PD630. ACS Synth Biol 8: 1921–1930

15. Liang Y, Yu H (2021) Genetic toolkits for engineering *Rhodococcus* species with versatile applications. Biotechnol Adv 49:107748
16. Hu YL, Zhang Q, Liu SH, Sun JL, Yin FZ, Wang ZR, Shi J, Jiao RH, Ge HM (2023) Building *Streptomyces albus* as a chassis for synthesis of bacterial terpenoids. Chem Sci 14: 3661–3667
17. Kang DH, Ko SC, Heo YB, Lee HJ, Woo HM (2022) RoboMoClo: a robotics-assisted modular cloning framework for multiple gene assembly in biofoundry. ACS Synth Biol 11: 1336–1348
18. Pollak B, Matute T, Nuñez I et al (2019) Universal loop assembly: open, efficient and cross-kingdom DNA fabrication. Synth Biol 5:1–13
19. Round JW, Robeck LD, Eltis LD (2021) An integrative toolbox for synthetic biology in *Rhodococcus*. ACS Synth Biol 10:2383–2395
20. Jansen Z, Reilly SR, Lieber-Kotz M, Li AZ, Wei Q, Kulhanek DL, Gilmour AR, Thyer R (2023) Interrogating the function of Bicistronic translational control elements to improve consistency of gene expression. ACS Synth Biol 12:1608–1615
21. Cobb RE, Wang Y, Zhao H (2015) High-efficiency multiplex genome editing of *Streptomyces* species using an engineered CRISPR/Cas system. ACS Synth Biol 4:723–728
22. Zhang J, Zhang D, Zhu J, Liu H, Liang S, Luo Y (2020) Efficient multiplex genome editing in *Streptomyces via* engineered CRISPR-Cas12a systems. Front Bioeng Biotechnol 8:1–14
23. Tong Y, Whitford CM, Robertsen HL, Blin K, Jørgensen TS, Klitgaard AK, Gren T, Jiang X, Weber T, Lee SY (2019) Highly efficient DSB-free base editing for streptomycetes with CRISPR-BEST. Proc Natl Acad Sci USA 116: 20366–20375
24. Nakashima N, Tamura T (2004) Isolation and characterization of a rolling-circle-type plasmid from *Rhodococcus erythropolis* and application of the plasmid to multiple-recombinant- protein expression. Appl Environ Microbiol 70: 5557–5568

Chapter 22

Golden Gate Assembly of Transcriptional Unit Libraries into a Rearrangeable Gene Cluster

Amanda L. Hughes and Lars M. Steinmetz

Abstract

Both regulatory sequences and genome organization contribute to the production of diverse transcript isoforms, which can influence how genes, or sets of genes, are expressed. An efficient, modular approach is needed to generate the combinatorial complexity required to empirically test many combinations of different regulatory sequences and different gene orders. Golden Gate assembly provides such a tool for seamless one-pot cleavage and ligation, by using type IIS restriction enzymes, which cleave outside of their recognition site. In addition to reducing the number of steps, this one-pot reaction can improve correct assemblies by the continued cleavage of self-ligation products that retain the recognition site. Switching the specific restriction enzyme used between steps allows for modular assembly of several units. A protocol to perform modular assemblies with two type IIS restriction enzymes, namely BsaI-v2-HF and BsmBI-v2, is described here. This protocol includes a description for generating destination vectors that add loxPsym sites between transcriptional units, allowing for diversification of gene order, orientation, and spacing.

Keywords Seamless assembly, Multi-fragment assembly, One-pot restriction-ligation, SCRaMbLE

1 Introduction

Clustering of metabolic genes is thought to play a role in their coregulation via transcriptional mechanisms [1, 2]. These mechanisms may include bidirectional transcription, transcriptional interference, antisense transcription, and chromatin alterations, as seen at the *GAL7, 10, 1* gene cluster in *Saccharomyces cerevisiae* [3, 4]. To systematically test the contributions of gene order and promoter regulatory logic on the expression of metabolic pathways, a modular strategy that can produce all possible combinations is required. Golden Gate cloning provides the opportunity to efficiently generate transcriptional unit (TU) libraries with diverse promoter sequences and further assemble these into clusters of genes separated by loxPsym sites, allowing for rearrangement of the cluster using the SCRaMbLE system of the Sc2.0 genome [5].

Daniel Schindler (ed.), *Golden Gate Cloning: Methods and Protocols*, Methods in Molecular Biology, vol. 2850, https://doi.org/10.1007/978-1-0716-4220-7_22,

Golden Gate cloning takes advantage of the unique properties of type IIS restriction endonucleases. While type II restriction endonucleases cleave DNA within symmetrical recognition sites, type IIS restriction endonucleases cleave outside of asymmetrical recognition sites. These properties of type IIS restriction endonucleases create unique cohesive ends, allowing for the ordered, directional assembly of multiple cleaved fragments. Many of the frequently used enzymes generate 4 bp (e.g., BbsI, BsaI, BsmBI) or 3 bp (e.g., BspQI) overhangs, creating up to 256 or 64 unique ends, respectively. However, to avoid self-assembly and the creation of concatemers, these overhang sequences must be asymmetrical in design. For example, a "CCGG" overhang could self-assemble and generate spurious products.

Unlike cloning with type II enzymes, intermolecular interactions do not compete with intramolecular interaction, so ligation reactions may be performed with equimolar amounts of insert and vector. Additionally, the products of type IIS restriction enzymes are not marked by recognitions sites, containing only the desired sequence. Furthermore, owing to the removal of recognition sites during cloning, the final product is not cleaved by the type IIS restriction endonuclease, allowing for efficient one-pot assembly, in which restriction digestion and ligation are performed in the same reaction mix [6]. By designing overhangs with high ligation fidelity, many pieces may be assembled together in a single step. Benefitting from data-optimized assembly design tools, Pryor et al. have successfully assembled 52 pieces into a 40 kb phage genome [7]. By switching between different endonucleases flanking complementary overhangs and different antibiotic resistance markers on the destination vectors, the process can be repeated to assemble larger and larger pieces. Assembly of DNA "parts," such as promoter, terminator, and CDS sequences, into transcriptional units (TUs), which can, themselves, be assembled into multi-TU genetic circuits or regulons, as described here, has been standardized [8–10]. By pre-cloning loxPsym sites into the TU destination vectors, this protocol generates a metabolic circuit punctuated by loxPsym sites between each gene and allows for genetic rearrangement, or SCRaMbLEing. The example shown here uses the *GAL7*, *GAL10*, and *GAL1* genes of the yeast galactose metabolic pathway, but this approach could be used to optimize expression of heterologous pathways in yeast.

2 Materials

2.1 Enzymes

1. T4 polynucleotide kinase (10 U/μL).
2. T4 DNA ligase (400 U/μL).
3. 2x KAPA HiFi master mix.

4. Restriction enzymes: SalI-HF, BbsI-HF, XbaI-HF, PstI, BamHI-HF, SbfI-HF, NotI-HF, BsmBI-v2, BsaI-HFv2.
5. rSAP (10 U/μL).
6. Plasmid-safe ATP-dependent DNase.
7. NEBridge BsaI.

2.2 Cells and Reagents

1. 1x Tris-Borate-EDTA buffer (TBE): 90 mM Tris base, 90 mM boric acid, 2 mM ethylenediaminetetraacetic acid.
2. Agarose.
3. Competent cells (for cloning large constructs, RecA-).
4. *ccdB* survival cells (for maintenance of acceptor vectors with *ccdB* toxin gene).
5. Luria-Bertani broth medium: 1% tryptone, 0.5% yeast extract, 1% sodium chloride.
6. Ampicillin (100 μg/mL final concentration).
7. Kanamycin (50 μg/mL final concentration).
8. Chloramphenicol (33 μg/mL final concentration).

2.3 Oligonucleotides and Plasmids

1. Primers to generate acceptor vectors are listed in Table 1.
2. Primer for TU "parts" are listed in Table 2.
3. Primers to confirm library composition are listed in Table 3.
4. Plasmids (*see* **Note 1**) are listed in Table 4.

2.4 Consumables

1. PCR reaction strips/plates.
2. Solid-phase reversible immobilization (SPRI) paramagnetic beads.
3. Miniprep kit plasmid purification.
4. Maxiprep kit plasmid purification.
5. Dialysis membrane discs.
6. 0.1 cm electrode gap cuvette.
7. Oxford Nanopore Ligation Sequencing Kit and Native Barcoding Expansion (*see* **Note 2**).
8. FLO-MIN106 flow cell (*see* **Note 3**).

2.5 Equipment and Software

1. Thermocycler.
2. Electroporation system.
3. Qubit fluorometer.
4. Agilent 2100 Bioanalyzer Instrument.
5. Illumina MiSeq sequencer.
6. GridION/MinION.
7. Minimap2.

Table 1
Primers to generate acceptor vectors

primer	sequence	purpose
MCStop	GCATGCCTGCAGAATCATGTCGACGGATCC	circularise HCKan vector and add multiple cloning site
MCS bottom	GGATCCGTCGACATGATTCTGCAGGCATGC	circularise HCKan vector and add multiple cloning site
KanRsdmBsmBIF	AGGCTCGCTCAGGCGC	site-directed mutagenesis of KanR BsmBI
KanRsdmBsmBIR	AAATACGCGATCGCCGT	site-directed mutagenesis of KanR BsmBI
pUC19sdmBsaIF	CGCGGTATCATTGCAGCACT	site-directed mutagenesis of AmpR BsaI
pUC19sdmBsaIR	GCTACCACGCTCACCGGCTC	site-directed mutagenesis of AmpR BsaI
pSMARTseqFWD2	TTCTACCGAAGAAAGGCCCA	HCKan sequencing
pSMARTseqREV2	GGAGGCTTTTGACTTTCTGC	HCKan sequencing
HCKanREV2	TGGAATTTAATCGCGGCCTC	HCKan sequencing
GAGGccdBF	CTACTGAAGACAGGAGGCGAGACG*TGATCG* *GCACGTAAGAGGTTCC*	*ccdB* cassette amplification
TAGCccdBR	CTACTGTCGACGGTCTCTTAGCGGCTCGAGA CG*CCGACTGGCTGTGTATAAGGGAG*	*ccdB* cassette amplification
TACGccdBR	CTACTGTCGACGGTCTCTTACGGGCTCGAGA CG*CCGACTGGCTGTGTATAAGGGAG*	*ccdB* cassette amplification
GGCTccdBF	CTACTGAAGACAGGGCTCGAGACG*TGATCG* *GCACGTAAGAGGTTCC*	*ccdB* cassette amplification
AAAAccdBR	CTACTGTCGACGGTCTCTAAAAGAGGCGAGA CG*CCGACTGGCTGTGTATAAGGGAG*	*ccdB* cassette amplification
CAGT-loxPsymF	CTAGGGTCTCTCAGT**ATAACTTCGTATAATGT** **ACATTATACGAAGTTAT**	loxPsym dsDNA
CAGT-loxPsymR	CCTC**ATAACTTCGTATAATGTACATTATACGAA** **GTTAT**ACTGAGAGACC	loxPsym dsDNA
GCTA-loxPsymF	CTAGGGTCTCTGCTA**ATAACTTCGTATAATGT** **ACATTATACGAAGTTAT**	loxPsym dsDNA

GCTA-loxPsymR	CCTC**ATAACTTCGTATAATGTACATTATACGAAGTTAT**TAGCAGAGACC	loxPsym dsDNA
CGTA-loxPsymF	CTAGGGTCTCTCGTA**ATAACTTCGTATAATGTACATTATACGAAGTTAT**	loxPsym dsDNA
CGTApro-loxPsymR	AGCC**ATAACTTCGTATAATGTACATTATACGAAGTTAT**TACGAGAGACC	loxPsym dsDNA
jLloxPsym_ccdBF	GGTCTCTGCGA**ATAACTTCGTATAATGTACATTATACGAAGTTAT**CGAGACG*TGATCGGCACGTAAGAGGTTCC*	loxPsym:ccdB amplification
bLloxPsym_ccdBF	GGTCTCTAGGC**ATAACTTCGTATAATGTACATTATACGAAGTTAT**CGAGACG*TGATCGGCACGTAAGAGGTTCC*	loxPsym:ccdB amplification
dLloxPsym_ccdBF	GGTCTCTCACT**ATAACTTCGTATAATGTACATTATACGAAGTTAT**CGAGACG*TGATCGGCACGTAAGAGGTTCC*	loxPsym:ccdB amplification
aRloxPsym_ccdBR	GGTCTCTAGGT**ATAACTTCGTATAATGTACATTATACGAAGTTAT**CGAGACG*GTGAGACCGACTGGCTGTGT*	loxPsym:ccdB amplification
cRloxPsym_ccdBR	GGTCTCTGGCA**ATAACTTCGTATAATGTACATTATACGAAGTTAT**CGAGACG*GTGAGACCGACTGGCTGTGT*	loxPsym:ccdB amplification
eRloxPsym_ccdBR	GGTCTCTCGAC**ATAACTTCGTATAATGTACATTATACGAAGTTAT**CGAGACG*GTGAGACCGACTGGCTGTGT*	loxPsym:ccdB amplification
spacer4F	TTATATCTGAGATGAACATCTATTTCACAGAATCTCACTTCCAATTCAAGTTTTTACGAAAAGGTATAGACGTATACTACAGTCGTGCGTTTTATAGCAGTGTC	spacer for HCKanLox:Lox insert
spacer4R	TTATGACACTGCTATAAAACGCACGACTGTAGTATACGTCTATACCTTTTCGTAAAAACTTGAATTGGAAGTGAGATTCTGTGAAATAGATGTTCATCTCAGAT	spacer for HCKanLox:Lox insert
spacer17F	TTATAGAATCGCATATGGGATTTGATGCGAACATTTAGCAAAAACGAAGAGGCATATGAAACTGTAAGTACCTGGTCAGCACTTTTTCAGTGCTTTGTGGGCGG	spacer for HCKanLox:Lox insert
spacer17R	TTATCCGCCCACAAAGCACTGAAAAAGTGCTGACCAGGTACTTACAGTTTCATATGCCTCTTCGTTTTTGCTAAATGTTCGCATCAAATCCCATATGCGATTCT	spacer for HCKanLox:Lox insert

spacer18F	TTATTAATTTTCAGATGTTGGATTATTGAACCAGTTATTGAAGGACGTTTGTCAGTTACTATTAGTCAACAGTAGCTCAACCGAGCCCTACATAGGTTGTAAGG	spacer for HCKanLox:Lox insert
spacer18R	TTATCCTTACAACCTATGTAGGGCTCGGTTGAGCTACTGTTGACTAATAGTAACTGACAAACGTCCTTCAATAACTGGTTCAATAATCCAACATCTGAAAATTA	spacer for HCKanLox:Lox insert
spacer13aF	ACCTAAATCATGGGAACCACGTCATGAAAAGTAAGAGTTGAAGAGTCCTAGCTATACAAAATTAGAGTGTTTTCAATTACATGAGCAAATACTAGATCTTAACA	spacer for HCAmp assembly
spacer13bR	GCCTTGTTAAGATCTAGTATTTGCTCATGTAATTGAAAACACTCTAATTTTGTATAGCTAGGACTCTTCAACTCTTACTTTTCATGACGTGGTTCCCATGATTT	spacer for HCAmp assembly
spacer19cF	TGCCAAGTCAACTGGGTTACGTACCAGTATGAAGGGTTGGTATCCCAAAGAAACTATCTCATGTTATCGATCGGTATAGTAAACTACATTCTTTCCTTCAGAGT	spacer for HCAmp assembly
spacer19dR	AGTGACTCTGAAGGAAAGAATGTAGTTTACTATACCGATCGATAACATGAGATAGTTTCTTTGGGATACCAACCCTTCATACTGGTACGTAACCCAGTTGACTT	spacer for HCAmp assembly
ddrmR	CAGCTGAAGCTTCGTACGC	amplify HygR
BclI_ddrmF	CTACTTGATCAGCATAGGCCACTAGTGGATCTG	amplify HygR
BamHI5loxPspaceF2	CTAGTGGATCCCCTGCAGAATCATGTCCG	amplify 5LoxSpacer array
BamHI5loxPspaceR	CTAGTGGATCCGTCCGTCTCGGCGAATAACTTCG	amplify 5LoxSpacer array
SalI-SbfI_GAL7hr	GCATGTCGACCCTGCAGGAATTGTTTCCCCCTAGCATTG	amplify HR1 (upstream of GAL7)
PstI_GAL7hr	GCATCTGCAGCGGTGTCGTTATCGATGG	amplify HR1 (upstream of GAL7)
BamHi_GAL1hr	GCATGGATCCGACAGAAAAATTCCGATGGAC	amplify HR2 (downstream of GAL1)
NotI_GAL1hr	GCATGCGGCCGCCGATGCCTCATAAACTTCGG	amplify HR2 (downstream of GAL1)

Altered bases in site-directed mutagenesis are red. BsaI and BsmBI sites are underlined. Sequence for *ccdB* amplification is in italics. LoxPsym sites are in bold

Table 2
Primers for TU "parts"

primer	sequence	purpose
SbfI-BsmBI_xProF	ACGTGCCTGCAGGCGTCTCGGGCTXXXXXXXXXXXXXXXXXXXX	promoter amplification
SbfI-BsmBI_xProR	ACGTGCCTGCAGGCGTCTCGCATCXXXXXXXXXXXXXXXXXXXX	promoter amplification
SbfI-BsmBi_GAL7terF	ACGTGCCTGCAGGCGTCTCGTAACAAAGAAAGTGGAATATTCATTCATATCATATTTT	GAL7 terminator amplification
SbfI-BsmBi_GAL10terF	ACGTGCCTGCAGGCGTCTCGTAACTTTGCCAGCTTACTATCCTTCTTG	GAL10 terminator amplification
SbfI-BsmBi_GAL1terF	ACGTGCCTGCAGGCGTCTCGTAACGTATACTTCTTTTTTTACTTTGTTCAGAACAA	GAL1 terminator amplification
SbfI-BsmBi_GAL7terR	ACGTGCCTGCAGGCGTCTCGGAGGACAATGAGCCTTGCTGCAA	GAL7 terminator amplification
SbfI-BsmBi_GAL10terR	ACGTGCCTGCAGGCGTCTCGGAGGTTAAGGAAAATGACAGAAAATATATCTAATGAG	GAL10 terminator amplification
SbfI-BsmBi_GAL1terR	ACGTGCCTGCAGGCGTCTCGGAGGATTTCTTTTCCTCCTCGCGC	GAL1 terminator amplification
BsmBi_SCT1proFex	AGCGTGCGTCTCGGGCTCACGTCTAGATAATGGCAGTAAAC	left-external primer for promoter amplification
BsmBi_VMA1proFex	AGCGTGCGTCTCGGGCTCAGCTACCTAGATACCTTAAACTTG	left-external primer for promoter amplification
BsmBi_ILV5proFex	AGCGTGCGTCTCGGGCTCGCTGTCACTGAACTAAAACAA	left-external primer for promoter amplification
BsmBi_MSE1proFex	AGCGTGCGTCTCGGGCTAAGGTATGGACTCTAGAAATGAAAGG	left-external primer for promoter amplification
BsmBi_NMA2proFex	AGCGTGCGTCTCGGGCTTAGGTTCAGGAGTACCAAGAATG	left-external primer for promoter amplification
BsmBi_LCB4proFex	AGCGTGCGTCTCGGGCTAACCAGAACGACCCCTCG	left-external primer for promoter amplification
BsmBi_THI72proFex	AGCGTGCGTCTCGGGCTGTATTTTGAGCGCTAGGCGT	left-external primer for promoter amplification

BsmBi_BNA3proFex	AGCGTGCGTCTCGGGCTAGAGCAGATTGTTTTGAGTAGGA	left-external primer for promoter amplification
BsmBi_ATP5proFex	AGCGTGCGTCTCGGGCTGTAAGATCTGGCTGCAAACCT	left-external primer for promoter amplification
BsmBi_GND2proFex	AGCGTGCGTCTCGGGCTAAGAAGAAGGGAAAAAAAGAAAGGAG	left-external primer for promoter amplification
BsmBi_SCT1proRex	AGCGTGCGTCTCGCATCTTTTTACCCCGATGTAAACGC	right-external primer for promoter amplification
BsmBi_VMA1proRex	AGCGTGCGTCTCGCATCTGTTTCTTTTCTTCTCTATTTGCTTT	right-external primer for promoter amplification
BsmBi_ILV5proRex	AGCGTGCGTCTCGCATCGTTTTATTTTTTACTTATATTGCTGGTAGG	right-external primer for promoter amplification
BsmBi_MSE1proRex	AGCGTGCGTCTCGCATCATTGTTATTTTATAGTTCTTCTGTACCAG	right-external primer for promoter amplification
BsmBi_NMA2proRex	AGCGTGCGTCTCGCATCTACGTAATAAGTGCACTTTTTCTTGC	right-external primer for promoter amplification
BsmBi_LCB4proRex	AGCGTGCGTCTCGCATCTATCTATAGTAGATTCTTCGCGTACTT	right-external primer for promoter amplification
BsmBi_THI72proRex	AGCGTGCGTCTCGCATCAATGGTGTAATGCTAGAGTTTTTTTTG	right-external primer for promoter amplification
BsmBi_BNA3proRex	AGCGTGCGTCTCGCATCTGAGAGCAAGTAACAATGGATTCT	right-external primer for promoter amplification
BsmBi_ATP5proRex	AGCGTGCGTCTCGCATCGATTGTCCACGGTTTAAAAATAATGC	right-external primer for promoter amplification
BsmBi_GND2proRex	AGCGTGCGTCTCGCATCTCTGTTCCTCGTGTTTTTTTAATTGT	right-external primer for promoter amplification

SCT1proLin	AGCCATAGTTGTTGAGAAGCCGATGGATGAGG	left-internal primer for promoter amplification
VMA1proLin	CAGCTTTTAATCCAGTGTGAAGCCCCTTTGAGCG	left-internal primer for promoter amplification
ILV5proLin	GGTACCGGAGGAAGCCGCTATAACCG	left-internal primer for promoter amplification
MSE1proLin	GAAGACATTTTTGCTAGATGAAGCCAAACGTCTTGGTTATAAAC	left-internal primer for promoter amplification
NMA2proLin	CAATTTACATTGTTATTTTTCTTGGAAGCGACACCGTCACATATG	left-internal primer for promoter amplification
LCB4proLin	AGAAGGCGCAGAAGAAGCGGAAACCAATAAAGTCA	left-internal primer for promoter amplification
THI72proLin	CAACCAGCAGGAAGCGCCTATTGTTAAGC	left-internal primer for promoter amplification
BNA3proLin	AGTAATTATCTTGGCATAGGAAGCGATAAATGCCATCGTTC	left-internal primer for promoter amplification
ATP5proLin	GCCAGCACCGCTTCTCGAAAGTGTCC	left-internal primer for promoter amplification
GND2proLin	GAAACGTTCAATTTCTGACGCTTCTACTTTTTCATCGAGAGC	left-internal primer for promoter amplification
SCT1proRin	CCTCATCCATCGGCTTCTCAACAACTATGGCT	right-internal primer for promoter amplification
VMA1proRin	CGCTCAAAGGGGCTTCACACTGGATTAAAAGCTG	right-internal primer for promoter amplification
ILV5proRin	CGGTTATAGCGGCTTCCTCCGGTACC	right-internal primer for promoter amplification

MSE1proRin	GTTTATAACCAAGACGTTTGGCTTCATCTAGCAA AAATGTCTTC	right-internal primer for promoter amplification
NMA2proRin	CATATGTGACGGTGTCGCTTCCAAGAAAAATAAC AATGTAAATTG	right-internal primer for promoter amplification
LCB4proRin	TGACTTTATTGGTTTCCGCTTCTTCTGCGCCTTCT	right-internal primer for promoter amplification
THI72proRin	GCTTAACAATAGGCGCTTCCTGCTGGTTG	right-internal primer for promoter amplification
BNA3proRin	GAACGATGGCATTTATCGCTTCCTATGCCAAGAT AATTACT	right-internal primer for promoter amplification
ATP5proRin	GGACACTTTCGAGAAGCGGTGCTGGC	right-internal primer for promoter amplification
GND2proRin	GCTCTCGATGAAAAAGTAGAAGCGTCAGAAATT GAACGTTTC	right-internal primer for promoter amplification
BsmBi_GAL7f	AGCGTGCGTCTCGGATGACTGCTGAAGAATTTG ATTTTTC	GAL7 ORF amplification
BsmBi_GAL7r	AGCGTGCGTCTCGGTTACAGTCTTTGTAGATAAT GAATCTGAC	GAL7 ORF amplification
BsmBi_GAL10f	AGCGTGCGTCTCGGATGACAGCTCAGTTACAAA GTGAAAG	GAL10 ORF amplification
BsmBi_GAL10r	AGCGTGCGTCTCGGTTAGGAAAATCTGTAGACA ATCTTGG	GAL10 ORF amplification
BsmBi_GAL1f	AGCGTGCGTCTCGGATGACTAAATCTCATTCAGA AGAAGTG	GAL1 ORF amplification
BsmBi_GAL1r	AGCGTGCGTCTCGGTTATAATTCATATAGACAGC TGCCCA	GAL1 ORF amplification

BsmBI sites are underlined. User-defined sequences for promoter amplification are represented by italicized Xs. Altered bases in overlap extension PCRs are indicated in red

Table 3
Primers to confirm library composition

Primer	Sequence	Purpose
GAL7-10ampliseqF1	**TCGTCGGCAGCGTCAGATGTGTATAAGAGACAG**TGCCTGCAGAACATGTCGACG	Ampliseq forward primer for GAL7 and GAL10 TU libraries
GAL7-10ampliseqF2	**TCGTCGGCAGCGTCAGATGTGTATAAGAGACAG**NTGCCTGCAGAACATGTCGACG	Ampliseq forward primer for GAL7 and GAL10 TU libraries
GAL7-10ampliseqF3	**TCGTCGGCAGCGTCAGATGTGTATAAGAGACAG**NNTGCCTGCAGAACATGTCGACG	Ampliseq forward primer for GAL7 and GAL10 TU libraries
GAL7-10ampliseqF4	**TCGTCGGCAGCGTCAGATGTGTATAAGAGACAG**NNNTGCCTGCAGAACATGTCGACG	Ampliseq forward primer for GAL7 and GAL10 TU libraries
GAL7ampliseqR1	**GTCTCGTGGGCTCGGAGATGTGTATAAGAGACAG**TGGATTGTAACGTCTATGGGAATGGC	Ampliseq reverse primer for GAL7 TU library
GAL7ampliseqR2	**GTCTCGTGGGCTCGGAGATGTGTATAAGAGACAG**NTGGATTGTAACGTCTATGGGAATGGC	Ampliseq reverse primer for GAL7 TU library
GAL7ampliseqR3	**GTCTCGTGGGCTCGGAGATGTGTATAAGAGACAG**NNTGGATTGTAACGTCTATGGGAATGGC	Ampliseq reverse primer for GAL7 TU library
GAL7ampliseqR4	**GTCTCGTGGGCTCGGAGATGTGTATAAGAGACAG**NNNTGGATTGTAACGTCTATGGGAATGGC	Ampliseq reverse primer for GAL7 TU library
GAL10ampliseqR1	**GTCTCGTGGGCTCGGAGATGTGTATAAGAGACAG**CCAGCACCACCTGTAACCAAAC	Ampliseq reverse primer for GAL10 TU library
GAL10ampliseqR2	**GTCTCGTGGGCTCGGAGATGTGTATAAGAGACAG**NCCAGCACCACCTGTAACCAAAC	Ampliseq reverse primer for GAL10 TU library
GAL10ampliseqR3	**GTCTCGTGGGCTCGGAGATGTGTATAAGAGACAG**NNCCAGCACCACCTGTAACCAAAC	Ampliseq reverse primer for GAL10 TU library
GAL10ampliseqR4	**GTCTCGTGGGCTCGGAGATGTGTATAAGAGACAG**NNNCCAGCACCACCTGTAACCAAAC	Ampliseq reverse primer for GAL10 TU library

(continued)

Table 3
(continued)

Primer	Sequence	Purpose
GAL1ampliseqF1	**TCGTCGGCAGCGTCAGATGTGTATAAGAGACAG**TGAGCGATATCTAGGGTCTCtCGT	Ampliseq forward primer for GAL1 TU library
GAL1ampliseqF2	**TCGTCGGCAGCGTCAGATGTGTATAAGAGACAG**NTGAGCGATATCTAGGGTCTCtCGT	Ampliseq forward primer for GAL1 TU library
GAL1ampliseqF3	**TCGTCGGCAGCGTCAGATGTGTATAAGAGACAG**NNTGAGCGATATCTAGGGTCTCtCGT	Ampliseq forward primer for GAL1 TU library
GAL1ampliseqF4	**TCGTCGGCAGCGTCAGATGTGTATAAGAGACAG**NNNTGAGCGATATCTAGGGTCTCtCGT	Ampliseq forward primer for GAL1 TU library
GAL1ampliseqR1	**GTCTCGTGGGCTCGGAGATGTGTATAAGAGACAG**TTGCGCTAGAATTGAACTCAGGTAC	Ampliseq reverse primer for GAL1 TU library
GAL1ampliseqR2	**GTCTCGTGGGCTCGGAGATGTGTATAAGAGACAG**NTTGCGCTAGAATTGAACTCAGGTAC	Ampliseq reverse primer for GAL1 TU library
GAL1ampliseqR3	**GTCTCGTGGGCTCGGAGATGTGTATAAGAGACAG**NNTTGCGCTAGAATTGAACTCAGGTAC	Ampliseq reverse primer for GAL1 TU library
GAL1ampliseqR4	**GTCTCGTGGGCTCGGAGATGTGTATAAGAGACAG**NNNTTGCGCTAGAATTGAACTCAGGTAC	Ampliseq reverse primer for GAL1 TU library
ONTII_NB01_SbfIt	AAGGTTAA*CACAAAGACACCGACAACTTTCTT*CAGCACCtgca	ONT NB01, SbfI ligation
ONTII_NB01_SbfIb	GGTGCTG*AAGAAAGTTGTCGGTGTCTTTGTG*TTAACCTTAGCAAT	ONT NB01, SbfI ligation
ONTII_NB01_NotIt	ggccGGTGCTG*AAGAAAGTTGTCGGTGTCTTTGTG*TTAACCTTAGCAAT	ONT NB01, NotI ligation
ONTII_NB01_NotIb	AAGGTTAA*CACAAAGACACCGACAACTTTCTT*CAGCACC	ONT NB01, NotI ligation
ONTII_NB02_SbfIt	AAGGTTAA*ACAGACGACTACAAACGGAATCGA*CAGCACCtgca	ONT NB02, SbfI ligation

ONTII_NB02_SbfIb	GGTGCTG*TCGATTCCGTTTGTAGTCGTCTGT*TTAACCTTAGCAAT	ONT NB02, SbfI ligation
ONTII_NB02_NotIt	ggccGGTGCTG*TCGATTCCGTTTGTAGTCGTCTGT*TTAACCTTAGCAAT	ONT NB02, NotI ligation
ONTII_NB02_NotIb	AAGGTTAA*ACAGACGACTACAAACGGAATCGA*CAGCACC	ONT NB02, NotI ligation
ONTII_NB03_SbfIt	AAGGTTAA*CCTGGTAACTGGGACACAAGACTC*CAGCACCtgca	ONT NB03, SbfI ligation
ONTII_NB03_SbfIb	GGTGCTG*GAGTCTTGTGTCCCAGTTACCAGG*TTAACCTTAGCAAT	ONT NB03, SbfI ligation
ONTII_NB03_NotIt	ggccGGTGCTG*GAGTCTTGTGTCCCAGTTACCAGG*TTAACCTTAGCAAT	ONT NB03, NotI ligation
ONTII_NB03_NotIb	AAGGTTAA*CCTGGTAACTGGGACACAAGACTC*CAGCACC	ONT NB03, NotI ligation
ONTII_NB04_SbfIt	AAGGTTAA*TAGGGAAACACGATAGAATCCGAA*CAGCACCtgca	ONT NB04, SbfI ligation
ONTII_NB04_SbfIb	GGTGCTG*TTCGGATTCTATCGTGTTTCCCTA*TTAACCTTAGCAAT	ONT NB04, SbfI ligation
ONTII_NB04_NotIt	ggccGGTGCTG*TTCGGATTCTATCGTGTTTCCCTA*TTAACCTTAGCAAT	ONT NB04, NotI ligation
ONTII_NB04_NotIb	AAGGTTAA*TAGGGAAACACGATAGAATCCGAA*CAGCACC	ONT NB04, NotI ligation

Nextera sequencing handles for amplicon sequencing are in bold. Oxford Nanopore ligation sequencing barcodes are italicized and overhangs for SbfI and NotI ligation are in lowercase

Table 4:
Plasmids used and generated in this study

Name	Purpose	Parental	Source
pSMART®-HCKan	Generation of TU acceptor plasmids, high-copy		Lucigen
pSMART®-HCAmp	Generation of TU accepter plasmids, high-copy		Lucigen
pUC-IIs	ORF domestication	pUC19	This work
pAV115	Golden Gate assembly of regulon via BsaI, mRFP marker		[12]
pAG32	Hygromycin drug resistance marker		[13]
pTEI122	*ccdB* cassette		[14]
HCKan:MCS	Generation of TU acceptor plasmids, high-copy, KanR	pSMART-HCKan	This work
HCKan:Lox:ccdB	Golden Gate cloning of TUs via BsmBI, release by BsaI	HCKan: MCS	This work
HCKan:LoxSpacer	100 bp flanked by loxPsym sites, release by BsaI	HCKan: Lox: ccdB:Lox	This work
HCAmp:5LoxSpacer	Array of 5, 100 bp units flanked loxPsym sites	HCAmp: MCS	This work
pGALhr5Lox (HygR)	Regulon destination vector for Golden Gate cloning of multiple TUs via BsaI, regions of homology for integration into yeast *GAL7, 10, 1* gene cluster, HygR for selection	pAV115	This work

3 Methods

3.1 Prepare Acceptor Vectors

Vectors for Golden Gate cloning contain type IIS restriction enzymes flanking overhang sequences used for ligation. If other standard vectors are used (*see* **Note 1**), the overhang sequences present in these vectors should be used for assembly. The inclusion of monomeric red fluorescent protein (mRFP) or a negative selection marker, such as the *ccdB* toxin gene, which is removed during restriction digestion, improves selection for correct Golden Gate assemblies. Preparing custom vectors as below allows for control over the sequence and number of overhangs used in assembly as well as for the inclusion of loxPsym sites, allowing for subsequent rearrangement of gene order and orientation in the regulon.

3.1.1 TU Assembly Vectors

The vectors described here are generated from pSMART®HCKan and pSMART®HCAmp, which allow for high-copy maintenance in bacteria. Other vectors may be used, as long as type IIS restriction sites are removed from the backbone. In this case, a BsmBI site is removed via site-directed mutagenesis of the circular HCKan vector. The pSMART®HCKan and pSMART®HCAmp Blunt Cloning

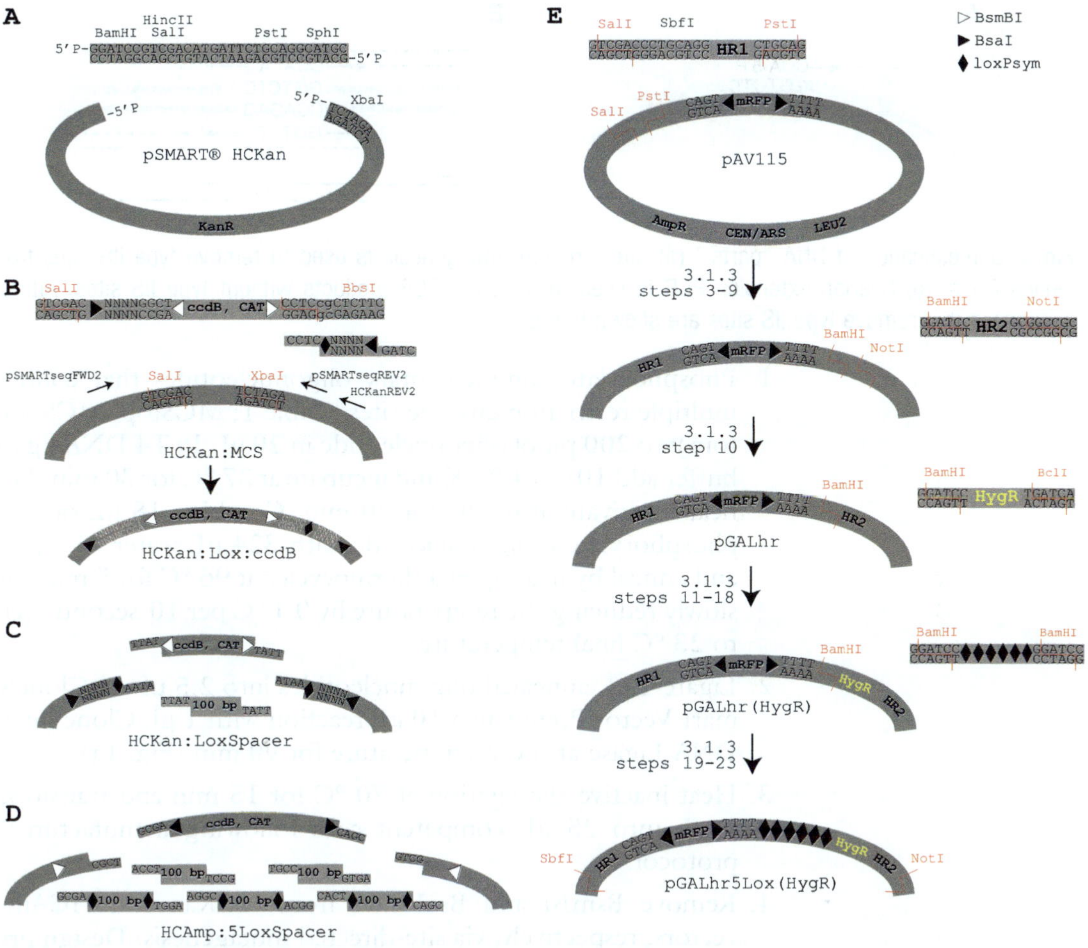

Fig. 1 Generation of acceptor vectors for Golden Gate assembly with loxPsym sites. (**a**) A multiple cloning site (MCS) is ligated in the pSMART-HCKan vector to allow for subsequent addition of a *ccdB* toxin gene and loxPsym site, creating HCKan:Lox:ccdB vectors (**b**) for TU assembly. (**c**) 100 bp spacers are cloned with flanking loxPsym sites and are assembled with 100 bp ds DNA to generate an array of loxPsym-spacers (**d**). (**e**) The regulon acceptor vector, pGALhr5Lox(HygR), is generated from pAV115 to provide regions of homology for integration into the yeast genome, a hygromycin resistance marker for selection, and a loxPsym-spacer array for altering gene spacing after Cre-lox induced rearrangement of the regulon

kits come as linearized vectors, which must first be circularized via ligation. By performing an intermolecular ligation with DNA encoding a multiple cloning site (MCS) to generate HCKan:MCS (Fig. 1a) and HCAmp:MCS, several restriction sites, specifically SalI and XbaI, are available for cloning a *ccdB* cassette, loxPsym site, and Golden Gate overhangs flanked by BsmBI and BsaI sites. To generate this HCKan:Lox:ccdB vector, two DNA fragments are simultaneously cloned into the backbone: (1) the *ccdB* cassette with BsaI and BsmBI sites flanking 4 bp sequences used for regulon and TU assembly and 5′ SalI recognition site on the 5′ side and BsmBI site with a 4 bp TU assembly sequence followed by a BbsI recognition site on the 3′ side and (2) a loxPsym site with a 4 bp sequence for regulon assembly flanked by a 3′ BsmBI site (Fig. 1b).

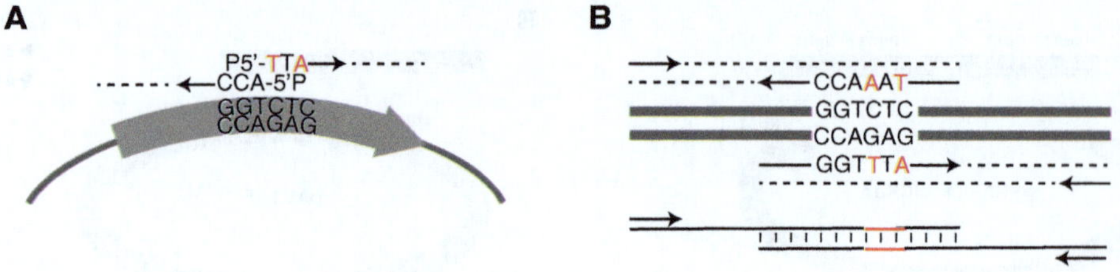

Fig. 2 Domestication of DNA "parts." (**a**) Site-directed mutagenesis is used to remove type IIS sites from cloned ORFs. (**b**) Overlap extension PCR is used to generate PCR products without type IIS sites. Altered sequences that remove type IIS sites are shown in red

1. Phosphorylate complementary oligonucleotides that contain multiple restriction enzyme sites (Table 1; MCStop, MCSbottom): to 200 pmol oligonucleotide in 20 μL 1x T4 DNA ligase buffer add 10U T4 PNK and incubate at 37 °C for 30 min then heat inactivate at 65 °C for 20 min. Combine 18 μL of each phosphorylated oligonucleotide with 324 μL water (0.5 μM) and anneal by heating in a thermocycler at 96 °C for 5 min and slowly reducing the temperature by 0.1 °C per 10-second cycle to 23 °C final temperature.
2. Ligate 1 μL annealed oligonucleotides into 2.5 μL 4x CloneSmart Vector Premix in a 10 μL reaction with 1 μL CloneSmart DNA Ligase at room temperature for 90 min (Fig. 1a).
3. Heat inactive the ligation at 70 °C for 15 min and transform 1 μL into 25 μL competent cells following manufacturer's protocol.
4. Remove BsmBI and BsaI sites from HCKan and HCAmp vectors, respectively, via site-directed mutagenesis. Design primers with abutting 5′ ends, in which the restriction site is removed (Fig. 2a; Table 1, KanRsdmBsmBIF, KanRsdmBsmBIR, pUC19sdmBsaIF, pUC19sdmBsaIR).
5. Phosphorylate primers (5 μM each) in a 40 μL reaction with 1x T4 DNA ligase buffer and 1μL T4 polynucleotide kinase (10 U/μL) at 37 °C for 30 min, then heat inactivate at 65 °C for 20 min.
6. Amplify the full-length plasmid (~5 ng) with the phosphorylated primers, using a high-fidelity DNA polymerase, for a maximum of 16 cycles. Check 5 μL of amplification on a 1% agarose gel for full-length product (~2 kb).
7. Digest original (methylated) plasmid: Add 5 μL 10x CutSmart buffer and treat with 1 μL DpnI for 30 min at 37 °C, followed by a 20-min heat inactivation at 65 °C.
8. Use 2 μL of DpnI-digested DNA in a 20 μL ligation reaction with 1x T4 DNA ligase buffer and 400U T4 DNA ligase at 16 °C overnight. Heat inactivate ligation at 65 °C for 10 min.

9. Transform 2 μL of the ligation into 25 μL competent cells, using the manufacturer's protocol.
10. Miniprep plasmid and confirm removal of restriction site by Sanger sequencing with pSMARTseq primers (Table 1).
11. Design primers to amplify *ccdB* cassette from pTEI122. The forward primer contains a TU assembly sequence flanked internally by a BsmBI site and externally by a BbsI site. The reverse primer contains the sequences for regulon and TU assembly flanked by BsaI and BsmBI, respectively, with a 5′ XbaI site for cloning into the MCS (Table 1, Fig. 1b).
12. Use 5 ng of pTEI122 plasmid as the template for a high-fidelity PCR reaction with these primers.
13. Clean-up PCR reactions with 1.2 volumes SPRI paramagnetic beads and elute in 20 μL 10mM Tris-HCl.
14. Design reverse complementary oligonucleotides encoding the loxPsym sequence, such that one end has a SalI-compatible overhang and the other has an overhang compatible with the TU assembly overhang generated by BbsI cleavage of the *ccdB* PCR product (**step 11**) (Table 1, Fig. 1b).
15. Phosphorylate 100 pmol each oligonucleotide in 40 μL 1x T4 DNA ligase buffer with 1.5 μL T4 PNK at 37 °C for 30 min, followed by heat inactivation at 65 °C for 20 min.
16. Dilute oligonucleotides 1:5 (0.5 μM) and anneal to create overhang sequences for ligation by heating at 96 °C for 5 min in a thermocycler and then slowly cooling down by 1 °C every 10 s until 23 °C.
17. Cleave 1 μg of HCKan:MCS with 1 μL SalI-HF and 1 μL XbaI for 1 h at 37 °C in a 50 μL reaction with 1x CutSmart buffer and dephosphorylate with 1μL rSAP; heat inactivate the reaction at 65 °C for 20 min.
18. Cut 1 μg *ccdB* PCR product with 1 μL each XbaI and BbsI-HF in 50 μL 1x CutSmart buffer for 1 h at 37 °C and heat inactivate at 65 °C for 20 min.
19. Ligate *ccdB* PCR product and corresponding loxPsym dsDNA into digested HCKan:MCS in a 20 μL reaction at 16 °C for 12 h.

T4 DNA ligase buffer	2 μL
T4 DNA Ligase (400 U/μL)	1 μL
HCKan: MCS, SalI/XbaI (**step 17**)	2 μL (0.03 pmol)
ccdB, XbaI/BbsI (**step 18**)	4 μL (0.09 pmol)
loxPsym dsDNA (**step 16**)	2 μL 1:10 diluted (0.09 pmol)

20. Heat inactivate the ligation at 65 °C for 10 min.
21. Transform 2 μL into *ccdB* survival cells, plating on LB containing kanamycin and chloramphenicol (*see* **Note 4**).
22. Miniprep and confirm via Sanger sequencing with pSMARTseq primers HCKan:Lox:ccdB (Fig. 1b).

3.1.2 Clone loxPsym-Spacer Array

To avoid cloning with multiple loxPsym fragments (*see* **Note 5**), loxPsym sites are first incorporated via PCR on either side of a *ccdB* cassette. BsmBI sites cleave on the internal side of the loxPsym sites, allowing for replacement of the *ccdB* cassette by various spacer sequences (Fig. 1c). These units are then assembled via Golden Gate cloning with additional 100 bp spacer sequences to produce an array of spacer units punctuated by loxPsym sites (Fig. 1d).

1. Design primers to amplify *ccdB* cassette with 5′-BsaI-loxPsym-BsmBI tails (Table 1).
2. Phosphorylate 5 μM each primer in 40 μL 1x T4 DNA ligase buffer with 1.5 μL T4 PNK at 37 °C for 30 min, then heat inactivate at 65 °C for 20 min.
3. Use the phosphorylated primer mix to amplify the *ccdB* cassette from ~5 ng pTEI122 with a high-fidelity polymerase.
4. Purify PCR products with 1.8 volumes SPRI paramagnetic beads.
5. Cut 500 ng HCKan:MCS in 50 μL 1x CutSmart buffer with 1 μL HincII and simultaneously treat with 1 μL rSAP at 37 °C for 1 h, then heat inactivate at 65 °C for 20 min.
6. Ligate Lox:ccdB PCR into HCKan:MCS in a 20 μL reaction volume:

T4 DNA ligase buffer	2 μL
T4 DNA ligase (400 U/μL)	1 μL
HCKan:MCS, HincII	2.5 μL (0.02 pmol)
Lox:ccdB PCR	0.06 mol

7. Incubate at 16 °C overnight (≥12 h). Heat inactivate at 65 °C for 10 min
8. Transform 2 μL into 25 μL competent cells.
9. Miniprep and confirm via M13 sequencing HCKanLox:ccdB:Lox
10. Design 100 bp reverse complementary oligonucleotides with random sequence of ~32% GC content (consistent with yeast genome) and lacking Gal4p binding sites, including 5′-TTAT sequence (Table 1).

11. Phosphorylate 2.5 μM spacer oligonucleotides in 40 μL 1x T4 DNA ligase buffer with 1 μL T4 PNK at 37 °C for 30 min. Heat inactivate at 65 °C for 20 min.
12. Dilute oligonucleotide mix to 0.5 μM with water and anneal by heating at 96 °C for 5 min and then cooling slowly (1 °C every 10 s) to 23 °C in a thermocycler.
13. Cut 500 ng HCKanLox:ccdB:Lox vector in 25 μL 1x NEB r3.1 buffer with 1 μL BsmBI-v2 at 55 °C for 1 h. Lower temperature to 37 °C and treat with 1 μL rSAP for 30 min. Heat inactivate at 80 °C for 20 min.
14. Ligate dsDNA spacer DNA into HCKan backbone flanked by loxPsym sites:

T4 DNA ligase buffer	2 μL
T4 DNA ligase (400 U/μL)	1 μL
HCKanLox:ccdB:Lox, BsmBI	2 μL (0.02 pmol)
dsDNA spacer, 1:10 diluted	1 μL (0.06 pmol)

15. Incubate at room temperature for 1 h, then heat inactivate at 65 °C for 10 min.
16. Transform 2 μL into 25 μL competent cells.
17. Miniprep HCKan:LoxSpacer (Fig. 1c) and confirm sequence via M13 sequencing.
18. Design reverse complementary 100 bp oligonucleotides with overhangs corresponding to the sequence generated by BsaI-cleavage of the loxPsym-spacer units (Table 1).
19. Phosphorylate 2.5 μM each oligonucleotide in 40 μL 1x T4 DNA ligase buffer with 1 μL T4 PNK at 37 °C for 30 min, then heat inactivate at 65 °C for 20 min.
20. Dilute to 0.5 μM with water and anneal by heating at 96 °C for 5 min then slowly cooling by 1 °C per 10 second cycle in thermocycler until 23 °C.
21. Golden Gate assembly of 3 loxPsym-spacer units (**step 17**) and 2 dsDNA spacers (**step 20**) into HCAmp:MCS (Subheading 3.1.1, **step 10**) in a 20 μL reaction.

T4 DNA Ligase buffer	2 μL
T4 DNA Ligase (400 U/μL)	0.25 μL
BsmBI-v2 (10 U/μL)	0.35 μL
HCAmp:ccdB(je)	0.05 pmol
HCKan:jaLoxPsym:spacer4	0.05 pmol
HCKan:bcLoxPsym:spacer17	0.05 pmol

(continued)

HCKan:deLoxPsym:spacer18	0.05 pmol
Spacer13ab dsDNA	0.05 pmol
Spacer19cd dsDNA	0.05 mol

22. Cycle thirty times between 42 and 16 °C for 5 min each. Heat inactivate ligase at 65 °C for 10 min, then hold at 10 °C.
23. Reduce background by treating Golden Gate reaction with Plasmid-Safe DNase in tandem with BsmBI-v2. Add ATP to 2 mM and 0.25 μL each Plasmid-Safe DNase and BsmBI-v2 and heat at 37 °C for 30 min then heat inactivate at 80 °C for 20 min.
24. Transform 2 μL Golden Gate reaction into 25 μL competent cells.
25. Miniprep HCAmp:5LoxPspacer plasmid (Fig. 1d) and confirm via Sanger sequencing with M13 primers.

3.1.3 Prepare Regulon Assembly Vector

As the vectors to assembly TUs have a kanamycin resistance marker, regulon assembly requires a different marker. The pAV115 vector, which encodes ampicillin resistance and RFP flanked by BsaI sites, is used here. Removal of RFP provides visual confirmation of assembly. The sequences used for assembly into this plasmid are "CAGT" and "TTTT." Other plasmids may be used, paying attention to the sequence overhangs generated by BsaI cleavage. Engineering a hygromycin resistance marker (or other available marker) surrounded by regions of homology (here, the sequence surrounding the yeast *GAL7, 10, 1* gene cluster) into this backbone allows for positive selection of integration into the yeast genome. Additionally, loxPsym sites flanking 100 bp sequences are added to allow for alteration of gene spacing upon Cre-lox-induced recombination. The homology arms are cloned outside of the BsaI sites, which must remain intact for Golden Gate assembly. The homology arm cloned upstream of the "CAGT" assembly sequence includes an SbfI recognition site to allow for isolation of the assembled regulon along with a NotI cleavage site, which is encoded downstream of the "TTTT" assembly sequence in the vector. The SbfI and NotI recognition sites are 8 bp and do not cleave in the sequences used for GAL regulon assembly. If this is not the case for targeted sequences, other sites may be used.

1. Design primers to amplify regions of homology surrounding the yeast *GAL7, 10, 1* gene cluster. For cloning upstream of the "CAGT" site, the forward primer includes SalI and SbfI recognition sites and the reverse primer has a PstI tail. For cloning downstream of the "TTTT" site, the forward primer includes a BamHI site and the reverse contains a NotI site (*see* **Note 6**) (Table 1).
2. Amplify homology arms from yeast genomic DNA, using a high-fidelity polymerase and clean-up PCR products with SPRI paramagnetic beads.

3. Cut pAV115 with SalI-HF and PstI-HF, treating the vector with rSAP, at 37 °C for 1 h.
4. Cut upstream homology region with SalI-HF at 37 °C for 30 min, add PstI-HF, and continue digestion for an additional 30 min (*see* **Note 7**).
5. Clean-up digestions with SPRI paramagnetic beads to remove PstI, which cannot be heat inactivated.
6. For a 20 μL ligation reaction, mix:

T4 DNA ligase buffer	2 μL
T4 DNA ligase (400 U/μL)	1 μL
Upstream HR, SalI/PstI (**step 4**)	0.06 pmol
pAV115, SalI/PstI (**step 3**)	0.02 mol

7. Incubate at room temperature for 1 h, then heat inactivate the ligation at 65 °C for 10 min.
8. Transform 2 μL into 25 μL competent cells, plating on LB +ampicillin plates.
9. Miniprep colonies and confirm insertion of homology region via sequencing with M13 forward primer.
10. Clone downstream homology arm similarly to upstream one (**steps 3–9**), using BamHI-HF and NotI-HF restriction enzymes. The M13 reverse primer can be used to confirm the insert sequence.
11. Amplify hygromycin resistance gene (HygR) from pAG32 using ddrmF and ddrmR oligos, including a BclI site in the tail of the forward oligo (Table 1). This BclI site and a BamHI site in the PCR product can be used to clone into a BamHI site in the pGALhr backbone.
12. Cleave the HygR gene product with BamHI-HF and BclI-HF.
13. Cut the pGALhr vector with BamHI-HF.
14. Purify digestion products (**steps 12** and **13**) with 1.8 volumes SPRI paramagnetic beads to remove BamHI-HF, as it cannot be heat inactivated.
15. For a 20 μL ligation reaction, combine:

T4 DNA ligase buffer	2 μL
T4 DNA ligase (400 U/μL)	1 μL
HygR, BamHI/BclI (**step 12**)	0.06 pmol
pGALhr, BamHI (**step 13**)	0.02 mol

16. Ligate at room temperature for 30 min, then heat inactivate at 65 °C for 10 min.
17. Transform 2 μL into 25 μL competent cells, plating on LB +ampicillin plates.
18. Miniprep colonies and confirm insertion via Sanger sequencing with M13 reverse primer. As BamHI is used to clone the loxPsym array, select plasmids with the BamHI site maintained on the side proximal to the BsaI assembly sites (Fig. 1e).
19. Amplify spaced loxPsym array (Subheading 3.1.2, Fig. 1d) with primers containing BamHI recognition sites, using a high-fidelity polymerase (Table 1). Purify product with SPRI paramagnetic beads.
20. Digest pGALhr(HygR) vector and loxPsym PCR products with BamHI-HF, treating the vector with rSAP in parallel, at 37 °C for 1 h. Purify products to remove BamHI-HF.
21. Ligate 0.06 pmol loxPsym array into 0.02 pmol vector in 20 μL 1x T4 DNA ligase buffer with 400 U T4 DNA ligase in a 20 μL reaction:

T4 DNA ligase buffer	2 μL
T4 DNA Ligase (400 U/μL)	1 μL
pGAHLhr(HygR), BamHI	0.02 pmol
loxPsym array PCR, BamHI	0.06 pmol

22. Incubate at room temperature for 30 min, then heat inactivate at 65 °C for 10 min.
23. Transform 2 μL into 25 μL competent cells, plating on LB +ampicillin.
24. Miniprep and confirm full sequence of pGALhr5Lox(HygR) regulon assembly vector (homology arms, hygromycin resistance gene, and loxPsym array) via full-length plasmid sequencing service (Fig. 1e).

3.2 Domesticate Open Reading Frames

1. Phosphorylate BsmBI-tailed primers: to 200 pmoles primers in 35 μL, add 4 μL 10x T4 DNA ligase buffer and 1 μL T4 PNK (10 U/μL). Incubate at 37 °C for 30 min and then heat inactivate at 65 °C for 20 min.
2. Amplify open reading frame (ORF) sequences from genomic DNA with a high-fidelity DNA polymerase. Purify PCR reactions with 1.8 volumes SPRI paramagnetic beads and elute in 50 μL 10mM Tris-HCl.
3. Remove BsaI site from ampicillin resistance gene in pUC19 (as Subheading 3.1.1, **steps 3–7**) to generate pUC-IIs vector (Table 1).

4. Cut 1 μg pUC-IIs vector with 20U EcoRV-HF in 50 μL 1x CutSmart buffer, simultaneously treating with 1 μL rSAP.
5. To clone ORF into pUC-IIs, combine:

T4 DNA ligase buffer	2 μL
T4 DNA Ligase (400 U/μL)	1 μL
pUC-IIs, EcoRV (**step 4**)	2 μL (0.02 pmol)
ORF PCR product (**steps 1** and **2**)	0.06 pmol

6. Incubate overnight (≥12 h) at 16 °C. Heat inactivate ligase at 65 °C for 10 min.
7. Transform 2 μL ligation into 25 μL competent cells via the manufacturer's protocol.
8. Perform site-directed mutagenesis to remove type IIS restriction sites. Design primers with abutting 5′ ends, in which the restriction site is removed (Fig. 2a).
9. Phosphorylate primers (5 μM each) in 40 μL reaction with 1x T4 DNA ligase buffer and 1 μL T4 polynucleotide kinase (10 U/μL) at 37 °C for 30 min, then heat inactivate at 65 °C for 20 min.
10. Use phosphorylated primers in a high-fidelity PCR reaction to amplify ~5 ng pUC-IIs:ORF plasmid, cycling a maximum of 16 times. Check 5 μL on 1% agarose gel for full-length product.
11. Add 5 μL 10x CutSmart buffer and treat with 1 μL DpnI to digest original plasmid at 37 °C for 30 min, then heat inactivate at 65 °C for 20 min.
12. Use 2 μL in 20 μL ligation reaction with 1x T4 DNA ligase buffer and 1 μL T4 DNA ligase (400 U/μL) at 16 °C overnight (≥12 h).
13. Heat inactivate and transform 2 μL reaction into 25 μL competent cells, using the manufacturer's protocol.
14. Miniprep pUC-IIs:ORF plasmid and confirm removal of restriction site by Sanger sequencing.

3.3 Amplify Promoters/Terminators (See Note 8)

3.3.1 Sequences without Type IIS Restriction Sites

1. Design primers with inward-oriented BsmBI sites and the appropriate 4 bp overhang sequence (Table 2) with 6 bp upstream to improve cleavage efficiency close to the end of a DNA fragment.
2. Amplify promoter/terminator sequences from ~50 ng genomic DNA with a high-fidelity DNA polymerase.
3. Purify the PCR products with 1.8 volumes SPRI paramagnetic beads, eluting in 20 μL 10mM Tris-HCl.

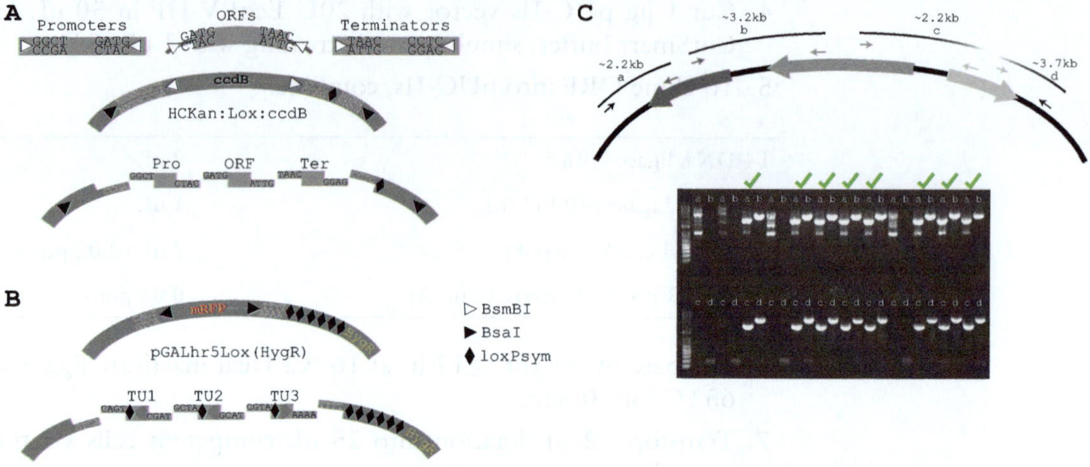

Fig. 3 Golden Gate assembly. (**a**) Assembly of TUs from promoter, ORF, and terminator sequences via BsmBI Golden Gate cloning into the HCKan:Lox:ccdB vector. (**b**) Release of the assembled TUs with a flanking loxPsym site is performed with BsaI, allowing for assembly of the regulon. (**c**) Colony PCR across junctions of the TUs confirms correct assemblies

3.3.2 Sequences with Type IIS Restriction Site

1. Design two sets of primers—left-external × left-internal and right-internal × right-external—so that the internal primers overlap and lack the restriction site (Fig. 2b). The external primers should include inward-oriented BsmBI recognition sequences and appropriate 4 bp overhang sequences (Table 2), with 6 bp upstream to improve cleavage efficiency close to the end of a DNA fragment.
2. Amplify the left and right products in separate high-fidelity PCR reactions, and purify the PCR products with 1.8 volumes SPRI paramagnetic beads, eluting in 20 μL 10mM Tris-HCl.
3. Combine ~2 nM each left and right PCR products with 0.5 μM external primers for the full-length PCR. Purify the full-length PCR product with 1.2 volumes SPRI paramagnetic beads and elute in 20 μL 10mM Tris-HCl.

3.4 Assemble TU Libraries

3.4.1 Golden Gate Assembly (BsmBI-v2)

Promoter libraries with "GGCT"/"CATC" overhangs and terminators with "TAAC"/"GAGG" overhangs are assembled in HCKan:ccdB vectors with "GATG"/"GTTA" flanked open reading frames to generate TU libraries with promoter combinatorial complexity (Fig. 3a). Each TU is cloned into an HCKan:ccdB vector with specific BsaI-derived overhangs for the subsequent ordered assembly of multiple TUs (Fig. 3b).

1. For Golden Gate cloning of TUs, combine:

T4 DNA ligase buffer	1 μL
T4 DNA ligase (400 U/μL)	0.25 μL
BsmBI-v2 (20 U/μL)	0.35 μL

(continued)

HCKan:ccdB	0.05 pmol
Pro mix	0.1 pmol
Ter	0.1 pmol
pUC-IIs:ORF	0.1 pmol

2. Perform 30 cycles of 42 °C then 16 °C for 5 min each, followed by 55 °C for 50 min, then 65 °C for 10 min. These final steps digest the remaining unassembled plasmid to reduce background and heat inactivate the ligase, respectively.
3. To further reduce background, treat the reaction with Plasmid-Safe DNase in tandem with BsmBI-v2. Add ATP to 2 mM and 0.25 μL each Plasmid-Safe DNase and BsmBI-v2 and heat at 37 °C for 30 min then heat inactivate at 80 °C for 20 min.
4. Transform 2 μL of the reaction mixture into 25 μL competent cells via the manufacturer's protocol. Plate ~125 μL transformed cells, or enough to obtain high-density single colonies, on several LB +Kanamycin plates (~9 cm diameter).
5. Scrape plate with ~3 mL LB and miniprep plasmid libraries from the mixture using ≥2 columns.

3.4.2 Assess Promoter Library Coverage (Illumina AmpliSeq)

1. Design sets of 4 primers surrounding the promoter insert (one upstream in the vector and the other at the 5′ end of the ORF) with handles for Nextera sequencing adapters (Table 3). Each set of 4 should contain 0, 1, 2, or 3 random nucleotides in between the handle and primer sequence to stagger the insert sequence and increase complexity during sequencing (Fig. 4a).

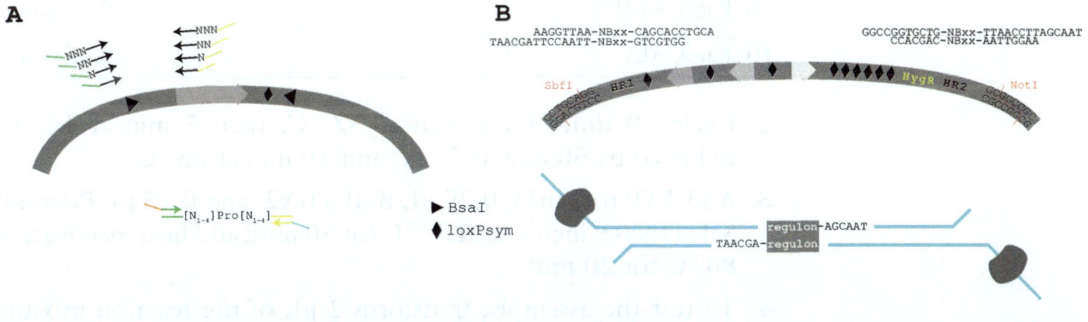

Fig. 4 Confirming library composition. (**a**) Amplicon sequencing of the cloned TU libraries is performed using primers that anneal just upstream of the assembly site in the TU acceptor vector and at the 5′-end of the ORF. Primers contain Nextera handles (green and yellow), which are extended (orange and blue) in a second PCR for Illumina sequencing. (**b**) Composition of the assembled regulon promoter libraries is confirmed via long-read sequencing. Barcoded adapters designed for compatibility with Oxford Nanopore Technologies native adapters are ligated to SbfI/NotI cleaved regulon and are then ligated to the native adapters for nanopore sequencing

2. Amplify promoter sequences from GAL TU libraries with a high-fidelity DNA polymerase for 10 cycles. Clean up PCRs with 0.8 volumes SPRI paramagnetic beads and elute in 11 μL 10mM Tris-HCl.
3. Use 1 μL of this PCR product in a second PCR that extends the Nextera handles for sequencing. Perform 12–15 cycles in a 10 μL high-fidelity PCR reaction with 1 μL Nextera indexing oligos.
4. Clean up PCR with 0.6 volumes SPRI paramagnetic beads and elute in 10 μL 10 mM Tris-HCl.
5. Quantitate 1 μL of 1:10 dilution with Qubit HS dsDNA and check size distribution with Agilent Bioanalyzer HS DNA. Combine in equimolar amounts for Illumina 150 bp single-end sequencing with a dedicated barcode read on a MiSeq machine (*see* **Note 9**).

3.5 Assemble Regulon

3.5.1 Golden Gate Assembly (BsaI-HFv2)

Transcriptional units with complementary overhangs are assembled into pGALhr5Lox(HygR) via "ACTG" and "TTTT" overhangs from BsaI-HFv2 cleavage (Fig. 3b). Homology arms and the hygromycin resistance marker will allow for selection of the integrated regulon.

1. To assemble multiple TUs into a regulon, combine:

T4 DNA ligase buffer	1 μL
NEBridge BsaI enzyme mix	0.5 μL
pGALhr5Lox(HygR)	0.05 pmol
HCKan:GAL7	0.05 pmol
HCKan:GAL10	0.05 pmol
HCKan:GAL1	0.05 pmol

2. Cycle 30 times for 5 min at 37 °C, then 5 min at 16 °C, followed by 50 min at 37 °C and 10 min at 65 °C.
3. Add ATP to 2 mM, 0.25 μL BsaI-HFv2, and 0.25 μL Plasmid-Safe DNase; incubate at 37 °C for 30 min; and heat inactivate at 80 °C for 20 min.
4. To test the assembly, transform 2 μL of the reaction mixture into 25 μL competent cells. Plate 150 μL transformed cells on LB +ampicillin plates.
5. Once correct assemblies are confirmed by colony PCR, perform high-efficiency transformation of Golden Gate assembly into electrocompetent cells.

6. Dialyze assembly reaction using pre-wetted (~15 min) dialysis membrane floating on ~100 mL water for 30 min to remove salts.
7. Add 2 μL dialyzed reaction into 25 μL electrocompetent cell suspension thawed on ice. Transfer suspension with DNA into chilled 0.1 cm cuvette on ice. Pulse at 25 μF, 200 Ω, and 2.0 kV; this usually results in 4.3–4.4 ms time constant.
8. Immediately recover cell suspension in 950 μL recovery media for 1 h at 37 °C. Plate 10 μL of 1:1000 diluted transformation mixture on LB Ampicillin plate (to estimate number of colonies) and the remainder across several large (245 × 245 mm^2) plates.
9. Scrape plates with ~15 mL LB each or enough to collect all cells from plates. Prepare plasmid using a Maxiprep kit.

3.5.2 Colony PCR Confirmation

1. Pick individual colonies into 20 μL water.
2. Use 2 μL of colony:water in PCR reactions with primers that amplify across junctions of assembled units and acceptor vector (Fig. 3c). Check PCR products on gel.

3.5.3 Assess Library Coverage (Oxford Nanopore Ligation Sequencing) (See **Note 2***)*

1. Cut assembled fragment from plasmid with SbfI-HF and NotI-HF and gel purify.
2. Prepare custom ONT native barcode adapters (Table 3). These adapters are designed to use the ONT native barcode sequences and to have an SbfI- or NotI-compatible overhang (Fig. 4b).
3. Phosphorylate 100 pmoles of each oligo in 40 μL of 1x T4 DNA ligase buffer with 10 units of T4 PNK.
4. Dilute phosphorylated oligos with 160 μL 10 mM Tris-HCl and heat in a thermocycler at 96 °C for 5 min, then decrease temperature 1 °C every 10 s until 23 °C.
5. Ligate custom ONT native barcode adapters to SbfI/NotI assembly fragment.

SbfI/NotI assembly fragment	~180 fmol
ONT_NBxx_SbfI dsDNA adapter	~460 fmol
ONT_NBxx_NotI dsDNA adapter	~460 fmol
5x Quick Ligase buffer (NEB)	10 μL
T4 DNA Ligase (2000 U/μL)	2.5 μL

6. Mix well and incubate at room temperature for 20 min.
7. Clean up the reaction to remove excess oligonucleotides: Dilute the reaction with 50 μL water and add 45 μL SPRI paramagnetic beads. Incubate on rotating wheel at room

temperature for ~10 min then wash twice with 200 μL 70% ethanol on a magnetic rack. Briefly dry the beads and elute with 31 μL water. Use Qubit HS dsDNA to measure DNA concentration.

8. Ligate ONT AMII adapter to barcoded DNA. Combine equimolar amounts of barcoded DNA to be sequenced on the same flow cell.

Barcoded DNA	200 fmol, total
AMII (ONT adapter)	5 μL
5x Quick Ligase buffer	20 μL
T4 DNA Ligase (2000 U/μL)	10 μL

9. Clean up reaction to remove excess adapters. Add 50 μL SPRI paramagnetic beads and rotate at room temperature for ~10 min. Wash beads twice with 250 μL LFB, resuspending the beads off of the magnet. Briefly dry the beads and elute with 16 μL EB.
10. Load the library on primed FLO-MIN106 flow cell (*see* **Note 3**). Run on GridION/MinION with settings for ligation sequencing with native barcode expansion, using live, high-accuracy base calling.
11. Align reads using minimap2 default (-x map-ont) to a genome file consisting of all possible promoter combinations for each TU. Sort mapped reads and keep only the longest mapping alignments for each TU. Calculate the percent of promoters covered for each TU and for all possible combinations; ideally, all promoters and possible combinations would be represented in the pool.
12. Comparing the area under the curve (AUC) of the cumulative distribution of promoter coverage to that of the line of equality can report on the equality of distribution via the Gini coefficient [11]. This can be calculated from the difference between the AUC of promoter coverage and the line of equality divided by the AUC of the line of equality. Numbers closer to zero indicate greater equality of promoter representation within the library.

4 Notes

1. There are many standardized Golden Gate cloning vectors available that can be used, but attention should be paid to ensure that the correct overhang sequences are used for assembly.

2. Oxford Nanopore Technologies (ONT) frequently updates its kits and software; it is advisable to check for updates on their website https://nanoporetech.com/. ONT currently offers a Native Barcoding Kit 96 with V14 chemistry rather than the older (V9) Native Barcoding Expansion coupled with their Ligation Sequencing Kit. In this newer version, the native adapter (NA) is used instead of the AMII adapter, both of which attach the motor protein-linked adapter for nanopore sequencing.
3. Depending on the number of samples combined and their complexity, different flow cells may be used. Here, 21952 potential combinations were possible (28 promoter for 3 TUs). For only 10x coverage, this would require an output of at least 200000 full-length (12000 bp) reads per sample, which is just within the theoretical maximum output of a Flongle flow cell (2.8 Gb). The theoretical maximum outputs for the MinION and PromethION flow cells are currently 50 Gb and 290 Gb, respectively.
4. Laboratory strains that contain the F' plasmid, such as NEB Stable, should also be able to grow with *ccdB*; however, *ccdB* survival cells, which are designed for growth with *ccdB*, produced far less background and fewer incomplete assemblies.
5. In general, assembling several individual loxPsym units in a single reaction appeared to decrease the efficiency of correct assembly. Incorporating these into the PCR product prior to cloning helped to reduce incorrect assemblies, but does require more stringent PCR conditions (higher Tm with selected polymerase) and reduced amplification cycles.
6. Homology arms to other desired loci may also be used but should be checked for restriction sites that do not cleave within the desired sequence. As an alternative, recombination sites orthogonal to Cre-lox could be cloned flanking the regulon assembly site.
7. SalI-HF digestion is performed first to decrease the efficiency of PstI cleavage at the SbfI site close to the end of the DNA, and PstI digestion time is kept to a minimum.
8. Promoter and terminator "parts" may be cloned for domestication; SbfI restriction sites are included in the primer tails for this purpose (Table 2). However, PCR products worked well for TU assembly.
9. If sequencing facilities are not available onsite, the amplicon (**step 2**) may be sent to a commercial sequencing provider.

Acknowledgments

This work was funded by Volkswagen Stiftung (94769 to L.M.S.).

References

1. Nützmann H-W, Scazzocchio C, Osbourn A (2018) Metabolic gene clusters in eukaryotes. Annu Rev Genet 52:159–183. https://doi.org/10.1146/annurev-genet-120417-031237
2. Mellor J, Woloszczuk R, Howe FS (2016) The interleaved genome. Trends Genet 32:57–71. https://doi.org/10.1016/j.tig.2015.10.006
3. Greger IH, Proudfoot NJ (1998) Poly (A) signals control both transcriptional termination and initiation between the tandem *GAL10* and *GAL7* genes of *Saccharomyces cerevisiae*. EMBO J 17:4771–4779. https://doi.org/10.1093/emboj/17.16.4771
4. Houseley J, Rubbi L, Grunstein M et al (2008) A ncRNA modulates histone modification and mRNA induction in the yeast GAL gene cluster. Mol Cell 32:685–695. https://doi.org/10.1016/j.molcel.2008.09.027
5. Dymond JS, Richardson SM, Coombes CE et al (2011) Synthetic chromosome arms function in yeast and generate phenotypic diversity by design. Nature 477:471–476. https://doi.org/10.1038/nature10403
6. Engler C, Kandzia R, Marillonnet S (2008) A one pot, one step, precision cloning method with high throughput capability. PLoS One 3: e3647. https://doi.org/10.1371/journal.pone.0003647
7. Pryor JM, Potapov V, Bilotti K et al (2022) Rapid 40 kb genome construction from 52 parts through data-optimized assembly design. ACS Synth Biol 11:2036–2042. https://doi.org/10.1021/acssynbio.1c00525
8. Weber E, Engler C, Gruetzner R et al (2011) A modular cloning system for standardized assembly of multigene constructs. PLOS ONE 6:e16765. https://doi.org/10.1371/journal.pone.0016765
9. Guo Y, Dong J, Zhou T et al (2015) YeastFab: the design and construction of standard biological parts for metabolic engineering in *Saccharomyces cerevisiae*. Nucleic Acids Res 43:e88. https://doi.org/10.1093/nar/gkv464
10. Martella A, Matjusaitis M, Auxillos J et al (2017) EMMA: an extensible mammalian modular assembly toolkit for the rapid design and production of diverse expression vectors. ACS Synth Biol 6:1380–1392. https://doi.org/10.1021/acssynbio.7b00016
11. O'Hagan S, Muelas MW, Day PJ et al (2018) GeneGini: assessment via the Gini coefficient of reference "housekeeping" genes and diverse human transporter expression profiles. Cell Syst 6:230–244.e1. https://doi.org/10.1016/j.cels.2018.01.003
12. Agmon N, Mitchell LA, Cai Y et al (2015) Yeast Golden Gate (yGG) for the efficient assembly of *S. cerevisiae* transcription units. ACS Synth Biol 4:853–859. https://doi.org/10.1021/sb500372z
13. Goldstein AL, McCusker JH (1999) Three new dominant drug resistance cassettes for gene disruption in *Saccharomyces cerevisiae*. Yeast Chichester Engl 15:1541–1553. https://doi.org/10.1002/(SICI)1097-0061(199910)15:14<1541::AID-YEA476>3.0.CO;2-K
14. Sharma M, Kretschmer C, Lampe C et al (2019) Targeting specificity of nuclear-encoded organelle proteins with a self-assembling split-fluorescent protein toolkit. J Cell Sci 132:jcs230839. https://doi.org/10.1242/jcs.230839

Chapter 23

Golden Gate Cloning for Efficient Biosynthesis of Lycopene in Synthetic Yeast

Yu Zhang, Ju Zheng, Xian Fu, and Yue Shen

Abstract

Golden Gate Assembly (GGA) represents a versatile method for assembling multiple DNA fragments into a single molecule, which is widely used in rapid construction of complex expression cassettes for metabolic engineering. Here we describe the GGA method for facile construction and optimization of lycopene biosynthesis pathway by the combinatorial assembly of different transcriptional units (TUs). Furthermore, we report the method for characterizing and improving lycopene production in the synthetic yeast chassis.

Key words Golden Gate Assembly, Biosynthesis pathway, SCRaMbLE, Lycopene, Synthetic yeast

1 Introduction

Golden Gate Assembly (GGA) is a widely used DNA assembly method that allows for the efficient construction of multiple DNA fragments into a single DNA molecule [1]. GGA utilizes type IIS restriction enzymes such as *Bsm*BI, *Bsa*I, *Bbs*I to cut DNA fragments and produce user-defined specified sticky ends outside recognition site on original fragments [2]. The resultant sticky end, a four-base overhang, could be used to connect modules in designed order and orientation. The GGA modules typically consist of a promoter (PRO), an open reading frame (ORF) and a terminator (TER) or transcriptional units (TUs). Since different overhang sequence can be specified for different GGA module, it would allow the high efficient assembly of all modules in a one-pot reaction [2, 3]. In comparison to conventional cloning, such as restriction enzyme digestion/ligation, GGA method is highly efficient, flexible, and accurate, which allows for the rapid and seamless assembly of more than 50 DNA fragments [4]. For example, Gibson Assembly is limited to linearized DNA substrates, and may fail

Daniel Schindler (ed.), *Golden Gate Cloning: Methods and Protocols*, Methods in Molecular Biology, vol. 2850,
https://doi.org/10.1007/978-1-0716-4220-7_23,

to assemble repetitive sequences. On the contrary, GGA method is versatile for assembling both linear and circular DNA fragments [2] and is well compatible with repetitive sequences [5].

With the aforementioned advantages, GGA is particularly useful for construction of metabolic pathways and optimization of gene expression via arbitrary combination of different TUs in heterogenous hosts [6, 7]. Engineering of biological pathways in various microorganisms is a promising direction for biotechnology to produce commodity chemicals. With the rapid development of synthetic genomics in recent years, whole genome synthesized microorganisms now represent as the ideal chassis for metabolic engineering applications owing to many useful design features [7–9]. For example, the synthetic yeast genome project (Sc2.0) introduced an inducible genome shuffling system called Synthetic Chromosome Rearrangement and Modification by LoxP-mediated Evolution (SCRaMbLE) into synthetic yeast [8, 10], which allows systematic rearrangement of the yeast genome. The resultant cell pools could be used to screen for high-production strain and provide insights into the compatibility between host chassis and biosynthesis pathway [7, 11, 12].

Here we describe the application of this method for readily constructing lycopene biosynthesis pathway. Subsequently, we apply GGA cloning to efficiently optimize the biosynthesis pathway by combinatorial assembly of TUs with different promoters and terminators. Finally, we integrated the GGA-optimized lycopene pathway into a SCRaMbLEd-optimized synthetic yeast to achieve increased production of lycopene.

2 Materials

2.1 Polymerase Chain Reaction (PCR)

1. Genomic DNA from BY4741 strain (*MAT*a *his3Δ1 leu2Δ0 met15Δ0 ura3Δ0*) or pRS416 vector with *Bsa*I site removal.
2. High-fidelity proofreading DNA polymerase and buffer (*see* **Note 1**).
3. 2.5 mM deoxynucleotide triphosphate (dNTP) mixture.
4. 10 μM forward and reverse primers.
5. Agarose.
6. 1 × Tris-acetate-EDTA (TAE) buffer: 40 mM Tris, 20 mM acetic acid, 1 mM EDTA.
7. Gel and PCR purification kit.

2.2 GGA

2.2.1 GGA Cloning with HcKan Vectors

1. HcKan_P, HcKan_O, HcKan_T vectors [6].
2. Purified PCR products or pUC57-ORF plasmids (*see* **Note 2**).
3. 10 × T4 DNA ligase buffer.

4. T4 DNA ligase (400 U/μL).
5. *Bsa*I-HF (20 U/μL).

2.2.2 GGA Cloning with POT Vectors

1. POT2, POT4, POT6, POT7 vectors [6].
2. PRO-HcKan_P, ORF-HcKan_O, TER-HcKan_T plasmids.
3. 10 × T4 DNA ligase buffer.
4. 10 × BSA (1 mg/mL).
5. T4 DNA ligase (400 U/μL).
6. *Bsm*BI (10 U/μL).

2.2.3 GGA Cloning with pRS416 Vectors

1. Purified PCR products of pRS416 linear vector.
2. TU-POT plasmids.
3. 10 × T4 DNA ligase buffer.
4. 10 × BSA (1 mg/mL).
5. T4 DNA Ligase (400 U/μL).
6. *Bsa*I-HF (20 U/μL).

2.3 Primers

2.3.1 Primers Used for the Amplification of PROs or TERs

1. pTEF2-F: 5′-AGCGTGGGTCTCAGGCTGGGCGCCATAACCAAG-3′.
2. pTEF2-R: 5′-GTGCTGGGTCTCGCATCTTTAGTTAATTATAGTTCGTTGACCGTATATTC-3′.
3. pTDH3-F: 5′- AGCGTGGGTCTCAGGCTTCATTATCAATACTGCCATTTC-3′.
4. pTDH3-R: 5′- GTGCTGGGTCTCGCATCTTTGTTTGTTTATGTGTGTTTATTCG-3′.
5. pCCW12-F: 5′- AGCGTGGGTCTCAGGCTCTATTGGCGTCTGATTTCCG-3′.
6. PCCW12-R: 5′- GTGCTGGGTCTCGCATCTTATTGATATAGTGTTTAAGCGAATGAC-3′.
7. pPGK1-F: 5′-AGCGTGGGTCTCAGGCTGTTTGCAAAAAGAACAAAACTG-3′.
8. pPGK1-R: 5′- GTGCTGGGTCTCGCATCCATTGTTTATATTTGTTGTAAAAAGT-3′.
9. pCYC1-F: 5′- AGCGTGGGTCTCAGGCTTTCATTTGGCGAGCGTTGG-3′.
10. pCYC1-R: 5′- GTGCTGGGTCTCGCATCTTAGTGTGTGTATTTGTGTTTGTGTGTC-3′.
11. pRPL18b-F: 5′-AGCGTGGGTCTCaGGCTTTAGAACGCTCCTGTAAC-3′.
12. pRPL18b-R: 5′-GTGCTGGGTCTCgCATCTTTTGTTTTTTGTTTTCTTCTAATTG-3′.

13. tADH1-F: 5′-AGCGTGGGTCTCATAGCCGAATTTCTTA TGATTTATGAT-3′.
14. tADH1-R: 5′-GTGCTGGGTCTCGGAGGCCGGTAGAGG TGTGG-3′.
15. tCIT1-F: 5′-AGCGTGGGTCTCTTAGCTAGCAGTGCTTT TAACTAAGAATTATTAGTC-3′.
16. tCIT1-R: 5′-GTGCTGGGTCTCGGAGGAAAAAAAAGCC ACGCGTG-3′.
17. tFUM1-F: 5′-AGCGTGGGTCTCTTAGCTAGCGGAAAAT TTGATTTTTGATTCAGG-3′.
18. tFUM1-R: 5′-GTGCTGGGTCTCGGAGG CGGAAAAATT TTGTTTTCTATTCATTGATA TC-3′.
19. tERP2-F: 5′-AGCGTGGGTCTCTTAGC AGAACTTTTCA ATCTACGAAAAATATATGTC CG-3′.
20. tERP2-R: 5′-GTGCTGGGTCTCGGAGGGCAAGACAAAT GACCAGATATAAACG-3′.

2.3.2 Primers Used for the Amplification of pRS416 Linear Backbone

1. pRS416-F: 5′-GGTCTCTAGGTAGAGAGAGACGGGATA A CGCAGGAAAGAACATGTG AGC-3′.
2. pRS416-R: 5′-GGTCTCATGAGACCTAGAGACGTATTAC CGCCTTTGAGTGAGCTGAT ACC-3′.

2.3.3 Primers Used for Plasmid Verification or Sequencing

1. HcKan-seq-F: 5′-GATCCTTTGATTTTCTACCG-3′.
2. HcKan-seq-R: 5′-CTCGATAACTCAAAAAATACG-3′.
3. POT-seq-F: 5′-CTTTCCTGCGTTATCCCCTGATTC-3′.
4. POT-seq-R: 5′-GATGTGCTGCAAGGCGATTAAG-3′.
5. pRS416-seq-F: 5′-CGGAGCCTATGGAAAAACGCC-3′.
6. pRS416-seq-R: 5′-TCTTCCGCTTCCTCGCTCAC-3′.
7. CrtE-seq-F: 5′-GCTTTCAACTACTGGTTGGACG-3′.
8. CrtE-seq-R: 5′-CAATGGGATGTCAGCCAACT-3′.
9. CrtI-seq-F: 5′-GCTAAGTTCAACTTCAACGCTC-3′.
10. CrtI-seq-R: 5′-GAAAGCCAAAACACCAACAGATC-3′.
11. CrtB-seq-F: 5′-ATGAATAATCCGTCGTTACTCAATC-3′.
12. CrtB-seq-R: 5′-CTACTAGAGCGGGCGCT-3′.
13. IDI-seq-F: 5′-ATGCAAACGGAACACGTCAT-3′.
14. IDI-seq-R: 5′-TAATTGTGCTGCGCGAAAGC-3′.

2.3.4 Primers Used for the Construction of Genomic Integrated Fragments

1. rDNA25-HoL-F: 5′- ACAAATCAGACAACAAAGGCT-3′.
2. rDNA25-HoL-KANMX-R: 5′-ATTCTGGGCCTCCATGTC AGCAGAATTCGGTAAGCG-3′.
3. KANMX-F: 5′-CGCTTACCGAATTCTGCTGACATGGAG GCCCAGAAT-3′.
4. KANMX-CCW12p-R: 5′-CAAAACGGAAATCAGACGCCA ATAGAGCCAGGTAGAGCA.
   GTATAGCGACCAGCATTCAC-3′.
5. ADH1t-rDNA25-HoR-F: 5′-TCTTATTGACCACACCTCT ACCGGCCTCTGAGACCTG.
   CTGGCCCAGTGAAATGC-3′.
6. rDNA25-HoR-R: 5′-GTTTGACCTCAAATCAGGTAGG-3′.

2.4 Escherichia coli (E. coli) Transformation and Verification

2.4.1 E. coli Transformation

1. *E. coli* DH5α competent cells.
2. GGA reaction mixture or plasmids.
3. Luria-Bertani (LB) medium: 10 g/L Tryptone, 5 g/L Yeast Extract, 10 g/L NaCl. For solid medium, agar is added to 1.5%.
4. LB solid medium containing 100 μg/mL Kanamycin (LB^{Kan+} plate) or 50 μg/mL Carbenicillin (LB^{Carb+} plate).
5. 20 mg/mL 5-Bromo-4-chloro-3-indole-β-D-galactoside (X-Gal).
6. 50 mg/mL isopropyl-β-D-thiogalactoside (IPTG).

2.4.2 Plasmid Extraction from E. coli

1. LB^{Kan+} liquid medium, LB^{Carb+} liquid medium, or 2 × YT^{Carb+} medium: 16 g/L tryptone, 10 g/L yeast extract, 5 g/L NaCl, 50 μg/mL carbenicillin.
2. Plasmid purification kit (*see* **Note 3**).
3. 75% ethanol.

2.4.3 Verification of Digested Plasmids or PCR Products

1. *Bsa*I-HF (20 U/μL) and 10 × NEBuffer CutSmart or *Bsm*BI (10 U/μL) and 10 × NEBuffer 3.1.
2. 2 × Taq PCR Mastermix.
3. 10 μM Forward and reverse verification primers.
4. Agarose.
5. 1 × TAE buffer.

2.5 Yeast Transformation and Verification

2.5.1 Yeast Transformation

1. Synthetic synII diploid strains (*MAT*a/*MATα his3Δ1/his3Δ1 leu2Δ0/leu2Δ0 LYS2/lys2Δ0 met15Δ0/MET15 ura3Δ0/ura3Δ0*) or BY4741 strain.
2. SC-His−Leu−Ura medium: 20 g/L glucose, 5 g/L ammonium sulfate, 1.7 g/L yeast nitrogen base, 0.735 g/L synthetic complete drop-out –His, −Leu, –Ura. For solid medium, agar is added to 3%.

3. 100 × histidine (6.3 g/L).
4. 50 × leucine (13 g/L).
5. 0.1 M lithium acetate (LiAc).
6. 1 M LiAc.
7. DNA plasmids or fragments.
8. 10 mg/mL salmon sperm DNA (ssDNA). ssDNA needs to be boiled at 99 °C for 10 min and immediately place on ice.
9. 50% (w/v) polyethylene glycol (PEG) 3350.
10. Dimethyl sulfoxide (DMSO).
11. 5 mM $CaCl_2$.
12. SC–Ura plate.

2.5.2 Verification of Yeast Colonies

1. 20 mM NaOH.
2. 2 × Taq PCR Mastermix.
3. 10 μM forward and reverse verification primers.
4. Agarose.
5. 1 × TAE buffer.

2.5.3 Yeast Plasmid Extract

1. SC–Ura liquid medium.
2. Yeast plasmid purification kit.
3. Lyticase (12.5 KU/mL).
4. 1 M Sorbitol buffer: 182.2 g/L sorbitol, 10% 1 M ethylenediaminetetraacetic acid (EDTA) (pH 7.5).

2.6 SCRaMbLE

1. pRS413-Cre/EBD plasmid [7].
2. SC–His–Ura liquid medium.
3. 1 mM β-Estradiol, dissolved in ethanol.
4. SC–Ura plate.

3 Methods

3.1 Golden Gate Cloning of Lycopene Biosynthesis Pathway in Yeast

3.1.1 Clone Biological Part into Plasmids

1. Amplify PROs and TERs from BY4741 genomic DNA using designed primers in Subheading 2.3.1.

   Incorporate prefixes into primers, for promoters, use prefixes below:

   Prefix of forward primer: AGCGTGGGTCTCAGGCT
   Prefix of reverse primer: GTGCTGGGTCTCGCATC
   For terminators, use prefixes below:
   Prefix of forward primer: AGCGTGGGTCTCTTAGC
   Prefix of reverse primer: GTGCTGGGTCTCGGAGG

2. Then, purify the PCR products by agarose gel (1%) electrophoresis following standard protocol provided by gel and PCR purification kit to obtain PROs or TERs fragments.
3. Extract each ORF-pUC57 plasmid manually from *E.coli* using a standard plasmid purification kit protocol, respectively. The ORFs involved in lycopene biosynthesis pathway include geranylgeranyl diphosphate synthase (CrtE), 15-*cis*-phytoene synthase (CrtB), phytoene desaturase (CrtI), and isopentenyl-diphosphate delta-isomerase (IDI) (Fig. 1).

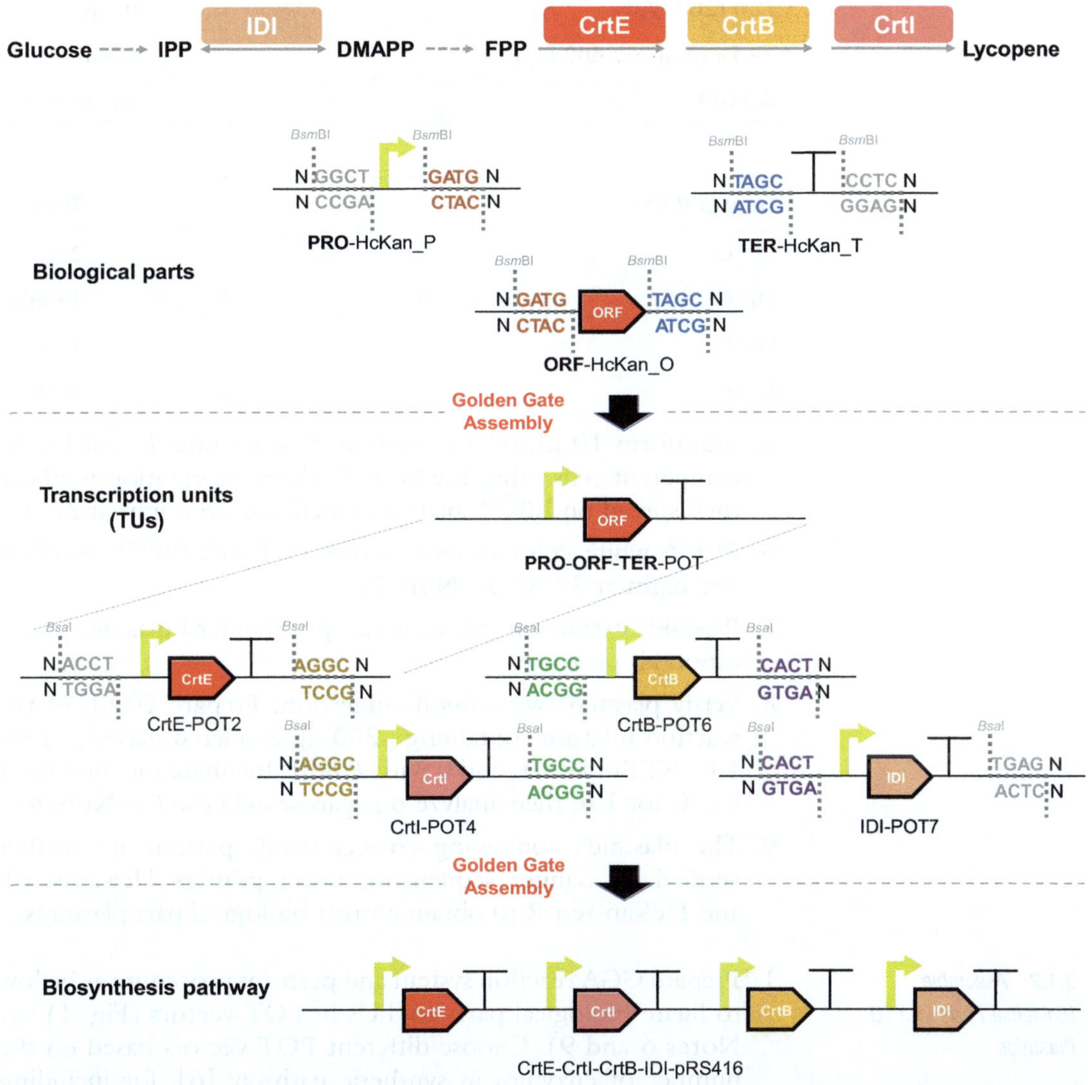

Fig. 1 Construction of lycopene biosynthesis pathway *via* GGA. The biological parts PROs, ORFs, or TERs were generated by PCR or de novo synthesized. These three parts were assembled into TUs *via* the GGA method. Then several TUs were assembled in sequence to the biosynthesis pathway *via* the GGA method. Four nucleotide bases flanked to parts or TUs indicated recognition sites, and dashed line indicated cutting sites of *Bsm*BI or *Bsa*I

4. Prepare GGA reaction system and perform procedure as follows on PCR instrument to generate biological part plasmids. Purified PROs and TERs or ORFs are ligated to HcKan_P, HcKan_T, HcKan_O vectors, respectively (Fig. 1) (*see* **Notes 4–6**).

Component	Amount
10 × T4 DNA ligase buffer	1 μL
HcKan_P/HcKan_T/HcKan_O	40 ng
PRO/TER PCR products or ORF Plasmids	40 ng
*Bsa*I (20 U/μL)	0.5 μL
T4 DNA ligase (400 U/μL)	0.5 μL
ddH_2O	Up to 10 μL

Temperature	Time
37 °C	2 h
50 °C	15 min
80 °C	15 min
12 °C	Store

5. Transform 10 μL of the reaction product into *E. coli* DH5α competent cells using the heat shock transformation method, then spread on LB^{Kan+} plate and incubate overnight at 37 °C.
6. Pick 6 white colonies and culture in 1 mL LB^{Kan+} medium overnight at 37 °C (*see* **Note 7**).
7. Plasmids extraction manually using a standard plasmid purification kit protocol.
8. Verify plasmids with *Bsm*BI digestion. Prepare 10 μL of the reaction mixture containing ~300 ng extracted plasmid, 1 μL 10 × NEBuffer 3.1, and 0.3 μL *Bsm*BI. Incubate the mixture at 55 °C for 1 h, then analyze on agarose gel (1%) (*see* **Note 8**).
9. The plasmids containing correct bands pattern are further verified by Sanger sequencing using primers HcKan-seq-F and HcKan-seq-R to obtain correct biological part plasmids.

3.1.2 Assemble Biological Parts into TU Plasmids

1. Prepare GGA reaction system and perform procedure as bellow to ligate biological parts to TUs in POT vectors (Fig. 1) (*see* **Notes 6** and **9**). Choose different POT vectors based on the number of enzymes in synthetic pathway [6], for including 4 TUs (CrtE, CrtI, CrtB and IDI), they are ligated to POT2, POT4 POT6 and POT7 vectors, respectively (*see* **Note 10**).

Component	Amount
10 × T4 DNA ligase buffer	1 μL
POT2/POT4/POT6/POT7	40 ng
pTEF2-HcKan_P	40 ng
CrtE-HcKan_O/CrtB-HcKan_O/CrtI-HcKan_O/IDI-HcKan_O	40 ng
tADH1-HcKan_T	40 ng
*Bsm*BI (10 U/μL)	0.5 μL
T4 DNA Ligase (400 U/μL)	0.5 μL
ddH_2O	Up to 10 μL

Temperature	Time
55 °C	1 h
Pause and add 0.5 μL T4 DNA Ligase, mix well with pipette	
25 °C	1 h
50 °C	15 min
80 °C	15 min
12 °C	Store

2. Transform 10 μL of the reaction product into *E. coli* DH5α competent cells, and spread on LB^{Carb+} plate at 37 °C for overnight.
3. Pick 6 white colonies for plasmids extraction (*see* **Note 11**).
4. Verify plasmids with *Bsa*I digestion. Prepare 10 μL of the reaction mixture containing ~300 ng extracted plasmid, 1 μL 10 × NEBuffer CutSmart, and 0.3 μL *Bsa*I-HF. Incubate the mixture at 37 °C for 1 h, then analyze on agarose gel (1%) (*see* **Note 12**).
5. Verify plasmids by Sanger sequencing using primers POT-seq-F and POT-seq-R, to obtain correct TU plasmids.

3.1.3 Assemble TUs into Biosynthesis Pathway Plasmids

1. Amplify the pRS416 linear backbone (*see* **Note 13**) from the pRS416 vector using designed primers pRS416-F and pRS416-R, then purify the PCR products.
2. Prepare the GGA reaction system and perform the procedure as below for biosynthesis pathway construction. 4 TUs are assembled into pRS416 linear backbone (*see* **Notes 6** and **14**).

Component	Amount
10 × T4 DNA ligase buffer	2 μL
pRS416 backbone PCR product	70 ng
CrtE-POT2	70 ng
CrtI-POT4	70 ng
CrtB-POT6	70 ng
IDI-POT7	70 ng
*Bsa*I (20 U/μL)	1 μL
T4 DNA Ligase (400 U/μL)	1 μL
ddH_2O	Up to 20 μL

Temperature	Time	Cycles
37 °C 16 °C	2 min 5 min	50 cycles
50 °C	5 min	1 cycle
80 °C	5 min	1 cycle
12 °C	Store	

3. Transform 10 μL of the reaction product into *E. coli* DH5α competent cells, spread on LB^{Carb+} plate with 40 μL X-Gal and 5 μL IPTG, then incubate overnight at 37 °C.
4. Pick 6 ~ 24 blue colonies from a mixture of blue and white transformants (*see* **Notes 15–16**) for plasmids extraction.
5. Verify the extracted plasmids by *Bsm*BI digestion reaction (*see* **Note 17**).
6. Further verify the correct band patterned plasmids by Sanger sequencing using primers pRS416-seq-F, pRS416-seq-R, and verifying primers of each ORF.

3.1.4 Obtain Lycopene Producing Synthetic Yeast

1. Transform ~10 ng of the lycopene biosynthesis pathway plasmid into synthetic *synII* diploid yeast using PEG-LiAc transformation method [13], plate the transformed cells on SC–Ura plate and incubate at 30 °C for 3 days.
2. Pick red yeast colonies to verify pathway plasmid by PCR reaction (*see* **Note 18**), then store correct clones in 25% glycerol at −80 °C.

3.2 Combinatorial Golden Gate Cloning to Optimize Lycopene Pathway in Yeast

3.2.1 Generate TU Library

1. Mix 6 promoter plasmids with various activities and 4 terminator plasmids in equimolar amounts, along with a single ORF and a single POT vector within one tube (Fig. 2). Perform the combinatorial GGA reaction as below, assembling ORF with random promoters and terminators to create a TU library.

Component	Amount
10 × T4 DNA ligase buffer	1 μL
10 × BSA (1 mg/mL)	1 μL
POT2/POT4/POT6/POT7	40 ng
pTDH3-HcKan_P, pCCW12-HcKan_P, pTEF2-HcKan_P, pPGK1-HcKan_P, pCYC1-HcKan_P, pRPL18b-HcKan_P	To equimolar with backbone
CrtE-HcKan_O/CrtB-HcKan_O/CrtI-HcKan_O/IDI-HcKan_O	To equimolar with backbone
tCIT1-HcKan_T, tFUM1-HcKan_T, tERP2-HcKan_T, tADH1-HcKan_T	To equimolar with backbone
*Bsm*BI (10 U/μL)	0.5 μL
ddH_2O	Up to 10 μL

Temperature	Time
55 °C	1 h
Pause and add 0.5 μL T4 DNA Ligase, mix well with pipette	
25 °C	1 h
50 °C	15 min
80 °C	15 min
12 °C	Forever

2. Transform the reaction product into *E. coli* DH5α competent cells, spread on LB^{Carb+} plate and incubate overnight at 37 °C.
3. Pick 96 white colonies to 96-deep well plate with 1 mL 2 × YT^{Carb+} medium (*see* **Note 19**), and culture overnight at 37 °C, 220 rpm. Then perform high-throughput plasmids extraction using the workstation of Hamilton Microlab STAR Liquid Handling System.
4. Verify plasmids with *Bsa*I digestion in 96-well PCR plate.
5. Verify the plasmids containing the TU-specific band by Sanger sequencing to identify promoters and terminators on TU plasmids.

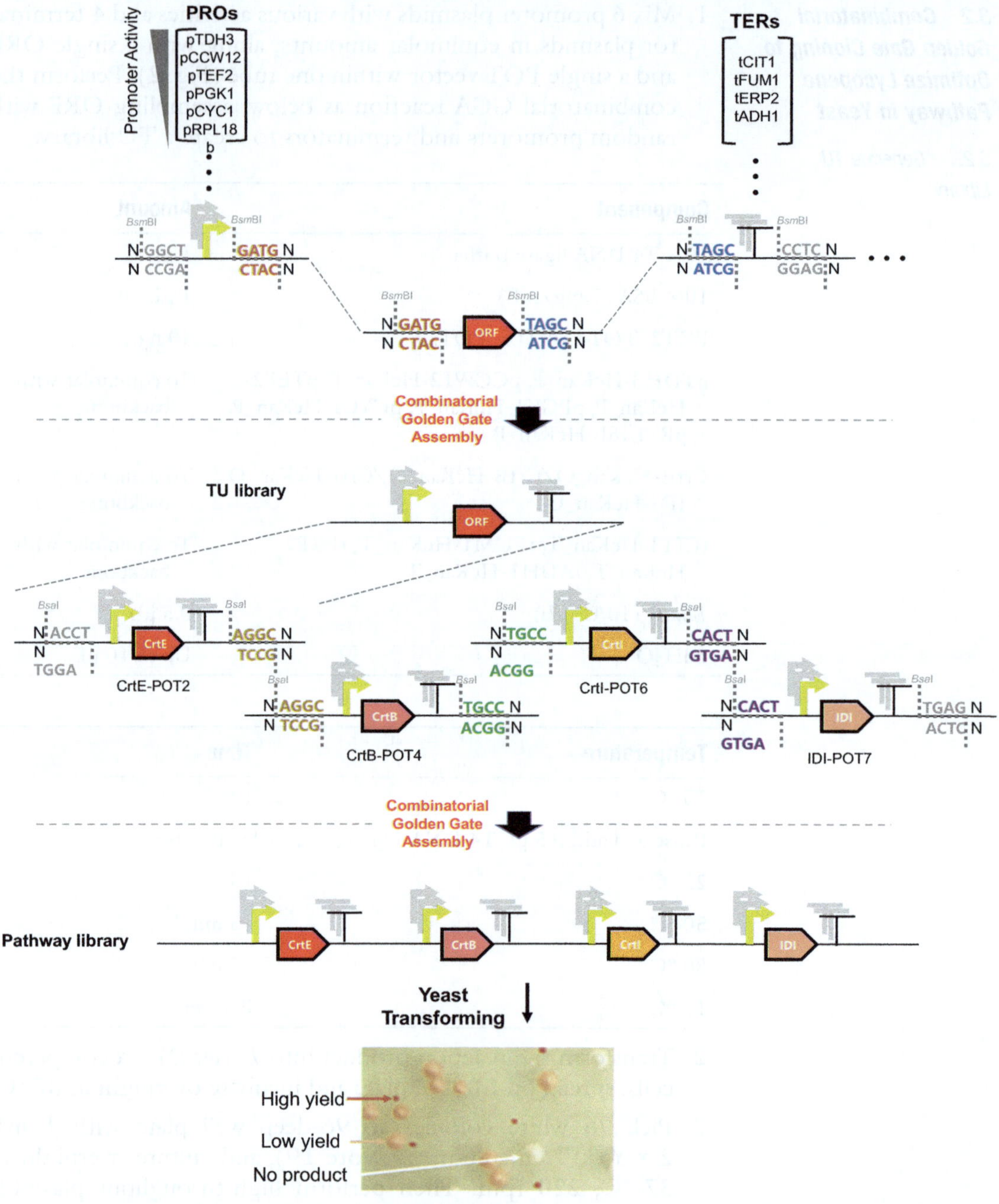

Fig. 2 Scheme depicting the construction of lycopene biosynthesis pathway *via* combinatorial GGA. ORFs are assembled with diverse PROs and TERs to generate a TU library, and multiple TUs libraries are further assembled into a pathway library. The resulting yeasts transformed with pathway library plasmids display varying color intensities

3.2.2 Generate Pathway Library

1. Mix equimolar amounts of obtained TUs of each ORF, along with the pRS416 linear backbone, in a single tube (Fig. 2). Perform the combinatorial GGA reaction as below to assemble TUs together to generate lycopene pathway library plasmids.

Component	Volume
10 × T4 DNA ligase buffer	2 μL
10 × BSA (1 mg/mL)	2 μL
pRS416 backbone PCR product	70 ng
CrtE-POT2 plasmids with different promoters or terminators	To equimolar with backbone each
CrtB-POT4 plasmids with different promoters or terminators	To equimolar with backbone each
CrtI-POT6 plasmids with different promoters or terminators	To equimolar with backbone each
IDI-POT7 plasmids with different promoters or terminators	To equimolar with backbone each
*Bsa*I-HF (20 U/μL)	1 μL
T4 DNA Ligase (400 U/μL)	1 μL
ddH_2O	Up to 20 μL

Temperature	Time	Cycles
37 °C 16 °C	2 min 5 min	50 cycles
50 °C	5 min	1 cycle
80 °C	5 min	1 cycle
12 °C	Forever	

2. Transform reaction product into *E. coli* DH5α competent cells, spread on LB^{Carb+} plate with 40 μL X-Gal, 5 μL IPTG, then incubate overnight at 37 °C.
3. Pick 96 blue *E. coli* colonies to check integrity of biosynthesis pathway by PCR reaction using verifying primers of each ORF.
4. Extract plasmids containing all ORF bands from *E. coli* strains.

3.2.3 Screen the Optimal Pathway

1. Mix 10 ng of obtained pathway plasmids into a single tube, transform the mixture into BY4741 yeast strain (Fig. 2), spread on SC–Ura plate, and incubate at 30 °C for 3 days.
2. Select darker red colonies with high-yield lycopene for further verifying pathway plasmid by PCR (Fig. 2) (*see* **Note 18**).

3. Extract plasmids from these lycopene high-yield yeasts following the standard protocol provided by yeast plasmid purification kit.
4. Verify by Sanger sequencing to identify the PROs and TERs in the optimal lycopene pathway plasmid.
5. Meanwhile, transform 10 ng plasmid into *E. coli* DH5α competent cells, then store clones in 25% glycerol at −80 °C.

3.3 Characterization of Optimal Lycopene Pathway in Synthetic Yeast

3.3.1 SCRaMbLE to Obtain Improved Chassis for Lycopene Production

1. Transform ~10 ng of the plasmid pRS413-Cre/EBD into the above synthetic yeast containing lycopene pathway, spread on SC–Ura–His plate and incubate at 30 °C for 3 days.
2. Randomly pick colonies and inoculate them in 3 mL SC–Ura–His liquid medium, culture the yeast overnight at 30 °C, 220 rpm.
3. Transfer the precultures into 50 mL fresh liquid medium with an initial $OD_{600} = 0.1$.
4. Add 50 μL of β-Estradiol to achieve a final concentration of 1 μM.
5. Cultivate the mixture at 30 °C, 220 rpm to induce SCRaMbLE for 24 h (Fig. 3).
6. Plate 1 mL of the culture on several SC–Ura plates and incubate at 30 °C for 3 days (*see* **Note 20**).
7. Select darker red strains with high-yield lycopene for further integration of optimal lycopene pathway (Fig. 3) (*see* **Note 18**).

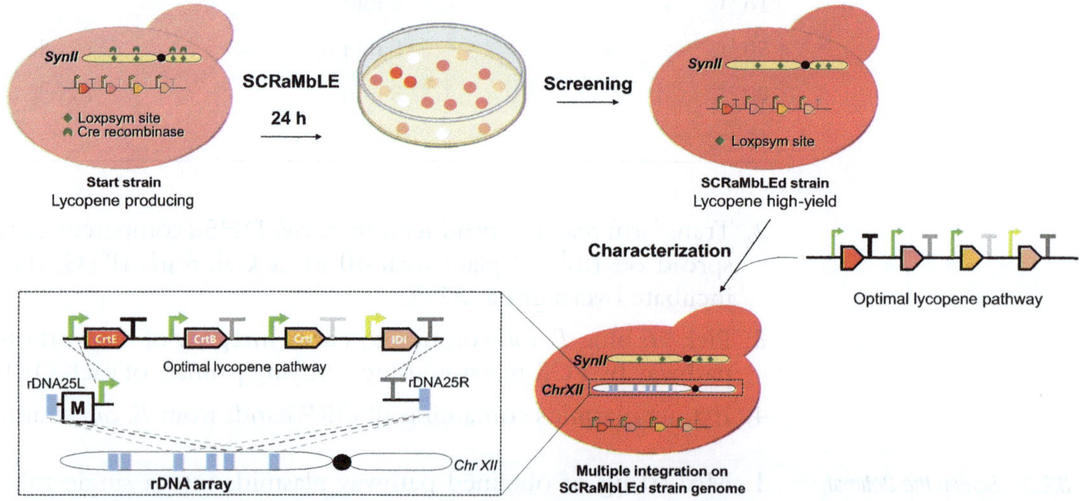

Fig. 3 Characterization of optimal lycopene synthetic pathway in SCRaMbLEd yeast chassis. The synthetic yeast strain harboring *synII* chromosome with improved ability to produce lycopene *via* SCRaMbLE served as the chassis. The optimal lycopene synthetic pathway constructed by GGA was integrated into the rDNA loci for further characterization

3.3.2 Test the Optimal Pathway in SCRaMbLEd Chassis

1. Prepare the PCR reaction system to amplify the 500 bp upstream and downstream sequence of genomic rDNA25 locus from BY4741 yeast genome, and pCCW12 promoter, tADH1 terminator, KanMX marker from plasmid using primers in Subheading 2.3.4, then purify the PCR products.
2. Perform overlap extension-PCR (OE-PCR) to assemble the upstream and downstream genomic integrated fragments (Fig. 3).
3. Prepare 50 μL of the *Bsm*BI digestion reaction mixture to cut off the optimal pathway from plasmid, then purify the digestion products.
4. Transform 500 ng of the purified PCR products of upstream/downstream genomic integrated fragment, and 500 ng of the purified digestion product of the optimal lycopene pathway into the SCRaMbLEd yeast strain (Fig. 3) (*see* **Note 21**). Plate on SC–Ura plate with G418 (500 μg/mL) and incubate at 30 °C for 3 days.
5. Pick darker red colonies and strike on a fresh plate at 30 °C for 2 ~ 3 days to obtain single clones (*see* **Note 18**). Store these strain clones in 25% glycerol at −80 °C for further analysis.

4 Notes

1. The Q5 High-Fidelity DNA Polymerase (NEB, M0491L) was used to amplify PROs, TERs, and the PrimeSTAR HS DNA Polymerase (Takara, R040) was used to amplify pRS416 linear backbone. You can choose other high-fidelity polymerases to replace them.
2. The ORFs were codon optimized according to *Saccharomyces cerevisiae*, then synthesized in pUC57-simple vector by GCAT-bio Co. Ltd. Cat. LS-GS-00003. To enable ORFs ligate into HcKan_O vector subsequently, the following prefixes are introduced besides of ORFs when synthesize or *via* PCR amplification.

   5′ end of ORF: AGCGTGGGTCTCAG

   3′ end of ORF: GTGCTGGGTCTCGGCTA
3. The TIANprep Mini Plasmid Kit (TIANGEN, DP103) was used for plasmid extraction manually, and the MagPure Plasmid LQ Kit (Magen, P1811) was used for plasmid purification on liquid transfer workstation.
4. The efficiency of GGA can be significantly affected by the presence of *Bsa*I or *Bsm*BI recognition sites in biological parts. For ORF, these recognition sites can be removed by

synonymous codon substitution. In the case of PRO or TER, it is necessary to increase the amount of the plasmids in reaction or substituting them with similar parts.

5. Forty nanograms of HcKan_P, HcKan_O, HcKan_T vectors facilitate the desired number of transformants on the plate.
6. T4 DNA ligase buffer contains ATP, which may degrade after undergoing multiple freeze-thaws. To ensure reaction efficiency, it is recommended to use fresh buffer.
7. The white transformants are the ones in which target parts have replaced the red fluorescent protein (RFP) gene of HcKan vectors successfully.
8. The correct bands pattern at 632 bp, 1110 bp, and different sizes for specific PROs, TERs or ORFs. When HcKan_P, HcKan_O or HcKan_T vectors remain uncut or experience self-ligated, the band pattern is 632 bp, 1110 bp and 1130 bp.
9. Forty nanograms of POT vectors are appropriate for the transformants number on plate.
10. The unique 4 bp overhangs of different POTs enable ligation in numerical order to pathway vector (Figs. 1 and 2), so the specific POT is chosen based on the number of enzymes on the pathway: for one enzyme, assemble its gene into POT1; for two, assemble them into POT2 and POT3; for three, assemble them into POT2, POT4 and POT5; for four, assemble them into POT2, POT4, POT6, and POT7, and so on.
11. The white transformants are the ones in which target parts have replaced the RFP gene of POT vectors successfully.
12. The correct bands pattern at 4889 bp, 22 bp, 21 bp and different sizes for specific TU. In cases where POT vectors remain uncut or experience self-ligated, the band pattern is 21 bp, 22 bp, 1055 bp, and 4889 bp.
13. The pRS416 was used due to *URA* selection marker, alternative pathway vectors such as pRS415 for *LEU* marker or pRS413 for *HIS* marker also can be chosen.
14. Seventy nanograms of pRS416 linear vector are appropriate for the transformants number on plate.
15. The blue transformants are the ones in which TU transfer to pRS416 vector successfully. The colony turns blue because there is functional β-Galactosidases gene *lacZ* on pRS416 vector but absent on POT vector.
16. The efficiency of GGA decreases significantly occurs as the number of TU increase, which needs to verify more transformants.
17. The correct bands pattern at 2428 bp, 2426 bp, 42 bp and specific size for biosynthesis pathway.

18. The product of lycopene synthetic pathway imparts a red color to the strain; the intensity of this color can serve as an indicator of lycopene yield (Fig. 2).
19. 2 × YT medium with higher nutrition is used to increase plasmid concentration in high-throughput plasmid extraction.
20. To prevent further rearrangement of strain after SCRaMbLE, it is necessary to remove pRS413-Cre/EBD plasmid by supplementation of histidine in medium.
21. The lycopene pathway may have multiple copies within genome, due to the presence of over 200 rDNA sites on chromosome *XII* [14] (Fig. 3).

Acknowledgments

This work was supported by the National Key Research and Development Program of China (No. 2021YFC2100800), Tip-top Scientific and Technical Innovative Youth Talents of Guangdong Special Support Program (2019TQ05Y876).

References

1. Engler C, Kandzia R, Marillonnet S (2008) A one pot, one step, precision cloning method with high throughput capability. PLoS One 3(11):e3647
2. Bird JE, Marles-Wright J, Giachino A (2022) A user's guide to golden gate cloning methods and standards. ACS Synth Biol 11(11): 3551–3563
3. Kotera I, Nagai T (2008) A high-throughput and single-tube recombination of crude PCR products using a DNA polymerase inhibitor and type IIS restriction enzyme. J Biotechnol 137(1–4):1–7
4. Pryor JM, Potapov V, Bilotti K, Pokhrel N, Lohman GJ (2022) Rapid 40 kb genome construction from 52 parts through data-optimized assembly design. ACS Synth Biol 11(6):2036–2042
5. Scior A, Preissler S, Koch M, Deuerling E (2011) Directed PCR-free engineering of highly repetitive DNA sequences. BMC Biotechnol 11(1):1–9
6. Guo Y, Dong J, Zhou T, Auxillos J, Li T, Zhang W, Wang L, Shen Y, Luo Y, Zheng Y (2015) YeastFab: the design and construction of standard biological parts for metabolic engineering in Saccharomyces cerevisiae. Nucleic Acids Res 43(13):e88–e88
7. Zhang Y, Chiu T-Y, Zhang J-T, Wang S-J, Wang S-W, Liu L-Y, Ping Z, Wang Y, Chen A, Zhang W-W (2021) Systematical engineering of synthetic yeast for enhanced production of lycopene. Bioengineering 8(1): 14
8. Richardson SM, Mitchell LA, Stracquadanio G, Yang K, Dymond JS, DiCarlo JE, Lee D, Huang CLV, Chandrasegaran S, Cai Y (2017) Design of a synthetic yeast genome. Science 355(6329): 1040–1044
9. Zhang H, Fu X, Gong X, Wang Y, Zhang H, Zhao Y, Shen Y (2022) Systematic dissection of key factors governing recombination outcomes by GCE-SCRaMbLE. Nat Commun 13(1): 5836. https://doi.org/10.1038/s41467-022-33606-0
10. Dymond JS, Richardson SM, Coombes CE, Babatz T, Muller H, Annaluru N, Blake WJ, Schwerzmann JW, Dai J, Lindstrom DL (2011) Synthetic chromosome arms function in yeast and generate phenotypic diversity by design. Nature 477(7365):471–476
11. Liu W, Luo Z, Wang Y, Pham NT, Tuck L, Pérez-Pi I, Liu L, Shen Y, French C, Auer M (2018) Rapid pathway prototyping and engineering using in vitro and in vivo synthetic

genome SCRaMbLE-in methods. Nat Commun 9(1):1936

12. Blount B, Gowers GF, Ho J, Ledesma-Amaro-R, Jovicevic D, McKiernan R, Xie Z, Li B, Yuan Y, Ellis T (2018) Rapid host strain improvement by in vivo rearrangement of a synthetic yeast chromosome. Nat Commun 9(1):1932
13. Gietz RD, Schiestl RH, Willems AR, Woods RA (1995) Studies on the transformation of intact yeast cells by the LiAc/SS-DNA/PEG procedure. Yeast 11(4):355–360. https://doi.org/10.1002/yea.320110408
14. Lopes TS, Klootwijk J, Veenstra AE, van der Aar PC, van Heerikhuizen H, Raúe HA, Planta RJ (1989) High-copy-number integration into the ribosomal DNA of Saccharomyces cerevisiae: a new vector for high-level expression. Gene 79(2):199–206. https://doi.org/10.1016/0378-1119(89)90202-3

Chapter 24

YeastFab Cloning of Toxic Genes and Protein Expression Optimization in Yeast

Stefan A. Hoffmann

Abstract

YeastFab is a Golden Gate–based cloning standard and parts repository. It is designed for modular, hierarchical assembly of transcription units and multi-gene assemblies for expression in *Saccharomyces cerevisiae*. This makes it a suitable toolbox to optimize the expression strength of heterologous genes in yeast. When cloning heterologous coding sequences into YeastFab vectors, in several cases we have observed toxicity to the cloning host *Escherichia coli*. The provided protocol details how to clone such toxic genes from multiple synthetic DNA fragments while adhering to the YeastFab standard. The presented cloning strategy includes a C-terminal FLAG tag that allows screening for constructs with a desired protein expression in yeast by western blot. The design allows scarlessly removing the tag through a Golden Gate reaction to facilitate cloning of expression constructs with the native, untagged transgene.

Key words Cloning of toxic genes, Heterologous expression, YeastFab, Modular assembly, Removable FLAG tag

1 Introduction

The yeast *Saccharomyces cerevisiae* is one of the most well-studied model organisms, not least thanks to its traditional use in ethanol fermentation and bread making. With substantial knowledge about its physiology, cell biology, and genetics and the availability of a broad toolbox for genetic engineering, it is oftentimes the organism of choice for metabolic engineering [1] or heterologous expression [2]. Developed to facilitate such efforts, YeastFab is both an assembly standard based on Golden Gate cloning and an *S. cerevisiae* genetic parts repository of promoters and terminators [3, 4]. It allows modular construction of transcription units from these parts for expression in yeast and their assembly into multi-gene constructs. YeastFab has been proven to be a useful toolbox to clone heterologous metabolic pathways and optimize metabolite production in yeast [3, 5].

Daniel Schindler (ed.), *Golden Gate Cloning: Methods and Protocols*, Methods in Molecular Biology, vol. 2850,
https://doi.org/10.1007/978-1-0716-4220-7_24,

Following the YeastFab standard, genetic parts, either promoters (PRO), open reading frames (ORF), or terminators (TER), are first cloned into respective entry vectors. This step is also called domestication. In the next level, transcription units are assembled from these parts in so-called POT plasmids (PRO-ORF-TER). The assembly standard results in a cloning scar of only a single base pair between promoter and start codon, and the stop codon and terminator, as the start codon ATG and the stop codon TAG themselves are part of the 4 bp overhang used for Golden Gate assembly. Transcription units can be further assembled into multi-gene constructs.

Cloning can be challenging for recombinant genes with products that are toxic to the cloning host, such as certain nucleases, channel membrane proteins, kinases, or proteins interfering with DNA replication. Readthrough from upstream promoters or cryptic transcription can result in enough gene product being made to exert toxicity on the cloning host *Escherichia coli* even in the absence of dedicated bacterial promoters in front of the gene of interest, as in the YeastFab ORF and POT plasmids. In cases of toxicity to the cloning host, apparent cloning efficiencies will be noticeably low, with cloned genes frequently or always having loss-of-function mutations. This can severely obstruct cloning, both for domestication in an entry vector and for transcription unit assembly in YeastFab. Likewise, the production host can be subject to a gene product's toxicity, which can be tackled by controlling protein expression though use of weak or inducible yeast promoters. In this case, it is desirable to query protein abundance to choose a promoter giving sufficiently high expression without causing a strong growth defect.

Here, we provide a step-by-step protocol to clone a gene with significant toxicity to the cloning host using the YeastFab standard, screen for a suitable yeast promoter with the help of a C-terminal protein tag, and scarlessly remove the tag in a Golden Gate reaction. For reference, an overview of the respective Golden Gate cloning steps is given in Fig. 1.

2 Materials

2.1 Standard Equipment, Consumables, and Software

Standard equipment for molecular cloning (including a plasmid extraction kit and a Qubit fluorometer or similar), SDS-PAGE, and western blot are required. The use of molecular biology software such as SnapGene, Geneious, or Benchling is highly recommended.

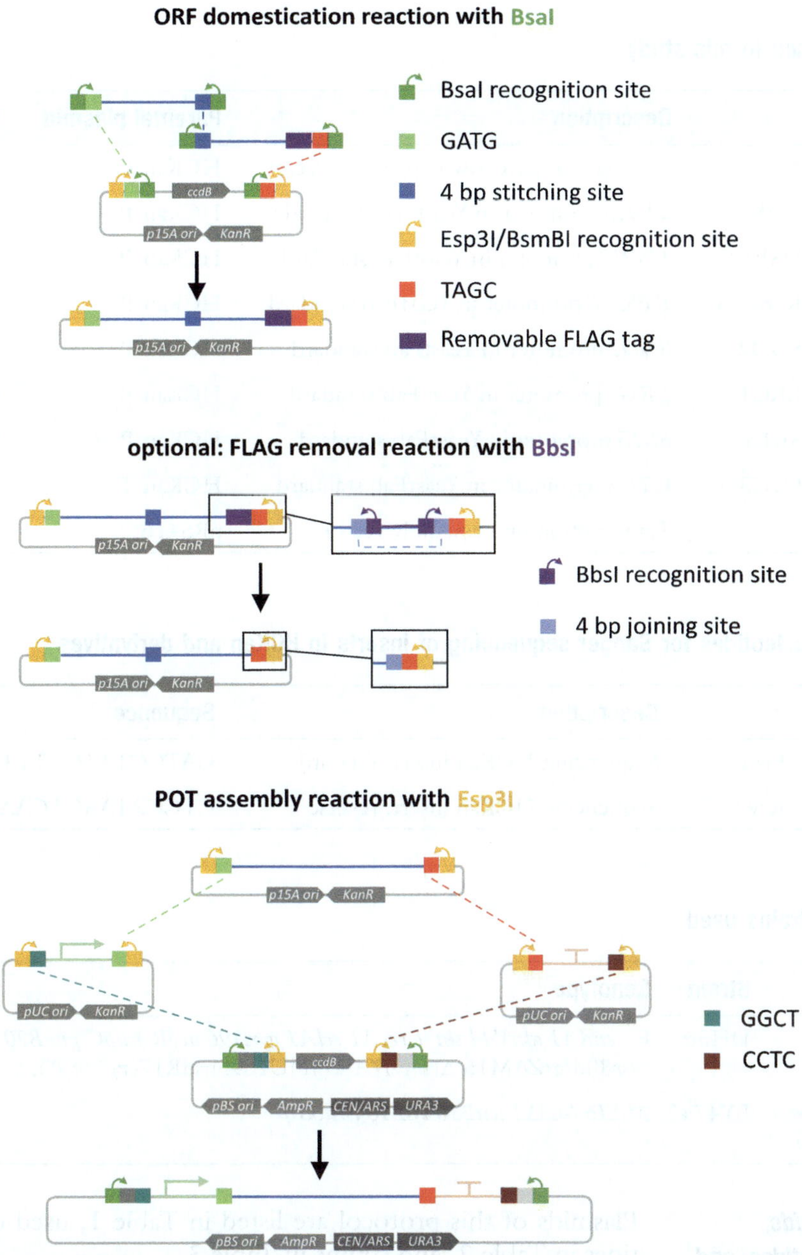

Fig. 1 Overview of Golden Gate cloning steps described. The coding sequence of a large, potentially toxic gene can be domesticated from multiple parts into a low-copy vector compatible with the YeastFab standard. The presented design includes a FLAG tag that can be scarlessly removed from the domesticated coding sequence with a Golden Gate elimination reaction. The coding sequence can then be assembled into a transcription unit for expression in yeast. Promoters and terminators can originate from the YeastFab parts repository or have been cloned according to the same standard

Table 1
Plasmids used in this study

Name	Description	Parental plasmid	References
p15A-Kan-O	Coding sequence low-copy entry vector	HCKan-O	This study
HCKan-P-TDH3	*TDH3* promoter in YeastFab standard	HCKan-P	[3]
HCKan-P-HSP12	*HSP12* promoter in YeastFab standard	HCKan-P	Unpublished
HCKan-P-RPS27B	*RPS27B* promoter in YeastFab standard	HCKan-P	Unpublished
HCKan-P-SOD2	*SOD2* promoter in YeastFab standard	HCKan-P	Unpublished
HCKan-P-ERG1	*ERG1* promoter in YeastFab standard	HCKan-P	Unpublished
HCKan-P-SET5	*SET5* promoter in YeastFab standard	HCKan-P	Unpublished
HCKan-T-CYC3	*CYC3* terminator in YeastFab standard	HCKan-T	Unpublished
POT4	Transcription unit assembly vector	pRS416	[3]

Table 2
DNA oligonucleotides for Sanger sequencing of inserts in HCKan and derivatives

Name	Description	Sequence
HCKan chc Fwd	Sequencing HCKan insert, forward	GATCCTTTGATTTTCTACCG
HCKan chc Rev	Sequencing HCKan insert, reverse	CTCGATAACTCAAAAAATACG

Table 3
Microbial strains used

Species	Strain	Genotype	References
Escherichia coli	DH5α	F^- *endA1 glnV44 thi-1 recA1 relA1 gyrA96 deoR nupG purB20* φ80d*lacZ*ΔM15 Δ(*lacZYA-argF*)U169, hsdR17($r_K^- m_K^+$), λ^-	[11]
Saccharomyces cerevisiae	BY4742	*MATα his3Δ1 leu2Δ0 lys2Δ0 ura3Δ0*	[12]

2.2 Plasmids, Oligonucleotides, and Strains

Plasmids of this protocol are listed in Table 1, used oligonucleotides in Table 2, and strains in Table 3.

2.3 Media

- Luria-Bertani broth (LB): 1% (w/v) tryptone, 0.5% (w/v) yeast extract, 1% (w/v) sodium chloride, and pH 7.
- Synthetic Complete uracil drop-out medium (SC-Ura): 0.17% (w/v) yeast nitrogen base without amino acids, 0.2% (w/v) yeast synthetic drop-out medium supplements without uracil (Sigma-Aldrich), 0.5% (w/v) ammonium sulfate, and 2% glucose.

- Solid media with the above compositions (LB agar and SC-Ura agar) with additional 2% agar.

2.4 Enzymes

- BsaI-HFv2 (NEB).
- Esp3I.
- BbsI-HF (NEB).
- T4 DNA ligase and buffer.

2.5 Buffers

- Tris-acetate ethylenediaminetetraacetic acid (TAE) buffer: 40 mM Tris base, 20 mM acetic acid, and 1 mM ethylenediaminetetraacetic acid (commercially available as 50× stock solution).
- SDS-PAGE sample buffer: 60 mM Tris-HCl pH 6.8, 4% β-mercaptoethanol, 5% glycerol, 2% sodium dodecyl sulfate, and 0.0025% bromophenol blue.
- SDS-PAGE running buffer: 25 mM Tris base, 192 mM glycine, and 0.1% SDS (commercially available as 10× stock solution).
- Tris-buffered saline Tween (TBST) buffer: 20 mM Tris base, 150 mM NaCl, and 0.1% (w/v) Tween 20 (commercially available as 10× stock solution).

2.6 Other Chemicals

- Kanamycin (50 mg/mL stock in water).
- Carbenicillin or ampicillin (100 mg/mL stock in water).
- Molecular biology grade agarose.
- Glycerol (sterile 50% solution).
- Skim milk powder.

2.7 Consumables Western Blot

- SDS-PAGE gels.
- Nitrocellulose membrane.
- Mouse anti-FLAG antibody M2.
- Mouse anti-glyceraldehyde-3-phosphate dehydrogenase (anti-GAPDH) or anti-β-actin or anti-α-tubulin.
- Horseradish peroxidase (HRP)-coupled rabbit anti-mouse antibody.
- Enhanced chemiluminescence (ECL) substrate.

3 Methods

3.1 Design of Insert

Here it is assumed that genes to be cloned into the ORF entry vector ("domestication") are obtained as synthetic, linear, double-stranded DNA fragments ("gene fragments"). Typically, those gene fragments are substantially cheaper compared to cloned genes and can be purchased from a variety of DNA synthesis

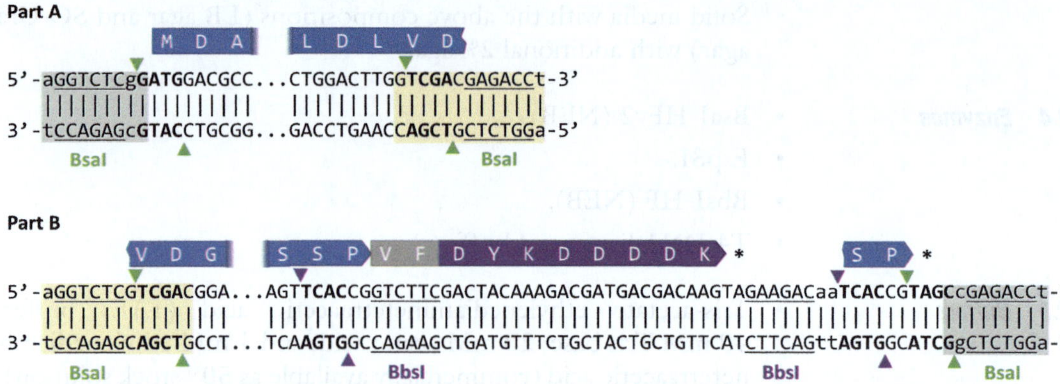

Fig. 2 Design to domesticate vaccinia *D5R* coding sequence gene with a removable FLAG tag from two gene fragments. Recognition sites for type IIs enzymes BsaI and BbsI are underlined, and the cutting sites indicated by matching colors. Translation of the *D5R* coding sequence is shown in blue, of the FLAG tag in purple. Domestication prefix and suffix sequences are highlighted in gray, the sequences used for stitching in yellow

providers. While synthetic gene fragments typically have a maximal length of 2 kb or less, genes larger than that can be cloned as well in a single reaction with a multi-part insert. The gene fragments can be designed to assemble into the full-length gene during the domestication reaction. Further, the design strategy presented here includes a removable FLAG tag as a fully compatible extension of the YeastFab standard. While YeastFab enables easy, modular cloning of transcription units with different promoters and terminators, the FLAG tag allows direct screening for desired protein expression level from those constructs. The tag can be scarlessly removed from the coding sequence through a single Golden Gate reaction. This facilitates cloning of the native, untagged gene without having to amplify and re-domesticate the coding sequence. Most design steps outlined below are easy to achieve with a molecular cloning software such as SnapGene, Geneious, or Benchling.

As an example, the design to clone vaccinia *D5R* with a removable FLAG tag from two gene fragments is illustrated in Fig. 2. It is followed by a generalized procedure detailing the respective design steps of the synthetic DNA fragments for domestication.

1. Obtain the native coding sequence of interest (*see* **Note 1**).
2. Codon optimize the coding sequence for *Saccharomyces cerevisiae*. An open-source codon optimization tool can be found here: cuba.genomefoundry.org/sculpt_a_sequence. Most DNA synthesis providers also provide a free online tool for this purpose but usually require creating an account to access it.
3. If the stop codon is TAA or TGA, change it to TAG.
4. Check for recognition sites for the type IIs restriction enzymes BsaI, BsmBI/Esp3I, and BbsI. Remove all these sites through

synonymous recoding. Preferably change to a codon of a similar frequency in *S. cerevisiae* (*see* **Note 2**).

5. Add the prefix aGGTCTCg**G** directly in front of the **ATG** start codon (5′-aGGTCTCg**GATG**-3′) and the suffix **c**cGAGACCt directly after the **TAG** stop codon (5′-**TAGC**cGAGACCt-3′). The BsaI recognition sites GGTCTC and GAGACC are used for Golden Gate cloning into the entry vector. BsaI cleavage results in four-base overhangs covering the start and stop codons. Identities of nucleotides in lowercase can be changed.
6. For the addition of a removable FLAG tag, add in the sequence shown in bold between the TAG stop codon and the two directly preceding codons (NNNNNN):

NNNNNN-**GTCTTC*GACTACAAAGACGATGACGACAAGTAG*AAGACaaNNNNNN**-TAG

BbsI recognition sites are underlined, and the FLAG tag including its stop codon is shown in italics. Note that the two codons (**NNNNNN**) now preceding the original stop codon are the same sequence as NNNNNN. This is to create a compatible overhang for scarless removal of the FLAG tag by BbsI.

7. If the coding sequence is too long to be ordered by the supplier of choice as noncloned linear DNA fragment, it can be broken up into smaller pieces to be assembled during the domestication reaction. If this is required, choose a 6 bp stretch (N*NNNN*N) at which to separate the sequence. (Note that the 6 bases NNNNNN here designate a different part of the sequence than in the previous step) The central 4 bp of this stretch will create the overhang for joining the fragments. This 4 bp sequence must not be GATG or TAGC, as these are used as overhangs for cloning into the entry vector. If joining multiple fragments, the 4 bp overhangs of the different fusion sites need to be dissimilar from each other, too. Separate the coding sequence at the chosen position(s). Include the 6 bp stretch N*NNNN*N in both the upstream and the downstream fragments. Add 5′-GAGACCt-3′ at the 3′ end on the upstream fragment and 5′-aGGTCTC-3′ at the 5′ end of the downstream fragment (n_L is the desired sequence left of the 6 bp stretch, n_R is the sequence right of it):
   - Full-length sequence: 5′-n_L-N*NNNN*N-n_R-3′
   - Upstream fragment: 5′-n_L-N*NNNN*N-GAGACCt-3′
   - Downstream fragment: 5′-aGGTCTC-N*NNNN*N-n_R-3′
8. It is recommended to perform the domestication reaction (with BsaI) and the reaction for FLAG tag removal (with BbsI) in silico to check the design. This also provides reference sequences to check sequencing results against.

3.2 Domestication Reaction

In the domestication reaction, the coding sequence of interest is introduced into a YeastFab-compatible entry vector for subsequent POT assembly of transcription units with promoter and terminator. The domestication can be done with a single DNA piece as insert or multiple pieces that have been designed to be concurrently assembled in the Golden Gate reaction. In the example given here (vaccinia *D5R*), the coding sequence is being assembled from two gene fragments. In reactions with more than a single insert, DNA components should be added in equimolar ratios.

Correct clones of vaccinia *D5R* could not be obtained using the standard high-copy YeastFab vector for coding sequence HCKan-O, likely due to cryptic transcription of this gene known to be toxic to *E. coli* [6]. To domesticate toxic genes, a variant of the domestication vector was constructed by swapping the origin of replication for the low-copy p15A origin (p15A-Kan_O). The use of a low-copy domestication vector and growing transformed *E. coli* at 30 °C instead of 37 °C is recommended when toxicity to the cloning host is evident or deemed likely.

1. Add the following reaction components to a PCR tube and mix the components by flicking or brief vortexing and spin the liquid down in a microcentrifuge (*see* **Note 3**).

Component	Volume/amount
Acceptor vector p15A-Kan_O	~80 ng (~50 fmol)
Insert part A	~50 fmol
Insert part B	~50 fmol
10× T4 ligase buffer	1 μL
T4 DNA ligase 400 CEU/μL	0.5 μL
BsaI-HFv2 (NEB) 20 U/μL	0.5 μL
Nuclease-free water	Up to 10 μL

2. Put the PCR tube into a thermocycler with a heated lid and run the following Golden Gate program optimized for multipart assemblies.

Step	Temperature	Cycles	Time
Initial reaction	37 °C	1	1 h
Cycling digest and ligation	37 °C 16 °C	30	5 min 10 min
Final ligation	16 °C	1	20 min
Final digest	37 °C	1	30 min

(continued)

Step	Temperature	Cycles	Time
Inactivation	80 °C	1	5 min
	8 °C		Hold

3. Transform the reaction into a CcdB-sensitive *E. coli* cloning strain. For instance, DH5alpha cells made chemically competent with an RbCl method [7] work well. For their transformation, mix 5 μL of the Golden Gate domestication reaction with 50 μL of the cell suspension of competent cells, incubate for 30 min on ice, heatshock for 45 s at 37 °C, and incubate on ice for another 5 min. Then add 500 μL Luria-Bertani broth (LB) and incubate for 45 min at 37 °C while shaking (*see* **Note 4**).
4. Plate 100 μL of the transformation outgrowth on LB agar with 50 μg/mL kanamycin with a spreader or glass beads. Pellet the remaining cells by centrifugation for 3 min at 6000× *g*. Tip away the supernatant, resuspend the pellet in the remaining liquid (~50 μL), and plate on a second plate of LB agar with kanamycin.
5. Incubate the plates at 30 °C for 2–3 days, after which colonies should be clearly visible. If the insert is toxic to the cloning host, colony sizes might be heterogenous, with correct clones likely forming smaller colonies. Pick colonies of different sizes into 5 mL LB with 50 μg/mL kanamycin each and incubate in a shaking incubator at 30 °C until cultures are dense (typically 1–2 days).
6. Extract plasmid DNA from 4 mL of each of the cultures using a silica column-based plasmid extraction kit according to the manufacturer's instructions. Store the rest of each culture at 4 °C.
7. Perform a test digest on the isolated plasmid DNA to check the insert size. Digest about 200–400 ng of DNA in 10 μL with 5 U of Esp3I in the appropriate buffer. Incubate for about 1 h at 37 °C.
8. Run the digest reactions alongside a suitable DNA ladder on a 1% agarose/TAE gel. A correct clone should give two bands, one from the p15A-Kan_O at 1.8 kb and another one at the size of the assembled insert.
9. Verify candidates by Sanger sequencing. For insert longer than about 1.5 kb, it is recommended to send plasmid DNA for whole plasmid sequencing. Measure DNA concentrations with a fluorescence-based method and prepare samples adjusted to the concentration required by the sequencing service provider (*see* **Note 5**).

10. Compare the sequencing result to the expected sequence. Prepare a cryo-culture according to standard procedure for long-term storage of a correct clone.

3.3 Removal of FLAG Tag

With the presented design, a single Golden Gate reaction allows to generate tag-free versions of domesticated coding sequences. This obviates the need to PCR amplify or re-order and re-clone the coding sequence. The type IIs enzyme used for this purpose (BbsI) is orthogonal to the ones used by the YeastFab standard (BsaI, Esp3I/BsmBI).

1. Add the following reaction components to a PCR tube and mix the components by flicking or brief vortexing and spin the liquid down in a microcentrifuge.

Component	Volume/amount
Coding sequence with FLAG in p15A-Kan_O	~20 ng
10× T4 ligase buffer	1 μL
T4 DNA ligase 400 CEU/μL	0.5 μL
BbsI-HF (NEB) 10 U/μL	0.5 μL
Nuclease-free water	Up to 10 μL

2. Put the PCR tube into a thermocycler with a heated lid and run the Golden Gate program as in Subheading 3.2, **step 2** (*see* **Note 6**).
3. Transform into *E. coli* as in Subheadings 3.2, **steps 3** and **4**.
4. Incubate the plates for 2 days at 30 °C and pick two or more colonies into 5 mL LB with 50 μg/mL kanamycin, again making sure to include small colonies in case of size heterogeneity.
5. Incubate in a shaking incubator at 30 °C until cultures are dense and isolate plasmid DNA from 4 mL of fresh cultures, storing cultures at 4 °C until sequence verification.
6. Perform a test digest on the isolated plasmid DNA to check the insert size. Digest about 200–400 ng of DNA in 10 μL with 5 U of Esp3I in the appropriate buffer. Incubate for about 1 h at 37 °C.
7. Run the digest reactions alongside a suitable DNA ladder on a 1% agarose/TAE gel. A correct clone should give two bands, one from the backbone and one from the insert, like in the domestication reaction (*see* **Note 7**).
8. For sequence confirmation, send prepared plasmid DNA for Sanger sequencing with the primer "HCKan chc Rev," covering the 3′ part of the coding sequence, or send for whole

plasmid sequencing for larger inserts. The sequencing is to confirm removal of the FLAG tag and that the gene has not been mutationally altered. Prepare a cryo-culture according to standard procedure for long-term storage of a correct clone.

3.4 Assembly of Transcription Units

Once the coding sequence has been successfully domesticated in p15A-Kan_O, it can be assembled in POT reactions with promoters and terminators according to the YeastFab standard. Facile construction of different expression constructs including the protein tag allows screening for desired expression behavior, balancing protein expression strength and potential fitness defects. Combinations of promoters and terminators giving desirable behavior can be used to construct transcriptional units without FLAG tag from the p15A-Kan_O construct with the tag-less coding sequence created in "Removal of FLAG tag" using the same protocol.

In the given example, the domesticated *D5R* coding sequence is assembled in the YeastFab standard in POT4 with a *CYC3* terminator and different promoters. Promoters ranging from very strong to fairly weak were chosen (reported gene transcript counts [8] in their endogenous context in brackets). Namely, the 500 bp upstream sequences of *TDH3* (15,909), *HSP12* (56), *RPS27B* (5356), *SOD2* (267), *ERG1* (307), and *SET5* (60) are tested for expression of *D5R*. However, transcription is known to be highly context dependent [9]. For instance, we have previously observed the 500 bp upstream of *HSP12* to be a rather strong promoter (unpublished) in the YeastFab reporter system [4].

1. Adjust the concentration of DNA parts to be assembled with the domesticated coding sequence, promoters in HCKan-P, terminators in HCKan-T, and the POT vector of choice (*see* **Note 8**), to 50 ng/μL each. This allows setting up multiple reactions with different components without having to adjust volumes.
2. Add the following reaction components to a PCR tube.

Component	Volume/Amount
Promoter in HCKan-P 50 ng/μL	1 μL
Terminator in HCKan-T 50 ng/μL	1 μL
POT vector 50 ng/μL	2 μL
Coding sequence in p15A-Kan_O	~30 fmol
10× T4 ligase buffer	1 μL
T4 DNA ligase 400 CEU/μL	0.5 μL
Esp3I 10 U/μL	0.5 μL
Nuclease-free water	Up to 10 μL

3. Proceed with the Golden Gate reaction and subsequent transformation into *E. coli*, as done in Subheadings 3.2, **steps 3, 4** and **5**, but plate on LB agar with 100 μg/mL carbenicillin or ampicillin.
4. Incubate the plates for 2 days at 30 °C and pick several colonies into 5 mL LB with 100 μg/mL carbenicillin or ampicillin, again making sure to include small colonies in case of size heterogeneity.
5. Incubate in a shaking incubator at 30 °C until cultures are dense and isolate plasmid DNA from 2 mL of fresh cultures, storing cultures at 4 °C until sequence verification.
6. Perform a test digest of the isolated plasmid DNA to check the insert size. Digest about 200–400 ng of DNA in 10 μL with 10 U of BsaI-HFv2. Incubate for about 1 h at 37 °C.
7. Run the digest reactions alongside a suitable DNA ladder on a 1% agarose/TAE gel. A correct clone should give two bands, one at 4.9 kb from the POT vector backbone, and another one at the size corresponding to the coding sequence plus the promoter length (500 bp in YeastFab standard parts) and terminator length (200 bp in YeastFab standard parts).
8. For sequence confirmation, send plasmid DNA of clones with correct digest patterns for whole plasmid sequencing. Prepare a cryo-culture according to standard procedure for long-term storage of a correct clone.

3.5 Assaying Protein Expression in Yeast

Transcription units with FLAG tag assembled in a POT vector can be screened for their protein expression. Constructs are first transformed into yeast. Transformed strains are grown to prepare SDS-PAGE samples, which are then assayed for protein expression using a western blotting procedure. The steps are briefly outlined here. Consult protocols describing SDS-PAGE and western blotting for detailed instructions on the techniques (e.g., [10]).

Assessing the *D5R* constructs with different promoters by western blot showed a range of expression strengths of D5 protein, covering strong (*pTDH3*, *pRPS27B*), medium (*pHSP12*), weak (*pSOD2*, *pERG1*), and very weak or absent (*pSET5*) expression (Fig. 3). None of the constructs imparted noticeable toxicity on the yeast host.

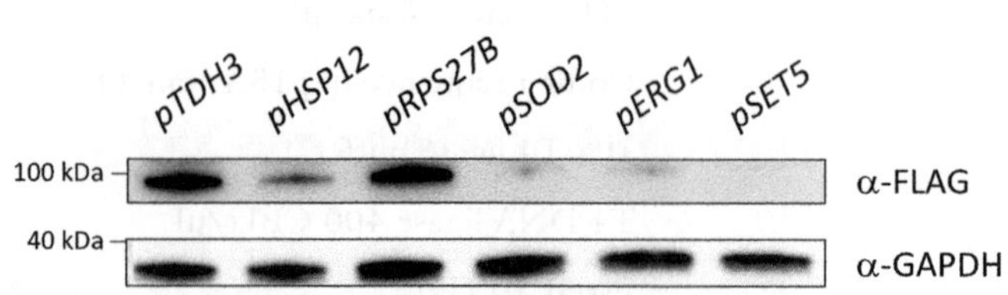

Fig. 3 Anti-FLAG tag western blot against D5-FLAG expressed in yeast from assembled *D5R* transcriptional units with indicated promoters. GAPDH was used as loading control

1. From a dense overnight culture of *S. cerevisiae* BY4742 in yeast extract peptone dextrose medium (YPD), inoculate a fresh culture 1:50 in YPD. For every transformation, a culture volume of 5 mL is needed. Grow the culture for 4 h at 30 °C with orbital shaking.
2. Pellet the cells (4000× *g* for 5 min) and discard the supernatant. Wash with half the original volume of sterile distilled water by resuspending and pelleting. Repeat the wash step once.
3. Resuspend the pellet in 1 mL 100 mM lithium acetate and pellet in a microcentrifuge tube at 10,000× *g* for 15 s. Remove the supernatant by pipetting and resuspend pellet in 100 mM lithium acetate with 50 μL for every 5 mL starting culture.
4. Denature 10 mg/mL salmon sperm DNA by incubating 5 min at 100 °C and placing on ice immediately after. For each transformation, 12.5 μL is needed.
5. For each transformation, to 50 μL of the cell suspension in 100 mM lithium acetate, add 266.67 μL of 50% polyethylene glycol 3350, 40 μL of 1 M lithium acetate, 12.5 μL 10 mg/mL denatured salmon sperm DNA, and about 500 ng of the POT construct. Make up to a total volume 400 μL with sterile distilled water.
6. Vortex mixture for 1 min. Incubate shaking at 30 °C for 1 h.
7. Heat shock for 20 min at 42 °C in a dry bath or water bath.
8. Pellet cells (4000× *g* for 3 min), remove supernatant, and resuspend cells in 200 μL sterile distilled water. Spread 50 μL of the cell suspension on SC-Ura agar using sterile glass beads or a cell spreader.
9. Incubate plates at 30 °C for 2–3 days until colonies appear. Plates with colonies can be stored for up to a month at 4 °C.
10. Inoculate 10 mL of SC-Ura for each transformation from single colonies and incubate shaking overnight at 30 °C.
11. From dense overnight cultures, inoculate fresh cultures 1:50 in 10 mL SC-Ura. Grow the culture for 4 h at 30 °C with orbital shaking.
12. Measure the optical density at 600 nm for each culture. Collect the volume equivalent to 1 optical density unit for each; 1 optical density unit equals 1 mL of a culture at an optical density of 1.
13. Pellet cells (10,000× *g* for 5 min), remove supernatant, and resuspend cells in 500 μL cold distilled water. Pellet again, remove supernatant by pipetting, and resuspend 200 μL 100 mM NaOH.
14. Incubate for 10 min at room temperature for alkaline lysis of the cell walls. Pellet cells again (6000× *g* for 5 min), remove

supernatant by pipetting, and resuspend each in 50 μL SDS-PAGE sample buffer. Samples can be stored at −20 °C until use.

15. Run 10 μL of each sample on a gradient SDS-PAGE gel for 40 min at 200 V along with a pre-stained protein ladder.
16. Blot the gel onto a nitrocellulose or polyvinylidene fluoride membrane with a method of your choice.
17. Assess successful blotting by staining membrane with Ponceau S. After staining, wash away excess dye with water and image the membrane. Afterward, wash with TBST until the membrane is mostly destained.
18. Block membrane by incubation in 5% skim milk powder in TBST for 1 h at room temperature.
19. Incubate with mouse anti-FLAG M2 antibody (Sigma), wash membrane three times with TBST, and incubate with a secondary HRP-coupled anti-mouse antibody. Wash again three times and develop with ECL.
20. As loading control, choose a suitable housekeeping protein (e.g., GAPDH, β-actin, α-tubulin) with a molecular weight different enough to be clearly separated from the protein of interest. Develop with a primary mouse antibody and a secondary HRP-coupled anti-mouse antibody analogous to **step 18**.

4 Notes

1. Coding sequences can for instances be retrieved from files from NCBI's GenBank (www.ncbi.nlm.nih.gov/genbank), a vast database containing annotated nucleotide sequences.
2. The *Saccharomyces cerevisiae* nuclear genome codon usage table can be found here: https://www.kazusa.or.jp/codon/cgi-bin/showcodon.cgi?species=4932&aa=1&style=N
3. A DNA molarity calculator can facilitate calculation of required DNA masses from desired molar amounts and fragment lengths. Such a tool can for instance be found here: www.bioline.com/media/calculator/01_07.html
4. Other protocols for competent cell preparation and transformation can be used, including commercially available kits for preparation or ready-made competent *E. coli*. In each case, a transformation outgrowth step must be included to allow cells to build resistance against kanamycin.
5. To verify the sequence of shorter inserts, Sanger sequencing can be used. Suitable sequencing primers are "HCKan chc Fwd" and "HCKan chc Rev." Both are priming on the plasmid

backbone of all HCKan vectors and the p15A-Kan_O derivative near the cloning site, pointing toward the insert.

6. For the elimination reaction a shorter Golden Gate reaction scheme with a reduced number of cycles can also be used.
7. Toxic genes are frequently inactivated through insertions of transposons. This will be detectable through an increased size of the insert.
8. The choice of the POT vector is relevant if multi-gene assemblies are to be carried out subsequently and informs the position of the transcriptional unit in those higher-order assemblies [3].

Acknowledgments

The author would like to thank Yizhi Cai for supporting this work. Plasmids with promoters in YeastFab standard were a gift from Junbiao Dai, while HCKan-T-CYC3 was a gift from Yizhi Cai.

References

1. Nielsen J, Larsson C, van Maris A, Pronk J (2013) Metabolic engineering of yeast for production of fuels and chemicals. Curr Opin Biotechnol 24(3):398–404. https://doi.org/10.1016/j.copbio.2013.03.023
2. Romanos MA, Scorer CA, Clare JJ (1992) Foreign gene expression in yeast: a review. Yeast 8(6):423–488. https://doi.org/10.1002/yea.320080602
3. Guo Y, Dong J, Zhou T, Auxillos J, Li T, Zhang W, Wang L, Shen Y, Luo Y, Zheng Y, Lin J, Chen GQ, Wu Q, Cai Y, Dai J (2015) YeastFab: the design and construction of standard biological parts for metabolic engineering in Saccharomyces cerevisiae. Nucleic Acids Res 43(13):e88. https://doi.org/10.1093/nar/gkv464
4. Garcia-Ruiz E, Auxillos J, Li T, Dai J, Cai Y (2018) YeastFab: high-throughput genetic parts construction, measurement, and pathway engineering in yeast. Methods Enzymol 608: 277–306. https://doi.org/10.1016/bs.mie.2018.05.003
5. Zhou Y, Li G, Dong J, Xing XH, Dai J, Zhang C (2018) MiYA, an efficient machine-learning workflow in conjunction with the YeastFab assembly strategy for combinatorial optimization of heterologous metabolic pathways in Saccharomyces cerevisiae. Metab Eng 47:294–302. https://doi.org/10.1016/j.ymben.2018.03.020
6. Evans E, Klemperer N, Ghosh R, Traktman P (1995) The vaccinia virus D5 protein, which is required for DNA replication, is a nucleic acid-independent nucleoside triphosphatase. J Virol 69(9):5353–5361. https://doi.org/10.1128/JVI.69.9.5353-5361.1995
7. Green R, Rogers EJ (2013) Transformation of chemically competent E. Coli. Methods Enzymol 529:329–336. https://doi.org/10.1016/B978-0-12-418687-3.00028-8
8. Lipson D, Raz T, Kieu A, Jones DR, Giladi E, Thayer E, Thompson JF, Letovsky S, Milos P, Causey M (2009) Quantification of the yeast transcriptome by single-molecule sequencing. Nat Biotechnol 27(7):652–658. https://doi.org/10.1038/nbt.1551
9. Brooks AN, Hughes AL, Clauder-Munster S, Mitchell LA, Boeke JD, Steinmetz LM (2022) Transcriptional neighborhoods regulate transcript isoform lengths and expression levels. Science 375(6584):1000–1005. https://doi.org/10.1126/science.abg0162
10. Hnasko TS, Hnasko RM (2015) The Western blot. Methods Mol Biol 1318:87–96. https://doi.org/10.1007/978-1-4939-2742-5_9
11. Grant SG, Jessee J, Bloom FR, Hanahan D (1990) Differential plasmid rescue from transgenic mouse DNAs into Escherichia coli methylation-restriction mutants. Proc Natl Acad Sci USA 87(12):4645–4649. https://doi.org/10.1073/pnas.87.12.4645

12. Brachmann CB, Davies A, Cost GJ, Caputo E, Li J, Hieter P, Boeke JD (1998) Designer deletion strains derived from Saccharomyces cerevisiae S288C: a useful set of strains and plasmids for PCR-mediated gene disruption and other applications. Yeast 14(2):115–132. https://doi.org/10.1002/(SICI)1097-0061(19980130)14:2<115::AID-YEA204>3.0.CO;2-2

Chapter 25

Golden Gate Cloning for the Standardized Assembly of Gene Elements with Modular Cloning in *Chlamydomonas*

Peter Emelin, Sarah Abdul-Mawla, and Felix Willmund

Abstract

Modern synthetic biology requires fast and efficient cloning strategies for the assembly of new transcription units or entire pathways. Modular Cloning (MoClo) is a standardized synthetic biology workflow, which has tremendously simplified the assembly of genetic elements for transgene expression. MoClo is based on Golden Gate Assembly and allows to combine genetic elements of a library through a hierarchical syntax-driven pipeline. Here we describe the assembly of a genetic cassette for transgene expression in the single-celled model alga *Chlamydomonas reinhardtii*.

Key words Golden Gate Assembly, Modular cloning, *Chlamydomonas reinhardtii*, Synthetic biology

1 Introduction

Over several decades, transgene expression in model organisms or biotechnology-relevant species was achieved through traditional molecular cloning approaches that used various restriction endonucleases to fuse a gene of interest (GOI) into a convenient vector. Such cloning approaches are often laborious and expression results are not always of the desired quality. For example, the single-celled model alga *Chlamydomonas reinhardtii* suffered from highly inefficient transgene expression, and many different vectors, promoters, and expression strategies were tested that had varying success for different genes [1]. However, there is a tremendous need to find new solutions with synthetic biology, in order to exploit the biology's potential for expression of therapeutic high-value products, to think of new solutions of depleting the increased atmospheric CO_2 levels (or capture released CO_2), and for making plants more robust against the challenges of severe weather fluctuations of global warming. The success of modern synthetic biology

Authors Peter Emelin and Sarah Abdul-Mawla have equally contributed to this chapter.

Daniel Schindler (ed.), *Golden Gate Cloning: Methods and Protocols*, Methods in Molecular Biology, vol. 2850,
https://doi.org/10.1007/978-1-0716-4220-7_25,

approaches depends on the efficient introduction of new genetic elements and quick iterative cycles of design–construction–testing–learning, to optimize synthetic genetic elements. An adequate cloning approach, termed modular cloning (MoClo), was established in 2011 that uses syntax-driven assembly of standardized genetic elements [2]. Parameters used in this approach are also the basis for the *Chlamydomonas* MoClo tool kit described hereafter. MoClo is based on Golden Gate Assembly, which uses type IIS restriction endonucleases and directed assembly of multiple genetic parts in one single reaction [3]. Type IIS restriction endonucleases cleave DNA outside of their recognition site and can thus be removed from a DNA fragment of interest, if placed upstream and/or downstream of the respective DNA sequence (see Fig. 1a). Most importantly, type IIS enzymes generate 5′ overhangs

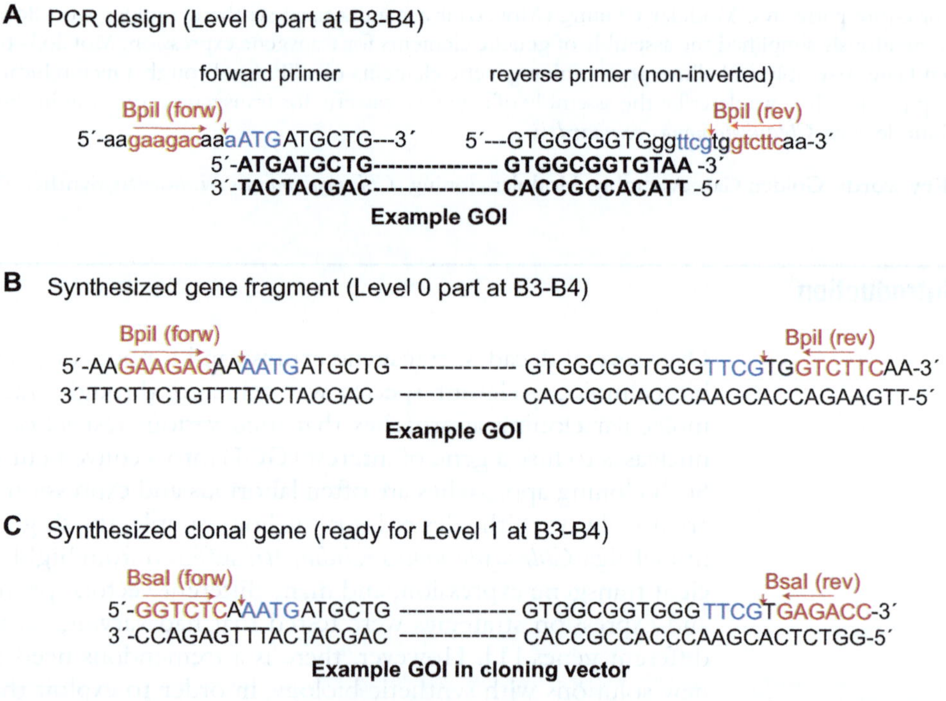

Fig. 1 Examples for designing a gene element for Level 0 cloning. In all three examples, fusion sites correspond to cloning of the respective gene part into Level 0 acceptor vector pAGM1287, which generates the fusion sites AATG and TTCG for position B3 to B4, a typical position for a coding sequence (see Fig. 2a). (**a**) Example for designing primers to amplify a gene of interest via polymerase chain reaction (PCR). First three codons and last three codons of an example gene are given for the sense and antisense strands ("Example GOI," in bold). Primer should anneal to at least 20 base pairs of the target gene (partially shown here) and contain a forward-oriented BpiI recognition site for the forward primer, and a reverse-oriented BpiI recognition site for the reverse primer. These sites will be removed after restriction digest and generate the correct terminal fusion sites. Make sure to skip the stop codon (here TAA), which should not be presented in the Level 0 part. (**b**) Example how to design the termini when ordering double-stranded gene fragments. (**c**) Example for designing the 5′ and 3′ flanking regions if clonal genes are ordered, which will be delivered in a cloning vector

on the antisense strands that can be custom designed to generate scar-less fusions for newly assembled DNA fragments [3]. MoClo takes advantage of these overhangs, termed "fusion sites" [2], to specifically define these overhangs in its syntax [2]. Here, gene elements are categorized in "promoters", "5′ untranslated regions (UTRs)", "signal sequences", "coding sequence (CDS)", and "3′-UTR with terminator". The defined fusion sites then direct assembly from individual gene parts to entire transcription units. The first, and most important step is to generate a Level 0 plasmid library (the collection of basic parts), which contains the individual gene elements at their respective positions (Fig. 2a). Typical Level 0 plasmids are based on pUC19 vectors [4] and confer Spectinomycin resistance during propagation in *Escherichia coli*. In addition, Level 0 plasmids contain *lacZ* genes within their respective cloning site. *LacZ* encodes for β-Galactosidase, an enzyme that converts 5-Brom-4-Chlor-3-Indolyl-β-D-Galactopyranoside ("X-Gal," which is added to the medium) into the blue product 5-bromo-4-chloro-indoxyl. By this, resulting clones are blue, if the *lacZ* gene was not replaced by the GOI, or white, if cloning was successful (blue/white selection). In the MoClo tool kit described here [2], insertion of the gene fragment is achieved via BpiI type IIS

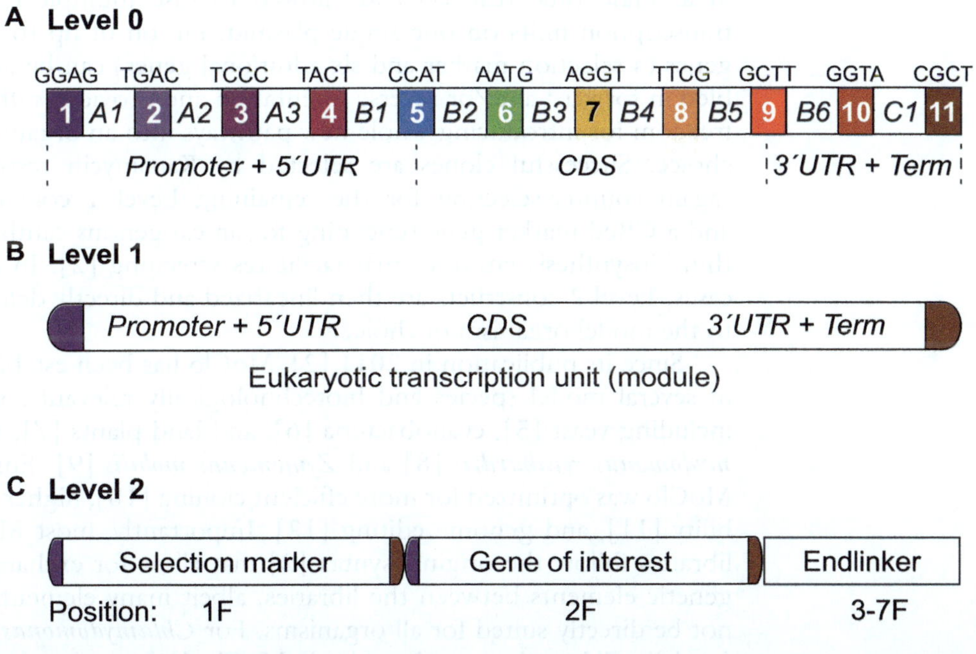

Fig. 2 Syntax of *Chlamydomonas* MoClo. (**a**) Positions and fusion sites for the respective Level 0 MoClo parts. Various commonly used and new promoters, 5′-UTRs, tags, signal sequences, reporters, and terminators are already provided by the published Level 0 library [8]. (**b**) Typical structure of a Level 1 module, which comprises a full eukaryotic transcription unit. Vector is not shown. (**c**) Example of a final Level 2 construct, which contains two modules (a selection marker and a GIO) in position 1F and 2F, respectively. Endlinker 2 for bridging positions 3F to 7F is shown (provided by plasmid [pICH41744, see Table 1])

restriction enzymes (forward orientation: 5′-GAAGAC(N)$_2$↓-3′; arrow indicating cleavage), and recognition sites are removed upon insertion of the desired gene fragment when successfully cloned. The final Level 0 clones contain the gene element that is now flanked by inward-directed BsaI type IIS recognition sites that generate the defined fusion sites after cleavage (5′-GGTCTC (N)$_1$↓-3′; arrow indicating cleavage), without being present on the released fragment (Figs. 1a–c and 2a). In the second step, Level 1 plasmids (the transcription units) are assembled. Here, Level 0 plasmids comprising promoter, 5′-UTRs, coding sequences, tags, and terminator elements are combinedly digested and overhangs define the order of subsequent assembly (Fig. 2b). T4 ligation is performed parallel to the restriction digests, since the use of type IIS restriction digest (BsaI for Level 1) facilitates the directionality of the reaction. Level 1 destination vectors carry Carbenicillin/Ampicillin resistance (thus counter-selecting for the remaining Level 0 constructs), and a *LacZ* fragment for blue/white selection. Once Level 1 clones are successfully cloned, BpiI recognition sites will again flank the gene assembly in order to generate specific fusion sites. Here, BpiI defines new fusion sites for directional cloning of multiple genes into Level 2 vectors (the higher-order gene assemblies, Fig. 1c). Level 2 constructs are again cloned in a single tube reaction and allowed to fuse multiple desired transcription units on one single plasmid. Fusion of up to seven genes (a selection marker and six additional genes) can be assembled in forward and/or reverse orientation and provides sufficient freedom for introducing whole new pathways into an organism of choice. Successful clones are selected by Kanamycin resistance (again counter-selecting for the remaining Level 1 constructs) and a CRed marker gene (encoding for an exogenous canthaxanthin biosynthesis enzyme) that facilitates screening [2]. In many cases, Level 2 constructs are then linearized and directly delivered to the model organism of choice.

Since its publication in 2011 [2], MoClo has been established in several model species and biotechnologically relevant species, including yeast [5], cyanobacteria [6], and land plants [7], *Chlamydomonas reinhardtii* [8] and *Zymomonas mobilis* [9]. Further, MoClo was optimized for more efficient cloning [10], higher flexibility [11], and genome editing [12]. Importantly, most MoClo libraries follow the original syntax [2] that allow for exchange of genetic elements between the libraries, albeit many elements will not be directly suited for all organisms. For *Chlamydomonas reinhardtii* (*Chlamydomonas* hereafter), MoClo has greatly advanced transgene expression for basic research [e.g., 13,14], establishment of new technologies [15, 16], and biotechnology [17]. Since general protocols for MoClo exist [18, 19], we will here focus on MoClo with *Chlamydomonas* and describe aspects that are of importance for cloning the genes for this organism. The

Chlamydomonas MoClo library started with 119 Level 0 genetic elements that provide a tremendous flexibility for expression of reporter genes, the use of selection markers, signal sequences, and gene silencing [8]. Importantly, the number of Level 0 constructs is constantly increasing, and many vectors are available through the *Chlamydomonas* resource center (https://www.chlamycollection.org).

2 Materials

2.1 Lab Equipment and Consumables

To carry out the MoClo assembly for *Chlamydomonas* genes, the following standard laboratory equipment and consumables are required.

1. Equipment for gel electrophoresis.
2. NanoDrop for determining DNA concentrations.
3. Thermocycler.
4. PCR reaction tubes.
5. Petri dishes.
6. Micro reaction tubes.
7. Glass tubes.
8. Laboratory incubators at 37 °C.

2.2 Strains, Vectors, and Preparation of Selection Media

For transformation, chemically competent Top10 and DH5α *Escherichia coli* strains are needed. All relevant vectors mentioned in this protocol are available as single plasmids or plasmid collections (Table 1). Backbone vectors for standard Levels 0–2 cloning are available via Addgene (https://www.addgene.org/kits/marillonnet-moclo/), Kit #1000000044. Many Level 0 and 1 parts and direct Level 2 cloning can be received from the *Chlamydomonas* resource center (https://www.chlamycollection.org).

Lysogeny Broth (LB)-medium:

Tryptone	10 g/L
Yeast extract	50 g/L
Sodium chloride	10 g/L

Solid media for plates are prepared with 1.5% agar

Level 0 and Level 1 cloning will be screened by blue/white selection. Thus, LB agar plates need to be prepared, which are supplemented with X-Gal and IPTG:

Prepare LB medium, autoclave on liquid cycle, and let the medium cool down to 50 °C. Add Isopropyl β-D-1-thiogalactopyranoside (IPTG) to a final concentration of 1 mM and 1 mL

Table 1
Relevant cloning vectors for *Chlamydomonas* MoClo

Name	Gene fragment to be inserted	Position	Resistance	Reference
Level 0 vectors				
pAGM1251	Promoter + 5′-UTR	A1-B1	Spectinomycin	[2]
pAGM1263	5′-UTR	B1	Spectinomycin	
pAGM1276	N-terminal tag, signal sequence	B2	Spectinomycin	
pAGM1287	Coding sequence	B3-B4	Spectinomycin	
pAGM1299	Coding sequence	B4	Spectinomycin	
pAGM1301	C-terminal tag	B5	Spectinomycin	
pICH41233	Promoter	A1-A3	Spectinomycin	
pICH41246	5′-UTR or 5-UTR and N-terminal tag	B1-B2	Spectinomycin	
pICH41258	N-terminal tag 2/signal sequence	B3	Spectinomycin	
pICH41264	Coding sequence	B4-B5	Spectinomycin	
pICH41276	3′-UTR and terminator	B6-C1	Spectinomycin	
pICH41295	Promoter +5′-UTR	A1-B2	Spectinomycin	
pICH41308	Coding sequence	B3-B5	Spectinomycin	
pICH53388	3′-UTR	B6-C1	Spectinomycin	
pICH53399	Terminator	C1	Spectinomycin	
Level 1 vectors				
pICH47732	Gene 1 (e.g., selection marker)	1 forward	Ampicillin	[2]
pICH47742	Gene 2 (e.g., transgene 1)	2 forward	Ampicillin	
pICH47751	Gene 3 (e.g., transgene 2)	3 forward	Ampicillin	
pICH47761	Gene 4 (e.g., transgene 3)	4 forward	Ampicillin	
pICH47772	Gene 5 (e.g., transgene 4)	5 forward	Ampicillin	
pICH47781	Gene 6 (e.g., transgene 5)	6 forward	Ampicillin	
pICH47791	Gene 7 (e.g., transgene 6)	7 forward	Ampicillin	
pICH47802	Gene 1 (e.g., selection marker)	1 reverse	Ampicillin	
pICH47811	Gene 2 (e.g., transgene 1)	2 reverse	Ampicillin	
pICH47822	Gene 3 (e.g., transgene 2)	3 reverse	Ampicillin	
pICH47831	Gene 4 (e.g., transgene 3)	4 reverse	Ampicillin	
pICH47841	Gene 5 (e.g., transgene 4)	5 reverse	Ampicillin	
pICH47852	Gene 6 (e.g., transgene 5)	6 reverse	Ampicillin	
pICH47861	Gene 7 (e.g., transgene 6)	7 reverse	Ampicillin	
Level 2 vectors				
pAGM4673	Level 2 acceptor	–	Kanamycin	[2]
pICH41722	Endlinker 1, bridging none-used positions	2 to 7	Spectinomycin	
pICH41744	Endlinker 2, bridging none-used positions	3 to 7	Spectinomycin	
pICH41766	Endlinker 3, bridging none-used positions	4 to 7	Spectinomycin	
pICH41780	Endlinker 4, bridging none-used positions	5 to 7	Spectinomycin	
pICH41800	Endlinker 5, bridging none-used positions	6 to 7	Spectinomycin	
pICH41822	Endlinker 6, bridging none-used positions	7	Spectinomycin	
pMBS807	Level 2 with *aadA* cassette (Spec^R) 1 forward	GOI at	pAGM4673	[10]
pMBS808	Level 2 with *aphVIII* cassette (Paro^R)	2 to 7	pAGM4673	
pMBS809	1 forward	GOI at	pAGM4673	
pMBS810	Level 2 with nat1 cassette (NTC^R) 1 forward	2 to 7	pAGM4673	
pMBS811	Level 2 with *aphVII* cassette (Hygro^R)	GOI at	pAGM4673	
		2 to 7		

(continued)

Table 1
(continued)

Name	Gene fragment to be inserted	Position	Resistance	Reference
	1 forward Level 2 with *ble* cassette ($Bleo^R$) 1 forward	GOI at 2 to 7 GOI at 2 to 7		

X-Gal stock per liter of media. Add Spectinomycin or Ampicillin for medium of Level 0 and Level 1 constructs, respectively. For Level 2 constructs, only Kanamycin needs to be added to the LB medium. Pour approximately 25 mL per petri dish and let the media solidify for approximately 30 min. Store the plates at 4 °C, avoid light exposure.

2.3 Enzymes and Antibiotics

The following enzymes are necessary to conduct the outlined protocol.

1. BsaI-HFv2 (isoschizomer of Eco31I; 20,000 units/mL).
2. BbsI-HF (isoschizomer of BpiI, Bso31I, and BspTNI; 20,000 units/mL).
3. T4 DNA ligase (400,000 units/mL) (*see* **Note 1**).

Generally, the buffers are supplied with the enzymes.

All antibiotics are dissolved in double-distilled H_2O and subsequently sterilized with a 0.2 μm filter. To reach the working concentration, dilute 1:1000, respectively.

1. 100 mg/mL Spectinomycin (stock concentration).
2. 100 mg/mL Ampicillin (stock concentration).
3. 50 mg/mL Kanamycin (stock concentration).

2.4 Chemicals and Buffer Components

1. X-Gal: 100 mg/mL in Dimethyl sulfoxide (DMSO) (stock).
2. IPTG: 1 M (stock).
3. DNA ladder (e.g., GeneRuler 1 kb DNA Ladder, ready-to-use).
4. DNA Gel Loading Dye.
5. Midori Green DNA Dye.

TAE-Buffer 1×

EDTA (Ethylenediaminetetraacetic acid)	0.01 M
Tris-acetate (pH 8.3)	0.4 M

Agarose gels are prepared with 1% agarose and Midori Green DNA Dye (1:25,000)

2.5 Chlamydomonas Strains

Any *Chlamydomonas* strain can be used for transformation of MoClo constructs. However, the best transgene expression is currently achieved with the UVM4 and UVM11 cell wall deficient expression strains [20].

3 Methods

3.1 Gene of Interest Design for Level 0 Cloning

For retrieving a *Chlamydomonas* DNA sequence for a gene of interest or respective gene elements (promoter, UTRs, signal sequence, terminator), the most recent gene models are found at the *Chlamydomonas* CC-4532 v6.1 genome portal (https://phytozome-next.jgi.doe.gov). For endogenous and foreign genes, several things should be considered during the design of MoClo Level 0 parts:

1. Domestication: Endogenous BsaI and BpiI type IIS recognition sites need to be removed prior to MoClo. This can be achieved during gene synthesis, by site-directed mutagenesis, or by fragmenting a GOI and mutating the sites by respective primer design. Care needs to be taken that mutated sites are silent and do not alter amino acid composition in CDSs.
2. According to the most recent *Chlamydomonas* genome, alternative splicing is frequent in *Chlamydomonas*, leading to multiple transcript variants. In addition, predicted gene models are not always correct. Our recent ribosome profiling data can be used to browse a *Chlamydomonas* GOI, to validate a gene model and check which splice variant is the most prevalent form [21, 22]. Sequences should be corrected if needed.
3. Codon usage: Foreign genes should be optimized for *Chlamydomonas* codon usage according to the table offered by the Kazusa DNA Research Institute (https://www.kazusa.or.jp/codon/cgi-bin/showcodon.cgi?species=3055). If needed, the DNA sequence can be reverse-translated from amino acid sequence via http://www.bioinformatics.org/sms2/rev_trans.html [23], based on the respective codon usage table.
4. Introns: Expression can be improved by replacing endogenous introns with introns that confer intron-mediated enhancement (IME). Introns should be introduced approximately every 400 bp in a codon sequence. The three *RBCS2* introns are most frequently used to increase expression and are placed at CAG/G or AAG/G sequences (slash indicates intron insertion position) [24, 25]. These flanking sequences represent canonical insertion sites for *Chlamydomonas* introns. Additional IME-inducing introns were recently described [26].
5. Start codon site: Expression seems to improve if the folding energy of mRNA secondary structures surrounding the start

codon is increased [27]. Such mutations can be introduced through synonymous mutations for the GOI.

Prior to ordering synthesized genes and/or primers, it is strongly advised to generate the desired plasmid in silico for the analysis of suitable restriction sites, sequencing results, and for archiving. Furthermore, the respective genomic DNA segments should be amplified and sequenced from the used laboratory strain, since mutations might have occurred that can interfere with cloning.

3.2 Generation of Gene Elements for Level 0 Cloning

Generate your gene fragment by one of the following ways (*see* **Note 2**):

1. Small gene parts can be generated by aligning two complementary oligos that are offset by four nucleotides on both sides.
2. Examples for oligo design for amplifying a gene element via PCR are given in Fig. 1a.
3. Gene fragment synthesis facilitates Level 0 cloning. For designing the fusion sites, see Fig. 1b.
4. Clonal gene synthesis: Genes can be ordered that are already cloned in vectors suitable for subsequent Level 1 cloning (Fig. 1c). Some companies might not offer cloning vectors with Spectinomycin selection marker. Alternatively, any other selection markers can be used; however, it is advised to maintain a Level 0 library with consistent selection markers. Avoid using Ampicillin selection marker, since these will interfere with Level 1 cloning.

3.3 Generating a Level 0 Plasmid

1. Use a 100 μL PCR tube for the reaction. Ligate your gene fragment (generated by one of the chosen methods from above) into Level 0 vector pAGM1287 by mixing 40 fmol of the DNA and plasmid, respectively, along with 2 μL T4 DNA ligase buffer, 0.5 μL BbsI (10U), and 0.5 μL T4 DNA ligase (200U) in a total volume of 20 μL.
2. Place the tube into a thermocycler set to the following parameters for the digest/ligation reaction:
   - 37 °C for 5 h
   - 50 °C for 5 min
   - 80 °C for 10 min
   - Infinite hold at 4 °C
3. Transform the construct into competent cells as described below and plate on LB agar plates containing X-Gal, IPTG, and Spectinomycin.

4. On the next day, pick several (four to eight) white colonies and grow them in liquid LB medium, supplemented with Spectinomycin.
5. Perform a plasmid mini-prep isolation, according to the instructions of the isolation kit.
6. Find an appropriate restriction enzyme with two cutting sites in silico (e.g., BsaI, which releases the part of interest from the cloning vector) and perform a test digest to check for putatively correct clones. Sequence the insertion by using forward and/or reverse sequencing oligos (ideally aligning to sequences parts upstream and downstream of the insert) (*see* **Note 3**).

3.4 Generating a Level 1 Plasmid

Level 1 constructs are full *Chlamydomonas* transcription units that are assemblies of Level 0 parts, containing a desired promotor and terminator, the CDS of interest, optional protein tags N-terminally or C-terminally fused to the coding sequence, optional reporter sequences, or other tools. Here, a combination of Level 0 parts and synthesized gene fragments or PCR fragments (containing the correct syntax) is also possible. The procedure is almost identical to the Level 0 plasmid generation. For Level 0, the plasmid pICH47742 is used as a destination vector, which confers resistance against Ampicillin, and BsaI is used as the restriction enzyme. The desired parts are mixed in an equimolar amount of 40 fmol per gene element (*see* **Note 4**). Transformed *E. coli* cells are plated onto LB agar plates supplemented with X-Gal, IPTG, and Ampicillin. Restriction digest and sequencing are again necessary to verify the correct ligation of the different parts.

1. Use a 100 μL PCR tube for the reaction. Ligate your gene fragment (generated by one of the chosen methods from above) into Level 1 vector pICH47742 by mixing 40 fmol of the desired Level 0 parts (*see* **Note 4**) along with 2 μL T4 DNA ligase buffer, 0.5 μL BsaI (10U), and 0.5 μL T4 DNA ligase (200U) in a total volume of 20 μL.
2. Place the tube into a thermocycler set to the following parameters for the digest/ligation reaction:
   - 37 °C for 5 h
   - 50 °C for 5 min
   - 80 °C for 10 min
   - Infinite hold at 4 °C
3. Transform the construct into competent cells as described below and plate on LB agar plates containing X-Gal, IPTG, and Ampicillin.

4. On the next day, pick several (four to eight) white colonies and grow them in liquid LB medium supplemented with Ampicillin.
5. Perform a plasmid mini-prep isolation, according to the instructions of the isolation kit.
6. Find an appropriate restriction enzyme with two cutting sites in silico and perform a test digest to check for putatively correct clones. Sequence the insertion by using forward and/or reverse sequencing oligos (ideally aligning to sequences parts upstream and downstream of the insert) (*see* **Note 3**).

3.5 Generating a Level 2 Plasmid

Level 2 plasmids are higher-order multi-gene constructs that are comprised of the desired number of Level 1 parts and a resistance gene for the target organism. Several Level 1 transcription units that can be used as resistance gene are already provided by the published MoClo library [8]. The procedure is again almost identical to the Level 0 reaction. The desired Level 1 parts are ligated into the destination vector pAGM4673 (containing the CRed gene, flanked by the BpiI restriction sites, for red/white selection). BpiI is used as the restriction enzyme. Also here, a combination of Level 1 plasmids and synthesized gene fragments or PCR fragments is possible (if the correct syntax is considered). Level 2 acceptor plasmids contain positions for seven transcription units that can be assembled in one construct. If only positions 1 and 2 are used (as frequently applied for expressing just the selection marker and a GOI), the unused positions 3–7 are bridged by the addition of an "end-linker" fragment (here Endlinker 2, see Table 1) (see Fig. 2c).

Upon transformation of the MoClo reactions into *E. coli*, cells are plated onto LB agar plates containing only Kanamycin and correct clones are screened by red/white selection. White colonies are selected for further test digest (*see* **Note 5**).

1. Use a 100 μL PCR tube for the reaction. Ligate your Level 1 parts into Level 2 vector pAGM4673 by mixing 40 fmol of the desired Level 1 parts along with 2 μL T4 DNA ligase buffer, 0.5 μL BpiI (10U), and 0.5 μL T4 DNA ligase (200U) in a total volume of 20 μL.
2. Place the tube into a thermocycler set to the following parameters for the digest/ligation reaction:
   - 37 °C for 5 h
   - 50 °C for 5 min
   - 80 °C for 10 min
   - Infinite hold at 4 °C
3. Transform the construct into competent cells as described below and plate on LB agar plates containing Kanamycin.

4. On the next day, pick several (four to eight) white colonies and grow them in liquid LB medium supplemented with Kanamycin.
5. Perform a plasmid mini-prep isolation, according to the instructions of the isolation kit.
6. Find an appropriate restriction enzyme with two cutting sites in silico and perform a test digest to check for putatively correct clones. Sequence the insertion by using forward and/or reverse sequencing oligos (ideally aligning to sequences parts upstream and downstream of the insert) (*see* **Note 3**).

3.6 Generating a Level 2 Plasmid by Skipping Level 1 (Fast Track)

Recently, a series of new Level 2 destination vectors were created, which allows for the direct ligation of Level 0 plasmids into final Level 2 constructs, omitting the creation of Level 1 plasmids [10]. These vectors (pMBS807–pMBS811, based on the destination vector pAGM4673) greatly reduce the time needed to construct a suitable Level 2 plasmid, if only one GOI is needed.

The procedure is identical to the creation of Level 1 plasmids, where BsaI is used to integrate the desired Level 0 plasmids. Screening of transformed *E. coli* strains is done on LB agar plates supplemented with Kanamycin using red/white selection.

1. Use a 100 μL PCR tube for the reaction. Ligate your Level 0 parts into the desired Level 2 vector pMBS807–pMBS811 by mixing 40 fmol of the Level 0 parts along with 2 μL T4 DNA ligase buffer, 0.5 μL BsaI (10U), and 0.5 μL T4 DNA ligase (200U) in a total volume of 20 μL.
2. Place the tube into a thermocycler set to the following parameters for the digest/ligation reaction:
   - 37 °C for 5 h
   - 50 °C for 5 min
   - 80 °C for 10 min
   - Infinite hold at 4 °C
3. Transform the construct into competent cells as described below and plate on LB agar plates containing Kanamycin.
4. On the next day, pick several (four to eight) white colonies and grow them in liquid LB medium supplemented with Kanamycin.
5. Perform a plasmid mini-prep isolation, according to the instructions of the isolation kit.
6. Find an appropriate restriction enzyme with two cutting sites in silico and perform a test digest to check for putatively correct clones. Sequence the insertion by using forward and/or reverse sequencing oligos (ideally aligning to sequences parts upstream and downstream of the insert) (*see* **Note 3**).

3.7 Transformation into Chemically Competent E. coli Top10/DH5α Strain

After each step (Levels 0–2), cloning reactions are transformed into chemically competent *E. coli* strains (*see* **Note 6**).

1. Take an aliquot of your competent cells from the –80 °C freezer and briefly thaw them until all ice is melted (*see* **Note 7**).
2. Add 10 of your cloning reaction (Level 0, 1, or 2) to 50 μL of competent cells and mix gently by flicking the tube. Immediately place back on ice.
3. Incubate on ice for 30 min.
4. Place tubes into a preheated water bath or heat block set to 42 ° C and incubate for 60–90 s.
5. Transfer the tube back on ice and let recover for 2–5 min.
6. Add 600–1000 μL LB medium without antibiotics and incubate cells for 45 min at 37 °C with shaking (300 rpm).
7. In the meantime, take two respective LB antibiotic plates per construct and let them warm up to room temperature.
8. Plate 50 μL of the cell suspension on one plate, briefly centrifuge the remaining cells in a tabletop centrifuge at romm temperature (RT) and 14,000 × g, resuspend in approximately 50 μL, and plate on the second plate (this serves as backup if transformation efficiency was low).
9. Dry the remaining liquid under the sterile bench and incubate plates overnight in an incubator at 37 °C.
10. For Level 0 and Level 1, putative positive clones will be white, whereas colonies with negative clones show a blue color. For Level 2, positive will be white, whereas colonies with a negative clone have a red color.

4 Notes

1. If a buffer without ATP is chosen, a working concentration of 1 mM ATP needs to be supplemented to the MoClo reaction.
2. For the design of CDS, it is recommended to remove the Stop codon for Level 0, hence providing maximal freedom for future MoClo design. If no C-terminal fusion is needed, the existing MoClo library provides Stop codon elements that replace the endogenous Stop codon [8].
3. Public deposition of successful Level 0 constructs will foster the research community and increase the impact of your study. For *Chlamydomonas*, most parts are deposited without charge at the *Chlamydomonas* resource center (https://www.chlamycollection.org).

4. To simplify the ligation process and save time during Level 1 and Level 2 assembly, the desired Level 0 or Level 1 parts do not necessarily have to be mixed in an equimolar amount. The concentration of the purified plasmids from *E. coli* can be used as an approximation:
   - If the plasmid concentration is below 300 ng/μL, two microliters are added to the ligation mix.
   - If the plasmid concentration is above 300 ng/μL, one microliter is added to the ligation mix.
5. If multiple genes are assembled, try to avoid repetitive use of similar promoters and terminators for the individual genes. Different promoters are available in the *Chlamydomonas* MoClo kit [8].
6. Alternatively, transformation can be performed with standard electroporation protocols.
7. Cells should be thawed directly prior to transformation since competence might be lost for longer time on ice.

References

1. Schroda M (2019) Good news for nuclear transgene expression in *Chlamydomonas*. Cells 8(12). https://doi.org/10.3390/cells8121534
2. Weber E, Engler C, Gruetzner R, Werner S, Marillonnet S (2011) A modular cloning system for standardized assembly of multigene constructs. PLoS One 6(2):e16765. https://doi.org/10.1371/journal.pone.0016765
3. Engler C, Kandzia R, Marillonnet S (2008) A one pot, one step, precision cloning method with high throughput capability. PLoS One 3(11):e3647. https://doi.org/10.1371/journal.pone.0003647
4. Yanisch-Perron C, Vieira J, Messing J (1985) Improved M13 phage cloning vectors and host strains: nucleotide sequences of the M13mp18 and pUC19 vectors. Gene 33(1):103–119. https://doi.org/10.1016/0378-1119(85)90120-9
5. Lee ME, DeLoache WC, Cervantes B, Dueber JE (2015) A highly characterized yeast toolkit for modular, multipart assembly. ACS Synth Biol 4(9):975–986. https://doi.org/10.1021/sb500366v
6. Vasudevan R, Gale GAR, Schiavon AA, Puzorjov A, Malin J, Gillespie MD, Vavitsas K, Zulkower V, Wang B, Howe CJ, Lea-Smith DJ, McCormick AJ (2019) CyanoGate: a modular cloning suite for engineering cyanobacteria based on the plant MoClo syntax. Plant Physiol 180(1):39–55. https://doi.org/10.1104/pp.18.01401
7. Engler C, Youles M, Gruetzner R, Ehnert TM, Werner S, Jones JD, Patron NJ, Marillonnet S (2014) A golden gate modular cloning toolbox for plants. ACS Synth Biol 3(11):839–843. https://doi.org/10.1021/sb4001504
8. Crozet P, Navarro FJ, Willmund F, Mehrshahi P, Bakowski K, Lauersen KJ, Perez-Perez ME, Auroy P, Gorchs Rovira A, Sauret-Gueto S, Niemeyer J, Spaniol B, Theis J, Trösch R, Westrich LD, Vavitsas K, Baier T, Hübner W, de Carpentier F, Cassarini M, Danon A, Henri J, Marchand CH, de Mia M, Sarkissian K, Baulcombe DC, Peltier G, Crespo JL, Kruse O, Jensen PE, Schroda M, Smith AG, Lemaire SD (2018) Birth of a photosynthetic chassis: a MoClo toolkit enabling synthetic biology in the microalga *Chlamydomonas reinhardtii*. ACS Synth Biol 7(9):2074–2086. https://doi.org/10.1021/acssynbio.8b00251
9. Behrendt G, Frohwitter J, Vlachonikolou M, Klamt S, Bettenbrock K (2022) Zymo-parts: a golden gate modular cloning toolbox for heterologous gene expression in *Zymomonas mobilis*. ACS Synth Biol 11(11):3855–3864. https://doi.org/10.1021/acssynbio.2c00428
10. Niemeyer J, Schroda M (2022) New destination vectors facilitate modular cloning for Chlamydomonas. Curr Genet 68(3-4):

531–536. https://doi.org/10.1007/s00294-022-01239-x

11. Klein CA, Emde L, Kuijpers A, Sobetzko P (2019) MoCloFlex: a modular yet flexible cloning system. Front Bioeng Biotechnol 7: 271. https://doi.org/10.3389/fbioe.2019.00271
12. Hahn F, Korolev A, Sanjurjo Loures L, Nekrasov V (2020) A modular cloning toolkit for genome editing in plants. BMC Plant Biol 20(1):179. https://doi.org/10.1186/s12870-020-02388-2
13. Trösch R, Ries F, Westrich LD, Gao Y, Herkt C, Hoppstädter J, Heck-Roth J, Mustas M, Scheuring D, Choquet Y, Räschle M, Zoschke R, Willmund F (2022) Fast and global reorganization of the chloroplast protein biogenesis network during heat acclimation. Plant Cell 34(3):1075–1099. https://doi.org/10.1093/plcell/koab317
14. Westrich LD, Gotsmann VL, Herkt C, Ries F, Kazek T, Trosch R, Armbruster L, Mühlenbeck JS, Ramundo S, Nickelsen J, Finkemeier I, Wirtz M, Storchova Z, Räschle M, Willmund F (2021) The versatile interactome of chloroplast ribosomes revealed by affinity purification mass spectrometry. Nucleic Acids Res 49(1): 400–415. https://doi.org/10.1093/nar/gkaa1192
15. Niemeyer J, Scheuring D, Oestreicher J, Morgan B, Schroda M (2021) Real-time monitoring of subcellular H2O2 distribution in *Chlamydomonas reinhardtii*. Plant Cell 33(9): 2935–2949. https://doi.org/10.1093/plcell/koab176
16. Kreis E, Konig K, Misir M, Niemeyer J, Sommer F, Schroda M (2023) TurboID reveals the proxiomes of Chlamydomonas proteins involved in thylakoid biogenesis and stress response. Plant Physiol. https://doi.org/10.1093/plphys/kiad335
17. Fabris M, Abbriano RM, Pernice M, Sutherland DL, Commault AS, Hall CC, Labeeuw L, McCauley JI, Kuzhiuparambil U, Ray P, Kahlke T, Ralph PJ (2020) Emerging technologies in algal biotechnology: toward the establishment of a sustainable, algae-based bioeconomy. Front Plant Sci 11:279. https://doi.org/10.3389/fpls.2020.00279
18. Bird JE, Marles-Wright J, Giachino A (2022) A user's guide to golden gate cloning methods and standards. ACS Synth Biol 11(11): 3551–3563. https://doi.org/10.1021/acssynbio.2c00355
19. Marillonnet S, Grutzner R (2020) Synthetic DNA assembly using Golden Gate cloning and the hierarchical modular cloning pipeline. Curr Protoc Mol Biol 130(1):e115. https://doi.org/10.1002/cpmb.115
20. Neupert J, Karcher D, Bock R (2009) Generation of *Chlamydomonas* strains that efficiently express nuclear transgenes. Plant J 57(6): 1140–1150. https://doi.org/10.1111/j.1365-313X.2008.03746.x
21. Gotsmann VL, Ting MKY, Haase N, Rudorf S, Zoschke R, Willmund F (2023) Utilizing high-resolution ribosome profiling for the global investigation of gene expression in *Chlamydomonas*. Plant J. https://doi.org/10.1111/tpj.16577
22. Gotsmann VL, Ting AY, Haase N, Rudorf S, Zoschke R, Willmund F (2023) Utilizing high resolution ribosome profiling for the global investigation of gene expression in Chlamydomonas. BioRxiv. https://doi.org/10.1101/2023.02.13.528309
23. Stothard P (2000) The sequence manipulation suite: JavaScript programs for analyzing and formatting protein and DNA sequences. Biotechniques 28(6):1102–1104. https://doi.org/10.2144/00286ir01
24. Eichler-Stahlberg A, Weisheit W, Ruecker O, Heitzer M (2009) Strategies to facilitate transgene expression in *Chlamydomonas reinhardtii*. Planta 229(4):873–883. https://doi.org/10.1007/s00425-008-0879-x
25. Baier T, Wichmann J, Kruse O, Lauersen KJ (2018) Intron-containing algal transgenes mediate efficient recombinant gene expression in the green microalga *Chlamydomonas reinhardtii*. Nucleic Acids Res 46(13): 6909–6919. https://doi.org/10.1093/nar/gky532
26. Baier T, Jacobebbinghaus N, Einhaus A, Lauersen KJ, Kruse O (2020) Introns mediate post-transcriptional enhancement of nuclear gene expression in the green microalga *Chlamydomonas reinhardtii*. PLoS Genet 16(7): e1008944. https://doi.org/10.1371/journal.pgen.1008944
27. Weiner I, Atar S, Schweitzer S, Eilenberg H, Feldman Y, Avitan M, Blau M, Danon A, Tuller T, Yacoby I (2018) Enhancing heterologous expression in *Chlamydomonas reinhardtii* by transcript sequence optimization. Plant J 94(1):22–31. https://doi.org/10.1111/tpj.13836

Chapter 26

Assembling DNA Plasmids with the Multi-Kingdom (MK) Cloning System

Sarina Scoville and David M. Chiasson

Abstract

The Golden Gate cloning technique is used to assemble DNA parts into higher-order assemblies. Individual parts containing compatible overhangs generated by type IIS restriction enzymes are joined together using DNA ligase. The technique enables users to assemble custom transcription units (TUs) for a wide array of experimental assays. Several Golden Gate cloning systems have been developed; however, they are typically used with a narrow range of organisms. Here we describe the Multi-Kingdom (MK) cloning system that allows users to generate DNA plasmids for use in a broad range of organisms.

Key words Golden Gate cloning, DNA assembly, Synthetic biology, DNA parts, Modular cloning

1 Introduction

Golden Gate is a modular technique for assembling custom DNA plasmids using a library of DNA parts [1]. The DNA assembly method utilizes type IIS restriction enzymes such as BsaI (Eco31I), BpiI (BbsI), Esp3I (BsmBI), AarI (PaqCI), and SapI (BspQI) or their isoschizomers (shown in brackets). These enzymes recognize either a 6 bp or 7 bp non-palindromic DNA motif. A typical type II restriction enzyme cleaves both DNA strands within its recognition site, leaving either a staggered or blunt overhang. In contrast, type IIS (shifted cleavage) restriction enzymes cleave DNA outside of their recognition site. Since the overhangs resulting from the staggered cut are independent of the recognition site, researchers can engineer custom DNA overhangs. This capability allows molecular biologists to design DNA cloning systems that are built upon custom overhangs flanking DNA parts. Several Golden Gate cloning systems have been independently developed for bacteria [2–6], fungi [7–10], mammalian cells [11, 12], and plants [13–18]. These cloning systems utilize type IIS restriction enzymes; however, the choice of enzyme, overhangs,

Daniel Schindler (ed.), *Golden Gate Cloning: Methods and Protocols*, Methods in Molecular Biology, vol. 2850,
https://doi.org/10.1007/978-1-0716-4220-7_26,

backbones, and overall design may differ between systems. Further, most Golden Gate cloning systems were originally designed for experimental work in a limited range of organisms such as bacteria, yeast, animals, or plants. To circumvent this issue, the Multi-Kingdom (MK) cloning system was developed to enable users to assemble DNA plasmids for experimental work in a broad range of organisms [19].

The MK cloning system [19] was built using the architecture of an existing Golden Gate cloning system for plants [15]. Since a typical molecular biology laboratory works with model organisms from different kingdoms, the MK cloning system was designed to streamline DNA plasmid assembly. The MK cloning system utilizes type IIS enzymes to assemble Level 1 (LI) entry clones (BpiI), Level 2 (LII) transcription units (TUs; BsaI), and Level 3 (LIII) higher-order assembly of transcription units (BpiI). All DNA parts are normally cleared of both BsaI and BpiI recognition sites by introducing silent mutations through polymerase chain reaction (PCR) or by ordering synthesized DNA parts with a modified sequence. The MK system uses the universal LI acceptor plasmid p641-BpiI LI (Fig. 1a) and a series of six LII (Fig. 2) and LIII backbones (Fig. 3). Four of the LII and LIII backbones (p421,

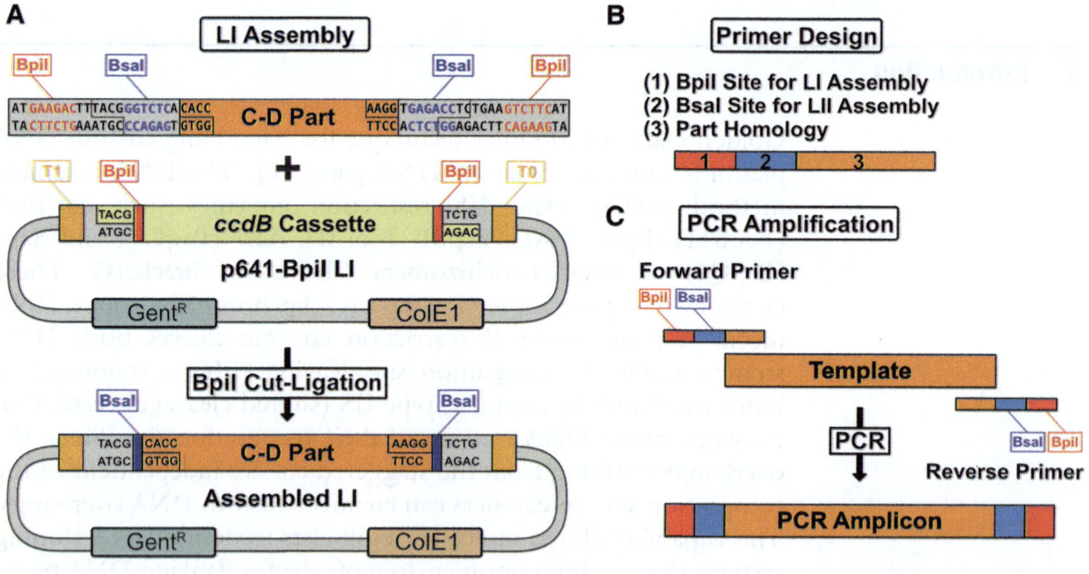

Fig. 1 Generating LI clones for the MK system. (**a**) A DNA part (C-D position example) is initially inserted into the universal p641-BpiI LI backbone. Flanking BsaI and BpiI restriction enzyme sites and overhangs are added by PCR or during the synthesis of the part. The DNA part and p641-BpiI LI are mixed in a BpiI cut-ligation reaction to generate an LI clone. p641-BpiI LI contains flanking T1 and T0 terminator sequences, a gentamicin resistance cassette (Gent^R), a ColE1 origin of replication, and *ccdB* negative selection. (**b**) To create a DNA part by PCR, the primers contain: (1) BpiI restriction site for directional LI cloning, (2) BsaI restriction site for LII cloning, and (3) Homology to the part template. (**c**) A DNA part can be amplified by PCR using DNA primers to append the BpiI and BsaI restriction enzyme sites and custom overhangs

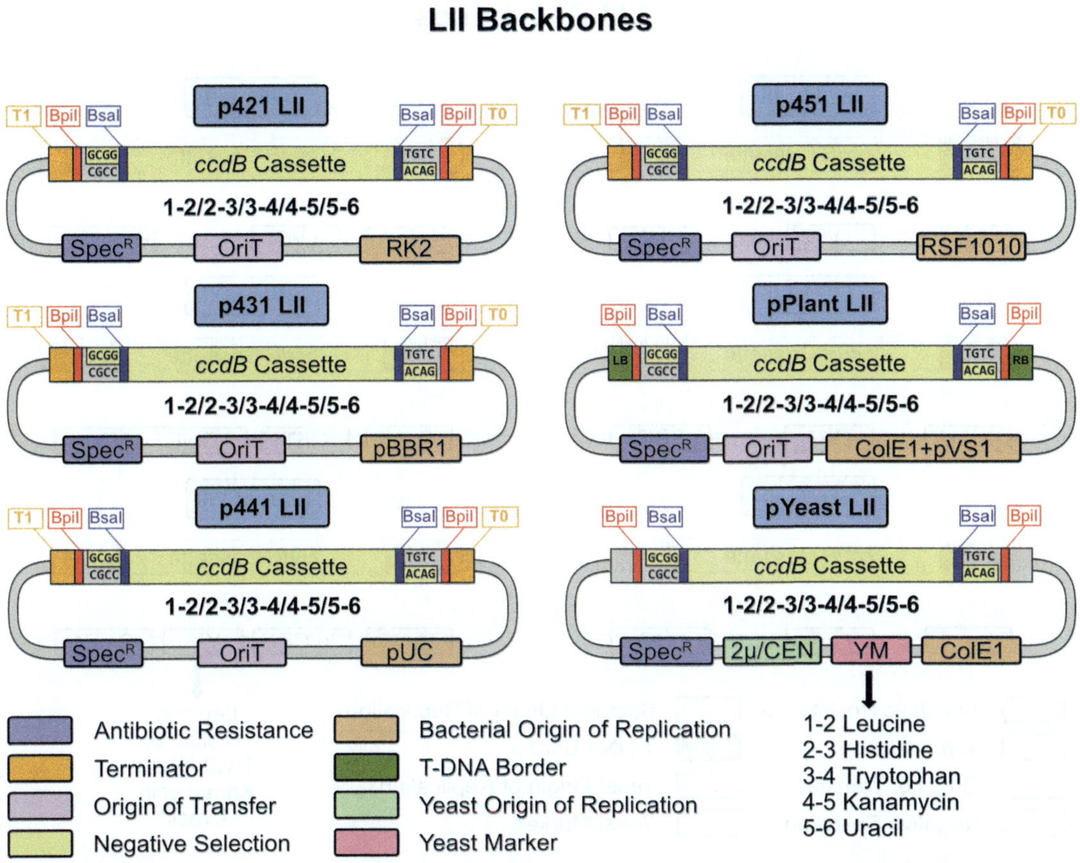

Fig. 2 Overview of the LII MK backbones. The system contains six unique LII plasmid backbone series (positions 1-2, 2-3, 3-4, 4-5, 5-6) with a *ccdB* negative-selection cassette and spectinomycin resistance ($Spec^R$). The p421 (RK2), p431 (pBBR1), p441 (ColE1), and p451 (RSF1010) series contain different origins of replication for gram-negative bacteria, an origin of transfer (OriT), and flanking terminators (T1, T0). Plant backbones contain a double origin of replication for *E. coli* (ColE1) and *Agrobacterium* (pVS1) along with an OriT and T-DNA borders for genomic integration (LB, left border; RB, right border). Yeast backbones contain a ColE1 origin of replication for bacteria, an origin of replication for yeast (high copy 2 μ or low copy CEN6/ARS4), and an auxotrophic or dominant marker

p431, p441, p451) contain origins of replication for bacteria (RK2, pBBR1, pUC, and RSF1010) with differing host ranges and copy numbers obtained from the Standard European Vector Architecture (SEVA) collection [20]. These plasmids also contain T0 and T1 terminators flanking cloned cargo and an origin of transfer (OriT) for conjugation. The pPlant series contains an origin of replication for both *Escherichia coli* (ColE1) and *Agrobacterium* sp. (pVS1), along with the transfer-DNA (T-DNA) borders required to generate transgenic plants. The pYeast series contains a ColE1 origin of replication along with either the 2 μ (high copy) or CEN6/ARS4 (low copy) origin or replication for yeast (*Saccharomyces cerevisiae*) obtained from a modular yeast toolkit [7]. In

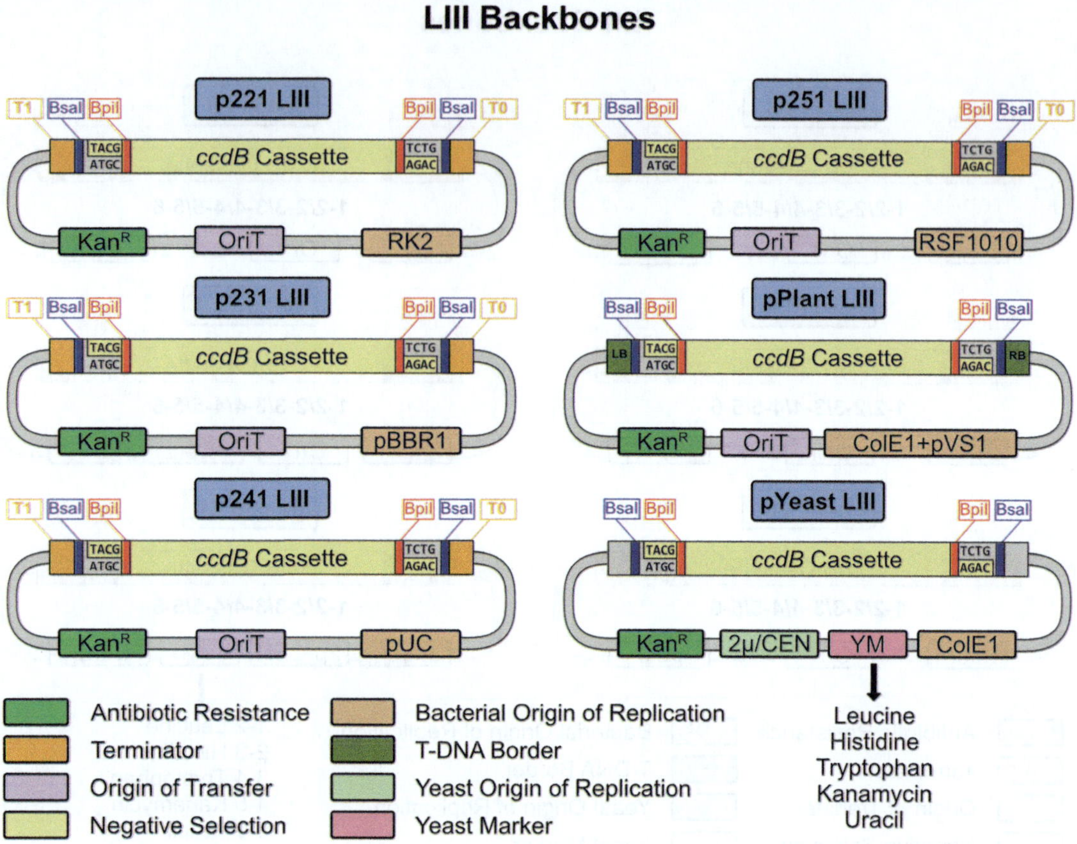

Fig. 3 Overview of the LIII MK backbones. LIII plasmid backbones are identical to LII backbones except that they contain a kanamycin resistance cassette (Kan^R) and the orientation of the BsaI and BpiI sites are reversed. Bacterial plasmids contain an origin of replication (RK2, pBBR1, ColE1, or RSF1010), an origin of transfer (OriT), and flanking terminators (T1, T0). Plant backbones contain a double origin of replication for *E. coli* (ColE1) and *Agrobacterium* (pVS1) along with an OriT and T-DNA borders for genomic integration (LB, left border; RB, right border). Yeast backbones contain a ColE1 origin of replication for bacteria, an origin of replication for yeast (high copy 2 µ or low copy CEN6/ARS4), and an auxotrophic or dominant marker

addition, each pYeast plasmid series contains cassettes for either auxotrophic complementation (leucine, histidine, tryptophan, or uracil) or drug selection (kanamycin) in yeast.

An LI DNA part is initially generated with flanking BsaI recognition sites and unique 4 bp overhang sequences corresponding to the positions of either A-B (promoter), B-C (N-terminal tag), C-D (gene of interest), D-E (C-terminal tag), E-F (terminator), or F-G (flexible position). These flanking sequences are added during PCR amplification (Fig. 1b) or DNA synthesis. The target DNA sequence is then cloned into p641-BpiI LI using BpiI and T4 ligase. Since p641-BpiI LI and all LII and LIII plasmid backbones contain the *ccdB* negative-selection cassette, all colonies on agar plates contain backbone plasmids with an inserted DNA fragment (*ccdB* cassette excised). Further, since each level of plasmids contain different antibiotic resistance cassettes (LI, gentamicin; LII, spectinomycin; LII, kanamycin), all colonies on agar plates must also contain the desired backbone and not donor plasmids used in the ligation reaction. LII transcription units are assembled with LI parts and an LII backbone using the enzyme BsaI and T4 ligase, while LIII higher-order assemblies are generated with BpiI and T4 ligase. With such a broad series of backbones, a laboratory can use the same type IIs restriction enzymes and DNA parts to generate custom plasmids for use in a wide array of organisms.

2 Materials

2.1 PCR Amplification

1. DNA oligonucleotides (*see* **Note 1**) dissolved at 10 mM in H_2O.
2. High-fidelity DNA polymerase with supplied reaction buffer (*see* **Note 2**).
3. Deoxynucleotide triphosphates (dNTPs): 10 mM in H_2O stored in aliquots at –20 °C.
4. DNA (genomic DNA, plasmid DNA, cDNA) template (*see* **Note 3**).
5. 0.2 mL PCR tubes and lids.
6. Thermocycler for PCR protocol.
7. DNA electrophoresis system.
8. 1% (w/v) agarose gel prepared in 1× TAE (40 mM Tris-Base, 20 mM acetic acid, 1 mM ethylenediaminetetraacetic acid) with a nucleic acid stain.
9. DNA ladder.
10. Gel imager for viewing DNA gels and collecting images.
11. Kit for PCR clean-up and/or DNA gel extraction kit (*see* **Note 4**).
12. Spectrophotometer for measuring DNA concentration.

2.2 MK Plasmid Assembly

1. Purified PCR product or synthesized DNA fragment and p641-Bpi LI plasmid for LI assembly (*see* **Note 5**).
2. LI plasmids with DNA parts and LII MK backbone for LII assembly.
3. Assembled LII plasmids and LIII MK backbone for LIII assembly.
4. 10 U/μL BpiI enzyme with supplied reaction buffer for LI and LIII assembly (*see* **Note 6**).
5. 20 U/μL BsaI enzyme with supplied reaction buffer for Level II assembly (*see* **Note 7**).

6. 400 U/μL T4 DNA ligase (*see* **Note 8**).
7. ATP: 10 mM in H_2O stored in aliquots at –20 °C.
8. Thermocycler for cut-ligation protocol.
9. Competent *Escherichia coli* cells such as TOP10 or NEB 10-beta (*see* **Note 9**).
10. Super Optimal broth with Catabolite repression (SOC) medium: 2% (w/v) Bacto-tryptone, 0.5% (w/v) yeast extract, 10 mM NaCl, 2.5 mM KCl, 10 mM $MgCl_2$, 10 mM $MgSO_4$, and 20 mM glucose.
11. ThermoMixer or a water bath.
12. Luria-Bertani (LB) medium: 1% (w/v) Bacto-tryptone, 0.5% (w/v) yeast extract, 1% (w/v) NaCl in deionized water, autoclave to sterilize. For plates, 1.5% (w/v) agar is added before autoclaving.
13. Antibiotics: Gentamicin: 15 mg/mL in H_2O; Spectinomycin: 50 mg/mL in H_2O; Kanamycin: 50 mg/mL in H_2O. Filter-sterilize stock solutions and store aliquots at –20 °C. Dilute antibiotics 1:1000 (final concentration of 15 μg/mL for Gentamicin, 50 μg/mL for Spectinomycin, 50 μg/mL for Kanamycin) for agar plates or liquid media.
14. Shaker-incubator for overnight bacterial cultures.
15. Kit for plasmid DNA purification (*see* **Note 10**).
16. Spectrophotometer for measuring DNA concentration.
17. Restriction enzymes to digest plasmids for verification (*see* **Note 11**).
18. Agarose gel, DNA electrophoresis system, and gel imager as outlined in Subheading 2.1 for visualizing restriction digests.

3 Methods

To introduce a new DNA part into the MK system, an LI part entry clone is generated in the p641-BpiI LI backbone (Fig. 1a). LI parts are either synthesized or amplified by PCR using primers containing BpiI and BsaI restriction sites and overhangs (Fig. 1c). The PCR amplicon is purified and inserted into p641-BpiI LI using a BpiI cut-ligation (alternating incubation temperatures of 37 °C and 16 °C) protocol. LI part plasmids are then verified by restriction digest and DNA sequencing.

3.1 Construct LI Parts by PCR

Visualize the part DNA sequence with computer software to identify any BpiI or BsaI restriction enzyme sites. If the sequence does not contain BpiI or BsaI sites, proceed with designing oligonucleotides for amplification. If the template contains BpiI or BsaI

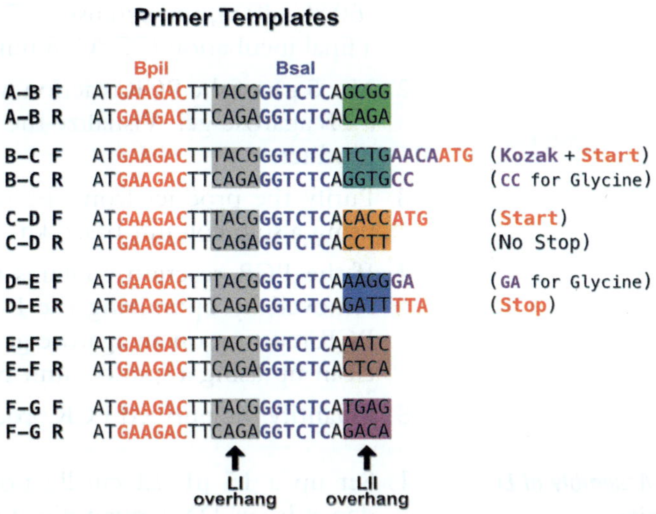

Fig. 4 Primer design for entry cloning. When amplifying a DNA part for LI cloning by PCR primer, conserved templates are added to the 5′ end of the oligo. A BpiI restriction site is included with the indicated LI overhang (gray) for directional cloning into p641-BpiI. Each primer also contains a BsaI restriction site along with the LII overhang (colored) to delineate the part position. A eukaryotic Kozak sequence or a ribosomal binding site can be added to 5′ to the start codon for B-C parts (N-terminal tags). The stop codon for a C-D part (gene of interest) is removed by default, while the reverse primer for D-E parts (C-terminal tag) includes a stop codon. Two nucleotides are added to the B-C reverse (CC) and D-E forward primers (GA) to maintain the reading frame when making protein fusions (adds a glycine residue)

restriction enzyme sites, modify the DNA sequence to remove the site(s) while preserving the triplet codons (silent mutations). If cloning by PCR, design overlapping primers to introduce the mutation(s) [15, 19]. If the sequence will be synthesized, change the appropriate base(s) manually. Add the appropriate restriction site overhangs to the 5′ and 3′ ends of the oligonucleotides (*see* **Note 1**) to include both BpiI and BsaI recognition sites and the corresponding overhang sequences (Fig. 4). Next, the primer includes ideally 20–22 bp of sequence that corresponds to the target amplicon with a GC content close to 50%. If the DNA sequence will be synthesized, manually add the BpiI and BsaI recognition sites and overhang sequences and proceed with ordering.

1. Set up a 50 μL PCR reaction in a tube containing both primers (2.5 μL each), 10 mM dNTPs (1 μL), 5X Q5 buffer (10 μL), Q5 DNA polymerase (0.5 μL), template (plasmid DNA, cDNA, or genomic DNA), and deionized water (to 50 μL). Run a protocol that includes an initial denaturation (98 °C, 30 s), 30–36 cycles of denaturation (98 °C, 10 s), annealing

(60 °C, 30 s), and extension (72 °C, 20–30 s/kb), followed by a final incubation (72 °C, 5 min) in a thermocycler.

2. Mix 5 μL of the PCR reaction with loading dye and separate on a 1% agarose gel. Visualize the gel and assess the purity of the product.
3. Purify the product from the remaining 45 μL PCR reaction using a spin column-based kit.
4. If the PCR reaction contains multiple bands that cannot be removed by optimizing the PCR, load the remaining 45 μL PCR reaction on an agarose gel, excise the band, and perform clean-up using a spin column-based kit or other method.
5. Quantify the purified PCR product by spectrophotometry.

3.2 Assembly of LI Plasmids

1. Set up a 15 μL LI cut-ligation by pipetting H_2O (amount depends on DNA quantities used), 1.5 μL of enzyme buffer (1X final concentration), 1.5 μL of 10 mM ATP, 0.75 μL BpiI, and 0.75 μL of T4 ligase into a PCR tube. Then pipette 100 ng of the purified PCR product (or synthesized DNA) and 100 ng of the p641-BpiI LI backbone plasmid into the tube.
2. Use a quick spin centrifuge to collect the contents and then vortex briefly. Repeat the quick spin to again collect the reaction at the bottom of the tube.
3. Set up the following cycling protocol in a thermocycler: 10 min at 37 °C, 10 min at 16 °C; repeat five times. The six cycles are followed by an incubation at 65 °C for 20 min to heat-kill the BpiI and T4 ligase enzymes.
4. Obtain a 50 μL aliquot (in a 1.5 mL tube) of frozen chemically competent TOP10 or NEB 10-beta cells and store on ice. Pipette 5 μL of the cut-ligation reaction into the tube and incubate on ice for 20–25 min.
5. Transfer the tube to a ThermoMixer block or water bath set at 42 °C for 1 min.
6. Place the tube back on ice for 3 min to recover.
7. Add 350 μL of SOC media and place back in the ThermoMixer block set at 37 °C with shaking at 850 revolutions per minute (RPM) for 60 min.
8. Pipette 100 μL of the cells onto an LB agar plate containing gentamicin.
9. Incubate at 37 °C overnight in an incubator.
10. Pick two colonies from the plate and transfer each to 3 mL of liquid LB media containing gentamicin (15 μg/mL) in a culture tube.
11. Incubate the cultures at 37 °C overnight with shaking (180 RPM).

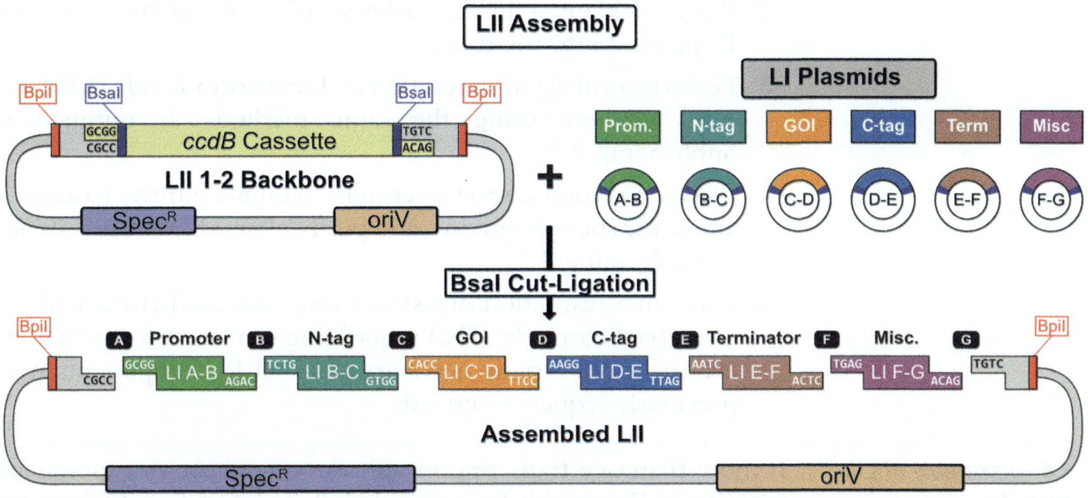

Fig. 5 Assembling LII plasmids. LI plasmids containing DNA parts (A-B, B-C, C-D, D-E, E-F, F-G positions) are mixed with an LII backbone in a BsaI cut-ligation to generate an LII plasmid. The MK system includes LI dummy parts to fill in any unused positions. All LII backbones contain spectinomycin resistance (Spec^R) and *ccdB* negative-selection cassettes

12. Isolate the plasmid DNA from the cultures and quantify the eluted plasmid by spectrophotometry.
13. Perform a restriction digest to verify that the plasmid contains the DNA part.
14. Sequence the inserted DNA part for final confirmation by Sanger sequencing or alternative platform.

3.3 Assembly of LII Plasmids

LI parts are assembled into transcription units using an LII backbone and a BsaI cut-ligation protocol (Fig. 5). A transcription unit of up to six components (positions A-B, B-C, C-D, D-E, E-F, F-G) can be assembled. Any positions that are not required can be filled with dummy parts containing short sequences of DNA[15]. Parts generally correspond to one position (e.g., a promoter in A-B); however, custom overhangs can also be used to bridge positions if desired (e.g., B-E). Typically, these plasmids are verified by restriction digest, then either used directly or combined with other LII plasmids to generate higher-order assemblies. Before proceeding with assembly, a software program is used to verify that the parts will assemble correctly in silico. This ensures that all parts contain compatible overhangs and will insert into the plasmid backbone correctly.

1. Set up a 15 μL LII cut-ligation by pipetting H_2O (amount depends on DNA quantities used), 1.5 μL of enzyme buffer (1X final concentration), 1.5 μL of 10 mM ATP, 0.75 μL BsaI, and 0.75 μL of T4 ligase into a PCR tube.

2. Pipette 100 ng of the LII backbone plasmid and 100 ng of each LI plasmid into the tube.
3. Perform a cut-ligation reaction and transform *E. coli* TOP10 or NEB 10-beta using the same methods as outlined in Subheading 3.2.
4. Pick two colonies, grow overnight in liquid cultures (using the antibiotic spectinomycin), and purify plasmid DNA as outlined in Subheading 3.2.
5. Perform a restriction digest to verify that the LII plasmid has assembled correctly. DNA sequencing can be performed; however, it is not strictly necessary since all LI components were previously sequence verified.

3.4 Assembly of LIII Plasmids

If more than one transcription unit is required for an experiment, up to five LII plasmids (positions 1-2, 2-3, 3-4, 4-5, 5-6) from any backbone series can be assembled in an LIII plasmid backbone using the enzyme BpiI (Fig. 6). Empty positions can be filled with LII dummy parts if less than five transcription units will be assembled [15]. The workflow to assemble an LIII plasmid is the same as assembling an LII plasmid. The only differences are that the enzyme BpiI (instead of BsaI) is used for the cut-ligation reaction and kanamycin (instead of spectinomycin) is added to agar plates and liquid media.

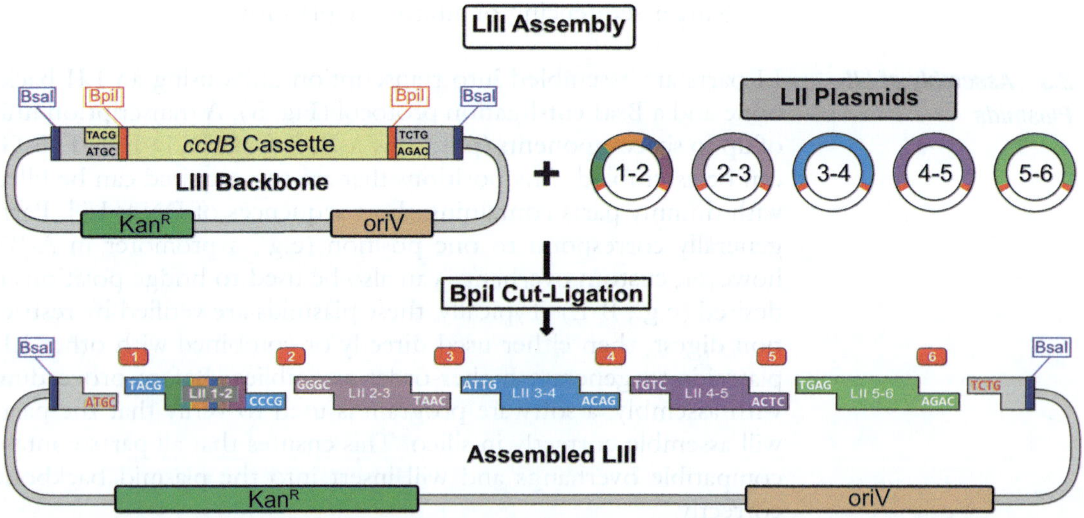

Fig. 6 Assembling LIII plasmids. To assemble an LIII plasmid, LII plasmids (1-2, 2-3, 3-4, 4-5, 5-6 positions) and an LIII backbone are mixed in a BpiI cut-ligation. The MK system includes LII dummy parts to fill in any unused positions. All LIII backbones contain kanamycin resistance (Kan^R) and *ccdB* negative-selection cassettes

4 Notes

1. DNA oligonucleotides are designed according to Fig. 4 for amplifying a DNA part from a template such as a plasmid, cDNA, or genomic DNA. Starting at the 5′ end of the oligo, two random bases (AT) are added such that the restriction enzyme does not cut at the very end of the DNA strand (less efficient). A BpiI recognition site is added (GAAGAC), followed by two bases (TT), and then a 4 bp custom overhang sequence for ligation into p641-BpiI LI (TACG for the forward primer, CAGA for the reverse primer). Each oligo then contains a BsaI recognition site (GGTCTC) followed by a random base (A) and then the custom 4 bp overhang sequence for the LI part position. The forward primer for a B-C part may contain a Kozak sequence for eukaryotes or a ribosome binding site (RBS) for bacteria to ensure efficient translation along with an ATG start sequence. The reverse primer for a B-C fusion tag and the forward primer for a D-E protein fusion tags contain two extra bases (CC for B-C R primer, GA for D-E F primer) to ensure the reading frame remains intact between fusion partners (since overhangs contain four bases). In general, C-D parts contain an ATG start sequence and no stop codon to allow for C-terminal fusions, although a stop codon can be added. If no stop codon is present and no C-terminal fusion is desired, a dummy part with stop codons is assembled in the D-E position. C-terminal tags cloned into the D-E position contain a stop codon (TAA, TGA, or TAG).
2. Any proofreading high-fidelity DNA polymerase can be used. *Taq* polymerase is not adequate for cloning parts due to the relatively high mutation rate.
3. PCR using a DNA plasmid as a template will yield a cleaner product, while a complex template such as genomic DNA may yield unwanted bands.
4. For relatively clean PCR products (single, expected-sized amplicon), a PCR clean-up kit can be used to purify the DNA. If unwanted bands are present in the PCR product, either optimize the PCR conditions (e.g., modify annealing temperature) or excise the correct product from the agarose gel and purify with a kit.
5. Empty LI, LII, and LIII backbone plasmids are propagated in a *ccdB*-resistant strain of *E. coli* such as DB3.1. DB3.1 is no longer commercially available from Thermo Fisher Scientific but can be obtained from the Belgian Coordinated Collection of Microorganisms (BCCM). We have found the *ccdB*-resistant *E. coli* strain *ccdB* Survival 2 T1 to be less consistent for propagating backbone plasmids.

6. BpiI (BbsI) from Thermo Fisher Scientific with 10X Buffer G is used for assembling LI and LIII plasmids. NEB now supplies BbsI-HF® with 10X rCutSmart buffer. Our initial testing has found that BbsI-HF also works well for assembly.
7. BsaI-HF v2 10X rCutSmart buffer from NEB is used for assembling LII plasmids. We have not tested BsaI from other suppliers.
8. T4 DNA ligase from NEB is used for ligating DNA parts. T4 ligase is active in most restriction enzyme buffers as long as 1 mM ATP is present. Therefore, the optimum buffer for the restriction enzyme is used in cut-ligation reactions rather than 10X T4 DNA ligase buffer.
9. The *ccdB*-susceptible *E. coli* strains TOP10 (Thermo Fisher Scientific), DH10β (Thermo Fisher Scientific), or NEB 10-beta (New England Biolabs) are commonly used for selecting assembled plasmids. These strains can be purchased or prepared with high competency for transformation and tolerate large DNA plasmids (>20 kb) such as assembled LIII clones.
10. We have used the NucleoSpin Plasmid Mini kit from Macherey-Nagel for purifying plasmids from *E. coli*. We have also had success using plasmid miniprep kits from other suppliers, such as NEB, Thermo Fisher Scientific, and Zymo Research.
11. Isolated plasmids are digested with restriction enzymes to verify assembly. BsaI can be used to digest LI and LIII plasmids and BpiI for LII plasmids. Further verification of the inserted DNA and backbone can be obtained by digesting with one or more restriction enzymes that cut both the backbone and the cargo.

Acknowledgment

This work was supported by a Discovery Grant from the Natural Sciences and Engineering Research Council (NSERC) to D.C. and an NSERC Undergraduate Student Research Award (USRA) to S. S.

References

1. Engler C, Kandzia R, Marillonnet S (2008) A one pot, one step, precision cloning method with high throughput capability. PLoS One 3(11):e3647
2. Andreou AI, Nakayama N (2018) Mobius Assembly: A versatile Golden-Gate framework towards universal DNA assembly. PLoS One 13(1):e0189892
3. Wu D, Schandry N, Lahaye T (2018) A modular toolbox for Golden-Gate-based plasmid assembly streamlines the generation of *Ralstonia solanacearum* species complex knockout

strains and multi-cassette complementation constructs. Mol Plant Pathol 19:1511–1522

4. Moore S, Lai J, Kelwick R, Chee S, Bell D, Polizzi K, Freemont P (2016) EcoFlex: A Multifunctional MoClo Kit for *E. coli* Synthetic Biology. ACS Synth Biol 5(10):1059–1069
5. Iverson SV, Haddock TL, Beal J, Densmore DM (2016) CIDAR MoClo: Improved MoClo Assembly Standard and New *E. coli* Part Library Enable Rapid Combinatorial Design for Synthetic and Traditional Biology. ACS Synth Biol 5(1):99–103
6. Blázquez B, León DS, Torres-Bacete J, Gómez-Luengo Á, Kniewel R, Martínez I, Sordon S, Wilczak A, Salgado S, Huszcza E, Popłoński J, Prieto A, Nogales J (2023) Golden Standard: a complete standard, portable, and interoperative MoClo tool for model and non-model proteobacteria. Nucleic Acids Res 51(19):e98
7. Lee ME, DeLoache WC, Cervantes B, Dueber JE (2015) A Highly Characterized Yeast Toolkit for Modular. Multipart Assembly. ACS Synth Biol 4(9):975–986
8. Agmon N, Mitchell LA, Cai Y, Ikushima S, Chuang J, Zheng A, Choi W-J, Martin JA, Caravelli K, Stracquadanio G, Boeke JD (2015) Yeast Golden Gate (yGG) for the Efficient Assembly of *S. cerevisiae* Transcription Units. ACS Synth Biol 4(7):853–859
9. Perez-Gonzalez A, Kniewel R, Veldhuizen M, Hemant K, Navarro-Rodriguez M, Rubio L, Caro E (2017) Adaptation of the GoldenBraid modular cloning system and creation of a toolkit for the expression of heterologous proteins in yeast mitochondria. BMC Biotechnol 17:80
10. Hernanz-Koers M, Gandía M, Garrigues S, Manzanares P, Yenush L, Orzaez D, Marcos JF (2018) FungalBraid: A GoldenBraid-based modular cloning platform for the assembly and exchange of DNA elements tailored to fungal synthetic biology. Fungal Genet Biol 116:51–61
11. Martella A, Matjusaitis M, Auxillos J, Pollard SM, Cai Y (2017) EMMA: An Extensible Mammalian Modular Assembly Toolkit for the Rapid Design and Production of Diverse Expression Vectors. ACS Synth Biol 6(7): 1380–1392
12. Duportet X, Wroblewska L, Guye P, Li Y, Eyquem J, Rieders J, Rimchala T, Batt G, Weiss R (2014) A platform for rapid prototyping of synthetic gene networks in mammalian cells. Nucleic Acids Res 42(21):13440–13451
13. Lampropoulos A, Sutikovic A, Wenzl C, Maegele I, Lohmann J, Forner J (2013) GreenGate - a novel, versatile, and efficient cloning system for plant transgenesis. PLoS One 8(12): e83043
14. Weber E, Engler C, Gruetzner R, Werner S, Marillonnet S (2011) A modular cloning system for standardized assembly of multigene constructs. PLoS One 6(2):e16765
15. Binder A, Lamber J, Morbitzer R, Popp C, Ott T, Lahaye T, Parniske M (2014) A modular plasmid assembly kit for multigene expression, gene silencing and silencing rescue in plants. PLoS One 9:e88218
16. Engler C, Youles M, Gruetzner R, Ehnert T-M, Werner S, Jones JDG, Patron N, Marillonnet S (2014) A golden gate modular cloning toolbox for plants. ACS Synth Biol 3(11): 839–843
17. Sarrion-Perdigones A, Vazquez-Vilar M, Palaci J, Castelijns B, Forment J, Ziarsolo P, Blanca J, Granell A, Orzaez D (2013) GoldenBraid 2.0: a comprehensive DNA assembly framework for plant synthetic biology. Plant Physiol 162(3):1618–1631
18. Gantner J, Ordon J, Ilse T, Kretschmer C, Gruetzner R, Lofke C, Dagdas Y, Burstenbinder K, Marillonnet S, Stuttmann J (2018) Peripheral infrastructure vectors and an extended set of plant parts for the Modular Cloning system. PLoS One 13(5):e0197185
19. Chiasson D, Giménez-Oya V, Bircheneder M, Bachmaier S, Studtrucker T, Ryan J, Sollweck K, Leonhardt H, Boshart M, Dietrich P, Parniske M (2019) A unified multi-kingdom Golden Gate cloning platform. Sci Rep 9(1):10131
20. Martínez-García E, Fraile S, Algar E, Aparicio T, Velázquez E, Calles B, Tas H, Blázquez B, Martín B, Prieto C, Sánchez-Sampedro L, Nørholm MHH, Volke DC, Wirth NT, Dvořák P, Alejaldre L, Grozinger L, Crowther M, Goñi-Moreno A, Nikel PI, Nogales J, de Lorenzo V (2023) SEVA 4.0: an update of the Standard European Vector Architecture database for advanced analysis and programming of bacterial phenotypes. Nucleic Acids Res 51(D1): D1558–D1567

Chapter 27

Advancements in Golden Gate Cloning: A Comprehensive Review

Jesús Laborda-Mansilla and Eva García-Ruiz

Abstract

Researchers have dedicated efforts to refining genetic part assembly techniques, responding to the demand for complex DNA constructs. The optimization efforts, targeting enhanced efficiency, fidelity, and modularity, have yielded streamlined protocols. Among these, Golden Gate cloning has gained prominence, offering a modular and hierarchical approach for constructing complex DNA fragments. This method is instrumental in establishing a repository of reusable parts, effectively reducing the costs and proving highly valuable for high-throughput DNA assembly projects. In this review, we delve into the main protocol of Golden Gate cloning, providing refined insights to enhance protocols and address potential challenges. Additionally, we perform a thorough evaluation of the primary modular cloning toolkits adopted by the scientific community. The discussion includes an exploration of recent advances and challenges in the field, providing a comprehensive overview of the current state of Golden Gate cloning.

Key words DNA assembly, Golden Gate cloning, Modular cloning, Synthetic biology, Type IIS restriction endonucleases (TIIS-RE)

1 Introduction

In molecular and synthetic biology, DNA assembly methods play a crucial role in generating diverse DNA constructs, from genes and expression systems to larger structures like artificial chromosomes and small genomes [1–3]. A plethora of available techniques like restriction digestion–ligation [4–6], homology-based assembly [2, 7, 8], and oligo-based assembly [9–11] offer a spectrum of strategies for assembling DNA fragments. Among these, the Golden Gate [12, 13] method (GG) stands out for its robustness, modular nature, speed, and ability to accommodate multiple fragments in a single reaction with minimal error rates, making it amenable to automation [14] and highly efficient in generating complex recombinant DNA constructs [1].

Daniel Schindler (ed.), *Golden Gate Cloning: Methods and Protocols*, Methods in Molecular Biology, vol. 2850,
https://doi.org/10.1007/978-1-0716-4220-7_27,

Since its inception in 2008, GG [12] cloning has become a fundamental standard cloning protocol. Its significance lies in its ability to assemble large DNA constructs through a one-pot reaction, combining smaller components. This streamlined approach, characterized by simultaneous digestion and ligation, provides a crucial advantage in terms of efficiency and speed, establishing GG cloning as an indispensable tool for researchers in the field.

This highly efficient method, conducted in a single reaction, relies on type IIS restriction endonucleases (TIIS-RE) [15] for digestion. These enzymes generate overhangs at a specified and directed distance from the recognition site, serving as fusion sites during assembly. The resulting DNA fragments, lacking the original recognition sites, can be ligated into a product without reconstructing the initial restriction site, thereby ensuring its accumulation over time. Importantly, the precise design of overhang sequences facilitates highly complex assemblies in a predetermined order with a low error rate [13, 16, 17].

The GG strategy demonstrates remarkable versatility, serving either as a "standalone" method or integrated within modular cloning toolkits. In the construction of complex synthetic pathways through the assembly of modular DNA parts [18], GG proves highly valuable. Its application extends to systematic protein engineering [19, 20], allowing for precise gene and protein modifications that introduce novel functionalities. Moreover, GG plays a pivotal role in combinatorial assembly, significantly contributing to the optimization of outcomes [21]. Its application also extends to *ad hoc* assembly of highly complex DNA constructs, exemplified in scenarios such as assembling small genomes [1].

When employed as a "standalone" method, critical factors require thorough considerations, particularly emphasizing the importance of a meticulous design of overhangs, ensuring the maximization of fidelity and efficiency throughout the assembly process [17, 22]. In contrast, when GG is utilized within a modular cloning kit, these kits function as repositories for libraries of domesticated biological parts housed in vectors. Here, the design of overhangs is bound by a pre-defined syntax, redirecting the attention toward a meticulous design of the assembly workflow.

The expanding variety of available modular cloning toolkits adds complexity, with each offering unique advantages and meeting specific needs. Navigating this diversity can be challenging when selecting the most suitable toolkit for a given project [23]. Our exploration extends beyond the fundamental protocol of GG, offering practical insights and recommendations for troubleshooting and optimizing reactions. These suggestions aim to streamline the design process and enhance the efficiency of Golden Gate cloning experiments. Additionally, we conduct a thorough evaluation of various modular cloning toolboxes, while also addressing recent advances and challenges in the field.

2 Stepwise Description of Golden Gate Cloning

In the following sections of this review, we will frequently mention terms such as "TIIS-RE," "TIIS-RE site," "TIIS-RE recognition site," "TIIS-RE cleavage site," "fusion site," "overhang," and "spacer." It is essential to note that "TIIS-RE" specifically refers to the restriction enzyme, while "TIIS-RE site" denotes the DNA sequence recognized, and cleaved by the TIIS-RE. The "TIIS-RE recognition site" refers specifically to the specific sequence recognized by the enzyme, where cleavage occurs outside of this sequence. The "TIIS-RE cleavage site" indicates the location where the enzyme cuts the double-stranded DNA[15]. After cleavage, a single-stranded DNA (ssDNA) fusion site, also known as an overhang or sticky end, is generated for assembly. Spacer nucleotides refer to the intervening nucleotides between the recognition site and the cleavage site.

Furthermore, to illustrate these concepts, we provide an example using the BbsI endonuclease, which recognizes the sequence 5′ GAAGAC N^6N^5 $\downarrow N^4N^3N^2N^1$ 3′ (the arrow indicates the place where the enzyme cuts the double-strand DNA, and N refers to any of the four possible nucleotides), consisting of a 6-nucleotide (6-nt) recognition sequence (GAAGAC), a 2-nt spacer (N^6N^5), and a 4-nt cleavage site ($\downarrow N^4N^3N^2N^1$). Following cleavage, a 4-nt ssDNA fusion site is generated for assembly. After ligation of the compatible ssDNA fusion sites, the restriction site and spacer nucleotides are not present in the final DNA construct.

We will utilize these terms consistently throughout the text to facilitate understanding of the Golden Gate cloning process.

2.1 Design of DNA Fragments

In the design of DNA fragments, an essential requirement for achieving the correct ligation involves the presence of compatible fusion sites at their ends. This necessitates incorporating the recognition site of TIIS-RE at both ends of each DNA fragment, ensuring the cleavage site occurs inward to the recognition sites (Fig. 1a). The distinctive advantage of TIIS-RE lies in its flexibility, allowing the cleavage site sequence to be any desired nucleotide. Consequently, the overhang at the end of one fragment can be designed to align with the overhang at the beginning of the adjacent fragment, thereby facilitating the assembly of multiple fragments at desired locations and directions (Fig. 1b).

2.2 Selection of the Restriction Enzyme

Ensuring a precise assembly process in GG cloning demands meticulous design, beginning with careful TIIS-RE selection. It is crucial to avoid additional TIIS-RE recognition sites in fragment sequences, preventing internal cleavage that hinders full-length construct formation. To address this, it is recommended to ensure the chosen TIIS-RE recognition sequence is absent in the desired

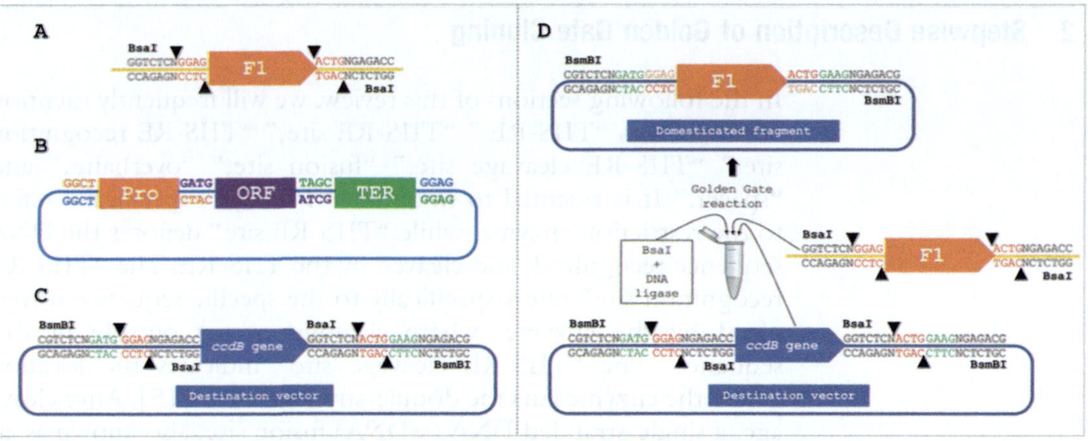

Fig. 1 Detailed representations of designed fusion sites and destination vectors in Golden Gate cloning. (**a**) Domestication of biological parts. The recognition site (in black) of the TIIS-RE will be incorporated at both ends of each DNA fragment (F1, Fragment1), ensuring that the cleavage site (in red) occurs inward to the recognition sites. (**b**) DNA fragments are assembled based on the compatibility of fusion site; the overhang at the end of one fragment can be designed to align with the overhang at the beginning of the adjacent fragment, facilitating the assembly of all fragments at the desired locations and directions in a destination vector. PRO, promoter; ORF, open reading frame; TER, terminator. (**c**) Destination vector containing a cloning marker such as *ccdB* gene flanked by two couple of TIIS-RE sites. With one site facing outward (BsaI), it aids in the integration of the assembled part into the destination vector, while the other, facing inward (BsmBI), facilitates the release of the assembled module for subsequent cloning steps. (**d**) Representation of Golden Gate assembly for standardization of parts. The overhangs represented here do not follow any particular syntax

assembly sequence. Alternatively, eliminate these recognition sites through mutagenesis [17] or optimize the sequence [22, 24] and synthesize it without the recognition sites.

Common GG workflows often employ TIIS-REs with a 6-nt recognition sequence and a 4-nt overhang, such as BsaI, BsmBI (with its isoschizomer Esp3I), BbsI (with its isoschizomer BpiI), and BtgZI. Additionally, 7-nt TIIS-REs, like AarI (with its isoschizomer PaqCI), and others such as SapI (with its isoschizomer BspQI), have 4-nt and 3-nt overhangs, respectively. These enzymes have proven successful in GG cloning [12, 25–27]. In the absence of a modular cloning toolkit where TIIS-RE and overhang sequences are pre-defined, there is flexibility in selecting these elements.

When selecting the TIIS-RE for non-modular toolkit assembly, consider that longer recognition sites decrease their probability within a desired DNA sequence. For instance, considering a 6-nt TIIS-RE recognition site, the estimated probability of its presence in a random 2-kb sequence of DNA is ~39% [22]. This probability decreases to ~12% with a 7-nt TIIS-RE site (based on the same equation used by [22]).

Considerations regarding overhang length involve a trade-off between achieving seamless assembly and maximizing combinatorial space. While a 3-nt overhang can facilitate seamless assembly, especially at positions like the start and stop codons, ensuring the preservation of the open reading frame [27, 28], it comes at the cost of reduced combinatorial space. However, despite this expected limitation, a recent study showcased the successful assembly of 13 fragments utilizing 3-nt overhangs, achieving >90% fidelity [17]. In contrast, 4-nt overhangs offer a broader spectrum of combinations compared to 3-nt overhangs. The utilization of 4-nt overhangs can approach a fidelity close to 90% when assembling 24 fragments [29]. Therefore, in scenarios involving a substantial number of fragments requiring accurate assembly, the recommendation leans toward the adoption of 4-nt overhangs.

2.3 Design of Overhangs

The selection of the sequence for GG fusion sites significantly impacts assembly success. Careful design of these nucleotides is critical to maximize fidelity and efficiency. Initially, a set of fundamental guidelines was established to minimize the risk of ligating imperfectly base-paired partners during an assembly reaction [30, 31].

These guidelines include avoiding self-complementary palindromic overhangs and prohibiting the same overhang within a single assembly reaction. Moreover, it is generally advised to maintain at least a minimum two-base pair difference between overhangs, underscoring that reduced similarity in the sequence correlates with a lower likelihood of ligation. Overhangs exclusively composed of A/T nucleotides or those with a G/C content of 25% are often avoided, as they are thought to possess weak joining abilities [17, 30].

Recent studies have significantly enhanced our understanding of the factors affecting a successful GG multifragment assembly [17, 22, 29]. Emerging computational web applications such as NEB webtool suite (NEBridge Golden Gate, https://goldengate.neb.com/, and NEBridge Ligase Fidelity, https://ligasefidelity.neb.com/) [17], iBioCAD GGA (https://ibiocad.igb.illinois.edu/) [22], and EGF CUBA (https://cuba.genomefoundry.org/) offer a breakthrough in crafting GG assemblies. These computational webtools facilitate the design of overhang combinations for the desired number of fragments to assemble that enhance fidelities and efficiencies [17, 22]. Moreover, these web applications can integrate the elimination of TIIS-RE recognition sites present in the sequence of interest, all within one single step.

Illustrating the capabilities of one of these computational online tools (NEB webtool suite), Pryor et al. effectively assembled 52 DNA fragments in a one-pot assembly [1]. The study evaluated the assembly of a 52-fragment 4.9 kb lactose operon and the assembly of a 52-fragment 40 kb T7 phage genome. Detection of correct assemblies relied on the functionality of the assembled DNA. For the T7 phage genome assembly, a rigorous selection method identified the correct assemblies. Functional T7 phage infectious particles were exclusively produced solely in the case of correct assemblies, resulting in the formation of discernible phage plaques on an *Escherichia coli* lawn. Importantly, incorrect assemblies failed to produce infectious particles, preventing any inadvertent selection. In this scenario, the assembling fidelity was notably high (~100%), thanks to the stringent selection approach employed [1].

In contrast, the 52-fragment lactose operon assembly employed a less stringent selection. Correctly assembled lactose operon into a destination vector produced blue *E. coli* colonies, while white colonies served as indicators of assembly errors. The fidelity in this scenario varied widely between experimental replicates, ranging from 10% to 50% [1]. In a similar study, a 35-fragment lactose operon assembly achieved a fidelity near 70% [17]. The findings led to the suggestion that assemblies exceeding 50 fragments might be approaching the limits of GG cloning capabilities using TIIS-RE with 4-bp fusion sites, even with the improvements from the use of computational webtools. To enhance the fidelity of GG assembly, it is recommended to either reduce the number of fragments or implement a stringent selection approach, as demonstrated in the generation of infectious particles [1].

Overall, using personalized overhang selection with computational online design tools is expected to greatly improve assembly success. This approach is favored over general guidelines, especially when dealing with complex and large assemblies involving multiple fragments [1, 22, 32].

2.4 Addition of TIIS-RE Sites

Given the recent drop in DNA synthesis costs and delivery times, it is feasible to economically synthesize designed DNA fragments on demand. Another option is to leverage one of the many available modular cloning toolkits where many pre-existing parts already cloned in vectors are readily available for immediate use. Although it may be necessary to domesticate additional DNA sequences of interest. Whether adding new elements to a modular cloning kit or assembling a multifragment construct using non-modular cloning kit components, it is essential to include TIIS-RE sites at both ends of each DNA fragment, as explained in the "*Design of DNA Fragments*" section. This can be achieved through polymerase chain reaction (PCR) by incorporating primers containing these sites to ensure proper orientation. Direct use of PCR products in assembly is discouraged due to potential errors from residual DNA polymerase, deoxinucleotides triphosphates (dNTPs), buffer salts, primers, and primer dimers. This issue may arise from the polymerase's capacity to refill the generated overhangs during the GG reaction. Moreover, primer dimers, sharing the same TIIS-RE sites as PCR products, have the potential to be cloned, leading to inaccuracies in the resulting constructs. Similarly, inadequately designed primers may lead to the amplification of unintended regions of the template DNA alongside the target sequence, which contains the TIIS-RE sites [32, 33]. Collectively, these factors emphasize the need for a purification step before utilizing the PCR product in the GG reaction.

2.5 Golden Gate Reaction: Digestion–Ligation in a One-Pot Single Step

In the GG reaction, TIIS-RE, DNA ligase, DNA fragments, and the destination vector are combined in a single tube for simultaneous cutting and ligation. DNA is added in equimolar concentrations of each fragment, preventing preferential ligation and low efficiencies. Accurate quantification of each DNA part is therefore advisable [32].

The widely used protocol entails cycling incubation steps, typically lasting 5–10 min, at two distinct temperatures: 37 °C, optimal for DNA cutting, and 16 °C, optimal for ligation. Subsequently, a final incubation at 50 °C facilitates the cleavage of remaining uncut or re-circularized part-containing vectors [12], given that DNA ligase is inactivated at this temperature. Variations exist; initial molecular cloning toolkits like MoClo [13] performed the incubation at a constant 37 °C. However, subsequent adjustments optimizing temperatures for specific enzymes have proven beneficial, especially when assembling a substantial number of fragments, enhancing overall fidelity and efficiency [13, 32, 33]. Additionally, the efficiency of complex assemblies, such as the T7 phage genome, is influenced by the number of cycles, with a higher recommended number as the count of fragments to assemble increases [1].

Considerations regarding TIIS-RE should be noted. While many TIIS-RE employed in GG cloning demonstrate 100% activity at 37 °C, some may have different optimal temperatures. Ensuring compatibility with DNA ligases is crucial; for example, the commonly used BsmBI shows an optimal temperature of 50 °C, which may hinder ligase activity. In such cases, substituting with its isoschizomer Esp3I, which has a 37 °C optimal temperature, is recommended. In cases where an alternative isoschizomer enzyme is unavailable, finding a temperature that balances the trade-off between TIIS-RE activity and DNA ligase inactivation becomes necessary. Taking the example of BsmBI modular cloning, successful utilization has been achieved through alternating cycles of 42 °C and 16 °C [32, 34]. Additionally, certain TIIS-REs, such as AarI, require binding at two recognition sites, making it necessary to incorporate short sequences as enhancers usually provided by the supplier.

Regarding DNA ligases, a recent study conducted a thorough examination [29, 35]. The study focused on the widely employed T4 DNA ligase, known for its ability to ligate both sticky and blunt ends, and T7 DNA ligase, specialized in ligating only sticky ends. The study revealed that while T7 DNA ligase exhibited a modest improvement in fidelity compared to T4 DNA ligase, it demonstrated a significant rise in bias against specific overhangs, thereby hindering the final yield in multifragment DNA assembly. In summary, these findings emphasize a preference for sticking with T4 DNA ligase for optimal results in complex assemblies [29].

2.6 Verification

The GG reaction is directly transformed into a competent host organism, typically *Escherichia coli*, to obtain and propagate the assembled DNA. Upon confirmation of the correct assembly, the DNA is subsequently introduced into the ultimate host microorganism. However, relying solely on *E. coli* for the recovery of DNA assembled in the GG reaction poses a limitation. If the transcribed and translated DNA in *E. coli* induces toxicity, it can hinder its growth, potentially impeding the progress of the assembly process.

In addition, while the concentration of the construct achieved in the GG reaction is typically suitable for transformation and propagation in *E. coli*, it frequently does not reach the concentration needed for efficient transformation in diverse microorganisms, particularly when constructing a library. Actually, destination vectors in modular cloning kits are *E. coli* shuttle vectors, allowing for replication in both *E. coli* and other host organisms. However, few cases offer the possibility to bypass the use of *E. coli* by directly transforming the raw GG mix into the final host [36, 37]. For instance, when attempting to transform *Vibrio natriegens* with the GG mix, significantly lower transformation efficiency was observed compared to *E. coli* [37]. Developing highly efficient transformation protocols for other microorganisms is crucial, as it would expedite DNA fragment GG assembly, circumvent potential toxicity issues associated with *E. coli*, and broaden the applicability of the technique across diverse microbial hosts.

Finally, confirming the successful assembly typically involves employing established methods like colony PCR, restriction digestion, and sequencing. Sequencing is particularly important when generating initial DNA parts [38]. It is worth noting that once a DNA fragment is cloned into a vector, it can be confidently used for GG cloning, as the GG process itself does not introduce mutation.

After confirming multifragment assembly, the functionality of the constructed DNA is typically assessed within the final host organism to verify its integrity and effectiveness.

2.7 Other Considerations

Optimizing cloning efficiencies for accurate assemblies involves thoughtful design of destination vectors. These vectors should incorporate a counter-selection gene flanked by TIIS-RE sites, which is removed after digestion and ligation. Often, these genes encode for a fluorescent or visible chromophore, facilitating easy visual detection of colonies harboring undigested destination vectors. Alternatively, a more robust selection can be achieved using genes encoding toxicity, such as the *ccdB* gene or the mutant *E. coli pheS* gene, significantly reducing background clones containing uncut destination vectors [30, 39, 40].

Another effective strategy to reduce significantly the background of uncut vector involves adding a nuclease after the GG reaction. For instance, using ATP-dependent plasmid safe nuclease or exonuclease V, along with fresh TIIS-RE, can effectively linearize undigested vectors and eliminate all linear components [30, 40].

3 Modular Cloning Kits

Modular cloning toolkits based on GG are powerful tools in synthetic biology. These tools expedite the efficient and straightforward DNA assembly of multifragment vectors by harnessing a well-characterized library of genetic parts. The inherent versatility of these toolkits facilitates hierarchical assembly, leveraging reusable basic DNA parts in a rapid, reliable, and automation-friendly manner. This approach not only enhances predictability but also ensures an economically viable and practical process, underscoring their significance in the systematic programming of novel cellular behaviors [12, 13].

Modular hierarchical systems (reviewed elsewhere [23]) encompass multiple cloning levels, each pre-defining a set of fusion sites or overhangs. The initial stage involves the domestication of individual biological components like origins of replication, selectable markers, promoters, ribosome binding sites, protein tags, transcriptional terminators, and coding sequences, to name but a few. These elements are integrated into destination vectors flanked by TIIS-RE sites with distinctive fusion sites. The assembly of biological parts into destination vectors follows an organized, directional schedule, enabling effortless interchangeability. This initial step holds the advantage of providing immediate availability for the incorporation of constructed genetic parts into novel DNA constructs. It eliminates the need to initiate the assembly of components de novo, thereby minimizing additional efforts and cost, such as PCR amplification, purification, and verification.

This systematic approach results in standardized biological parts for various organisms, including bacteria (e.g., *Escherichia coli* [27, 41–43], *Vibrio natriegens* [37], and *Zymomonas mobilis* [44], cyanobacteria [45]), yeast (e.g., *Saccharomyces cerevisiae* [21, 34, 46, 47], *Pichia pastoris* [48], *Yarrowia lipolytica* [49], *Kluyveromyces marxianus* [50], and *Schizosaccharomyces pombe* [51]), filamentous fungi (e.g., *Penicillium digitatum* [52], *Trichoderma* [53], and *Aspergillus niger* [54]), algae (e.g., *Chlamydomonas reinhardtii* [55]), diatoms (e.g., *Phaeodactylum tricornutum* [56]), amoebae (*Dictyostelium discoideum* [57]), plants [13, 16, 18, 26, 58], insect cells [59], and mammalian cells

[60, 61]. Moreover, certain toolkits target specific organelles, such as yeast mitochondria [62] and chloroplast [63].

Following this, domesticated biological elements undergo a second GG cloning, often labeled as Level 1 assembly, where a set of parts is assembled into a new destination vector, generating a module that typically is represented by a transcriptional unit (TU) [13, 64].

The destination vectors feature a cloning marker surrounded by two unique TIIS-RE sites. One TIIS-RE site facing inward, it aids in the integration of the assembled module into the destination vector, while the other TIIS-RE site, facing outward, facilitates the release of the assembled module for subsequent cloning steps (Fig. 1c, d). This sequential process allows the assembled module to be utilized in a third cloning step, often labeled as Level 2 assembly, contributing to the generation of larger DNA constructs generally represented by a multigene pathway [13, 21]. Some toolkits support an infinite iterative GG cloning approach, involving alternation between destination vectors and TIIS-RE, thereby enabling the creation of extensive DNA constructs [25, 65].

Pioneering the field, the initial work on MoClo [13] and GoldenBraid2.0 [26] established foundational rules for usage and extensions, aiming to create a distinctive "syntax" [66]. These toolkits share fusion sites (partially) and employ BsaI for Level 1 assembly (generating modules or TUs), but differ in Level 2 assembly, using either BbsI (BpiI) [13] or BsmBI (Esp3I) [26]. Therefore, for newly added biological elements to be fully valuable to the scientific community and seamlessly integrated into these toolkits, adherence to this syntax is essential. Despite these efforts toward standardization and flexibility, subsequent versions and novel systems have introduced modifications to the cloning scheme to meet different purposes. The diversity between toolkits is notably observed in fusion site sequences [64], cloning markers [25], destination vectors [67], and enzyme sets at each assembly level [68].

Other adopted toolkits that maintain the same syntax as MoClo and GoldenBraid2.0 mainly focused on generating parts and destination vectors for engineering-specific organisms beyond plants, like CIDAR [42] for *E. coli*, for *S. cerevisiae* [62], the Marburg Collection [37] for *V. natriegens*, FungalBraid [52] for filamentous fungi, or CyanoGate [45] for cyanobacteria, to name but a few. However, numerous toolkits have been designed with their own syntax, limiting their adoption by the broader research community. Some, like Yeast Tool Kit (YTK) [47], have overcome this limitation and gained widespread acceptance. Being among the first toolkits published for yeast, along with YeastFab [21] and yGG [46], YTK's availability in published repositories and comprehensive characterization of biological parts for yeast, coupled with its inherent flexibility for destination vector design, have contributed to its broad recognition. This is exemplified by recent developments aimed at further expanding its utility for additional clustered regularly interspaced short palindromic repeats (CRISPR)/Cas9 applications [69] or extending its capabilities for multiplexed yeast engineering [34].

Considering the multitude of modular cloning toolkits available, scientists may encounter challenges in selecting the most suitable one for their projects [23]. Table 1 presents the key features of commonly adopted toolkits within the scientific community, providing valuable assistance in the decision-making process.

3.1 Recent Developments

Certain endeavors are directed toward developing a universal GG cloning system to streamline the assembly of DNA constructs for cross-kingdom experimental applications. Noteworthy efforts in this direction include Multi-Kingdom (MK) [70] Golden Gate, which has developed a platform based on the syntax of Modular Plant kit from Binder et al. [58]. This platform aims to facilitate the construction of genetic constructs for use in various species spanning the kingdoms of Fungi (yeast), Eubacteria (gram-negative bacteria), Protista (trypanosomes), Plantae (plants), and Animalia (human cells and frog oocytes). Similarly, uLoop [65] has devised a platform based on the syntax of Loop [68], focusing on generating multigene vectors applicable to diatoms, yeast, plants, and bacteria.

Recent advances emphasize the importance of simplifying recursive cloning for seamless addition of new fragments into existing assemblies. These toolkits are specially designed to iterate between two-level destination vectors, providing a framework for continuous and limitless DNA assembly. Mobius [25] adopts a strategic vector alternation—BsaI during Level 1 and AarI during Level 2 assembly—enhancing assembling capacity by promoting destination vector reuse. Following Level 2 assembly, constructs can iteratively combine into a Level 1 vector using BsaI, then into a Level 2 vector using AarI, creating a recursive process. In addition, here a rare cutting 7-bp TIIS-RE (AarI) diverges from the more prevalent 6-bp TIIS-RE, reducing part domestication demands [25]. Similarly, Loop [68] and its variant, uLoop [65], employ a comparable strategy where destination vectors alternate between BsaI and SapI (a rare cutting 7-bp TIIS-RE generating 3-nt overhangs instead of 4). This design facilitates the generation of larger constructs by iteratively switching between the two series of destination vectors. In contrast, MIDAS [39] introduces destination vectors featuring an additional cassette as a landing pad for multigene assembly. This innovative design permits the incorporation of new modules into an expanding multifragment vector through alternating AarI and BsmBI cloning steps [39]. Likewise, EcoFlex [41], an *E. coli* toolkit with a syntax distinct from MoClo and GoldenBraid2.0, introduced destination vectors containing a secondary module flanked by a third TIIS-RE site (BbsI), diverging from the TIIS-RE used in the main cloning scheme (BsaI and BsmBI) [41].

Table 1
List of Golden Gate modular cloning toolkits

GG toolkit	TIIS-RE L1	TIIS-RE L2	Overhang compatibility	Target organisms	Feature	References
MoClo	BsaI	BbsI=BpiI	Original	Plants	Series of destination vectors and parts.	[13]
Plant Parts Kit	BsaI	BbsI=BpiI	MoClo	Plants	It provides more new elements for plant engineering; also increases the number of destination vectors, allowing more fragments to be assembled.	[18]
GoldenBraid2.0	BsaI, alternative BtgZI	BsmBI=Esp3I	Original and partially compatible with MoClo	Plants	Derived from GoldenBraid, it has been adapted to partially align with MoClo syntax; collection of DNA parts and destination vectors for plant engineering.	[26]
Modular Plant kit	BsaI	BbsI=BpiI	Original	Plants	Collection of plant genetic parts.	[58]
Loop	BsaI	SapI	MoClo, GoldenBraid2.0	Plants	Recursive DNA assembly by alternating between two sets of destination vectors and TIIS-RE.	[68]
MoChlo	BsaI	BbsI=BpiI	MoClo	Chloroplast	It provides a set of parts for chloroplast engineering.	[63]
CIDAR	BsaI	BbsI=BpiI	MoClo	*Escherichia coli*	Extension of MoClo with a collection of DNA parts for *E. coli* and destination vectors.	[42]
EcoFlex	BsaI	BsmBI=Esp3I	Original	*E. coli*	Collection of genetic parts for a number of applications in *E. coli*; destination vectors feature a secondary module for expanded flexibility (addition of new fragments to an assembled construct) using BpiI (BbsI).	[41]

PODAC	BsaI	BsaI	Original	*E. coli*	It features methylated BsaI site to allow iterative cloning using only BsaI sites.	[71]
Mobius	BsaI	AarI	MoClo, GoldenBraid2.0	*E. coli*	Uses a rare cutter for part domestication; it allows iterative cloning by alternating between destination vectors and TIIS-RE.	[25]
Start-Stop	SapI	BsaI	Original	*E. coli*	It allows scarless cloning of open reading frame (ORFs) using 3-nt overhangs.	[27]
GoldenStandard	BsaI	BbsI=BpiI	MoClo, SEVA	Proteobacteria	It enables simple vector modification for increased compatibility between different proteobacteria groups.	[76]
The Marburg collection	BsaI	BsmBI=Esp3I	MoClo, Loop	*Vibrio natriegens*	Basic genetic parts for *V. natriegens* engineering; it enables customization of destination vectors.	[37]
Zymo-parts	BsaI	BbsI=BpiI	Original	*Zymomonas mobilis*	Basic genetic parts for *Z. mobilis* engineering.	[44]
CyanoGate	BsaI	BbsI=BpiI	MoClo	*Synechocystis* sp.	Basic genetic parts for cyanobacteria engineering.	[45]
Yeast Toolkit (YTK)	BsaI	BsmBI=Esp3I	Original	*Saccharomyces cerevisiae*	Collection of genetic parts for yeast engineering; it enables customization of destination vectors.	[47]
MTK	BsaI	BsmBI=Esp3I	YTK	*S. cerevisiae*	Expands the number of parts for yeast engineering; enables new genome integration sites.	[34]
YeastFab	BsaI	BsmBI=Esp3I	Original	*S. cerevisiae*	Collection of genetic parts for yeast engineering, featuring all promoters from yeast.	[40]

(continued)

Table 1 (continued)

GG toolkit	TIIS-RE L1	TIIS-RE L2	Overhang compatibility	Target organisms	Feature	References
yGG	BsaI or BsmBI	–	Original	*S. cerevisiae*	Collection of genetic parts for assembling transcriptional units (TUs).	[46]
Yeast mitochondria	BsaI	BsmBI=Esp3I	GoldenBraid2.0	*S. cerevisiae*	Collection of genetic parts for yeast engineering and expression of proteins in yeast mitochondria.	[62]
GG Yarrowia lipolytica	BsaI	–	Original	*Yarrowia lipolytica*	Collection of genetic parts for *Y. lipolytica* engineering.	[49]
GoldenPics	BbsI=BpiI	BsaI	GoldenMOCS	*Pichia pastoris*	Collection of genetic parts for *P. pastoris* engineering.	[48]
Kluyveromyces marxianus Golden Gate	BsaI	BsmBI=Esp3I	YTK	*Kluyveromyces marxianus*	Collection of genetic parts for *K. marxianus* engineering.	[50]
Schizosaccharomyces pombe	BsaI	–	Original	*Schizosaccharomyces pombe*	Collection of genetic parts for *S. pombe* engineering.	[51]
FungalBraid	BsaI, alternative BtgZI	BsmBI=Esp3I	GoldenBraid2.0	Filamentous fungi	Adaptation of GoldenBraid2.0 from plants to filamentous fungi.	[52]
MIDAS	BsaI	AarI	MoClo	Filamentous fungi	Destination vectors feature a cassette that permits the incorporation of new modules into an expanding multifragment vector through alternating AarI and BsmBI.	[39]
GoldenMOCS	BbsI=BpiI	BsaI	Original	*Aspergillus niger*	Basic genetic parts for filamentous fungi engineering.	[54]
TrichoGate	BsaI	–	Original	*Trichoderma* sp.	Basic genetic parts for *Trichoderma* sp. engineering.	[53]

Phaeodactylum tricornutum Golden Gate	BsaI	SapI	uLoop	*Phaeodactylum tricornutum*	Basic genetic parts for diatoms.	[56]
Dictyostelium discoideum Golden Gate	BsaI	BsmBI	GoldenBraid	*Dictyostelium discoideum*	Basic genetic parts for amoebae engineering.	[57]
GoldenBac	–	BsaI	Original	Insect cells	Uses the baculovirus expression vector system for expression in insect cells.	[59]
EMMA	BsmBI	BsaI	Original	Mammalian cells	Collection of genetic parts for mammalian cells engineering; it allows assembling up to 25 fragments in one reaction.	[60]
Mammalian Toolkit (MTK)	BsaI	BsmBI=Esp3I	YTK	Mammalian cells	Collection of characterized genetic parts for mammalian cell engineering; it allows the construction of multifragments with nine TUs.	[61]
JUMP	BsaI	BsmBI	PhytoBrick, BioBrick, SEVA	Adaptable to different organisms	It allows simple vector modification for increased compatibility; vectors feature secondary cloning sites flanked by AarI and BbsI.	[67]
uLoop	BsaI	SapI	Loop, MoClo, GoldenBraid2.0	Multi-kingdom: diatoms, yeast, plants, and bacteria	Recursive DNA assembly; it enables customization of host-specific vectors; it can generate destination vectors for the organism of choice.	[65]
Multi-Kingdom (MK)	BsaI	BbsI=BpiI	Modular Plant kit	Multi-kingdom: fungi, eubacteria, protista, plantae, and animalia	It enables parts assembly for the organism of choice using the same kit.	[70]

TIIS-RE L1 type IIS restriction endonuclease used in the assembly of domesticated parts in modules, *TIIS-RE L2* type IIS restriction endonuclease used in the assembly of several modules; "Original" denotes that the overhangs employed in the toolkit has been initially described within that context

A noteworthy strategy is employed in PODAC [71], where exclusively the BsaI TIIS-RE is utilized. The incorporation of a linear cassette containing a methylated BsaI site into the expanding multifragment construct enables iterative cloning using a single TIIS-RE. The underlying rationale is that methylated bases remain resistant to BsaI during the GG cloning reaction; however, during in vivo replication in *E. coli*, methylation is removed, exposing a new cloning site for a subsequent round of GG cloning [71]. Similarly, MetClo [72] uses DNA methylation to block TIIS-RE sites. In this approach, destination vectors contain TIIS-RE sites flanking the cloning marker, which partially overlap with the recognition site of a specific methylase. When the vector is transformed into an *E. coli* strain expressing the methylase, this can methylate the TIIS-RE site in vivo, thereby blocking the TIIS-RE activity. Conversely, if the vector is transformed into a standard *E. coli* strain lacking the methylase, the TIIS-RE site remains unmethylated. By employing different *E. coli* strains, the methylation of the TIIS-RE can be switched on or off [72].

Concerns persist about the universality and interchangeability of parts across various toolkits, irrespective of whether they share similar or different syntax. Addressing this, the JUMP [67] toolkit offers a range of destination vectors not only facilitating conventional modular cloning but also ensuring compatibility between standard parts from PhytoBricks [66] (common syntax for MoClo and GoldenBraid2.0), BioBricks [5], and the Standard European Vector Architecture (SEVA) [73–75]. It further enhances flexibility by incorporating two additional secondary cloning sites, enabling the integration of new functionalities [67]. Likewise, the recently introduced GoldenStandard [76] focuses on providing a toolkit based on SEVA vectors adapted to the common MoClo syntax, allowing the resulting constructs to be portable for use in various proteobacteria [76].

4 Conclusion

Golden Gate cloning emerges as a cornerstone in molecular and synthetic biology, offering a robust and versatile platform for DNA assembly. Over the years, significant improvements have transformed it into a method capable of effortlessly constructing large tailor-made DNA constructs, showcasing enhanced fidelity and efficiency. The streamlined process enables the successful assembly of over 50 fragments in a single step, significantly reducing effort and expediting various applications.

Despite these advancements, challenges persist. Enhancing the reaction yield holds the potential to liberate the method from reliance on *E. coli* cloning and propagation. This becomes particularly important when generating combinatorial libraries for diverse host organisms. The potential to directly perform functional analysis on the library within the final host organism, directly resulting from the GG reaction, could mark a groundbreaking advancement in the field.

On the other side, the abundance of modular cloning toolkits, while addressing specific requirements, introduces complexity in toolkit selection. This complexity underscores the growing need for comprehensive standardization, ensuring compatibility and portability between toolkits.

In conclusion, Golden Gate cloning is recognized as an essential and versatile tool, propelling research advancements across various fields, driving progress across diverse applications and organisms.

Acknowledgment

This work was supported by the Comunidad de Madrid grant Atraccion de Talento 2019-T1/BIO-13207.

References

1. Pryor JM, Potapov V, Bilotti K et al (2022) Rapid 40 kb Genome Construction from 52 Parts through Data-optimized Assembly Design. ACS Synth Biol 11:2036–2042. https://doi.org/10.1021/acssynbio.1c00525
2. Gibson DG, Young L, Chuang RY et al (2009) Enzymatic assembly of DNA molecules up to several hundred kilobases. Nat Methods 6: 343–345. https://doi.org/10.1038/nmeth.1318
3. Schindler D, Dai J, Cai Y (2018) Synthetic genomics: a new venture to dissect genome fundamentals and engineer new functions. Curr Opin Chem Biol 46:56–62. https://doi.org/10.1016/j.cbpa.2018.04.002
4. Cost GJ (2007) Enzymatic ligation assisted by nucleases: Simultaneous ligation and digestion promote the ordered assembly of DNA. Nat Protoc 2:2198–2202. https://doi.org/10.1038/nprot.2007.325
5. Shetty RP, Endy D, Knight TF (2008) Engineering BioBrick vectors from BioBrick parts. J Biol Eng 2:1–12. https://doi.org/10.1186/1754-1611-2-5
6. Xu P, Vansiri A, Bhan N, Koffas MAG (2012) EPathBrick: a synthetic biology platform for engineering metabolic pathways in *E. coli*. ACS Synth Biol 1:256–266. https://doi.org/10.1021/sb300016b
7. Li MZ, Elledge SJ (2007) Harnessing homologous recombination *in vitro* to generate recombinant DNA via SLIC. Nat Methods 4: 251–256. https://doi.org/10.1038/nmeth1010
8. Shao Z, Zhao H, Zhao H (2009) DNA assembler, an *in vivo* genetic method for rapid construction of biochemical pathways. Nucleic Acids Res 37:1–10. https://doi.org/10.1093/nar/gkn991
9. De KS, Stanton LH, Slaby T et al (2014) Rapid and reliable DNA assembly via ligase cycling reaction. ACS Synth Biol 3:97–106. https://doi.org/10.1021/sb4001992
10. Gibson DG (2009) Synthesis of DNA fragments in yeast by one-step assembly of overlapping oligonucleotides. Nucleic Acids Res 37:6984–6990. https://doi.org/10.1093/nar/gkp687
11. Liang J, Liu Z, Low XZ et al (2017) Twin-primer non-enzymatic DNA assembly: An efficient and accurate multi-part DNA assembly method. Nucleic Acids Res 45:e94. https://doi.org/10.1093/nar/gkx132
12. Engler C, Kandzia R, Marillonnet S (2008) A one pot, one step, precision cloning method with high throughput capability. PLoS One 3. https://doi.org/10.1371/journal.pone.0003647
13. Weber E, Engler C, Gruetzner R et al (2011) A modular cloning system for standardized assembly of multigene constructs. PLoS One 6. https://doi.org/10.1371/journal.pone.0016765

14. Kanigowska P, Shen Y, Zheng Y et al (2016) Smart DNA fabrication using sound waves: applying acoustic dispensing technologies to synthetic biology. J Lab Autom 21:49–56. https://doi.org/10.1177/2211068215593754
15. Szybalski W, Kim SC, Hasan N, Podhajska AJ (1991) Class-IIS restriction enzymes – a review. Gene 100:13–26. https://doi.org/10.1016/0378-1119(91)90345-C
16. Sarrion-Perdigones A, Falconi EE, Zandalinas SI et al (2011) GoldenBraid: an iterative cloning system for standardized assembly of reusable genetic modules. PLoS One 6. https://doi.org/10.1371/journal.pone.0021622
17. Pryor JM, Potapov V, Kucera RB et al (2020) Enabling one-pot Golden Gate assemblies of unprecedented complexity using data-optimized assembly design. PLoS One 15:1–19. https://doi.org/10.1371/journal.pone.0238592
18. Engler C, Youles M, Gruetzner R et al (2014) A Golden Gate modular cloning toolbox for plants. ACS Synth Biol 3:839–843. https://doi.org/10.1021/sb4001504
19. Püllmann P, Ulpinnis C, Marillonnet S et al (2019) Golden Mutagenesis: an efficient multi-site-saturation mutagenesis approach by Golden Gate cloning with automated primer design. Sci Rep 9:1–11. https://doi.org/10.1038/s41598-019-47376-1
20. Daffern N, Francino-Urdaniz IM, Baumer ZT, Whitehead TA (2024) Standardizing cassette-based deep mutagenesis by Golden Gate assembly. Biotechnol Bioeng 121:281–290. https://doi.org/10.1002/bit.28564
21. Guo Y, Dong J, Zhou T et al (2015) YeastFab: the design and construction of standard biological parts for metabolic engineering in *Saccharomyces cerevisiae*. Nucleic Acids Res 43:e88. https://doi.org/10.1093/nar/gkv464
22. Hamedirad M, Weisberg S, Chao R et al (2019) Highly efficient single-pot scarless Golden Gate assembly. ACS Synth Biol 8: 1047–1054. https://doi.org/10.1021/acssynbio.8b00480
23. Bird JE, Marles-Wright J, Giachino A (2022) A user's guide to Golden Gate cloning methods and standards. ACS Synth Biol 11:3551–3563. https://doi.org/10.1021/acssynbio.2c00355
24. Zulkower V, Rosser S (2020) DNA Chisel, a versatile sequence optimizer. Bioinformatics 36:4508–4509. https://doi.org/10.1093/bioinformatics/btaa558
25. Andreou AI, Nakayama N (2018) Mobius assembly: A versatile golden-gate framework towards universal DNA assembly. PLoS One 13:1–18. https://doi.org/10.1371/journal.pone.0189892
26. Sarrion-Perdigones A, Vazquez-Vilar M, Palací J et al (2013) Goldenbraid 2.0: a comprehensive DNA assembly framework for plant synthetic biology. Plant Physiol 162:1618–1631. https://doi.org/10.1104/pp.113.217661
27. Taylor GM, Mordaka PM, Heap JT (2019) Start-Stop Assembly: a functionally scarless DNA assembly system optimized for metabolic engineering. Nucleic Acids Res 47. https://doi.org/10.1093/nar/gky1182
28. Vries ST De, Kley L, Schindler D. Use of a Golden Gate plasmid set enabling scarless MoClo-compatible transcription unit assembly. https://doi.org/10.48550/arXiv.2402.03410
29. Potapov V, Ong JL, Kucera RB et al (2018) Comprehensive profiling of four base overhang ligation fidelity by T4 DNA ligase and application to DNA Assembly. ACS Synth Biol 7: 2665–2674. https://doi.org/10.1021/acssynbio.8b00333
30. Liang J, Chao R, Abil Z, et al (2014) Fairy-TALE: a high-throughput TAL effector synthesis platform. ACS Synth Biol 3(2):67–73. https://doi.org/10.1021/sb400109p
31. Engler C, Gruetzner R, Kandzia R, Marillonnet S (2009) Golden gate shuffling: a one-pot DNA shuffling method based on type ils restriction enzymes. PLoS One 4. https://doi.org/10.1371/journal.pone.0005553
32. Sikkema AP, Tabatabaei SK, Lee YJ et al (2023) High-complexity one-pot Golden Gate assembly. Curr Protoc 3:1–47. https://doi.org/10.1002/cpz1.882
33. Marillonnet S, Grützner R (2020) Synthetic DNA assembly using Golden Gate cloning and the hierarchical modular cloning pipeline. Curr Protoc Mol Biol 130:1–33. https://doi.org/10.1002/cpmb.115
34. Shaw WM, Khalil AS, Ellis T (2023) A multiplex MoClo toolkit for extensive and flexible engineering of *Saccharomyces cerevisiae*. ACS Synth Biol 12:3393–3405. https://doi.org/10.1021/acssynbio.3c00423
35. Potapov V, Ong JL, Langhorst BW et al (2018) A single-molecule sequencing assay for the comprehensive profiling of T4 DNA ligase fidelity and bias during DNA end-joining. Nucleic Acids Res 46:E79. https://doi.org/10.1093/nar/gky303
36. Rodd AM, Jawad A, Nano FE (2023) *Bacillus subtilis* small replicative plasmid collection incorporating five distinct origins of replication and compatible with Golden Gate DNA

assembly. FEMS Microbiol Lett 370. https://doi.org/10.1093/femsle/fnad039

37. Stukenberg D, Hensel T, Hoff J et al (2021) The Marburg collection: a Golden Gate DNA assembly framework for synthetic biology applications in *Vibrio natriegens*. ACS Synth Biol 10:1904–1919. https://doi.org/10.1021/acssynbio.1c00126
38. Ramírez Rojas AA, Brinkmann CK, Köbel TS, Schindler D (2023) DuBA.flow—a low-cost, long-read amplicon sequencing workflow for the validation of synthetic DNA constructs. ACS Synth Biol. https://doi.org/10.1021/acssynbio.3c00522
39. Van Dolleweerd CJ, Kessans SA, Van De Bittner KC et al (2018) MIDAS: a modular DNA assembly system for synthetic biology. ACS Synth Biol 7:1018–1029. https://doi.org/10.1021/acssynbio.7b00363
40. Garcia-Ruiz E, Auxillos J, Li T et al (2018) YeastFab: high-throughput genetic parts construction, measurement, and pathway engineering in yeast. In Methods Enzymol, vol. 608, pp 277–306. https://doi.org/10.1016/bs.mie.2018.05.003
41. Moore SJ, Lai HE, Kelwick RJR et al (2016) EcoFlex: a multifunctional MoClo kit for *E. coli* synthetic biology. ACS Synth Biol 5:1059–1069. https://doi.org/10.1021/acssynbio.6b00031
42. Iverson SV, Haddock TL, Beal J, Densmore DM (2016) CIDAR MoClo: improved MoClo assembly standard and new *E. coli* part library enable rapid combinatorial design for synthetic and traditional biology. ACS Synth Biol. https://doi.org/10.1021/acssynbio.5b00124
43. Schindler D, Milbredt S, Sperlea T, Waldminghaus T (2016) Design and assembly of DNA sequence libraries for chromosomal insertion in bacteria based on a set of modified MoClo vectors. ACS Synth Biol 5:1362–1368. https://doi.org/10.1021/acssynbio.6b00089
44. Behrendt G, Frohwitter J, Vlachonikolou M et al (2022) Zymo-Parts: a Golden Gate modular cloning toolbox for heterologous gene expression in *Zymomonas mobilis*. ACS Synth Biol 11:3855–3864. https://doi.org/10.1021/acssynbio.2c00428
45. Vasudevan R, Gale GAR, Schiavon AA et al (2019) Cyanogate: a modular cloning suite for engineering cyanobacteria based on the plant moclo syntax. Plant Physiol 180:39–55. https://doi.org/10.1104/pp.18.01401
46. Agmon N, Mitchell LA, Cai Y et al (2015) Yeast Golden Gate (yGG) for the efficient assembly of *S. cerevisiae* transcription units. ACS Synth Biol 4:853–859. https://doi.org/10.1021/sb500372z
47. Lee ME, DeLoache WC, Cervantes B, Dueber JE (2015) A highly characterized yeast toolkit for modular, multipart assembly. ACS Synth Biol 4:975–986. https://doi.org/10.1021/sb500366v
48. Prielhofer R, Barrero JJ, Steuer S et al (2017) GoldenPiCS: a Golden Gate-derived modular cloning system for applied synthetic biology in the yeast *Pichia pastoris*. BMC Syst Biol 11:1–14. https://doi.org/10.1186/s12918-017-0492-3
49. Celińska E, Ledesma-Amaro R, Larroude M et al (2017) Golden Gate assembly system dedicated to complex pathway manipulation in *Yarrowia lipolytica*. Microb Biotechnol 10:450–455. https://doi.org/10.1111/1751-7915.12605
50. Rajkumar AS, Varela JA, Juergens H et al (2019) Biological parts for *Kluyveromyces marxianus* synthetic biology. Front Bioeng Biotechnol 7:1–15. https://doi.org/10.3389/fbioe.2019.00097
51. Kakui Y, Sunaga T, Arai K et al (2015) Module-based construction of plasmids for chromosomal integration of the fission yeast *Schizosaccharomyces pombe*. Open Biol 5. https://doi.org/10.1098/rsob.150054
52. Hernanz-Koers M, Gandía M, Garrigues S et al (2018) FungalBraid: a GoldenBraid-based modular cloning platform for the assembly and exchange of DNA elements tailored to fungal synthetic biology. Fungal Genet Biol 116:51–61. https://doi.org/10.1016/j.fgb.2018.04.010
53. Nogueira-López G, Padilla-Arizmendi F, Inwood S et al (2019) TrichoGate: an improved vector system for a large scale of functional analysis of trichoderma genes. Front Microbiol 10:1–18. https://doi.org/10.3389/fmicb.2019.02794
54. Sarkari P, Marx H, Blumhoff ML et al (2017) An efficient tool for metabolic pathway construction and gene integration for *Aspergillus niger*. Bioresour Technol 245:1327–1333. https://doi.org/10.1016/j.biortech.2017.05.004
55. Crozet P, Navarro FJ, Willmund F et al (2018) Birth of a photosynthetic chassis: a MoClo toolkit enabling synthetic biology in the microalga *Chlamydomonas reinhardtii*. ACS Synth Biol 7:2074–2086. https://doi.org/10.1021/acssynbio.8b00251
56. Windhagauer M, Abbriano RM, Ashworth J et al (2021) Characterisation of novel regulatory sequences compatible with modular assembly in the diatom *Phaeodactylum*

tricornutum. Algal Res 53:102159. https://doi.org/10.1016/j.algal.2020.102159

57. Kundert P, Sarrion-Perdigones A, Gonzalez Y et al (2020) A GoldenBraid cloning system for synthetic biology in social amoebae. Nucleic Acids Res 48:4139–4146. https://doi.org/10.1093/NAR/GKAA185
58. Binder A, Lambert J, Morbitzer R et al (2014) A modular plasmid assembly kit for multigene expression, gene silencing and silencing rescue in plants. PLoS One 9. https://doi.org/10.1371/journal.pone.0088218
59. Neuhold J, Radakovics K, Lehner A et al (2020) GoldenBac: a simple, highly efficient, and widely applicable system for construction of multi-gene expression vectors for use with the baculovirus expression vector system. BMC Biotechnol 20:1–15. https://doi.org/10.1186/s12896-020-00616-z
60. Martella A, Matjusaitis M, Auxillos J et al (2017) EMMA: an extensible mammalian modular assembly toolkit for the rapid design and production of diverse expression vectors. ACS Synth Biol 6:1380–1392. https://doi.org/10.1021/acssynbio.7b00016
61. Fonseca JP, Bonny AR, Kumar GR et al (2019) A toolkit for rapid modular construction of biological circuits in mammalian cells. ACS Synth Biol 8:2593–2606. https://doi.org/10.1021/acssynbio.9b00322
62. Pérez-González A, Kniewel R, Veldhuizen M et al (2017) Adaptation of the GoldenBraid modular cloning system and creation of a toolkit for the expression of heterologous proteins in yeast mitochondria. BMC Biotechnol 17:1–11. https://doi.org/10.1186/s12896-017-0393-y
63. Occhialini A, Piatek AA, Pfotenhauer AC et al (2019) Mochlo: a versatile, modular cloning toolbox for chloroplast biotechnology. Plant Physiol 179:943–957. https://doi.org/10.1104/pp.18.01220
64. Lee ME, DeLoache WC, Cervantes B, Dueber JE (2015) A highly characterized yeast toolkit for modular, multipart assembly. ACS Synth Biol. https://doi.org/10.1021/sb500366v
65. Pollak B, Matute T, Nuñez I et al (2019) Universal loop assembly: open, efficient and cross-kingdom DNA fabrication. Synth Biol 5:1–13. https://doi.org/10.1093/synbio/ysaa001
66. Patron NJ, Orzaez D, Marillonnet S et al (2015) Standards for plant synthetic biology: a common syntax for exchange of DNA parts. New Phytol 208:13–19. https://doi.org/10.1111/nph.13532
67. Valenzuela-Ortega M, French C (2021) Joint universal modular plasmids (JUMP): a flexible vector platform for synthetic biology. Synth Biol 6. https://doi.org/10.1093/synbio/ysab003
68. Pollak B, Cerda A, Delmans M et al (2019) Loop assembly: a simple and open system for recursive fabrication of DNA circuits. New Phytol 222:628–640. https://doi.org/10.1111/nph.15625
69. Otto M, Skrekas C, Gossing M et al (2021) Expansion of the yeast modular cloning toolkit for CRISPR-based applications, genomic integrations and combinatorial libraries. ACS Synth Biol 10:3461–3474. https://doi.org/10.1021/acssynbio.1c00408
70. Chiasson D, Giménez-Oya V, Bircheneder M et al (2019) A unified multi-kingdom Golden Gate cloning platform. Sci Rep 9:1–12. https://doi.org/10.1038/s41598-019-46171-2
71. Van Hove B, Guidi C, De Wannemaeker L et al (2017) Recursive DNA assembly using protected oligonucleotide duplex assisted cloning (PODAC). ACS Synth Biol 6:943–949. https://doi.org/10.1021/acssynbio.7b00017
72. Lin D, O'Callaghan CA (2018) MetClo: methylase-assisted hierarchical DNA assembly using a single type IIS restriction enzyme. Nucleic Acids Res 46. https://doi.org/10.1093/nar/gky596
73. Martínez-García E, Goñi-Moreno A, Bartley B et al (2020) SEVA 3.0: An update of the Standard European Vector Architecture for enabling portability of genetic constructs among diverse bacterial hosts. Nucleic Acids Res 48:D1164–D1170. https://doi.org/10.1093/nar/gkz1024
74. Martínez-García E, Fraile S, Algar E et al (2023) SEVA 4.0: an update of the Standard European Vector Architecture database for advanced analysis and programming of bacterial phenotypes. Nucleic Acids Res 51:D1558–D1567. https://doi.org/10.1093/nar/gkac1059
75. Silva-Rocha R, Martínez-García E, Calles B et al (2013) The Standard European Vector Architecture (SEVA): a coherent platform for the analysis and deployment of complex prokaryotic phenotypes. Nucleic Acids Res 41:666–675. https://doi.org/10.1093/nar/gks1119
76. Blázquez B, León DS, Torres-Bacete J et al (2023) Golden Standard: a complete standard, portable, and interoperative MoClo tool for model and non-model proteobacteria. Nucleic Acids Res 51:E98–E98. https://doi.org/10.1093/nar/gkad758

Correction to: Combining Oligo Pools and Golden Gate Cloning to Create Protein Variant Libraries or Guide RNA Libraries for CRISPR Applications

Alicia Maciá Valero, Rianne C. Prins, Thijs de Vroet, and Sonja Billerbeck

Correction to:
Chapter 15 in: Daniel Schindler (ed.), *Golden Gate Cloning: Methods and Protocols*, Methods in Molecular Biology, vol. 2850,
https://doi.org/10.1007/978-1-0716-4220-7_15

The original version of the chapter had the incorrect author name as "Alicia Maciá" as the first name and "Valero" as the last.

It has been corrected as:
First name: "Alicia and *Last name*: Maciá Valero"

The updated version of this chapter can be found at
https://doi.org/10.1007/978-1-0716-4220-7_15

Daniel Schindler (ed.), *Golden Gate Cloning: Methods and Protocols*, Methods in Molecular Biology, vol. 2850,
https://doi.org/10.1007/978-1-0716-4220-7_28,

INDEX

Daniel Schindler (ed.), *Golden Gate Cloning: Methods and Protocols*, Methods in Molecular Biology, vol. 2850, https://doi.org/10.1007/978-1-0716-4220-7,

T

V

9781071640944